Studies in Big Data

Volume 162

Series Editor

Janusz Kacprzyk, Polish Academy of Sciences, Warsaw, Poland

The series "Studies in Big Data" (SBD) publishes new developments and advances in the various areas of Big Data- quickly and with a high quality. The intent is to cover the theory, research, development, and applications of Big Data, as embedded in the fields of engineering, computer science, physics, economics and life sciences. The books of the series refer to the analysis and understanding of large, complex, and/or distributed data sets generated from recent digital sources coming from sensors or other physical instruments as well as simulations, crowd sourcing, social networks or other internet transactions, such as emails or video click streams and other. The series contains monographs, lecture notes and edited volumes in Big Data spanning the areas of computational intelligence including neural networks, evolutionary computation, soft computing, fuzzy systems, as well as artificial intelligence, data mining, modern statistics and Operations research, as well as self-organizing systems. Of particular value to both the contributors and the readership are the short publication timeframe and the world-wide distribution, which enable both wide and rapid dissemination of research output.

The books of this series are reviewed in a single blind peer review process.

Indexed by SCOPUS, EI Compendex, SCIMAGO and zbMATH.

All books published in the series are submitted for consideration in Web of Science.

Pradeep Singh · Balasubramanian Raman

Deep Learning Through the Prism of Tensors

Springer

Pradeep Singh
Department of Computer Science
and Engineering
Indian Institute of Technology Roorkee
Roorkee, Uttarakhand, India

Balasubramanian Raman
Department of Computer Science
and Engineering
Indian Institute of Technology Roorkee
Roorkee, Uttarakhand, India

ISSN 2197-6503 ISSN 2197-6511 (electronic)
Studies in Big Data
ISBN 978-981-97-8021-1 ISBN 978-981-97-8019-8 (eBook)
https://doi.org/10.1007/978-981-97-8019-8

This Springer imprint is published by the registered company Springer Nature Singapore Pte Ltd.
The registered company address is: 152 Beach Road, #21-01/04 Gateway East, Singapore 189721, Singapore

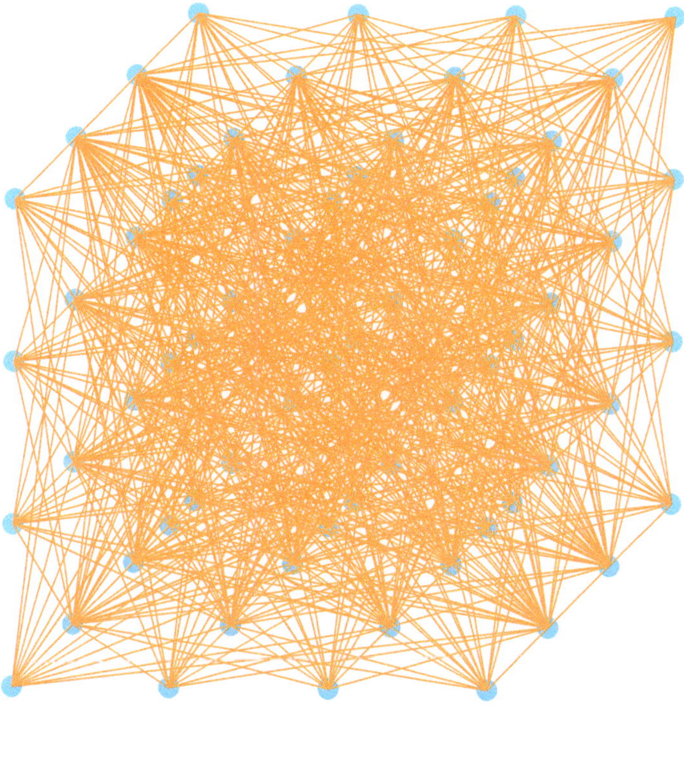

Dedicated to the Uncharted Territories of Mind and Machine

Preface

The pursuit of understanding complex systems and modeling them using mathematical structures has always been at the core of scientific research. In the recent years, this quest has led to the birth and rapid development of deep learning, a subset of machine learning that is transforming various fields of study. This includes, but is not limited to, computer vision and natural language processing in computer science; algorithmic trading and risk management in finance; medical imaging and drug discovery in medicine; materials science and quantum computing in physics; genomics and protein folding in biology; social network analysis and behavioral modeling in social sciences; and robotics and control systems in engineering.

At the heart of these advancements lies the mathematical construct of tensors, which provide the foundation for representing and manipulating high-dimensional data. This book, *Deep Learning Through the Prism of Tensors*, aims to offer a comprehensive and rigorous exploration of deep learning, unified by the powerful framework of tensor mathematics. Deep learning, with its intricate architectures and vast applications, can be daunting to both newcomers and seasoned practitioners. The goal of this book is to demystify deep learning by presenting its concepts through the lens of tensors. By grounding each topic in tensor operations and algebra, readers will gain a deeper and more intuitive understanding of how deep learning models function, how they are trained, and how they can be applied to solve real-world problems.

This book is designed for a diverse audience. It caters to students and educators seeking a thorough grounding in deep learning principles, researchers looking to deepen their understanding of the mathematical foundations of neural networks, and professionals eager to apply deep learning techniques in their work. Regardless of your background, this book provides the necessary tools to master deep learning through a rigorous, tensor-centric approach. Each chapter of this book is crafted to build on the previous one, ensuring a coherent and progressive learning experience. We begin with an introduction to deep learning, establishing the importance of tensors and their role in this domain. From there, we delve into the algebra and geometry of tensors, laying the mathematical groundwork essential for the subsequent chapters. The book covers a wide range of topics, from the foundational multilayer perceptrons and convolutional neural networks to the advanced architectures of transformers and generative models. Each topic is explained with a focus on the underlying tensor operations, accompanied by practical examples and exercises to reinforce the concepts.

The interdisciplinary nature of this book reflects the reality of modern deep learning research, where insights from mathematics, computer science, and engineering converge to drive innovation. By emphasizing the tensorial perspective, this book not only provides a robust understanding of existing models but also equips readers with the conceptual tools to explore and develop new architectures. In writing this book, we have drawn from our own experiences in research and teaching, striving to present complex ideas in a clear and accessible manner. The examples and exercises are designed to bridge the gap between theory and practice, helping readers to apply what they have learned to real-world challenges. The mathematical formalism in this book is rich and rigorous, reflecting the complex nature of the subject matter, yet it is presented with clarity and coherence to ensure that readers are not lost in abstraction.

As you embark on this journey through the fascinating world of tensors and deep learning, we hope that this book will become a valuable resource and guide, illuminating the paths that connect abstract mathematical concepts to tangible technological advancements. Whether you are a seasoned researcher or a curious enthusiast, may you find inspiration, insights, and intellectual joy in the pages that follow.

Roorkee, India
June 2024

Pradeep Singh
Balasubramanian Raman

Acknowledgements

We are immensely grateful to the numerous individuals and organizations that have made the completion of this book possible. First and foremost, we extend our heartfelt thanks to our home institution, Indian Institute of Technology (IIT) Roorkee, for providing an exceptional environment and the necessary resources for our work. We also deeply appreciate the support from various research bodies, particularly the Science and Engineering Research Board (SERB) and the Department of Science and Technology (DST), India, for their support. We would also like to acknowledge the National Board of Higher Mathematics (NBHM), Department of Atomic Energy (DAE), India, for their support.

We owe a profound debt of gratitude to our colleagues, whose insights, encouragement, and unwavering support have been invaluable throughout this journey. Their contributions have been a constant source of inspiration. Special thanks to the insightful discussions with research scholars and interns at the Machine Vision (MV) lab and the Machine Intelligence (MI) lab at the Department of Computer Science and Engineering, IIT Roorkee. It is a real privilege to witness the continuous blossoming of thoughts there.

We are also profoundly thankful for the boundless knowledge available through the internet. Special thanks to Google for its incredible search capabilities, which have been indispensable for fact-checking and finding relevant information quickly, and to Stack Exchange for its vast community of experts who have helped solve countless technical queries and provided deeper insights into complex topics. We acknowledge the role of Large Language Models in brainstorming and generating ideas. Open-source tools and libraries have been critical in the development of this book. We are deeply appreciative of the developers and contributors to libraries such as NumPy, SciPy, Pandas, Matplotlib, PyTorch, and TensorFlow, which provided the computational foundation for many of the examples and case studies presented. A special mention goes to GitHub for making collaboration and version control seamless, and to Jupyter for making interactive data exploration and visualization a delight.

To our colleagues and students at various institutions, thank you for your support, feedback, and encouragement. Your intellectual curiosity and dedication have been a source of motivation. Finally, we extend our deepest gratitude to our families and friends for their unwavering support and patience throughout this endeavor. Your encouragement has been a pillar of strength. This book is the result of collective knowledge and collaborative effort. To everyone who has contributed, directly or indirectly, thank you. Your contributions have made this work possible.

Introduction

In an era where data and computation are integral to nearly every aspect of human endeavor, the field of deep learning stands as a beacon of innovation and potential. This book, *Deep Learning Through the Prism of Tensors*, is a testament to the remarkable advancements and applications of deep learning that I have had the privilege to witness and contribute to.

Deep learning, at its core, is about harnessing the power of neural networks to process and make sense of vast amounts of data. The journey from rudimentary perceptrons to today's sophisticated architectures, such as transformers and graph neural networks, has been driven by the relentless pursuit of efficiency and accuracy in computational models. This book seeks to capture the essence of this journey through a comprehensive exploration of the tensorial perspective, which I believe is fundamental to understanding and advancing deep learning.

This book was fueled by a passion for mathematics and a deep appreciation for the elegance that tensors bring to the field. Tensors provide a powerful framework for representing and manipulating data, enabling the development of more efficient and scalable neural network architectures. By presenting deep learning through the lens of tensors, we aim to offer readers a unique and rigorous approach to understanding and applying these concepts. The structure of the book reflects our commitment to providing a thorough and accessible guide to deep learning. Starting with the foundational principles, we gradually introduce more advanced topics, ensuring that readers build a strong conceptual framework before delving into complex applications. Each chapter is designed to be both informative and practical, with numerous examples and exercises to reinforce learning.

One of the hallmarks of deep learning is its broad applicability across various domains. From image and sequence processing to the geometric understanding of data, the versatility of deep learning models continues to expand. This book not only covers the theoretical underpinnings but also showcases practical implementations using popular frameworks such as TensorFlow and PyTorch. I believe that this hands-on approach will equip readers with the skills necessary to tackle real-world problems.

As we navigate the rapidly evolving landscape of deep learning, it is important to remember that this field is still in its infancy. The potential for innovation is boundless, and I am excited to see how the concepts and techniques presented in this book will inspire future advancements. I encourage readers to approach this book with curiosity and an open mind, ready to explore the depths of deep learning through the prism of tensors. I hope that this book serves as a valuable resource for students, researchers, and practitioners alike. It is my sincere wish that you find inspiration and insight within these pages, and that you are empowered to contribute to the ever-growing body of knowledge in deep learning.

July 2024 Balasubramanian Raman

Contents

About the Authors

Dr. Pradeep Singh earned his Ph.D. and Master's degrees from the Indian Institute of Technology (IIT) Delhi, specializing in Dynamical Systems, and a Bachelor's degree in Data Science from IIT Madras. Currently, he is a Post-Doctoral Researcher and Principal Investigator at the Machine Intelligence Lab within the Department of Computer Science and Engineering at IIT Roorkee, where he is actively engaged in research at the intersection of Geometric Deep Learning, Neuro-symbolic AI, and Dynamical Systems. His research is supported by the National Post Doctoral Fellowship (N-PDF) from the Science and Engineering Research Board (SERB), Department of Science and Technology. Dr. Singh has earned multiple accolades, including All India Rank 1 in IIT GATE 2020, IIT JAM 2015, and CSIR NET 2019, along with prestigious fellowships such as the National Board for Higher Mathematics (NBHM) Masters, Doctoral, and Post Doctoral Fellowships from the Department of Atomic Energy, and the SPM Doctoral Fellowship from the Council of Scientific and Industrial Research.

Dr. Balasubramanian Raman received his Ph.D. from IIT Madras and his B.Sc. and M.Sc. in Mathematics from the University of Madras. He is a Professor and the Head of the Department of Computer Science and Engineering at IIT Roorkee, as well as the iHUB Divyasampark Chair Professor. He is also a Joint Faculty member in the Mehta Family School of Data Science and Artificial Intelligence at IIT Roorkee. With over 200 research papers published in reputed journals and conferences, his research interests span Machine Learning, Image and Video Processing, Computer Vision, and Pattern Recognition. Dr. Raman has served as a Guest Professor and Visiting Researcher at prestigious institutions such as Osaka Metropolitan University, Curtin University, the University of Cyberjaya, and the University of Windsor. He has held postdoctoral positions at Rutgers University and the University of Missouri-Columbia. Under his coaching, teams have achieved notable rankings in the ACM International Collegiate Programming Contest (ICPC) World Finals. He has been recognized with several awards, including the BOYSCAST Fellowship and the Ramkumar Prize for Outstanding Teaching and Research.

Abbreviations

AI	Artificial intelligence
BERT	Bidirectional Encoder Representations from Transformers
CNN	Convolutional neural network
EfficientNet	A family of convolutional neural networks
GAN	Generative Adversarial Network
GAT	Graph Attention Network
GNN	Graph neural network
GPT	Generative pre-trained transformer
GPU	Graphics Processing Unit
GRU	Gated Recurrent Unit
Inception	A deep convolutional architecture by Google
LSTM	Long short-term memory
Matplotlib	A plotting library for Python
MLP	Multi-layer perceptron
MobileNet	A lightweight deep neural network architecture
NLP	Natural language processing
NumPy	A library for numerical computation in Python
Pandas	A data analysis and manipulation library in Python
PyTorch	A deep learning framework
ReLU	Rectified Linear Unit
ResNet	Residual network
RNN	Recurrent neural network
SciPy	A library for scientific computation in Python
SGD	Stochastic gradient descent
SWIN	Shifted Window Transformer
TensorFlow	A deep learning framework
VGG	Visual Geometry Group
ViT	Vision Transformer

Chapter 1
A Tensorial Perspective to Deep Learning

Deep learning is a journey, and intelligence is the destination.

Yoshua Bengio, From AI to AGI

1.1 Introduction

Deep learning, a term that has buzzed through academic corridors and technology hubs, stands at the intersection of mathematical elegance and practical ingenuity. As a branch of artificial intelligence, it represents a cutting-edge approach to machine learning, heralding a new era where computers can grasp, interpret, and respond to the world in ways that were once the exclusive domain of human cognition. The evolution of deep learning over the past decade has been nothing short of meteoric. Its transformative power to handle vast and complex data structures, ranging from raw text and time-series to multidimensional images, has revolutionized fields as diverse as healthcare, finance, entertainment, and beyond. At its core, deep learning is a mathematical alchemy. It employs artificial neural networks, composed of intricately connected nodes, each a tiny mathematical oracle, unraveling patterns hidden within data. The journey of information through these networks is akin to a voyage of discovery, where raw, chaotic data is meticulously sculpted, layer by layer, into coherent insights and predictions.

The term 'deep' in deep learning may evoke images of profound complexity or philosophical wisdom. However, it's a technical reference to the architecture of the neural networks themselves. The 'depth' refers to the multitude of layers, or stages, that data traverses, each adding a new dimension of understanding, each unveiling a finer nuance of the pattern. Contrasting with shallow learning's superficial glance, deep learning delves into the very essence of data, navigating through tens or even hundreds of hidden layers. It's a symphony of non-linear transformations

P. Singh and B. Raman, *Deep Learning Through the Prism of Tensors*, Studies in Big Data 162, https://doi.org/10.1007/978-981-97-8019-8_1

that convert the unmanageable and abstract into the tangible and actionable. Deep learning's inspiration from the human brain's biological neural networks is poetic but not literal. The architectural resemblance is abstract, a metaphor that captures the spirit of connection and computation, but without claiming to model or replicate the exact functionality of the human mind. In this light, deep learning is not just a cold, mechanical process but a vibrant journey of exploration. It reflects a contemporary renaissance where mathematics, technology, and creativity converge, where numbers become images, texts sing, and algorithms dream.

The age we inhabit is not merely digital; it's drenched in data. From smartphones to satellites, every blink of technology generates a stream of information. Amidst this deluge, deep learning emerges not just as a tool, but as a transformative force, an alchemist's wand that transmutes raw data into wisdom. Unlike many traditional methods that stumble as data swells, deep learning thrives on abundance. The more data you pour into a deep learning model, the more nuanced and refined its insights potentially become. This thirst for information places deep learning on the throne of contemporary AI research, a discipline that doesn't plateau but ascends with every challenge. What sets deep learning apart is not just its efficiency but its intuition. In a world where traditional algorithms waited for human guidance, deep learning takes the lead. It automates what was once a tedious human task: feature engineering. No longer does a machine wait for an expert to tell it what to see; it learns to see for itself. Imagine a child, eyes wide with wonder, discovering the world one shape at a time. That's a deep learning model at its lower layers, recognizing edges, colors, and textures. Now picture that child growing, connecting shapes into objects, objects into scenes, scenes into stories. That's the model as it dives deeper, weaving a rich tapestry of understanding, recognizing faces in a crowd or storms in a satellite image.

1.1.1 What Is Deep Learning?

Deep learning is a subset of machine learning that focuses on neural networks with many layers, referred to as deep neural networks. These models are designed to learn hierarchical representations of data through multiple layers of abstraction (see Fig. 1.1). Each layer in a deep neural network transforms the input data into increasingly complex representations, capturing intricate patterns and features. Mathematically, a deep neural network can be expressed as a composition of functions, each representing a layer in the network. For example, consider a neural network with L layers, where the output of the lth layer is denoted as $\mathbf{h}^{(l)}$. The transformation performed by each layer can be represented as:

$$\mathbf{h}^{(l)} = \sigma(\mathbf{W}^{(l)}\mathbf{h}^{(l-1)} + \mathbf{b}^{(l)})$$

where $\mathbf{W}^{(l)}$ and $\mathbf{b}^{(l)}$ are the weight matrix and bias vector for the lth layer, respectively, and σ is a non-linear activation function. The composition of these functions forms a deep neural network:

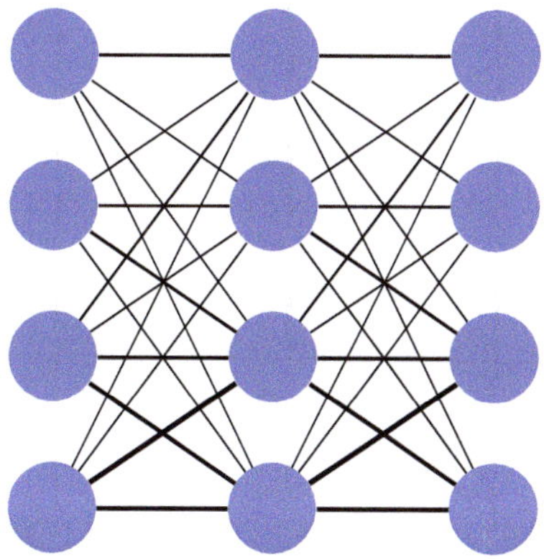

Fig. 1.1 From human cognition to algorithmic insights

$$f(\mathbf{x}) = \mathbf{h}^{(L)} = \sigma(\mathbf{W}^{(L)}(\sigma(\mathbf{W}^{(L-1)}(\cdots\sigma(\mathbf{W}^{(1)}\mathbf{x} + \mathbf{b}^{(1)})\cdots) + \mathbf{b}^{(L-1)})) + \mathbf{b}^{(L)})$$

This hierarchical structure enables deep neural networks to capture complex relationships in the data, making them powerful tools for various tasks.

The scope of deep learning extends across numerous domains, from computer vision and natural language processing to robotics and healthcare. In computer vision, deep learning models have achieved state-of-the-art performance in image classification, object detection, and image generation. For instance, convolutional neural networks (CNNs) have revolutionized image processing by leveraging spatial hierarchies through convolutional layers. In natural language processing, recurrent neural networks (RNNs) and transformers have enabled breakthroughs in tasks such as language translation, sentiment analysis, and text generation. RNNs process sequences of data by maintaining a hidden state that captures information from previous time steps. Transformers, on the other hand, utilize self-attention mechanisms to capture dependencies between words in a sentence without regard to their distance, significantly improving performance in language-related tasks.

The applications of deep learning are vast and impactful, transforming industries and driving advancements in technology. In healthcare, deep learning models are used for medical image analysis, aiding in the diagnosis of diseases such as cancer and diabetic retinopathy. These models can automatically detect abnormalities in medical images, providing faster and more accurate diagnoses than traditional methods. For example, a CNN trained to detect tumors in mammograms can highlight suspicious regions, assisting radiologists in their assessments.

In autonomous driving, deep learning algorithms process sensory data from cameras, LiDAR, and radar to perceive the environment and make driving decisions. These systems rely on CNNs for object detection and semantic segmentation, enabling vehicles to recognize and respond to pedestrians, other vehicles, and traffic signs. An example of a deep learning-based object detection system is YOLO (You Only Look Once), which predicts bounding boxes and class probabilities directly from full images in a single forward pass through the network:

$$(\mathbf{p}_c, b_x, b_y, b_w, b_h, \mathbf{c}) = \text{YOLO}(\mathbf{I})$$

where $\mathbf{p}_c$ is the probability of the presence of an object, (b_x, b_y, b_w, b_h) are the bounding box coordinates, and $\mathbf{c}$ are the class probabilities.

In finance, deep learning models are used for algorithmic trading, fraud detection, and risk assessment. These models analyze vast amounts of financial data to identify patterns and make predictions. For example, an LSTM network can be employed to predict stock prices based on historical price data:

$$\hat{y}_t = \text{LSTM}(\mathbf{x}_{t-1}, \mathbf{h}_{t-1}, \mathbf{c}_{t-1})$$

where $\hat{y}_t$ is the predicted stock price at time t, $\mathbf{x}_{t-1}$ is the input at the previous time step, and $\mathbf{h}_{t-1}$ and $\mathbf{c}_{t-1}$ are the hidden and cell states, respectively.

The impact of deep learning extends beyond specific applications, driving advancements in artificial intelligence (AI) and enabling the development of intelligent systems that can learn, reason, and adapt. Deep learning models have outperformed traditional machine learning algorithms in many tasks, leading to significant improvements in accuracy and efficiency. As these models continue to evolve, they hold the potential to solve even more complex problems, pushing the boundaries of what is possible in AI and beyond.

1.1.2 Importance of Tensors in Deep Learning

Tensors are the fundamental building blocks in deep learning, enabling the efficient representation and manipulation of multi-dimensional data (see Fig. 1.2). They extend the concept of scalars, vectors, and matrices to higher dimensions, allowing deep learning models to process complex data structures. The use of tensors facilitates various operations that are crucial for training and inference in neural networks.

In deep learning, data is often represented as tensors to handle multiple dimensions efficiently. A scalar is a tensor of rank 0, a vector is a tensor of rank 1, and a matrix is a tensor of rank 2. Higher-dimensional arrays, such as 3D images or sequences of vectors, are represented as tensors of rank 3 or higher. For example, consider a batch of grayscale images, each of size 28×28. This batch can be represented as a 3D tensor $\mathcal{X} \in \mathbb{R}^{N \times 28 \times 28}$, where N is the number of images in the batch. If the images are in color, an additional dimension is added to represent the color channels, resulting in a 4D tensor $\mathcal{X} \in \mathbb{R}^{N \times 28 \times 28 \times 3}$.

Tensors allow for the efficient storage and manipulation of large datasets. For instance, in natural language processing, a batch of sentences can be represented as a 3D tensor $\mathcal{X} \in \mathbb{R}^{N \times T \times d}$, where N is the batch size, T is the maximum sequence length, and d is the dimensionality of the word embeddings. This tensor representation enables deep learning models to perform parallel operations on the entire batch, significantly speeding up computations.

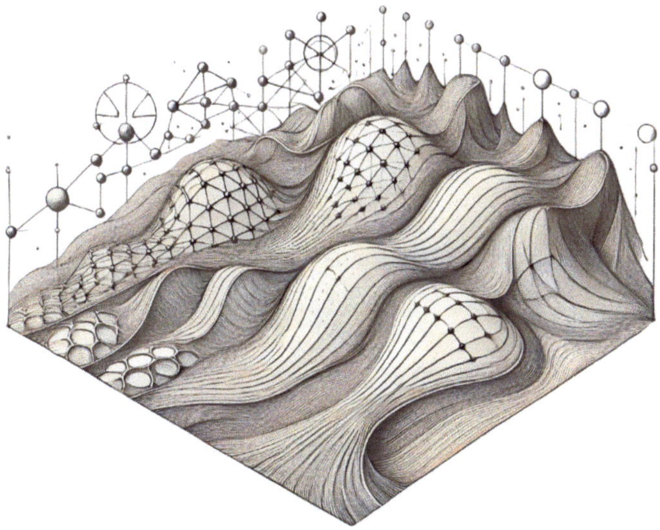

Fig. 1.2 Learning in the landscape of tensors: interwoven surfaces, tangent planes, and manifolds

Various operations on tensors are essential for the functioning of neural networks. These operations include element-wise operations, tensor reshaping, and tensor contractions, among others. Element-wise operations, such as addition, subtraction, multiplication, and division, are performed on corresponding elements of tensors. For example, given two tensors $\mathcal{A}, \mathcal{B} \in \mathbb{R}^{I \times J \times K}$, their element-wise sum $\mathcal{C}$ is computed as:

$$\mathcal{C}_{ijk} = \mathcal{A}_{ijk} + \mathcal{B}_{ijk}$$

for all i, j, k.

Tensor reshaping is another crucial operation, allowing tensors to be reorganized without changing their data. This is often required during data preprocessing and model building. For example, a 2D tensor representing a grayscale image of size 28×28 can be flattened into a 1D tensor of size 784 before feeding it into a fully connected layer:

$$\mathcal{X} \in \mathbb{R}^{28 \times 28} \rightarrow \mathbf{x} \in \mathbb{R}^{784}$$

Tensor contractions generalize the concept of matrix multiplication to higher dimensions. Given two tensors $\mathcal{A} \in \mathbb{R}^{I \times J}$ and $\mathcal{B} \in \mathbb{R}^{J \times K}$, their matrix product $\mathcal{C} \in \mathbb{R}^{I \times K}$ is computed as:

$$\mathcal{C}_{ik} = \sum_{j=1}^{J} \mathcal{A}_{ij} \mathcal{B}_{jk}$$

This operation can be extended to higher-dimensional tensors. For instance, given a 3D tensor $\mathcal{A} \in \mathbb{R}^{I \times J \times K}$ and a 2D tensor $\mathcal{B} \in \mathbb{R}^{K \times L}$, their contraction over the third dimension of $\mathcal{A}$ and the first dimension of $\mathcal{B}$ results in a 3D tensor $\mathcal{C} \in \mathbb{R}^{I \times J \times L}$:

$$\mathcal{C}_{ijl} = \sum_{k=1}^{K} \mathcal{A}_{ijk}\mathcal{B}_{kl}$$

Tensor operations are efficiently implemented in deep learning frameworks such as TensorFlow and PyTorch, which leverage hardware accelerations like GPUs and TPUs. For example, the convolution operation in CNNs is a tensor contraction that can be performed efficiently using these frameworks. Consider a 4D input tensor $\mathcal{X} \in \mathbb{R}^{N\times H\times W\times C}$ representing a batch of images, and a 4D kernel tensor $\mathcal{K} \in \mathbb{R}^{kH\times kW\times C\times F}$ representing the convolutional filters. The output tensor $\mathcal{Y} \in \mathbb{R}^{N\times H'\times W'\times F}$ is computed as:

$$\mathcal{Y}_{nhwf} = \sum_{i=1}^{kH}\sum_{j=1}^{kW}\sum_{c=1}^{C} \mathcal{X}_{n(h+i-1)(w+j-1)c}\mathcal{K}_{ijcf}$$

where H' and W' are the height and width of the output feature map.

1.2 Historical Development of Neural Networks

Early Beginnings

The development of neural networks dates back to the 1940s and 1950s, when researchers began to explore the possibility of creating machines that could mimic the human brain's learning processes (McCulloch and Pitts 1990). This early period laid the groundwork for what would eventually become the field of artificial neural networks. One of the earliest and most significant milestones in the history of neural networks is the invention of the perceptron by Rosenblatt (1958). The perceptron was inspired by the biological neuron and was designed as a simplified model of the brain's neural circuitry. A perceptron is a binary classifier that maps an input vector $\mathbf{x} \in \mathbb{R}^n$ to a binary output using a weight vector $\mathbf{w} \in \mathbb{R}^n$ and a bias term $b \in \mathbb{R}$. Mathematically, the output of a perceptron can be expressed as:

$$y = \phi(\mathbf{w}^T\mathbf{x} + b)$$

where ϕ is the activation function, typically the Heaviside step function:

$$\phi(z) = \begin{cases} 1 & \text{if } z \geq 0 \\ 0 & \text{otherwise} \end{cases}$$

The perceptron learning algorithm adjusts the weights and bias iteratively to minimize the classification error. Given a set of training examples $\{(\mathbf{x}_i, y_i)\}_{i=1}^{m}$, where $\mathbf{x}_i$ is the input vector and y_i is the desired output, the perceptron update rule is:

$$\mathbf{w} \leftarrow \mathbf{w} + \Delta\mathbf{w}$$

$$b \leftarrow b + \Delta b$$

where

$$\Delta\mathbf{w} = \eta(y_i - \hat{y}_i)\mathbf{x}_i$$

$$\Delta b = \eta(y_i - \hat{y}_i)$$

Here, η is the learning rate, and $\hat{y}_i$ is the predicted output. The weights are updated only if there is a misclassification ($y_i \neq \hat{y}_i$).

The perceptron demonstrated the potential of artificial neural networks to learn from data and make predictions, marking a significant step forward in the field of machine learning. Despite the initial success of the perceptron, early neural network models faced several limitations that hindered their development and broader adoption. One of the most critical limitations was the inability of the perceptron to solve non-linearly separable problems. This limitation was famously highlighted by Marvin Minsky and Seymour Papert in their 1969 book "Perceptrons," where they proved that a single-layer perceptron could not represent certain simple functions, such as the XOR function. The XOR function is defined as:

$$\text{XOR}(x_1, x_2) = \begin{cases} 1 & \text{if } x_1 \neq x_2 \\ 0 & \text{if } x_1 = x_2 \end{cases}$$

The XOR function is not linearly separable, meaning there is no straight line that can separate the positive examples (where XOR $= 1$) from the negative examples (where XOR $= 0$) in the input space.

To illustrate this limitation, consider a perceptron with inputs x_1 and x_2, weights w_1 and w_2, and bias b. The output is:

$$y = \phi(w_1x_1 + w_2x_2 + b)$$

No matter how the weights w_1 and w_2 and the bias b are chosen, the perceptron cannot correctly classify all possible input combinations of the XOR function. This fundamental limitation led to a temporary decline in interest and funding for neural network research, a period often referred to as the "AI winter." However, it also motivated researchers to explore more complex architectures that could overcome these limitations.

The introduction of multi-layer perceptrons (MLPs) and the backpropagation algorithm in the 1980s revitalized the field. MLPs, consisting of multiple layers of perceptrons, are capable of approximating any continuous function, given sufficient neurons and appropriate training. The backpropagation algorithm, introduced by Rumelhart et al. (1986), provided an efficient method for training these multi-layer networks by computing the gradient of the loss function with respect to the weights.

The backpropagation algorithm leverages the chain rule of calculus to propagate errors backward through the network, updating the weights layer by layer. For a network with a loss function L, the gradient of the loss with respect to a weight $w_{ij}^{(l)}$ in layer l is computed as:

$$\frac{\partial L}{\partial w_{ij}^{(l)}} = \delta_j^{(l)} h_i^{(l-1)}$$

where $\delta_j^{(l)}$ is the error term for neuron j in layer l, and $h_i^{(l-1)}$ is the activation of neuron i in the previous layer. The error term $\delta_j^{(l)}$ is computed recursively from the output layer to the input layer:

$$\delta_j^{(L)} = \frac{\partial L}{\partial h_j^{(L)}} \phi'(z_j^{(L)})$$

$$\delta_j^{(l)} = \left(\sum_k \delta_k^{(l+1)} w_{jk}^{(l+1)} \right) \phi'(z_j^{(l)})$$

where $z_j^{(l)}$ is the input to neuron j in layer l, and ϕ' is the derivative of the activation function.

Consider an MLP with L layers. Let $\mathbf{x}$ be the input vector, $\mathbf{y}$ the target vector, and $\hat{\mathbf{y}}$ the output vector. The loss function L measures the difference between $\mathbf{y}$ and $\hat{\mathbf{y}}$. For instance, in a regression task, the mean squared error (MSE) is a common choice:

$$L(\mathbf{y}, \hat{\mathbf{y}}) = \frac{1}{2} \sum_i (y_i - \hat{y}_i)^2$$

The output of each neuron in layer l is given by:

$$\mathbf{h}^{(l)} = \phi(\mathbf{W}^{(l)} \mathbf{h}^{(l-1)} + \mathbf{b}^{(l)})$$

where $\mathbf{W}^{(l)}$ is the weight matrix, $\mathbf{b}^{(l)}$ is the bias vector, and ϕ is the activation function. The backpropagation algorithm updates the weights and biases to minimize the loss function:

$$\mathbf{W}^{(l)} \leftarrow \mathbf{W}^{(l)} - \eta \frac{\partial L}{\partial \mathbf{W}^{(l)}}$$

$$\mathbf{b}^{(l)} \leftarrow \mathbf{b}^{(l)} - \eta \frac{\partial L}{\partial \mathbf{b}^{(l)}}$$

where η is the learning rate. The gradients $\frac{\partial L}{\partial \mathbf{W}^{(l)}}$ and $\frac{\partial L}{\partial \mathbf{b}^{(l)}}$ are computed as follows:

$$\frac{\partial L}{\partial \mathbf{W}^{(l)}} = \delta^{(l)} \mathbf{h}^{(l-1)T}$$

$$\frac{\partial L}{\partial \mathbf{b}^{(l)}} = \delta^{(l)}$$

where $\delta^{(l)}$ is the error term for layer l, computed recursively from the output layer to the input layer:

$$\delta^{(L)} = (\hat{\mathbf{y}} - \mathbf{y}) \odot \phi'(\mathbf{z}^{(L)})$$

$$\delta^{(l)} = (\mathbf{W}^{(l+1)T} \delta^{(l+1)}) \odot \phi'(\mathbf{z}^{(l)})$$

Here, $\odot$ denotes element-wise multiplication, and $\mathbf{z}^{(l)}$ is the pre-activation input to layer l.

Backpropagation enabled the training of deep neural networks, leading to significant advances in various applications, including pattern recognition, speech recognition, and early computer vision systems. The development of multi-layer perceptrons and the backpropagation algorithm marked a significant milestone in the evolution of neural networks, enabling them to solve complex, non-linear problems and laying the foundation for modern deep learning.

Convolutional Neural Networks (CNNs)

CNNs revolutionized the field of computer vision by introducing the concept of convolutional layers, which exploit the spatial structure of image data. Proposed by LeCun et al. (1998), CNNs became widely known with the success of LeNet-5 for handwritten digit recognition.

In a CNN, convolutional layers apply filters (kernels) to the input image, producing feature maps that capture local patterns such as edges and textures. Mathematically, the output of a convolutional layer is given by:

$$\mathbf{y}_{i,j} = \sum_{m=1}^{M} \sum_{n=1}^{N} \mathbf{W}_{m,n} \cdot \mathbf{x}_{i+m-1,j+n-1} + b$$

where $\mathbf{W}$ is the filter, $\mathbf{x}$ is the input, and b is the bias term. The filters are learned during training, allowing the network to adapt to specific features in the data.

CNNs typically include pooling layers, which downsample the feature maps, reducing their spatial dimensions while retaining important information. A common pooling operation is max-pooling, defined as:

$$\mathbf{y}_{i,j} = \max_{m,n} \mathbf{x}_{i+m,j+n}$$

CNNs also incorporate fully connected layers at the end of the network to perform classification based on the extracted features. The architecture of CNNs, with their hierarchical layers, makes them highly effective for image classification, object detection, and image segmentation tasks. Prominent architectures such as AlexNet, VGG, and ResNet have set benchmarks in image recognition competitions, showcasing the power of CNNs.

Recurrent Neural Networks (RNNs)
RNNs are designed to handle sequential data, making them suitable for tasks involving time series, natural language processing, and speech recognition (Hochreiter and Schmidhuber 1997). RNNs introduce loops in the network architecture, allowing information to persist across time steps. The hidden state in an RNN is updated at each time step t based on the current input $\mathbf{x}_t$ and the previous hidden state $\mathbf{h}_{t-1}$:

$$\mathbf{h}_t = \phi(\mathbf{W}_{hh}\mathbf{h}_{t-1} + \mathbf{W}_{xh}\mathbf{x}_t + \mathbf{b}_h)$$

The output at time step t can be computed as:

$$\mathbf{y}_t = \mathbf{W}_{hy}\mathbf{h}_t + \mathbf{b}_y$$

However, standard RNNs struggle with learning long-term dependencies due to the vanishing gradient problem. This issue is mitigated by advanced RNN variants such as Long Short-Term Memory (LSTM) networks and Gated Recurrent Units (GRUs). LSTMs introduce memory cells and gating mechanisms to control the flow of information. RNNs and their variants have been instrumental in advancing the state-of-the-art in tasks such as machine translation, speech recognition, and sentiment analysis.

Transformers and Attention Mechanisms
Transformers and attention mechanisms represent a significant paradigm shift in neural network architectures, particularly in natural language processing. Introduced by Vaswani et al. (2017) in their seminal paper "Attention is All You Need," transformers eschew the sequential nature of RNNs in favor of a fully attention-based approach, enabling parallel processing of input data. The core component of the transformer architecture is the self-attention mechanism, which allows each element of the input to attend to all other elements, capturing dependencies regardless of their distance in the sequence. The transformer architecture consists of an encoder-decoder structure, where both the encoder and decoder are composed of multiple layers of self-attention and feedforward neural networks. The encoder processes the input sequence to generate context-aware representations, while the decoder generates the output sequence, attending to both the input and previously generated outputs. Transformers have revolutionized natural language processing, leading to the development of pre-trained models such as BERT (Bidirectional Encoder Representations from Transformers) and GPT (Generative Pre-trained Transformer). These models leverage large-scale unsupervised pre-training followed by task-specific fine-tuning, achieving state-of-the-art performance on a wide range of NLP tasks. The success of transformers has extended beyond NLP, with applications in computer vision (Vision Transformers) and reinforcement learning, demonstrating their versatility and power in modeling complex dependencies in data.

Ergo, the development of backpropagation and MLPs, the introduction of CNNs, the evolution of RNNs and their variants, and the emergence of transformers and attention mechanisms represent significant milestones in the history of neural networks. These advancements have expanded the capabilities of neural networks, enabling them to tackle increasingly complex tasks across various domains.

1.3 Basics of Neural Networks

1.3.1 Neurons and Layers

The fundamental building block of a neural network is the artificial neuron, inspired by the biological neurons in the human brain. An artificial neuron, also known as a perceptron, is a computational unit that receives one or more inputs, processes them through a set of weights, applies a bias, and produces an output through an activation function. Mathematically, the operation of a single neuron can be described as follows. Let $\mathbf{x} = [x_1, x_2, \ldots, x_n]^T$ be the input vector, $\mathbf{w} = [w_1, w_2, \ldots, w_n]^T$ be the weight vector, and b be the bias term. The output y of the neuron is given by:

$$z = \mathbf{w}^T\mathbf{x} + b$$

$$y = \phi(z)$$

Here, z is the linear combination of the inputs weighted by $\mathbf{w}$ plus the bias b, and $\phi(z)$ is the activation function that introduces non-linearity into the model. Each neuron processes the input data, applies the weights and bias, and passes the result through the activation function to produce the output. This process allows the network to learn and represent complex patterns in the data.

Layered Architecture

A neural network consists of multiple layers of neurons arranged in a hierarchical structure. The layered architecture is what enables neural networks to learn and model complex relationships in the data. The primary layers in a neural network include the input layer, hidden layers, and the output layer.

Input Layer The input layer is the first layer in the network and consists of neurons that receive the input features. The number of neurons in the input layer corresponds to the number of features in the input data. For instance, if the input data is a grayscale image of size 28×28, the input layer will have 784 neurons, each representing a pixel.

Hidden Layers Hidden layers are the intermediate layers between the input and output layers. They consist of neurons that apply weights, biases, and activation functions to the input data. The role of hidden layers is to capture and transform the input data into higher-level representations. The output of each hidden layer serves as the input to the subsequent layer. The number of hidden layers and the number of neurons in each layer are hyperparameters that significantly impact the network's performance.

Output Layer The output layer is the final layer in the network and produces the prediction or classification based on the transformed input data. The number of neurons in the output layer depends on the task. For binary classification, the output layer typically has one neuron with a sigmoid activation function. For multi-class classification, the output layer has as many neurons as there are classes, often with a softmax activation function to produce a probability distribution over the classes:

$$\text{softmax}(z_i) = \frac{e^{z_i}}{\sum_j e^{z_j}}$$

The forward pass of the neural network involves computing the activations for each layer from the input layer to the output layer. For a network with L layers, the forward pass can be represented as:

$$\mathbf{h}^{(1)} = \phi(\mathbf{W}^{(1)}\mathbf{x} + \mathbf{b}^{(1)})$$

$$\mathbf{h}^{(2)} = \phi(\mathbf{W}^{(2)}\mathbf{h}^{(1)} + \mathbf{b}^{(2)})$$

$$\vdots$$

$$\mathbf{y} = \phi(\mathbf{W}^{(L)}\mathbf{h}^{(L-1)} + \mathbf{b}^{(L)})$$

In this formulation, $\mathbf{x}$ is the input vector, $\mathbf{h}^{(l)}$ is the activation of the lth layer, $\mathbf{W}^{(l)}$ and $\mathbf{b}^{(l)}$ are the weight matrix and bias vector for the lth layer, respectively, and $\mathbf{y}$ is the output vector.

The layered architecture of neural networks allows for the composition of simple linear transformations and non-linear activation functions to model complex functions. This hierarchical representation is key to the success of neural networks in various applications, from image recognition to natural language processing. For example, consider a neural network designed for handwritten digit recognition using the MNIST dataset. The network might consist of an input layer with 784 neurons (one for each pixel), two hidden layers with 128 and 64 neurons respectively, and an output layer with 10 neurons (one for each digit class). The network learns to transform the pixel values through the hidden layers into a representation that can accurately classify the digits.

Activation functions play a crucial role in neural networks by introducing non-linearity into the model. This non-linearity enables neural networks to approximate complex functions and model intricate patterns in data. Without activation functions, a neural network, regardless of its depth, would simply behave like a linear model, incapable of solving complex tasks. In this section, we explore the most common activation functions-Sigmoid, Tanh, and ReLU-as well as advanced activation functions that address specific limitations and improve network performance.

1.3.2 Forward and Backward Propagation

The core mechanics of training neural networks involve two critical processes: the forward pass and the backward pass. These processes allow the network to make predictions and iteratively adjust its weights to minimize the error, respectively.

Forward Pass

The forward pass in a neural network is the process of calculating the output predictions from the input data by passing the inputs through each layer of the network. During the forward pass, each layer transforms the input using a linear transformation followed by a non-linear activation function. This process allows the network to learn complex mappings from inputs to outputs.

Consider a neural network with L layers, where $\mathbf{x}$ is the input vector and $\mathbf{y}$ is the output vector. The weights and biases of the lth layer are denoted by $\mathbf{W}^{(l)}$ and $\mathbf{b}^{(l)}$, respectively. The activation function is denoted by ϕ. The forward pass can be described as follows:

Input layer The input vector $\mathbf{x}$ is fed into the network.
Hidden layers For each hidden layer $l = 1, 2, \ldots, L-1$, the output $\mathbf{h}^{(l)}$ is computed as:

$$\mathbf{z}^{(l)} = \mathbf{W}^{(l)}\mathbf{h}^{(l-1)} + \mathbf{b}^{(l)}$$

$$\mathbf{h}^{(l)} = \phi(\mathbf{z}^{(l)})$$

Here, $\mathbf{h}^{(0)} = \mathbf{x}$ is the input vector.
Output layer The final layer L produces the output vector $\mathbf{y}$:

$$\mathbf{z}^{(L)} = \mathbf{W}^{(L)}\mathbf{h}^{(L-1)} + \mathbf{b}^{(L)}$$

$$\mathbf{y} = \phi(\mathbf{z}^{(L)})$$

Each layer applies a linear transformation to its input, followed by a non-linear activation function. The output of the last layer is the network's prediction. For instance, in a classification task, the final activation function might be the softmax function, which converts the output logits into probabilities:

$$\text{softmax}(\mathbf{z})_i = \frac{e^{z_i}}{\sum_j e^{z_j}}$$

Example Consider a simple neural network with one hidden layer for binary classification. Let the input vector $\mathbf{x} \in \mathbb{R}^2$ be two-dimensional, and the hidden layer have three neurons. The weights and biases are $\mathbf{W}^{(1)} \in \mathbb{R}^{3\times 2}$, $\mathbf{b}^{(1)} \in \mathbb{R}^3$, $\mathbf{W}^{(2)} \in \mathbb{R}^{1\times 3}$, and $\mathbf{b}^{(2)} \in \mathbb{R}$. The forward pass is:

1. Compute the linear transformation for the hidden layer:

$$\mathbf{z}^{(1)} = \mathbf{W}^{(1)}\mathbf{x} + \mathbf{b}^{(1)}$$

2. Apply the activation function (e.g., ReLU) to obtain the hidden layer activations:

$$\mathbf{h}^{(1)} = \mathrm{ReLU}(\mathbf{z}^{(1)})$$

3. Compute the linear transformation for the output layer:

$$\mathbf{z}^{(2)} = \mathbf{W}^{(2)}\mathbf{h}^{(1)} + \mathbf{b}^{(2)}$$

4. Apply the sigmoid activation function to obtain the output:

$$\mathbf{y} = \sigma(\mathbf{z}^{(2)})$$

Backward Pass and Gradient Descent
The backward pass, or backpropagation, is the process of computing the gradients of the loss function with respect to the network's parameters (weights and biases) and updating the parameters to minimize the loss. This process uses the chain rule of calculus to propagate the error backward through the network, layer by layer.

Compute the loss The loss function $L(\mathbf{y}, \hat{\mathbf{y}})$ measures the difference between the predicted output $\hat{\mathbf{y}}$ and the true output $\mathbf{y}$. For example, in a binary classification task, the binary cross-entropy loss is:

$$L(\mathbf{y}, \hat{\mathbf{y}}) = -\left(y \log(\hat{y}) + (1 - y) \log(1 - \hat{y})\right)$$

Calculate the gradients The goal is to compute the gradients of the loss with respect to the weights and biases, $\frac{\partial L}{\partial \mathbf{W}^{(l)}}$ and $\frac{\partial L}{\partial \mathbf{b}^{(l)}}$, for each layer l.

For the output layer L:

$$\delta^{(L)} = \frac{\partial L}{\partial \mathbf{z}^{(L)}} = \hat{\mathbf{y}} - \mathbf{y}$$

$$\frac{\partial L}{\partial \mathbf{W}^{(L)}} = \delta^{(L)}\mathbf{h}^{(L-1)T}$$

$$\frac{\partial L}{\partial \mathbf{b}^{(L)}} = \delta^{(L)}$$

For the hidden layers $l = L - 1, L - 2, \ldots, 1$:

$$\delta^{(l)} = \left(\mathbf{W}^{(l+1)T}\delta^{(l+1)}\right) \odot \phi'(\mathbf{z}^{(l)})$$

$$\frac{\partial L}{\partial \mathbf{W}^{(l)}} = \delta^{(l)}\mathbf{h}^{(l-1)T}$$

$$\frac{\partial L}{\partial \mathbf{b}^{(l)}} = \delta^{(l)}$$

Here, $\delta^{(l)}$ represents the error term for layer l, and $\odot$ denotes element-wise multiplication.

Update the parameters The parameters are updated using gradient descent or its variants. In simple gradient descent, the weights and biases are updated as follows:

$$\mathbf{W}^{(l)} \leftarrow \mathbf{W}^{(l)} - \eta \frac{\partial L}{\partial \mathbf{W}^{(l)}}$$

$$\mathbf{b}^{(l)} \leftarrow \mathbf{b}^{(l)} - \eta \frac{\partial L}{\partial \mathbf{b}^{(l)}}$$

where η is the learning rate.

Example Consider the same simple neural network with one hidden layer for binary classification. The loss function is binary cross-entropy. The backward pass involves:

1. Compute the error at the output layer:

$$\delta^{(2)} = \hat{y} - y$$

2. Compute the gradients for the output layer:

$$\frac{\partial L}{\partial \mathbf{W}^{(2)}} = \delta^{(2)} \mathbf{h}^{(1)T}$$

$$\frac{\partial L}{\partial \mathbf{b}^{(2)}} = \delta^{(2)}$$

3. Propagate the error to the hidden layer:

$$\delta^{(1)} = (\mathbf{W}^{(2)T} \delta^{(2)}) \odot \text{ReLU}'(\mathbf{z}^{(1)})$$

4. Compute the gradients for the hidden layer:

$$\frac{\partial L}{\partial \mathbf{W}^{(1)}} = \delta^{(1)} \mathbf{x}^T$$

$$\frac{\partial L}{\partial \mathbf{b}^{(1)}} = \delta^{(1)}$$

By updating the weights and biases using these gradients, the network gradually learns to minimize the loss, improving its performance on the training data. In a nutshell, the forward pass computes the output predictions by passing the input data through the network, layer by layer. The backward pass computes the gradients of the loss with respect to the network's parameters and updates the parameters using gradient descent. This iterative process of forward and backward propagation enables neural networks to learn from data and improve their predictions.

1.4 Representation of Data as Tensors

1.4.1 Tensor Notation and Basics

Scalars, Vectors, Matrices, and Higher-Order Tensors

In deep learning, tensors are the fundamental data structures used to represent and manipulate multi-dimensional data. Tensors generalize the concepts of scalars, vectors, and matrices to higher dimensions, providing a flexible framework for representing complex datasets. Understanding tensor notation and their basic properties is essential for effectively working with deep learning models.

A scalar is a single numerical value, often represented as $a \in \mathbb{R}$. It is a tensor of rank 0, having no dimensions. Scalars are the simplest form of data and are commonly used to represent individual numbers, such as loss values or single pixel intensities in an image. A vector is an ordered collection of numbers, represented as $\mathbf{a} = [a_1, a_2, \ldots, a_n] \in \mathbb{R}^n$. It is a tensor of rank 1 with one dimension, where n is the length of the vector. Vectors are used to represent various types of data, such as feature vectors, word embeddings, and weights in neural networks. A matrix is a two-dimensional array of numbers, represented as $\mathbf{A} \in \mathbb{R}^{m \times n}$, where m and n are the number of rows and columns, respectively. A matrix is a tensor of rank 2. Matrices are widely used in linear algebra operations, such as matrix multiplication, and in representing data structures like images (in grayscale) and adjacency matrices in graphs. Higher-order tensors extend these concepts to more than two dimensions. A tensor of rank k can be represented as $\mathcal{A} \in \mathbb{R}^{I_1 \times I_2 \times \cdots \times I_k}$, where $I_1, I_2, \ldots, I_k$ are the dimensions of the tensor. For example, a rank-3 tensor (3D tensor) can be used to represent a batch of images with dimensions (N, H, W), where N is the batch size, H is the height, and W is the width of each image. In color images, an additional dimension for color channels can be included, resulting in a rank-4 tensor with dimensions (N, H, W, C), where C is the number of color channels (e.g., 3 for RGB images).

Tensor Shapes and Dimensions

The shape of a tensor is defined by the number of elements in each dimension. The shape of a tensor is crucial as it determines how operations can be performed on it. For example, the shape of a matrix $\mathbf{A} \in \mathbb{R}^{m \times n}$ is (m, n). The shape of a rank-3 tensor $\mathcal{A} \in \mathbb{R}^{I \times J \times K}$ is (I, J, K). Understanding tensor shapes and dimensions is essential for performing various operations such as reshaping, broadcasting, and tensor algebra. Reshaping allows tensors to be transformed into different shapes without changing their data. For instance, a vector of length 9 can be reshaped into a 3×3 matrix:

$$\mathbf{a} = [1, 2, 3, 4, 5, 6, 7, 8, 9] \in \mathbb{R}^9$$

can be reshaped into

$$\mathbf{A} = \begin{pmatrix} 1 & 2 & 3 \\ 4 & 5 & 6 \\ 7 & 8 & 9 \end{pmatrix} \in \mathbb{R}^{3\times 3}$$

Broadcasting is a technique used to perform element-wise operations on tensors of different shapes. Broadcasting automatically expands the smaller tensor along the missing dimensions to match the shape of the larger tensor. For example, adding a vector to each row of a matrix:

$$\mathbf{A} = \begin{pmatrix} 1 & 2 & 3 \\ 4 & 5 & 6 \\ 7 & 8 & 9 \end{pmatrix}, \quad \mathbf{b} = [1, 2, 3]$$

can be broadcasted as:

$$\mathbf{A} + \mathbf{b} = \begin{pmatrix} 1+1 & 2+2 & 3+3 \\ 4+1 & 5+2 & 6+3 \\ 7+1 & 8+2 & 9+3 \end{pmatrix} = \begin{pmatrix} 2 & 4 & 6 \\ 5 & 7 & 9 \\ 8 & 10 & 12 \end{pmatrix}$$

Tensors are foundational in deep learning frameworks such as TensorFlow and PyTorch, which provide optimized implementations of tensor operations, allowing efficient manipulation and computation. For instance, in PyTorch, a tensor can be defined and reshaped as follows:

```
import torch

# Define a vector
vector = torch.tensor([1, 2, 3, 4, 5, 6, 7, 8, 9])

# Reshape the vector into a 3x3 matrix
matrix = vector.view(3, 3)
```

1.4.2 Manipulating Data with Tensors

Tensor Operations in Neural Networks Tensors are central to the functionality of neural networks, enabling efficient representation and manipulation of multi-dimensional data. Various tensor operations are fundamental to neural network computations, including element-wise operations, broadcasting, matrix multiplication, and advanced tensor contractions. Understanding these operations is crucial for designing and optimizing neural networks.

Element-wise Operations Element-wise operations are performed independently on corresponding elements of tensors. Common element-wise operations include addition, subtraction, multiplication, and division. For two tensors $\mathcal{A}, \mathcal{B} \in \mathbb{R}^{I \times J}$, their element-wise sum $\mathcal{C}$ is computed as:

$$\mathcal{C}_{ij} = \mathcal{A}_{ij} + \mathcal{B}_{ij}$$

for all i, j. These operations are highly parallelizable and efficiently implemented in deep learning frameworks.

Broadcasting Broadcasting allows tensors of different shapes to be combined by automatically expanding the smaller tensor to match the dimensions of the larger tensor. This enables efficient computation without the need for explicit reshaping. For example, consider adding a vector $\mathbf{b} \in \mathbb{R}^J$ to each row of a matrix $\mathcal{A} \in \mathbb{R}^{I \times J}$:

$$\mathcal{C}_{ij} = \mathcal{A}_{ij} + \mathbf{b}_j$$

Matrix Multiplication Matrix multiplication is a core operation in neural networks, used in layers such as fully connected and convolutional layers. For two matrices $\mathbf{A} \in \mathbb{R}^{m \times n}$ and $\mathbf{B} \in \mathbb{R}^{n \times p}$, their product $\mathbf{C} \in \mathbb{R}^{m \times p}$ is computed as:

$$\mathbf{C}_{ik} = \sum_{j=1}^{n} \mathbf{A}_{ij} \mathbf{B}_{jk}$$

This operation can be extended to higher-dimensional tensors, enabling complex transformations in neural networks.

Tensor Contractions Tensor contractions generalize matrix multiplication to higher dimensions, allowing for efficient computation of inner products along specified axes. For example, consider two tensors $\mathcal{A} \in \mathbb{R}^{I \times J \times K}$ and $\mathcal{B} \in \mathbb{R}^{K \times L}$. Their contraction over the third dimension of $\mathcal{A}$ and the first dimension of $\mathcal{B}$ results in a tensor $\mathcal{C} \in \mathbb{R}^{I \times J \times L}$:

$$\mathcal{C}_{ijl} = \sum_{k=1}^{K} \mathcal{A}_{ijk} \mathcal{B}_{kl}$$

These operations are integral to the forward and backward passes in neural networks. During the forward pass, tensors flow through the network, undergoing various transformations. During the backward pass, gradients are computed and propagated backward to update the model parameters. Efficient implementation of tensor operations is crucial for the performance and scalability of neural networks.

Practical Examples with Tensor Libraries (TensorFlow, PyTorch) Modern deep learning frameworks, such as TensorFlow and PyTorch, provide optimized implementations of tensor operations, enabling efficient manipulation and computation. Here, we demonstrate practical examples of tensor operations using these libraries.

TensorFlow Example TensorFlow is a powerful library for numerical computation and deep learning. It provides a flexible and efficient way to perform tensor operations.

```
import tensorflow as tf

# Define tensors
A = tf.constant([[1, 2], [3, 4]], dtype=tf.float32)
B = tf.constant([[5, 6], [7, 8]], dtype=tf.float32)

# Element-wise addition
C = tf.add(A, B)

# Matrix multiplication
D = tf.matmul(A, B)

# Reshape tensor
E = tf.reshape(A, [1, 4])

# Print results
print("Element-wise addition:\n", C.numpy())
print("Matrix multiplication:\n", D.numpy())
print("Reshaped tensor:\n", E.numpy())
```

PyTorch Example PyTorch is a widely used deep learning library known for its dynamic computation graph and ease of use.

```
import torch

# Define tensors
A = torch.tensor([[1, 2], [3, 4]], dtype=torch.float32)
B = torch.tensor([[5, 6], [7, 8]], dtype=torch.float32)

# Element-wise addition
C = A + B

# Matrix multiplication
D = torch.matmul(A, B)

# Reshape tensor
E = A.view(1, 4)

# Print results
print("Element-wise addition:\n", C)
print("Matrix multiplication:\n", D)
print("Reshaped tensor:\n", E)
```

Example: Forward Pass in a Neural Network Consider a simple neural network with one hidden layer for binary classification. Using PyTorch, we can define and perform the forward pass as follows:

```
import torch
import torch.nn as nn

# Define the network
class SimpleNN(nn.Module):
    def __init__(self):
        super(SimpleNN, self).__init__()
        self.fc1 = nn.Linear(2, 3)
        self.fc2 = nn.Linear(3, 1)
```

```
    def forward(self, x):
        x = torch.relu(self.fc1(x))
        x = torch.sigmoid(self.fc2(x))
        return x

# Create the network
net = SimpleNN()

# Define an input tensor
input_tensor = torch.tensor([[0.5, 0.8]], dtype=torch.float32)

# Perform the forward pass
output = net(input_tensor)
print("Network output:\n", output)
```

In this example, the 'SimpleNN' class defines a neural network with two linear layers and ReLU and sigmoid activation functions. The 'forward' method performs the forward pass, transforming the input tensor through the layers to produce the output.

1.5 Practical Examples

1.5.1 Image Data Processing with Tensors

Processing image data with tensors is fundamental in computer vision tasks. Convolutional neural networks (CNNs) are particularly well-suited for this, as they leverage convolutional and pooling operations to extract hierarchical features from images. These operations are efficient and take advantage of the spatial structure of image data.

Convolutional Operations

Convolutional operations are the core building blocks of CNNs. A convolution operation applies a filter (also known as a kernel) to the input image to produce an output feature map. The filter is a small matrix of weights that slides over the input image, performing element-wise multiplications and summing the results to produce each element of the feature map. Mathematically, for a 2D input image $\mathbf{X} \in \mathbb{R}^{H \times W}$ and a filter $\mathbf{W} \in \mathbb{R}^{kH \times kW}$, where H and W are the height and width of the image, and kH and kW are the height and width of the filter, the convolution operation produces an output feature map $\mathbf{Y} \in \mathbb{R}^{(H-kH+1)\times(W-kW+1)}$ defined as:

$$\mathbf{Y}_{i,j} = \sum_{m=1}^{kH} \sum_{n=1}^{kW} \mathbf{X}_{i+m-1,j+n-1} \cdot \mathbf{W}_{m,n}$$

To illustrate this with a practical example, consider an input image **X** and a 3×3 filter **W**:

$$\mathbf{X} = \begin{pmatrix} 1 & 2 & 3 & 4 \\ 5 & 6 & 7 & 8 \\ 9 & 10 & 11 & 12 \\ 13 & 14 & 15 & 16 \end{pmatrix}, \quad \mathbf{W} = \begin{pmatrix} 1 & 0 & -1 \\ 1 & 0 & -1 \\ 1 & 0 & -1 \end{pmatrix}$$

The resulting feature map **Y** is computed by sliding the filter over the image and performing element-wise multiplications and summations:

$$\mathbf{Y} = \begin{pmatrix} (1 \cdot 1 + 2 \cdot 0 + 3 \cdot -1) + (5 \cdot 1 + 6 \cdot 0 + 7 \cdot -1) + (9 \cdot 1 + 10 \cdot 0 + 11 \cdot -1) & \cdots \\ \cdots & \cdots \end{pmatrix}$$

Continuing this calculation, we get:

$$\mathbf{Y} = \begin{pmatrix} -6 & -6 \\ -6 & -6 \end{pmatrix}$$

In practice, we often use additional techniques such as padding and strides to control the spatial dimensions of the output feature map. Padding adds a border of zeros around the input image, allowing the filter to be applied to the edges of the image. Strides control the step size of the filter as it slides over the image, effectively downsampling the output feature map.

TensorFlow Example: Convolutional Operation

```
import tensorflow as tf

# Define an input tensor (image)
input_tensor = tf.constant([[1, 2, 3, 4],
                            [5, 6, 7, 8],
                            [9, 10, 11, 12],
                            [13, 14, 15, 16]], dtype=tf.float32)
input_tensor = tf.reshape(input_tensor, [1, 4, 4, 1])  # Reshape to [batch, height,
width, channels]

# Define a filter tensor
filter_tensor = tf.constant([[1, 0, -1],
                             [1, 0, -1],
                             [1, 0, -1]], dtype=tf.float32)
filter_tensor = tf.reshape(filter_tensor, [3, 3, 1, 1]) # Reshape to [height, width,
 in_channels, out_channels]

# Apply the convolutional operation
output_tensor = tf.nn.conv2d(input_tensor, filter_tensor, strides=[1, 1, 1, 1],
 padding='VALID')
print("Convolution result:\n", tf.squeeze(output_tensor).numpy())
```

PyTorch Example: Convolutional Operation

```
import torch
import torch.nn.functional as F

# Define an input tensor (image)
input_tensor = torch.tensor([[1, 2, 3, 4],
                             [5, 6, 7, 8],
                             [9, 10, 11, 12],
                             [13, 14, 15, 16]], dtype=torch.float32)
input_tensor = input_tensor.view(1, 1, 4, 4)  # Reshape to [batch, channels, height,
 width]

# Define a filter tensor
filter_tensor = torch.tensor([[1, 0, -1],
                              [1, 0, -1],
                              [1, 0, -1]], dtype=torch.float32)
filter_tensor = filter_tensor.view(1, 1, 3, 3)  # Reshape to [out_channels, in_channels,
 height, width]

# Apply the convolutional operation
output_tensor = F.conv2d(input_tensor, filter_tensor, stride=1, padding=0)
print("Convolution result:\n", output_tensor.squeeze().numpy())
```

Pooling Operations

Pooling operations are used in CNNs to reduce the spatial dimensions of the feature maps, retaining important features while discarding redundant information. Pooling helps in making the network invariant to small translations and distortions in the input image. There are two common types of pooling operations: max-pooling and average pooling.

Max-Pooling Max-pooling selects the maximum value from each region of the feature map. For an input feature map $\mathbf{X} \in \mathbb{R}^{H \times W}$ and a pooling window of size $pH \times pW$, the max-pooling operation produces an output feature map $\mathbf{Y} \in \mathbb{R}^{\frac{H}{pH} \times \frac{W}{pW}}$ defined as:

$$\mathbf{Y}_{i,j} = \max_{m,n} \mathbf{X}_{i \cdot pH+m, j \cdot pW+n}$$

Average Pooling Average pooling computes the average value from each region of the feature map. For an input feature map $\mathbf{X} \in \mathbb{R}^{H \times W}$ and a pooling window of size $pH \times pW$, the average pooling operation produces an output feature map $\mathbf{Y} \in \mathbb{R}^{\frac{H}{pH} \times \frac{W}{pW}}$ defined as:

$$\mathbf{Y}_{i,j} = \frac{1}{pH \cdot pW} \sum_{m,n} \mathbf{X}_{i \cdot pH+m, j \cdot pW+n}$$

Example Consider a 4×4 input feature map and a 2×2 pooling window.

For max-pooling:

$$\mathbf{X} = \begin{pmatrix} 1 & 3 & 2 & 4 \\ 5 & 6 & 8 & 7 \\ 9 & 11 & 10 & 12 \\ 13 & 15 & 14 & 16 \end{pmatrix}$$

The resulting feature map after max-pooling is:

$$\mathbf{Y} = \begin{pmatrix} 6 & 8 \\ 15 & 16 \end{pmatrix}$$

For average pooling:

$$\mathbf{Y} = \begin{pmatrix} 3.75 & 5.25 \\ 12 & 13 \end{pmatrix}$$

TensorFlow Example: Pooling Operation

```
import tensorflow as tf

# Define an input tensor (feature map)
input_tensor = tf.constant([[1, 3, 2, 4],
                            [5, 6, 8, 7],
                            [9, 11, 10, 12],
                            [13, 15, 14, 16]], dtype=tf.float32)
input_tensor = tf.reshape(input_tensor, [1, 4, 4, 1])  # Reshape to [batch, height, width,
 channels]

# Apply max pooling
max_pooled = tf.nn.max_pool2d(input_tensor, ksize=2, strides=2, padding='VALID')
print("Max pooling result:\n", tf.squeeze(max_pooled).numpy())

# Apply average pooling
avg_pooled = tf.nn.avg_pool2d(input_tensor, ksize=2, strides=2, padding='VALID')
print("Average pooling result:\n", tf.squeeze(avg_pooled).numpy())
```

PyTorch Example: Pooling Operation

```
import torch
import torch.nn.functional as F

# Define an input tensor (feature map)
input_tensor = torch.tensor([[1, 3, 2, 4],
                             [5, 6, 8, 7],
                             [9, 11, 10, 12],
                             [13, 15, 14, 16]], dtype=torch.float32)
input_tensor = input_tensor.view(1, 1, 4, 4)  # Reshape to [batch, channels, height,
 width]

# Apply max pooling
max_pooled = F.max_pool2d(input_tensor, kernel_size=2, stride=2)
print("Max pooling result:\n", max_pooled.squeeze().numpy())

# Apply average pooling
avg_pooled = F.avg_pool2d(input_tensor, kernel_size=2, stride=2)
print("Average pooling result:\n", avg_pooled.squeeze().numpy())
```

1.5.2 Sequence Data Processing with Tensors

Processing sequence data, such as time series, text, and speech, is a critical task in many applications. RNNs and attention mechanisms are fundamental components for handling sequential data. These methods enable models to capture temporal dependencies and relationships within sequences, providing powerful tools for tasks such as language modeling, translation, and sequence prediction.

Recurrent Operations

RNNs are designed to process sequences by maintaining a hidden state that captures information from previous time steps. The hidden state is updated at each time step based on the current input and the previous hidden state. Mathematically, the operation of an RNN can be described as follows: Given an input sequence $\{\mathbf{x}_t\}$ where $t = 1, 2, \ldots, T$, the hidden state $\mathbf{h}_t$ at time step t is computed using the recurrent equation:

$$\mathbf{h}_t = \phi(\mathbf{W}_{hh}\mathbf{h}_{t-1} + \mathbf{W}_{xh}\mathbf{x}_t + \mathbf{b}_h)$$

where $\mathbf{W}_{hh}$ is the weight matrix for the hidden state, $\mathbf{W}_{xh}$ is the weight matrix for the input, $\mathbf{b}_h$ is the bias term, and ϕ is the activation function (often tanh or ReLU). The output $\mathbf{y}_t$ at time step t can be computed as:

$$\mathbf{y}_t = \mathbf{W}_{hy}\mathbf{h}_t + \mathbf{b}_y$$

where $\mathbf{W}_{hy}$ is the weight matrix for the output and $\mathbf{b}_y$ is the bias term.

Attention Mechanisms

Attention mechanisms have become a cornerstone in sequence processing, particularly in tasks requiring long-term dependencies and contextual understanding, such as machine translation and text summarization. Attention allows the model to focus on relevant parts of the input sequence when generating each part of the output sequence, effectively addressing the limitations of RNNs in capturing long-range dependencies.

Self-Attention Self-attention, or intra-attention, computes the representation of a sequence by relating each element to all other elements in the sequence. This mechanism enables the model to weigh the importance of each element based on its relevance to the others. The self-attention mechanism can be mathematically formulated as follows: Given an input sequence represented by a matrix $\mathbf{X} \in \mathbb{R}^{T\times d}$, where T is the sequence length and d is the dimensionality, the self-attention mechanism computes the query, key, and value matrices:

$$\mathbf{Q} = \mathbf{X}\mathbf{W}_Q$$

$$\mathbf{K} = \mathbf{X}\mathbf{W}_K$$

$$\mathbf{V} = \mathbf{X}\mathbf{W}_V$$

where $\mathbf{W}_Q, \mathbf{W}_K, \mathbf{W}_V \in \mathbb{R}^{d\times d_k}$ are learned weight matrices. The attention scores are computed as:

$$\mathbf{A} = \text{softmax}\left(\frac{\mathbf{Q}\mathbf{K}^T}{\sqrt{d_k}}\right)$$

where the softmax function ensures that the attention scores sum to one. The output of the self-attention mechanism is a weighted sum of the value vectors:

$$\mathbf{Z} = \mathbf{A}\mathbf{V}$$

Multi-Head Attention Multi-head attention extends the self-attention mechanism by allowing the model to attend to information from different representation subspaces at different positions. This is achieved by running multiple self-attention operations (heads) in parallel and concatenating their outputs. The multi-head attention mechanism is defined as:

$$\text{MultiHead}(\mathbf{X}) = \oplus(\text{head}_1, \ldots, \text{head}_h)\mathbf{W}_O$$

where each head is computed as:

$$\text{head}_i = \text{Attention}(\mathbf{Q}^i, \mathbf{K}^i, \mathbf{V}^i)$$

with:

$$\mathbf{Q}^i = \mathbf{X}\mathbf{W}^i_Q$$

$$\mathbf{K}^i = \mathbf{X}\mathbf{W}^i_K$$

$$\mathbf{V}^i = \mathbf{X}\mathbf{W}^i_V$$

Here, $\mathbf{W}^i_Q, \mathbf{W}^i_K, \mathbf{W}^i_V \in \mathbb{R}^{d \times d_k}$ are learned weight matrices for each head, and $\mathbf{W}_O \in \mathbb{R}^{hd_k \times d}$ is the output weight matrix. The attention mechanism for each head can be expressed as:

$$\text{Attention}(\mathbf{Q}^i, \mathbf{K}^i, \mathbf{V}^i) = \text{softmax}\left(\frac{\mathbf{Q}^i(\mathbf{K}^i)^\top}{\sqrt{d_k}}\right)\mathbf{V}^i$$

In this formulation, the input $\mathbf{X}$ is first projected into multiple subspaces to create the query, key, and value matrices for each head, allowing the model to attend to different aspects of the input simultaneously.

Example: Self-Attention Mechanism in PyTorch

```
import torch
import torch.nn.functional as F

def scaled_dot_product_attention(Q, K, V):
    d_k = Q.size(-1)
    scores = torch.matmul(Q, K.transpose(-2, -1)) / torch.sqrt(torch.tensor(d_k, dtype
        =torch.float32))
    attn_weights = F.softmax(scores, dim=-1)
    output = torch.matmul(attn_weights, V)
    return output, attn_weights

# Example
```

```
d_model = 64
seq_length = 10
batch_size = 32

Q = torch.randn(batch_size, seq_length, d_model)
K = torch.randn(batch_size, seq_length, d_model)
V = torch.randn(batch_size, seq_length, d_model)

output, attn_weights = scaled_dot_product_attention(Q, K, V)
print("Self-attention output shape:", output.shape)
print("Attention weights shape:", attn_weights.shape)
```

1.6 Exercises

1. Given two tensors $\mathbf{A} \in \mathbb{R}^{2\times 3}$ and $\mathbf{B} \in \mathbb{R}^{2\times 3}$:

$$\mathbf{A} = \begin{pmatrix} 1 & 2 & 3 \\ 4 & 5 & 6 \end{pmatrix}, \quad \mathbf{B} = \begin{pmatrix} 7 & 8 & 9 \\ 10 & 11 & 12 \end{pmatrix}$$

   (a) Compute $\mathbf{A} + \mathbf{B}$.
   (b) Compute the element-wise multiplication $\mathbf{A} \circ \mathbf{B}$.
   (c) Compute the matrix multiplication $\mathbf{A}\mathbf{B}^T$.

2. Consider a neural network with an input layer, one hidden layer with ReLU activation, and an output layer with a sigmoid activation function. Let the input $\mathbf{x} = \begin{pmatrix} 1 \\ 2 \end{pmatrix}$, the weights for the hidden layer $\mathbf{W}_1 \in \mathbb{R}^{2\times 2}$, and the weights for the output layer $\mathbf{W}_2 \in \mathbb{R}^{1\times 2}$:

$$\mathbf{W}_1 = \begin{pmatrix} 0.5 & -0.2 \\ 0.8 & 0.4 \end{pmatrix}, \quad \mathbf{W}_2 = \begin{pmatrix} 0.3 & -0.7 \end{pmatrix}$$

   Compute the output of the network for the given input $\mathbf{x}$.
3. Using the same network as in Exercise 2, let the true output $y = 1$. Calculate the gradients of the loss with respect to the weights $\mathbf{W}_1$ and $\mathbf{W}_2$ using the cross-entropy loss function.
4. Given a tensor $\mathbf{C} \in \mathbb{R}^{2\times 3\times 4}$ with random values, reshape it into a tensor of shape $(3, 4, 2)$. Verify that the total number of elements remains the same.
5. Consider a 3D tensor $\mathbf{D} \in \mathbb{R}^{4\times 4\times 4}$. Extract a sub-tensor containing the first two dimensions fixed at indices 1 and 2, respectively. Write down the resulting tensor's shape and values.
6. Let $\mathbf{E} \in \mathbb{R}^{3\times 4\times 5}$ and $\mathbf{F} \in \mathbb{R}^{4\times 5\times 2}$ be two third-order tensors with elements:

$$E_{ijk} = i + j + k, \quad F_{jkl} = j \cdot k \cdot l$$

   where i, j, k, l are the respective indices.

(a) Perform a tensor contraction on the third dimension of $\mathbf{E}$ and the second dimension of $\mathbf{F}$, resulting in a new tensor $\mathbf{G} \in \mathbb{R}^{3\times 4\times 2}$.
(b) Compute the Frobenius norm of the resulting tensor $\mathbf{G}$.

7. Given matrices $\mathbf{P} \in \mathbb{R}^{2\times 2}$ and $\mathbf{Q} \in \mathbb{R}^{2\times 2}$:

$$\mathbf{P} = \begin{pmatrix} 1 & 2 \\ 3 & 4 \end{pmatrix}, \quad \mathbf{Q} = \begin{pmatrix} 0 & 5 \\ 6 & 7 \end{pmatrix}$$

(a) Compute the Kronecker product $\mathbf{P} \otimes \mathbf{Q}$.
(b) Consider a tensor $\mathbf{R} \in \mathbb{R}^{2\times 2\times 2}$ where each slice $\mathbf{R}_{::k}$ is given by $\mathbf{P} \cdot k$ for $k = 1, 2$. Compute the tensor product $\mathbf{R} \times \mathbf{Q}$ along the second mode, resulting in a tensor $\mathbf{S} \in \mathbb{R}^{2\times 4\times 2}$.

8. Given a third-order tensor $\mathbf{T} \in \mathbb{R}^{2\times 2\times 2}$ with elements:

$$T_{ijk} = i + j - k$$

where $i, j, k \in \{1, 2\}$.

(a) Flatten the tensor $\mathbf{T}$ into a matrix $\mathbf{T}_{(1)}$ by unfolding along the first mode.
(b) Compute the eigenvalues and eigenvectors of the matrix $\mathbf{T}_{(1)}$.

9. Consider a tensor network where tensor $\mathbf{X} \in \mathbb{R}^{2\times 2\times 2}$ is passed through a series of transformations. Let:

$$X_{ijk} = i + 2j + 3k$$

The transformations are defined as follows: Apply a linear transformation $\mathbf{W} \in \mathbb{R}^{2\times 2}$ such that $\mathbf{Y} = \mathbf{W} * \mathbf{X}$, where $*$ denotes the tensor contraction along the second mode.

$$\mathbf{W} = \begin{pmatrix} 1 & 0 \\ 0 & 1 \end{pmatrix}$$

(a) Compute the resulting tensor $\mathbf{Y} \in \mathbb{R}^{2\times 2}$.
(b) Assume a loss function $L = \frac{1}{2} \sum_{ij} (Y_{ij} - \hat{Y}_{ij})^2$ with $\hat{Y}$ being the target tensor. Compute the gradients $\frac{\partial L}{\partial \mathbf{W}}$ and $\frac{\partial L}{\partial \mathbf{X}}$ using backpropagation through the tensor network.

Chapter 2
The Algebra and Geometry of Deep Learning

In the description of the physical world...there is no 'space' without time, there are only four-dimensional tensors.

Albert Einstein, Relativity: The Special and the General Theory

2.1 Introduction

Tensor algebra is a mathematical framework that extends the concepts of linear algebra to higher dimensions, enabling the representation and manipulation of multi-dimensional data. In deep learning, tensors are the primary data structures used to encode inputs, outputs, weights, and intermediate representations within neural networks. Understanding tensor algebra is essential for designing, implementing, and optimizing neural networks. A tensor is a multi-dimensional array that generalizes scalars (0D), vectors (1D), and matrices (2D) to higher dimensions. The order of a tensor refers to the number of dimensions it has. For instance, a scalar is a tensor of order 0, a vector is a tensor of order 1, a matrix is a tensor of order 2, and higher-order tensors extend this concept further. Mathematically, a tensor can be denoted as $\mathcal{A} \in \mathbb{R}^{I_1 \times I_2 \times \cdots \times I_N}$, where N is the order of the tensor and I_i are the dimensions. For example, a 3D tensor representing a batch of color images might have dimensions (N, H, W, C), where N is the batch size, H is the height, W is the width, and C is the number of color channels. Tensor operations extend the familiar operations of linear algebra to higher dimensions. Recall some fundamental tensor operations from Chap. 1:

1. **Element-wise Operations:** These operations are applied independently to each element of the tensors. For tensors $\mathcal{A}$ and $\mathcal{B}$ of the same shape, element-wise addition is defined as:

P. Singh and B. Raman, *Deep Learning Through the Prism of Tensors*, Studies in Big Data 162, https://doi.org/10.1007/978-981-97-8019-8_2

$$\mathcal{C}_{i_1,i_2,\ldots,i_N} = \mathcal{A}_{i_1,i_2,\ldots,i_N} + \mathcal{B}_{i_1,i_2,\ldots,i_N}$$

2. **Tensor Multiplication:** Tensor multiplication generalizes matrix multiplication. For a 2D matrix $\mathbf{A} \in \mathbb{R}^{m\times n}$ and a 2D matrix $\mathbf{B} \in \mathbb{R}^{n\times p}$, the product $\mathbf{C} = \mathbf{AB}$ is:

$$\mathbf{C}_{ik} = \sum_{j=1}^{n} \mathbf{A}_{ij}\mathbf{B}_{jk}$$

3. **Tensor Contraction:** This operation generalizes the inner product to tensors. For tensors $\mathcal{A} \in \mathbb{R}^{I_1\times I_2\times\cdots\times I_N}$ and $\mathcal{B} \in \mathbb{R}^{J_1\times J_2\times\cdots\times J_M}$, a contraction over specified dimensions sums over the products of the elements along those dimensions.
4. **Outer Product:** The outer product of two vectors $\mathbf{a} \in \mathbb{R}^n$ and $\mathbf{b} \in \mathbb{R}^m$ results in a matrix $\mathbf{C} \in \mathbb{R}^{n\times m}$:

$$\mathbf{C}_{ij} = \mathbf{a}_i\mathbf{b}_j$$

These operations are crucial for building and training neural networks. For example, the forward pass in a fully connected layer involves matrix multiplication between the input tensor and the weight tensor, followed by the addition of a bias tensor and the application of an activation function. Consider a simple neural network layer with input $\mathbf{x} \in \mathbb{R}^d$, weights $\mathbf{W} \in \mathbb{R}^{d\times h}$, and bias $\mathbf{b} \in \mathbb{R}^h$. The output $\mathbf{y} \in \mathbb{R}^h$ is computed as:

$$\mathbf{y} = \phi(\mathbf{W}^T\mathbf{x} + \mathbf{b})$$

where ϕ is an activation function such as ReLU. Tensor algebra provides a powerful and flexible framework for these computations, enabling efficient implementation of complex operations on multi-dimensional data.

Importance of Geometry in Deep Learning

Geometry plays a crucial role in deep learning, providing insights into the structure and behavior of neural networks. Geometric concepts help us understand how neural networks transform and represent data, leading to more effective models and training algorithms.

Geometric Interpretations of Neural Networks

1. **Feature Spaces:** Neural networks learn to map input data into high-dimensional feature spaces. Each layer of the network applies a geometric transformation, such as a linear mapping or non-linear activation, to project the data into new feature spaces. For example, a convolutional layer applies convolutional filters to extract local patterns, mapping the input image to a feature map that highlights relevant structures.
2. **Manifold Learning:** Data often lies on a low-dimensional manifold within the high-dimensional input space. Neural networks learn to identify and exploit these manifolds, effectively reducing the dimensionality of the data and focusing on

the most informative features. Techniques such as autoencoders explicitly aim to learn these manifolds, capturing the underlying structure of the data.

3. **Optimization Landscapes:** Training neural networks involves navigating complex optimization landscapes defined by the loss function. Geometric concepts such as gradients, curvature, and critical points help us understand and improve optimization algorithms. For instance, gradient descent follows the negative gradient of the loss function to find a local minimum, while advanced methods like Adam adaptively adjust the learning rate based on the geometry of the loss surface.

Example: Consider the training of a neural network using gradient descent. The weight update rule is given by:

$$\mathbf{W} \leftarrow \mathbf{W} - \eta \nabla_{\mathbf{W}} L(\mathbf{W})$$

where η is the learning rate and $\nabla_{\mathbf{W}} L(\mathbf{W})$ is the gradient of the loss function L with respect to the weights $\mathbf{W}$. The gradient provides a geometric direction for updating the weights, guiding the optimization process toward a minimum.

Transformations and Invariances Neural networks leverage geometric transformations to achieve invariance to certain properties of the input data. For example:

1. **Translation Invariance:** Convolutional layers achieve translation invariance by applying the same filter across different spatial locations in the input image. This allows the network to detect features regardless of their position.
2. **Rotation and Scale Invariance:** Advanced architectures and data augmentation techniques can introduce invariance to rotations and scaling. For instance, spatial transformer networks include modules that explicitly learn to perform geometric transformations, making the network robust to such variations.

Example: In image classification, data augmentation techniques such as random rotations, scaling, and translations are applied to the training images to improve the network's robustness to these transformations. This geometric manipulation of the training data helps the network generalize better to new, unseen images.

2.2 Fundamentals of Tensor Algebra

2.2.1 Definitions and Notation

Basic Tensor Concepts

Tensors are a generalization of scalars, vectors, and matrices to higher dimensions, and they form the foundation for data representation in many machine learning and deep learning applications. Understanding the basic concepts of tensors is crucial for effectively working with multi-dimensional data. A scalar is a single number and is considered a tensor of rank 0. For example, $a \in \mathbb{R}$ is a scalar. A vector is an ordered array of numbers and is a tensor of rank 1. For instance, $\mathbf{v} = [v_1, v_2, \ldots, v_n] \in \mathbb{R}^n$ is a vector. A matrix is a two-dimensional array of numbers and is a tensor of rank 2. For example, $\mathbf{A} \in \mathbb{R}^{m \times n}$ is a matrix with m rows and n columns:

$$\mathbf{A} = \begin{pmatrix} a_{11} & a_{12} & \cdots & a_{1n} \\ a_{21} & a_{22} & \cdots & a_{2n} \\ \vdots & \vdots & \ddots & \vdots \\ a_{m1} & a_{m2} & \cdots & a_{mn} \end{pmatrix}$$

A tensor is a multi-dimensional array that extends these concepts to higher dimensions. A tensor of rank 3, for example, can be represented as $\mathcal{A} \in \mathbb{R}^{I \times J \times K}$, where I, J, K are the dimensions of the tensor:

$$\mathcal{A} = \{a_{ijk}\}, \quad i \in \{1, \ldots, I\},\ j \in \{1, \ldots, J\},\ k \in \{1, \ldots, K\}$$

Higher-order tensors (rank 4, 5, etc.) are defined similarly. For example, a rank 4 tensor can be represented as $\mathcal{B} \in \mathbb{R}^{I \times J \times K \times L}$. Tensors can be used to represent complex data structures in various fields, including computer vision, natural language processing, and physics. In deep learning, tensors are used to represent input data, model parameters, and intermediate computations. For example, consider a batch of grayscale images, each of size 28×28 pixels. The batch can be represented as a 3D tensor $\mathcal{X} \in \mathbb{R}^{N \times 28 \times 28}$, where N is the number of images in the batch. If the images are in color (RGB), the batch can be represented as a 4D tensor $\mathcal{X} \in \mathbb{R}^{N \times 28 \times 28 \times 3}$, where the last dimension corresponds to the color channels.

Tensor Notation

Tensor notation provides a compact and efficient way to represent and manipulate multi-dimensional arrays. Understanding tensor notation is essential for working with tensor operations in both theoretical and practical settings.

Index Notation In index notation, the elements of a tensor are referenced by their indices. For example, the element in the i-th row and j-th column of a matrix $\mathbf{A}$ is denoted as a_{ij}. Similarly, the element at position (i, j, k) in a rank 3 tensor $\mathcal{A}$ is denoted as a_{ijk}.

Einstein Summation Convention The Einstein summation convention is a notational shorthand used to simplify tensor operations, particularly tensor contractions. According to this convention, repeated indices in a term imply summation over those indices. For example, the matrix multiplication $\mathbf{C} = \mathbf{AB}$ can be written as:

$$c_{ik} = \sum_j a_{ij} b_{jk}$$

Using the Einstein summation convention, this is written more compactly as:

$$c_{ik} = a_{ij} b_{jk}$$

2.2.2 *Basic Operations*

Tensor Addition

Tensor addition is an element-wise operation that applies to tensors of the same shape. This operation sums corresponding elements from the tensors to produce a new tensor. Given two tensors $\mathcal{A}$ and $\mathcal{B}$ of the same shape, their element-wise sum $\mathcal{C}$ is computed as:

$$\mathcal{C}_{i_1,i_2,\ldots,i_N} = \mathcal{A}_{i_1,i_2,\ldots,i_N} + \mathcal{B}_{i_1,i_2,\ldots,i_N}$$

where $\mathcal{A}, \mathcal{B}, \mathcal{C} \in \mathbb{R}^{I_1 \times I_2 \times \cdots \times I_N}$.

Example: Consider two matrices (rank-2 tensors) **A** and **B** each of shape 2×2:

$$\mathbf{A} = \begin{pmatrix} 1 & 2 \\ 3 & 4 \end{pmatrix}, \quad \mathbf{B} = \begin{pmatrix} 5 & 6 \\ 7 & 8 \end{pmatrix}$$

The element-wise sum $\mathbf{C} = \mathbf{A} + \mathbf{B}$ is:

$$\mathbf{C} = \begin{pmatrix} 1+5 & 2+6 \\ 3+7 & 4+8 \end{pmatrix} = \begin{pmatrix} 6 & 8 \\ 10 & 12 \end{pmatrix}$$

Tensor Multiplication

Tensor multiplication generalizes matrix multiplication to higher dimensions and can be performed in several ways, including element-wise multiplication, matrix multiplication, and higher-order tensor contractions. Here, we focus on the element-wise and matrix multiplication operations.

Element-wise Multiplication Element-wise multiplication of two tensors $\mathcal{A}$ and $\mathcal{B}$ of the same shape produces a tensor $\mathcal{C}$, where each element is the product of corresponding elements:

$$\mathcal{C}_{i_1,i_2,\ldots,i_N} = \mathcal{A}_{i_1,i_2,\ldots,i_N} \cdot \mathcal{B}_{i_1,i_2,\ldots,i_N}$$

Example: Consider two vectors (rank-1 tensors) **a** and **b**:

$$\mathbf{a} = \begin{pmatrix} 1 \\ 2 \\ 3 \end{pmatrix}, \quad \mathbf{b} = \begin{pmatrix} 4 \\ 5 \\ 6 \end{pmatrix}$$

Their element-wise product $\mathbf{c} = \mathbf{a} \odot \mathbf{b}$ is:

$$\mathbf{c} = \begin{pmatrix} 1 \cdot 4 \\ 2 \cdot 5 \\ 3 \cdot 6 \end{pmatrix} = \begin{pmatrix} 4 \\ 10 \\ 18 \end{pmatrix}$$

Matrix Multiplication Matrix multiplication involves the dot product of rows and columns of two matrices. Given matrices $\mathbf{A} \in \mathbb{R}^{m \times n}$ and $\mathbf{B} \in \mathbb{R}^{n \times p}$, their product $\mathbf{C} = \mathbf{AB} \in \mathbb{R}^{m \times p}$ is computed as:

$$c_{ij} = \sum_{k=1}^{n} a_{ik} b_{kj}$$

Example: Consider the matrices:

$$\mathbf{A} = \begin{pmatrix} 1 & 2 \\ 3 & 4 \end{pmatrix}, \quad \mathbf{B} = \begin{pmatrix} 5 & 6 \\ 7 & 8 \end{pmatrix}$$

Their matrix product $\mathbf{C} = \mathbf{AB}$ is:

$$\mathbf{C} = \begin{pmatrix} 1 \cdot 5 + 2 \cdot 7 & 1 \cdot 6 + 2 \cdot 8 \\ 3 \cdot 5 + 4 \cdot 7 & 3 \cdot 6 + 4 \cdot 8 \end{pmatrix} = \begin{pmatrix} 19 & 22 \\ 43 & 50 \end{pmatrix}$$

Tensor Contraction

Tensor contraction generalizes matrix multiplication to higher-order tensors. It involves summing over specified pairs of indices, reducing the dimensionality of the resulting tensor. For tensors $\mathcal{A} \in \mathbb{R}^{I \times J \times K}$ and $\mathcal{B} \in \mathbb{R}^{K \times L}$, contracting over the third dimension of $\mathcal{A}$ and the first dimension of $\mathcal{B}$ results in a tensor $\mathcal{C} \in \mathbb{R}^{I \times J \times L}$:

$$c_{ijl} = \sum_{k=1}^{K} a_{ijk} b_{kl}$$

Example: Consider a rank-3 tensor $\mathcal{A} \in \mathbb{R}^{2 \times 2 \times 2}$ and a rank-2 tensor $\mathcal{B} \in \mathbb{R}^{2 \times 2}$:

$$\mathcal{A}(:, :, 1) = \begin{pmatrix} 1 & 2 \\ 3 & 4 \end{pmatrix}, \quad \mathcal{A}(:, :, 2) = \begin{pmatrix} 5 & 6 \\ 7 & 8 \end{pmatrix}$$

$$\mathbf{B} = \begin{pmatrix} 9 & 10 \\ 11 & 12 \end{pmatrix}$$

The contraction over the third dimension of $\mathcal{A}$ and the first dimension of $\mathbf{B}$ produces $\mathcal{C} \in \mathbb{R}^{2 \times 2}$:

$$\mathcal{C}(:,:,1) = \sum_{k=1}^{2} \mathcal{A}(:,:,k) \cdot \mathbf{B}_{k,1} = \begin{pmatrix} 1 \cdot 9 + 2 \cdot 11 & 3 \cdot 9 + 4 \cdot 11 \\ 5 \cdot 9 + 6 \cdot 11 & 7 \cdot 9 + 8 \cdot 11 \end{pmatrix} = \begin{pmatrix} 31 & 63 \\ 99 & 141 \end{pmatrix}$$

Outer Products

The outer product of two vectors produces a matrix, and this concept extends to higher-order tensors. Given vectors $\mathbf{a} \in \mathbb{R}^n$ and $\mathbf{b} \in \mathbb{R}^m$, their outer product $\mathbf{C} \in \mathbb{R}^{n \times m}$ is defined as:

$$c_{ij} = a_i b_j$$

Example: Consider the vectors:

$$\mathbf{a} = \begin{pmatrix} 1 \\ 2 \end{pmatrix}, \quad \mathbf{b} = \begin{pmatrix} 3 \\ 4 \\ 5 \end{pmatrix}$$

Their outer product $\mathbf{C} = \mathbf{a} \otimes \mathbf{b}$ is:

$$\mathbf{C} = \begin{pmatrix} 1 \cdot 3 & 1 \cdot 4 & 1 \cdot 5 \\ 2 \cdot 3 & 2 \cdot 4 & 2 \cdot 5 \end{pmatrix} = \begin{pmatrix} 3 & 4 & 5 \\ 6 & 8 & 10 \end{pmatrix}$$

2.2.3 Advanced Operations

Kronecker Product

The Kronecker product is a tensor operation that generalizes the outer product to matrices. Given two matrices $\mathbf{A} \in \mathbb{R}^{m \times n}$ and $\mathbf{B} \in \mathbb{R}^{p \times q}$, their Kronecker product $\mathbf{C} \in \mathbb{R}^{(m \cdot p) \times (n \cdot q)}$ is defined as:

$$\mathbf{C} = \mathbf{A} \otimes \mathbf{B}$$

This operation produces a block matrix where each element a_{ij} of $\mathbf{A}$ is multiplied by the entire matrix $\mathbf{B}$:

$$\mathbf{C} = \begin{pmatrix} a_{11}\mathbf{B} & a_{12}\mathbf{B} & \cdots & a_{1n}\mathbf{B} \\ a_{21}\mathbf{B} & a_{22}\mathbf{B} & \cdots & a_{2n}\mathbf{B} \\ \vdots & \vdots & \ddots & \vdots \\ a_{m1}\mathbf{B} & a_{m2}\mathbf{B} & \cdots & a_{mn}\mathbf{B} \end{pmatrix}$$

Example: Consider two matrices $\mathbf{A} \in \mathbb{R}^{2 \times 2}$ and $\mathbf{B} \in \mathbb{R}^{2 \times 2}$:

$$\mathbf{A} = \begin{pmatrix} 1 & 2 \\ 3 & 4 \end{pmatrix}, \quad \mathbf{B} = \begin{pmatrix} 0 & 5 \\ 6 & 7 \end{pmatrix}$$

Their Kronecker product $\mathbf{C} = \mathbf{A} \otimes \mathbf{B}$ is:

$$\mathbf{C} = \begin{pmatrix} 1\begin{pmatrix} 0 & 5 \\ 6 & 7 \end{pmatrix} & 2\begin{pmatrix} 0 & 5 \\ 6 & 7 \end{pmatrix} \\ 3\begin{pmatrix} 0 & 5 \\ 6 & 7 \end{pmatrix} & 4\begin{pmatrix} 0 & 5 \\ 6 & 7 \end{pmatrix} \end{pmatrix} = \begin{pmatrix} 0 & 5 & 0 & 10 \\ 6 & 7 & 12 & 14 \\ 0 & 15 & 0 & 20 \\ 18 & 21 & 24 & 28 \end{pmatrix}$$

The Kronecker product is useful in various applications, such as signal processing, image processing, and constructing structured matrices in numerical linear algebra.

Tensor Decomposition

Tensor decomposition is a technique used to factorize a tensor into a set of simpler, often interpretable, components. This is analogous to matrix factorizations such as singular value decomposition (SVD) for matrices. Tensor decompositions are essential in applications like data compression, feature extraction, and understanding latent structures in high-dimensional data.

CANDECOMP/PARAFAC (CP) Decomposition The CP decomposition expresses a tensor as a sum of rank-1 tensors. For a tensor $\mathcal{X} \in \mathbb{R}^{I \times J \times K}$, the CP decomposition is given by:

$$\mathcal{X} \approx \sum_{r=1}^{R} \mathbf{a}_r \circ \mathbf{b}_r \circ \mathbf{c}_r$$

where $\mathbf{a}_r \in \mathbb{R}^I$, $\mathbf{b}_r \in \mathbb{R}^J$, and $\mathbf{c}_r \in \mathbb{R}^K$ are factor vectors, $\circ$ denotes the outer product, and R is the rank of the decomposition.

Tucker Decomposition The Tucker decomposition generalizes the CP decomposition by expressing a tensor as a core tensor multiplied by a matrix along each mode. For a tensor $\mathcal{X} \in \mathbb{R}^{I \times J \times K}$, the Tucker decomposition is given by:

$$\mathcal{X} \approx \mathcal{G} \times_1 \mathbf{A} \times_2 \mathbf{B} \times_3 \mathbf{C}$$

where $\mathcal{G} \in \mathbb{R}^{R_1 \times R_2 \times R_3}$ is the core tensor, $\mathbf{A} \in \mathbb{R}^{I \times R_1}$, $\mathbf{B} \in \mathbb{R}^{J \times R_2}$, and $\mathbf{C} \in \mathbb{R}^{K \times R_3}$ are factor matrices, and $\times_n$ denotes the mode-n product.

Example: Consider a tensor $\mathcal{X} \in \mathbb{R}^{2 \times 2 \times 2}$:

$$\mathcal{X}(:, :, 1) = \begin{pmatrix} 1 & 2 \\ 3 & 4 \end{pmatrix}, \quad \mathcal{X}(:, :, 2) = \begin{pmatrix} 5 & 6 \\ 7 & 8 \end{pmatrix}$$

A possible CP decomposition with rank 1 is:

$$\mathbf{a}_1 = \begin{pmatrix} 1 \\ 2 \end{pmatrix}, \quad \mathbf{b}_1 = \begin{pmatrix} 1 \\ 1 \end{pmatrix}, \quad \mathbf{c}_1 = \begin{pmatrix} 1 \\ 1 \end{pmatrix}$$

Thus,

$$\mathcal{X} \approx \mathbf{a}_1 \circ \mathbf{b}_1 \circ \mathbf{c}_1$$

Tensor Network Representations

Tensor networks provide a graphical and algebraic framework to represent and manipulate large tensors. They decompose high-dimensional tensors into networks of smaller tensors interconnected by contractions. Tensor networks are widely used in quantum physics, quantum chemistry, and machine learning due to their ability to efficiently represent complex data structures.

Matrix Product States (MPS) MPS is a type of tensor network that represents a high-dimensional tensor as a chain of low-dimensional tensors. For a tensor $\mathcal{X} \in \mathbb{R}^{I_1 \times I_2 \times \cdots \times I_N}$, the MPS representation is:

$$\mathcal{X}_{i_1 i_2 \ldots i_N} = \sum_{a_1, a_2, \ldots, a_{N-1}} \mathbf{A}^{[1]}_{i_1 a_1} \mathbf{A}^{[2]}_{a_1 i_2 a_2} \cdots \mathbf{A}^{[N]}_{a_{N-1} i_N}$$

where $\mathbf{A}^{[k]}$ are low-dimensional tensors.

Tensor Train (TT) The TT decomposition is another type of tensor network similar to MPS but emphasizes hierarchical structure. For a tensor $\mathcal{X} \in \mathbb{R}^{I_1 \times I_2 \times \cdots \times I_N}$, the TT decomposition is:

$$\mathcal{X}_{i_1 i_2 \ldots i_N} = \mathbf{G}^{[1]}_{i_1} \mathbf{G}^{[2]}_{i_2} \cdots \mathbf{G}^{[N]}_{i_N}$$

where $\mathbf{G}^{[k]}$ are core tensors. For example consider a tensor $\mathcal{X} \in \mathbb{R}^{2 \times 2 \times 2}$ with MPS representation:

$$\mathbf{A}^{[1]} = \begin{pmatrix} 1 & 0 \\ 0 & 1 \end{pmatrix}, \quad \mathbf{A}^{[2]} = \begin{pmatrix} 1 & 1 \\ 1 & 1 \end{pmatrix}, \quad \mathbf{A}^{[3]} = \begin{pmatrix} 1 \\ 2 \end{pmatrix}$$

The TT decomposition efficiently captures the structure of $\mathcal{X}$, allowing for scalable computations in higher dimensions.

2.3 Geometric Interpretation of Tensors

2.3.1 *Transformations and Rotations*

Geometric transformations are fundamental operations in both linear algebra and deep learning, where they are used to manipulate data and model complex relationships. Tensors, as generalizations of vectors and matrices, can undergo various geometric transformations, including affine and orthogonal transformations. Understanding these transformations provides insights into how neural networks process and transform data.

Affine Transformations

Affine transformations are a class of linear transformations that include translation, scaling, rotation, and shearing. An affine transformation can be represented as a combination of a linear transformation followed by a translation. Mathematically, an affine transformation $\mathbf{y}$ of a point $\mathbf{x}$ in n-dimensional space can be expressed as:

$$\mathbf{y} = \mathbf{A}\mathbf{x} + \mathbf{b}$$

where $\mathbf{A} \in \mathbb{R}^{n \times n}$ is a matrix representing the linear part of the transformation, and $\mathbf{b} \in \mathbb{R}^n$ is a vector representing the translation. For example, consider a 2D affine transformation involving scaling and translation. Let $\mathbf{A}$ be the scaling matrix and $\mathbf{b}$ be the translation vector:

$$\mathbf{A} = \begin{pmatrix} 2 & 0 \\ 0 & 3 \end{pmatrix}, \quad \mathbf{b} = \begin{pmatrix} 1 \\ 1 \end{pmatrix}$$

For a point $\mathbf{x} = \begin{pmatrix} x_1 \\ x_2 \end{pmatrix}$, the affine transformation is:

$$\mathbf{y} = \mathbf{A}\mathbf{x} + \mathbf{b} = \begin{pmatrix} 2 & 0 \\ 0 & 3 \end{pmatrix} \begin{pmatrix} x_1 \\ x_2 \end{pmatrix} + \begin{pmatrix} 1 \\ 1 \end{pmatrix} = \begin{pmatrix} 2x_1 + 1 \\ 3x_2 + 1 \end{pmatrix}$$

This transformation scales the x_1-coordinate by 2 and the x_2-coordinate by 3, then translates the point by (1, 1). Affine transformations are widely used in computer graphics and image processing. For instance, they can be used to augment training data by applying random translations, rotations, and scalings, improving the robustness and generalization of deep learning models.

Orthogonal Transformations

Orthogonal transformations are a special class of linear transformations that preserve the length and angle between vectors. These transformations include rotations and reflections, and they can be represented by orthogonal matrices. An orthogonal matrix $\mathbf{Q} \in \mathbb{R}^{n \times n}$ satisfies:

$$\mathbf{Q}^T\mathbf{Q} = \mathbf{Q}\mathbf{Q}^T = \mathbf{I}$$

where $\mathbf{I}$ is the identity matrix. Orthogonal transformations are essential in various applications, including computer vision, robotics, and physics, as they maintain the geometric properties of the data. For example, consider a 2D rotation matrix $\mathbf{Q}$ that rotates points by an angle θ:

$$\mathbf{Q} = \begin{pmatrix} \cos\theta & -\sin\theta \\ \sin\theta & \cos\theta \end{pmatrix}$$

For a point $\mathbf{x} = \begin{pmatrix} x_1 \\ x_2 \end{pmatrix}$, the rotated point $\mathbf{y}$ is given by:

$$\mathbf{y} = \mathbf{Q}\mathbf{x} = \begin{pmatrix} \cos\theta & -\sin\theta \\ \sin\theta & \cos\theta \end{pmatrix} \begin{pmatrix} x_1 \\ x_2 \end{pmatrix} = \begin{pmatrix} x_1\cos\theta - x_2\sin\theta \\ x_1\sin\theta + x_2\cos\theta \end{pmatrix}$$

If $\theta = \frac{\pi}{4}$ (45 degrees), the rotation matrix becomes:

$$\mathbf{Q} = \begin{pmatrix} \frac{\sqrt{2}}{2} & -\frac{\sqrt{2}}{2} \\ \frac{\sqrt{2}}{2} & \frac{\sqrt{2}}{2} \end{pmatrix}$$

For a point $\mathbf{x} = \begin{pmatrix} 1 \\ 0 \end{pmatrix}$, the rotated point $\mathbf{y}$ is:

$$\mathbf{y} = \mathbf{Q}\mathbf{x} = \begin{pmatrix} \frac{\sqrt{2}}{2} & -\frac{\sqrt{2}}{2} \\ \frac{\sqrt{2}}{2} & \frac{\sqrt{2}}{2} \end{pmatrix} \begin{pmatrix} 1 \\ 0 \end{pmatrix} = \begin{pmatrix} \frac{\sqrt{2}}{2} \\ \frac{\sqrt{2}}{2} \end{pmatrix}$$

This transformation rotates the point $(1, 0)$ by 45 degrees to $(\frac{\sqrt{2}}{2}, \frac{\sqrt{2}}{2})$.

Orthogonal transformations are also crucial in deep learning for tasks like data normalization and orthogonalization of weight matrices. For example, orthogonal initialization of weight matrices can help stabilize the training of neural networks by preserving the variance of the activations across layers.

2.3.2 Tensor Fields and Manifolds

Tensor fields and manifolds are central concepts in differential geometry, providing a framework for understanding the structure and behavior of data in high-dimensional spaces. In deep learning, these concepts help in analyzing and optimizing neural networks, particularly in understanding how data is transformed and how learning algorithms navigate the complex landscapes of high-dimensional parameter spaces.

Riemannian Geometry

Riemannian geometry (Chavel 2006; do Carmo 1992; Jost 2017; Lee 2018; Petersen 2016) is a branch of differential geometry that studies smooth manifolds equipped with a Riemannian metric, which defines the notion of distance and angles on the manifold. A Riemannian manifold (M, g) is a smooth manifold M endowed with a Riemannian metric g, a positive-definite inner product on the tangent space at each point. For a point p on the manifold M, the Riemannian metric g_p on the tangent space T_pM is a bilinear form:

$$g_p : T_pM \times T_pM \to \mathbb{R}$$

For tangent vectors $\mathbf{u}, \mathbf{v} \in T_p M$, the metric g_p provides the inner product $\langle \mathbf{u}, \mathbf{v} \rangle_p$, defining the length of vectors and angles between them. In coordinates, the Riemannian metric is often represented by a matrix G, whose components g_{ij} vary smoothly across the manifold. For example, consider the two-dimensional sphere S^2 embedded in $\mathbb{R}^3$. The Riemannian metric on S^2 can be derived from the Euclidean metric of $\mathbb{R}^3$. In spherical coordinates (θ, ϕ), the metric g is:

$$\mathrm{d}s^2 = \mathrm{d}\theta^2 + \sin^2\theta \, \mathrm{d}\phi^2$$

where θ is the polar angle and ϕ is the azimuthal angle. The corresponding metric tensor G in these coordinates is:

$$G = \begin{pmatrix} 1 & 0 \\ 0 & \sin^2\theta \end{pmatrix}$$

This metric tensor describes how distances are measured on the surface of the sphere.

Application in Deep Learning: Riemannian geometry is used in optimizing neural networks on manifolds, such as in the context of Riemannian gradient descent. For instance, the optimization of covariance matrices in statistical models can be performed on the manifold of positive-definite matrices using Riemannian metrics to ensure stability and convergence.

Geodesics and Distance Metrics

Geodesics are the generalization of straight lines to curved spaces, representing the shortest paths between points on a Riemannian manifold. The geodesic distance between two points $p, q \in M$ is the length of the shortest path connecting them, measured using the Riemannian metric. The geodesic equation, which describes geodesics on a manifold, is a second-order differential equation derived from the metric. For a curve $\gamma(t)$ parameterized by t, the geodesic equation is:

$$\frac{d^2\gamma^k}{\mathrm{d}t^2} + \Gamma^k_{ij} \frac{\mathrm{d}\gamma^i}{\mathrm{d}t} \frac{\mathrm{d}\gamma^j}{\mathrm{d}t} = 0$$

where Γ^k_{ij} are the Christoffel symbols, which depend on the metric tensor g_{ij}:

$$\Gamma^k_{ij} = \frac{1}{2} g^{kl} \left(\frac{\partial g_{il}}{\partial x^j} + \frac{\partial g_{jl}}{\partial x^i} - \frac{\partial g_{ij}}{\partial x^l} \right)$$

Example: On the two-dimensional sphere S^2, geodesics are great circles. For instance, the equator and meridians are geodesics. The geodesic distance $d(p, q)$ between two points p and q on S^2 can be calculated using the central angle θ between them:

$$d(p, q) = R\theta$$

where R is the radius of the sphere.

Application in Deep Learning: Geodesic distances and Riemannian metrics are useful in deep learning for tasks involving data on curved spaces. For example, in natural language processing, word embeddings can be modeled on Riemannian manifolds to capture complex semantic relationships. Additionally, the concept of geodesics is employed in manifold learning algorithms, such as Isomap, which aims to reduce dimensionality while preserving the intrinsic geometry of the data.

2.4 Tensors in Neural Networks

2.4.1 Weight Tensors and Operations

Tensors are fundamental in representing and manipulating data within neural networks. Weight tensors, in particular, are crucial as they store the parameters that the network learns during training. Operations on these tensors, such as linear transformations, convolutions, and pooling, form the backbone of neural network computations.

Linear Transformations in Neural Networks

Linear transformations are essential operations in neural networks, particularly in fully connected (dense) layers. A fully connected layer performs a linear transformation followed by a non-linear activation function. The linear transformation can be expressed as a matrix multiplication. Given an input vector $\mathbf{x} \in \mathbb{R}^n$, a weight matrix $\mathbf{W} \in \mathbb{R}^{m \times n}$, and a bias vector $\mathbf{b} \in \mathbb{R}^m$, the output vector $\mathbf{y} \in \mathbb{R}^m$ is computed as:

$$\mathbf{y} = \phi(\mathbf{W}\mathbf{x} + \mathbf{b})$$

where ϕ is an element-wise activation function such as ReLU, sigmoid, or tanh. For example, consider a simple fully connected layer with:

$$\mathbf{x} = \begin{pmatrix} x_1 \\ x_2 \end{pmatrix}, \quad \mathbf{W} = \begin{pmatrix} w_{11} & w_{12} \\ w_{21} & w_{22} \\ w_{31} & w_{32} \end{pmatrix}, \quad \mathbf{b} = \begin{pmatrix} b_1 \\ b_2 \\ b_3 \end{pmatrix}$$

The output $\mathbf{y}$ is:

$$\mathbf{y} = \phi\left(\begin{pmatrix} w_{11} & w_{12} \\ w_{21} & w_{22} \\ w_{31} & w_{32} \end{pmatrix}\begin{pmatrix} x_1 \\ x_2 \end{pmatrix} + \begin{pmatrix} b_1 \\ b_2 \\ b_3 \end{pmatrix}\right) = \phi\left(\begin{pmatrix} w_{11}x_1 + w_{12}x_2 + b_1 \\ w_{21}x_1 + w_{22}x_2 + b_2 \\ w_{31}x_1 + w_{32}x_2 + b_3 \end{pmatrix}\right)$$

The activation function ϕ is applied element-wise to the resulting vector.

Backpropagation: During training, backpropagation is used to update the weights $\mathbf{W}$ and biases $\mathbf{b}$ by computing gradients with respect to a loss function. The gradients of the loss L with respect to the weights and biases are given by:

$$\frac{\partial L}{\partial \mathbf{W}} = \frac{\partial L}{\partial \mathbf{y}} \cdot \frac{\partial \mathbf{y}}{\partial \mathbf{W}}, \quad \frac{\partial L}{\partial \mathbf{b}} = \frac{\partial L}{\partial \mathbf{y}} \cdot \frac{\partial \mathbf{y}}{\partial \mathbf{b}}$$

Using the chain rule, these gradients can be efficiently computed and used to update the parameters via gradient descent or other optimization algorithms.

Convolutions and Pooling

Convolutions are fundamental operations in convolutional neural networks (CNNs), used primarily for processing spatial data such as images. Convolutional layers apply filters (kernels) to the input data to detect local patterns and features.

Convolutional Operation: Given an input tensor $\mathcal{X} \in \mathbb{R}^{H \times W \times C}$ (height, width, channels) and a filter tensor $\mathcal{K} \in \mathbb{R}^{kH \times kW \times C \times F}$ (filter height, filter width, input channels, output channels), the convolution operation produces an output tensor $\mathcal{Y} \in \mathbb{R}^{H' \times W' \times F}$:

$$\mathcal{Y}_{i,j,f} = \sum_{m=1}^{kH} \sum_{n=1}^{kW} \sum_{c=1}^{C} \mathcal{X}_{i+m-1,j+n-1,c} \cdot \mathcal{K}_{m,n,c,f}$$

where H' and W' are the height and width of the output tensor, determined by the stride and padding of the convolution. For example, consider a grayscale image (single channel) with:

$$\mathcal{X} = \begin{pmatrix} 1 & 2 & 3 \\ 4 & 5 & 6 \\ 7 & 8 & 9 \end{pmatrix}, \quad \mathcal{K} = \begin{pmatrix} 1 & 0 \\ 0 & -1 \end{pmatrix}$$

The convolution operation with stride 1 and no padding produces:

$$\mathcal{Y} = \begin{pmatrix} (1 \cdot 1 + 2 \cdot 0 + 4 \cdot 0 + 5 \cdot -1) & (2 \cdot 1 + 3 \cdot 0 + 5 \cdot 0 + 6 \cdot -1) \\ (4 \cdot 1 + 5 \cdot 0 + 7 \cdot 0 + 8 \cdot -1) & (5 \cdot 1 + 6 \cdot 0 + 8 \cdot 0 + 9 \cdot -1) \end{pmatrix} = \begin{pmatrix} -4 & -4 \\ -4 & -4 \end{pmatrix}$$

Pooling Operation: Pooling is a downsampling operation that reduces the spatial dimensions of the input tensor, helping to control overfitting and reduce computational cost. Common pooling operations include max-pooling and average pooling.

Max-Pooling: Max-pooling selects the maximum value within a pooling window. Given an input tensor $\mathcal{X} \in \mathbb{R}^{H \times W \times C}$ and a pooling window of size $pH \times pW$, max-pooling produces an output tensor $\mathcal{Y} \in \mathbb{R}^{H' \times W' \times C}$:

$$\mathcal{Y}_{i,j,c} = \max_{m=1,\ldots,pH} \max_{n=1,\ldots,pW} \mathcal{X}_{i+m-1,j+n-1,c}$$

For example, consider a 2D input tensor with a pooling window of size 2×2:

$$\mathcal{X} = \begin{pmatrix} 1 & 3 & 2 & 4 \\ 5 & 6 & 8 & 7 \\ 9 & 11 & 10 & 12 \\ 13 & 15 & 14 & 16 \end{pmatrix}$$

Max-pooling with stride 2 produces:

$$\mathcal{Y} = \begin{pmatrix} \max(1, 3, 5, 6) & \max(2, 4, 8, 7) \\ \max(9, 11, 13, 15) & \max(10, 12, 14, 16) \end{pmatrix} = \begin{pmatrix} 6 & 8 \\ 15 & 16 \end{pmatrix}$$

Average Pooling: Average pooling calculates the average value within a pooling window. Given an input tensor $\mathcal{X} \in \mathbb{R}^{H \times W \times C}$ and a pooling window of size $pH \times pW$, average pooling produces an output tensor $\mathcal{Y} \in \mathbb{R}^{H' \times W' \times C}$:

$$\mathcal{Y}_{i,j,c} = \frac{1}{pH \cdot pW} \sum_{m=1}^{pH} \sum_{n=1}^{pW} \mathcal{X}_{i+m-1,j+n-1,c}$$

For example, consider the same 2D input tensor with a pooling window of size 2×2:

$$\mathcal{X} = \begin{pmatrix} 1 & 3 & 2 & 4 \\ 5 & 6 & 8 & 7 \\ 9 & 11 & 10 & 12 \\ 13 & 15 & 14 & 16 \end{pmatrix}$$

Average pooling with stride 2 produces:

$$\mathcal{Y} = \begin{pmatrix} \frac{1+3+5+6}{4} & \frac{2+4+8+7}{4} \\ \frac{9+11+13+15}{4} & \frac{10+12+14+16}{4} \end{pmatrix} = \begin{pmatrix} 3.75 & 5.25 \\ 12 & 13 \end{pmatrix}$$

In both max-pooling and average pooling, the stride determines how the pooling window moves across the input tensor, which affects the dimensions of the resulting output tensor.

2.4.2 *Practical Applications*

Implementing Tensors in Neural Network Layers

In a fully connected layer, each neuron receives input from all neurons in the previous layer. This layer performs a linear transformation followed by an activation function. The implementation involves matrix multiplication between the input tensor and the weight tensor, followed by the addition of a bias tensor. Given an input tensor

$\mathbf{X} \in \mathbb{R}^{N \times d}$ (where N is the batch size and d is the input dimension), a weight tensor $\mathbf{W} \in \mathbb{R}^{d \times h}$ (where h is the number of neurons in the layer), and a bias tensor $\mathbf{b} \in \mathbb{R}^{h}$, the output tensor $\mathbf{Y} \in \mathbb{R}^{N \times h}$ is computed as:

$$\mathbf{Y} = \phi(\mathbf{X}\mathbf{W} + \mathbf{b})$$

where ϕ is an activation function applied element-wise.

Example: Consider a dense layer with 2 input features and 3 output neurons. Let the input tensor be:

$$\mathbf{X} = \begin{pmatrix} x_{11} & x_{12} \\ x_{21} & x_{22} \\ \vdots & \vdots \\ x_{N1} & x_{N2} \end{pmatrix}$$

The weight tensor and bias tensor are:

$$\mathbf{W} = \begin{pmatrix} w_{11} & w_{12} & w_{13} \\ w_{21} & w_{22} & w_{23} \end{pmatrix}, \quad \mathbf{b} = \begin{pmatrix} b_1 & b_2 & b_3 \end{pmatrix}$$

The output tensor is:

$$\mathbf{Y} = \phi\left(\begin{pmatrix} x_{11} & x_{12} \\ x_{21} & x_{22} \\ \vdots & \vdots \\ x_{N1} & x_{N2} \end{pmatrix} \begin{pmatrix} w_{11} & w_{12} & w_{13} \\ w_{21} & w_{22} & w_{23} \end{pmatrix} + \begin{pmatrix} b_1 & b_2 & b_3 \end{pmatrix}\right)$$

Convolutional layers apply convolutional filters to the input tensor, detecting local patterns and features. The filters (kernels) are small tensors that slide over the input tensor, performing element-wise multiplications and summations. Given an input tensor $\mathcal{X} \in \mathbb{R}^{N \times H \times W \times C}$ (where N is the batch size, H and W are the height and width of the input, and C is the number of channels), a filter tensor $\mathcal{K} \in \mathbb{R}^{kH \times kW \times C \times F}$ (where kH and kW are the height and width of the filter, and F is the number of filters), the output tensor $\mathcal{Y} \in \mathbb{R}^{N \times H' \times W' \times F}$ is computed as:

$$\mathcal{Y}_{n,i,j,f} = \sum_{m=1}^{kH} \sum_{n=1}^{kW} \sum_{c=1}^{C} \mathcal{X}_{n,i+m-1,j+n-1,c} \cdot \mathcal{K}_{m,n,c,f}$$

Example: Consider a convolutional layer with a single 3×3 filter applied to a grayscale image (single channel). Let the input tensor be:

$$\mathcal{X} = \begin{pmatrix} 1 & 2 & 3 & 4 \\ 5 & 6 & 7 & 8 \\ 9 & 10 & 11 & 12 \\ 13 & 14 & 15 & 16 \end{pmatrix}$$

The filter tensor is:

$$\mathcal{K} = \begin{pmatrix} 1 & 0 & -1 \\ 1 & 0 & -1 \\ 1 & 0 & -1 \end{pmatrix}$$

The convolution operation with stride 1 and no padding produces:

$$\mathcal{Y} = \begin{pmatrix} (1 \cdot 1 + 2 \cdot 0 + 3 \cdot -1) + (5 \cdot 1 + 6 \cdot 0 + 7 \cdot -1) + (9 \cdot 1 + 10 \cdot 0 + 11 \cdot -1) & \cdots \\ \cdots & \cdots \end{pmatrix}$$

Pooling layers downsample the spatial dimensions of the input tensor, reducing the number of parameters and computational cost. Common pooling operations include max-pooling and average pooling. Max-pooling selects the maximum value within a pooling window. Given an input tensor $\mathcal{X} \in \mathbb{R}^{N \times H \times W \times C}$ and a pooling window of size $pH \times pW$, max-pooling produces an output tensor $\mathcal{Y} \in \mathbb{R}^{N \times H' \times W' \times C}$:

$$\mathcal{Y}_{n,i,j,c} = \max_{m=1,\ldots,pH} \max_{n=1,\ldots,pW} \mathcal{X}_{n,i+m-1,j+n-1,c}$$

Example: Consider a 2D input tensor with a pooling window of size 2×2:

$$\mathcal{X} = \begin{pmatrix} 1 & 3 & 2 & 4 \\ 5 & 6 & 8 & 7 \\ 9 & 11 & 10 & 12 \\ 13 & 15 & 14 & 16 \end{pmatrix}$$

Max-pooling with stride 2 produces:

$$\mathcal{Y} = \begin{pmatrix} \max(1, 3, 5, 6) & \max(2, 4, 8, 7) \\ \max(9, 11, 13, 15) & \max(10, 12, 14, 16) \end{pmatrix} = \begin{pmatrix} 6 & 8 \\ 15 & 16 \end{pmatrix}$$

Optimizing Tensor Operations

Optimizing tensor operations is crucial for efficient deep learning, especially when dealing with large datasets and complex models. Techniques for optimization include vectorization, efficient memory management, and leveraging hardware accelerations such as GPUs and TPUs.

Vectorization: Vectorization involves expressing tensor operations in terms of matrix and tensor operations, allowing for parallel execution. This approach minimizes the use of explicit loops and leverages optimized linear algebra libraries. For example, instead of using a loop to add two vectors, vectorized addition can be performed as:

```
import numpy as np

# Define two vectors
a = np.array([1, 2, 3, 4])
b = np.array([5, 6, 7, 8])

# Vectorized addition
c = a + b
```

Efficient Memory Management: Efficient memory management involves minimizing memory usage and avoiding unnecessary data copies. Techniques include in-place operations and careful management of data flow to reduce memory overhead. For example, in PyTorch, in-place operations can be performed using functions that modify the original tensor:

```
import torch

# Define a tensor
x = torch.tensor([1.0, 2.0, 3.0, 4.0])

# In-place addition
x.add_(2.0)
```

Hardware Acceleration: Leveraging hardware acceleration, such as GPUs and TPUs, can significantly speed up tensor operations. Deep learning frameworks like TensorFlow and PyTorch provide built-in support for these accelerators, enabling efficient computation. For example, in TensorFlow, operations can be performed on a GPU by specifying the device:

```
import tensorflow as tf

# Define a tensor on the GPU
with tf.device('/GPU:0'):
    x = tf.constant([1.0, 2.0, 3.0, 4.0])
    y = tf.constant([5.0, 6.0, 7.0, 8.0])
    z = x + y
```

2.5 Geometric Structures in Data

2.5.1 Point Clouds and Meshes

Geometric structures such as point clouds and meshes are essential representations for 3D data. They are widely used in computer graphics, computer vision, and machine learning applications, including 3D object recognition, reconstruction, and simulation. Understanding how to represent and manipulate these structures using tensors is crucial for developing algorithms that can effectively process and analyze 3D data.

Representing 3D Data with Tensors

Point Clouds A point cloud is a collection of points in a 3D space, typically represented by their coordinates. Each point can be described as $\mathbf{p}_i = (x_i, y_i, z_i)$, where x_i, y_i, and z_i are the coordinates of the i-th point. A point cloud with N points can be represented as a tensor $\mathcal{P} \in \mathbb{R}^{N\times 3}$. For example, consider a point cloud with four points:

$$\mathcal{P} = \begin{pmatrix} x_1 & y_1 & z_1 \\ x_2 & y_2 & z_2 \\ x_3 & y_3 & z_3 \\ x_4 & y_4 & z_4 \end{pmatrix}$$

This representation allows for efficient manipulation and processing of the 3D coordinates using tensor operations.

Meshes A mesh is a collection of vertices, edges, and faces that define the shape of a 3D object. The vertices are points in 3D space, similar to those in a point cloud. The edges connect pairs of vertices, and the faces are typically triangular or quadrilateral elements that define the surface of the object. A mesh can be represented by two tensors: one for the vertices and one for the faces. Let $\mathcal{V} \in \mathbb{R}^{N\times 3}$ represent the vertices and $\mathcal{F} \in \mathbb{R}^{M\times 3}$ represent the faces, where N is the number of vertices and M is the number of faces (assuming triangular faces). For example, consider a mesh with four vertices and two triangular faces:

Vertices:

$$\mathcal{V} = \begin{pmatrix} x_1 & y_1 & z_1 \\ x_2 & y_2 & z_2 \\ x_3 & y_3 & z_3 \\ x_4 & y_4 & z_4 \end{pmatrix}$$

Faces:

$$\mathcal{F} = \begin{pmatrix} 1 & 2 & 3 \\ 2 & 3 & 4 \end{pmatrix}$$

The face tensor $\mathcal{F}$ indicates that the first face is formed by vertices 1, 2, and 3, and the second face is formed by vertices 2, 3, and 4.

Geometric Operations on Point Clouds and Meshes

Geometric operations on point clouds include translation, rotation, and scaling. These operations can be efficiently implemented using tensor algebra. To translate a point cloud by a vector $\mathbf{t} = (t_x, t_y, t_z)$, we add $\mathbf{t}$ to each point in the point cloud tensor $\mathcal{P}$:

$$\mathcal{P}_{\text{translated}} = \mathcal{P} + \mathbf{t}$$

Example: Given a point cloud $\mathcal{P}$ and a translation vector $\mathbf{t} = (1, 2, 3)$:

$$\mathcal{P} = \begin{pmatrix} x_1 & y_1 & z_1 \\ x_2 & y_2 & z_2 \end{pmatrix}, \quad \mathbf{t} = \begin{pmatrix} 1 & 2 & 3 \end{pmatrix}$$

The translated point cloud is:

$$\mathcal{P}_{\text{translated}} = \begin{pmatrix} x_1 + 1 & y_1 + 2 & z_1 + 3 \\ x_2 + 1 & y_2 + 2 & z_2 + 3 \end{pmatrix}$$

To rotate a point cloud, we multiply each point by a rotation matrix $\mathbf{R} \in \mathbb{R}^{3\times 3}$. For a point cloud tensor $\mathcal{P}$:

$$\mathcal{P}_{\text{rotated}} = \mathcal{P}\mathbf{R}^T$$

Example: Given a point cloud $\mathcal{P}$ and a rotation matrix $\mathbf{R}$:

$$\mathcal{P} = \begin{pmatrix} x_1 & y_1 & z_1 \\ x_2 & y_2 & z_2 \end{pmatrix}, \quad \mathbf{R} = \begin{pmatrix} \cos\theta & -\sin\theta & 0 \\ \sin\theta & \cos\theta & 0 \\ 0 & 0 & 1 \end{pmatrix}$$

The rotated point cloud is:

$$\mathcal{P}_{\text{rotated}} = \mathcal{P}\mathbf{R}^T = \begin{pmatrix} x_1\cos\theta - y_1\sin\theta & x_1\sin\theta + y_1\cos\theta & z_1 \\ x_2\cos\theta - y_2\sin\theta & x_2\sin\theta + y_2\cos\theta & z_2 \end{pmatrix}$$

To scale a point cloud by a factor $\mathbf{s} = (s_x, s_y, s_z)$, we multiply each coordinate by the corresponding scaling factor:

$$\mathcal{P}_{\text{scaled}} = \mathcal{P} \cdot \mathbf{s}$$

Example: Given a point cloud $\mathcal{P}$ and a scaling factor $\mathbf{s} = (2, 2, 2)$:

$$\mathcal{P} = \begin{pmatrix} x_1 & y_1 & z_1 \\ x_2 & y_2 & z_2 \end{pmatrix}, \quad \mathbf{s} = \begin{pmatrix} 2 & 2 & 2 \end{pmatrix}$$

The scaled point cloud is:

$$\mathcal{P}_{\text{scaled}} = \begin{pmatrix} 2x_1 & 2y_1 & 2z_1 \\ 2x_2 & 2y_2 & 2z_2 \end{pmatrix}$$

Mesh Operations Geometric operations on meshes include similar transformations as those on point clouds, but they must also account for the connectivity of vertices. To translate a mesh, we apply the translation vector $\mathbf{t}$ to each vertex in the vertex tensor $\mathcal{V}$:

$$\mathcal{V}_{\text{translated}} = \mathcal{V} + \mathbf{t}$$

To rotate a mesh, we apply the rotation matrix $\mathbf{R}$ to each vertex:

$$\mathcal{V}_{\text{rotated}} = \mathcal{V}\mathbf{R}^T$$

To scale a mesh, we apply the scaling factor $\mathbf{s}$ to each vertex:

$$\mathcal{V}_{\text{scaled}} = \mathcal{V} \cdot \mathbf{s}$$

Example: Consider a mesh with vertex tensor $\mathcal{V}$ and face tensor $\mathcal{F}$. The mesh can be transformed similarly to point clouds, with the faces remaining unchanged since they are defined by vertex indices.

2.5.2 Graph Data and Networks

Graphs are powerful data structures used to model relationships and interactions in various domains, such as social networks, molecular structures, and knowledge graphs (Diestel 2010; West 2000). Representing and processing graph data using tensors enables the application of neural network techniques to graph-structured data, leading to the development of graph neural networks (GNNs). Two fundamental aspects of working with graph data are graph representation with tensors and graph convolutional networks (GCNs) (Kipf and Welling 2016; Bruna et al. 2014; Hamilton et al. 2017; Velickovic et al. 2017b).

Graph Representation with Tensors

A graph G is defined by a set of nodes (vertices) V and edges E, where each edge connects a pair of nodes. Formally, a graph is represented as $G = (V, E)$. In tensor notation, graphs can be represented using adjacency matrices, feature matrices, and possibly edge feature matrices. The adjacency matrix $\mathbf{A} \in \mathbb{R}^{N \times N}$ represents the connectivity of the graph, where N is the number of nodes. The element a_{ij} indicates the presence (and possibly the weight) of an edge between node i and node j:

$$a_{ij} = \begin{cases} 1 & \text{if there is an edge from node } i \text{ to node } j \\ 0 & \text{otherwise} \end{cases}$$

For undirected graphs, the adjacency matrix is symmetric ($a_{ij} = a_{ji}$). For example, consider a graph with 4 nodes and the following edges: (1, 2), (1, 3), and (3, 4). The adjacency matrix is:

$$\mathbf{A} = \begin{pmatrix} 0 & 1 & 1 & 0 \\ 1 & 0 & 0 & 0 \\ 1 & 0 & 0 & 1 \\ 0 & 0 & 1 & 0 \end{pmatrix}$$

The feature matrix $\mathbf{X} \in \mathbb{R}^{N \times F}$ represents the features of the nodes, where N is the number of nodes and F is the number of features per node. Each row $\mathbf{x}_i$ corresponds to the feature vector of node i. For example, consider a graph with 4 nodes, each having 2 features. The feature matrix is:

$$\mathbf{X} = \begin{pmatrix} x_{11} & x_{12} \\ x_{21} & x_{22} \\ x_{31} & x_{32} \\ x_{41} & x_{42} \end{pmatrix}$$

In some cases, the edges themselves have features, represented by an edge feature matrix $\mathbf{E} \in \mathbb{R}^{|E| \times F_e}$, where $|E|$ is the number of edges and F_e is the number of features per edge.

Graph Convolutional Networks

Graph convolutional networks (GCNs) are a class of neural networks designed to operate on graph-structured data. GCNs generalize the concept of convolution from grid-like structures (such as images) to graphs. The key idea is to aggregate feature information from neighboring nodes to learn node representations that capture the graph's structure. The graph convolution operation updates the feature vector of each node by aggregating the feature vectors of its neighbors. Mathematically, a single layer of a GCN can be expressed as:

$$\mathbf{H}^{(l+1)} = \sigma\left(\tilde{\mathbf{A}}\mathbf{H}^{(l)}\mathbf{W}^{(l)}\right)$$

where $\mathbf{H}^{(l)} \in \mathbb{R}^{N \times F_l}$ is the feature matrix at layer l, with F_l features per node, $\tilde{\mathbf{A}} = \mathbf{D}^{-1/2}\mathbf{A}\mathbf{D}^{-1/2}$ is the normalized adjacency matrix, where $\mathbf{D}$ is the degree matrix with $d_{ii} = \sum_j a_{ij}$, $\mathbf{W}^{(l)} \in \mathbb{R}^{F_l \times F_{l+1}}$ is the weight matrix of the l-th layer, and σ is an activation function, such as ReLU. For example, consider a GCN with an input feature matrix $\mathbf{X}$ and a single graph convolutional layer. The output feature matrix $\mathbf{H}^{(1)}$ is computed as:

$$\mathbf{H}^{(1)} = \sigma\left(\tilde{\mathbf{A}}\mathbf{X}\mathbf{W}^{(0)}\right)$$

where $\mathbf{W}^{(0)}$ is the weight matrix of the first layer.

Implementation: In practice, implementing GCNs involves defining the adjacency matrix, feature matrix, and using a deep learning framework to perform the graph convolution operations. Below is a simplified implementation using PyTorch and the PyTorch Geometric library:

```
import torch
import torch.nn.functional as F
from torch_geometric.nn import GCNConv
from torch_geometric.data import Data

# Define a simple graph
edge_index = torch.tensor([[0, 1, 2, 3],
                           [1, 0, 3, 2]], dtype=torch.long)
                           # Edges (source, target)
x = torch.tensor([[1, 2], [3, 4], [5, 6], [7, 8]],
        dtype=torch.float)  # Node features

# Create a graph data object
data = Data(x=x, edge_index=edge_index)

# Define a simple GCN model
class GCN(torch.nn.Module):
    def __init__(self):
        super(GCN, self).__init__()
        self.conv1 = GCNConv(2, 4)  # Input features = 2,
            Output features = 4
        self.conv2 = GCNConv(4, 2)  # Input features = 4,
            Output features = 2

    def forward(self, data):
        x, edge_index = data.x, data.edge_index
        x = self.conv1(x, edge_index)
        x = F.relu(x)
        x = self.conv2(x, edge_index)
        return x

# Initialize and apply the GCN model
model = GCN()
output = model(data)
print(output)
```

In this example, the GCN model consists of two graph convolutional layers. The first layer transforms the input features from 2 to 4 dimensions, and the second layer reduces them back to 2 dimensions. The ReLU activation function is applied after the first convolutional layer.

2.6 Visualizing Tensors and Geometric Structures

2.6.1 Techniques for Tensor Visualization

Visualizing tensors and geometric structures is essential for understanding the internal workings of deep learning models, diagnosing issues, and interpreting results. Effective visualization techniques can reveal patterns, highlight features, and provide insights that are not easily discernible from raw numerical data. Two fundamental techniques for tensor visualization are heatmaps and slices, and vector field visualization.

Heatmaps and Slices

Heatmaps are a powerful tool for visualizing the values within a tensor, especially when dealing with matrices or higher-order tensors. A heatmap represents data values through variations in color, making it easier to identify patterns, anomalies, and areas of interest. For a matrix $\mathbf{A} \in \mathbb{R}^{m\times n}$, a heatmap can visually represent each element a_{ij} by assigning a color based on its magnitude. The color scale is typically chosen such that higher values correspond to warmer colors (e.g., red) and lower values to cooler colors (e.g., blue). For example, consider a matrix:

$$\mathbf{A} = \begin{pmatrix} 1 & 2 & 3 \\ 4 & 5 & 6 \\ 7 & 8 & 9 \end{pmatrix}$$

A heatmap of **A** would show a gradient from blue (representing 1) to red (representing 9), allowing quick identification of the relative values. In Python, heatmaps can be generated using libraries like Matplotlib and Seaborn (see Fig. 2.1):

```
import matplotlib.pyplot as plt
import seaborn as sns
import numpy as np

# Define a matrix
A = np.array([[1, 2, 3], [4, 5, 6], [7, 8, 9]])

# Create a heatmap
sns.heatmap(A, annot=True, cmap='coolwarm')
plt.show()
```

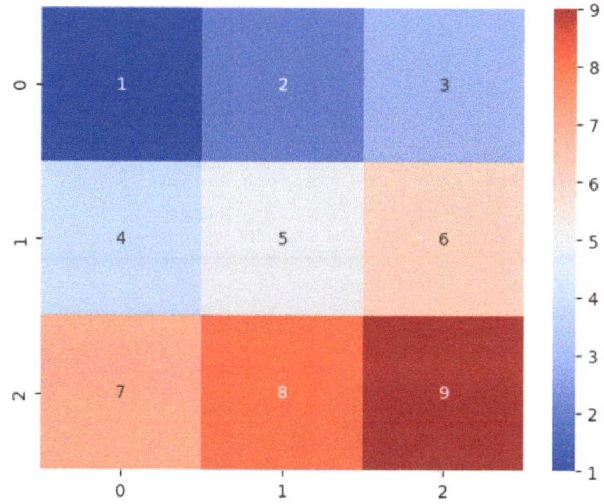

Fig. 2.1 An illustration of the heatmap for a 2D tensor

Slices of Higher-order Tensors For higher-order tensors, visualizing individual slices can be informative. A slice of a tensor is a lower-dimensional section obtained by fixing some of its indices. For a tensor $\mathcal{T} \in \mathbb{R}^{I \times J \times K}$, fixing the index k results in a matrix slice $\mathcal{T}_{:,:,k} \in \mathbb{R}^{I \times J}$. For example, consider a tensor $\mathcal{T} \in \mathbb{R}^{3 \times 3 \times 3}$:

$$\mathcal{T}(:, :, 1) = \begin{pmatrix} 1 & 2 & 3 \\ 4 & 5 & 6 \\ 7 & 8 & 9 \end{pmatrix}, \quad \mathcal{T}(:, :, 2) = \begin{pmatrix} 9 & 8 & 7 \\ 6 & 5 & 4 \\ 3 & 2 & 1 \end{pmatrix}$$

Visualizing each slice $\mathcal{T}(:, :, k)$ as a heatmap provides a clear view of the tensor's structure in different dimensions.

Vector Field Visualization Vector fields represent data where each point in space has a vector associated with it, often used to depict flow or directionality in data. Visualizing vector fields can be crucial for understanding gradients, forces, and movements within the data.

Quiver plots are commonly used to visualize vector fields. They display vectors as arrows, where the direction and length of each arrow represent the direction and magnitude of the vector at that point. For example, consider a 2D vector field defined by:

$$\mathbf{V}(x, y) = (y, -x)$$

This represents a rotational field. In Python, a quiver plot can be generated using Matplotlib (see Fig. 2.2):

```
import matplotlib.pyplot as plt
import numpy as np
```

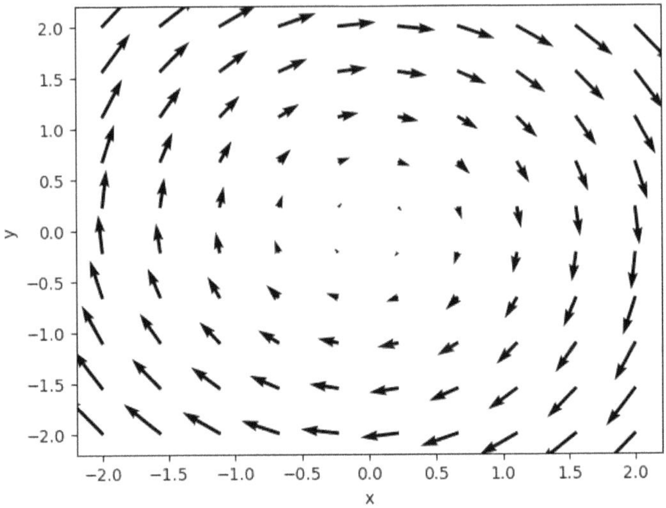

Fig. 2.2 Quiver plot of vector field

```
# Define a grid
x, y = np.meshgrid(np.linspace(-2, 2, 10),
np.linspace(-2, 2, 10))

# Define the vector field
u = y
v = -x

# Create a quiver plot
plt.quiver(x, y, u, v)
plt.xlabel('x')
plt.ylabel('y')
plt.title('Quiver Plot of Vector Field')
plt.show()
```

Streamline plots provide a continuous representation of vector fields, showing the paths that particles would follow under the influence of the vector field. This is particularly useful for visualizing flow patterns and understanding the overall behavior of the field.

Example: Continuing from the previous example, a streamline plot can be generated in Python (see Fig. 2.3):

```
import matplotlib.pyplot as plt
import numpy as np
```

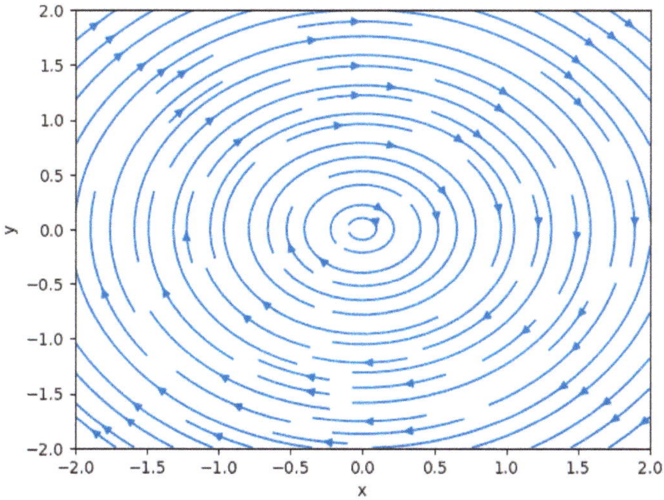

Fig. 2.3 Streamline plot of vector field

```
# Define a grid
x, y = np.meshgrid(np.linspace(-2, 2, 100), np.linspace(-2, 2,
 100))

# Define the vector field
u = y
v = -x

# Create a streamline plot
plt.streamplot(x, y, u, v)
plt.xlabel('x')
plt.ylabel('y')
plt.title('Streamline Plot of Vector Field')
plt.show()
```

2.6.2 *Geometric Visualization*

Geometric visualization involves representing high-dimensional data in lower dimensions, making it easier to understand and interpret complex structures. Two fundamental techniques in geometric visualization are manifold learning and embeddings and projections. These techniques enable us to visualize data while preserving its intrinsic geometric properties.

Manifold Learning

Manifold learning is a type of unsupervised learning that seeks to uncover the low-dimensional structure of high-dimensional data. The underlying assumption is that high-dimensional data points lie on a lower-dimensional manifold embedded within the higher-dimensional space. Manifold learning algorithms aim to discover this manifold and provide a meaningful low-dimensional representation of the data.

Principal Component Analysis (PCA) PCA is a linear manifold learning technique that finds the directions (principal components) along which the variance of the data is maximized. Mathematically, given a dataset $\mathbf{X} \in \mathbb{R}^{N\times d}$, PCA finds the eigenvectors and eigenvalues of the covariance matrix $\mathbf{C} = \frac{1}{N}\mathbf{X}^T\mathbf{X}$. The principal components are the eigenvectors corresponding to the largest eigenvalues. For example, consider a dataset with points in $\mathbb{R}^3$:

$$\mathbf{X} = \begin{pmatrix} x_{11} & x_{12} & x_{13} \\ x_{21} & x_{22} & x_{23} \\ \vdots & \vdots & \vdots \\ x_{N1} & x_{N2} & x_{N3} \end{pmatrix}$$

PCA reduces the dimensionality to k by projecting the data onto the top k principal components:

$$\mathbf{X}_{\text{reduced}} = \mathbf{X}\mathbf{W}_k$$

where $\mathbf{W}_k \in \mathbb{R}^{d\times k}$ contains the top k eigenvectors.

t-Distributed Stochastic Neighbor Embedding (t-SNE) t-SNE is a non-linear manifold learning technique that visualizes high-dimensional data by reducing it to two or three dimensions while preserving local structures (van der Maaten and Hinton 2008). t-SNE minimizes the divergence between two probability distributions: one representing pairwise similarities of the data points in the high-dimensional space and the other representing pairwise similarities in the low-dimensional space. For a dataset $\mathbf{X}$, t-SNE constructs two probability distributions P and Q and minimizes the Kullback-Leibler divergence between them:

$$\mathrm{KL}(P\|Q) = \sum_{i\neq j} p_{ij} \log \frac{p_{ij}}{q_{ij}}$$

where p_{ij} and q_{ij} are the probabilities of similarity between points i and j in the high-dimensional and low-dimensional spaces, respectively.

In Python, t-SNE can be implemented using the scikit-learn library (see Fig. 2.4):

```
import numpy as np
from sklearn.manifold import TSNE
```

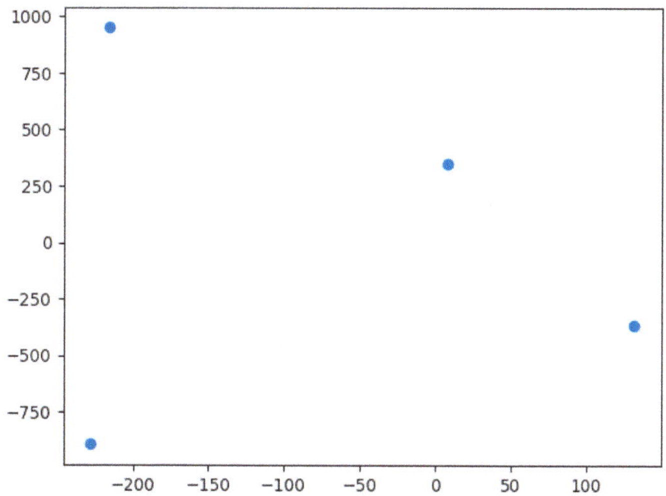

Fig. 2.4 t-SNE visualization

```
import matplotlib.pyplot as plt

# Define a dataset
X = np.array([[1, 2, 3], [4, 5, 6], [7, 8, 9], [10, 11, 12]])

# Apply t-SNE with a perplexity less than the number of samples
tsne = TSNE(n_components=2, perplexity=2)
X_embedded = tsne.fit_transform(X)

# Plot the results
plt.scatter(X_embedded[:, 0], X_embedded[:, 1])
plt.title('t-SNE Visualization')
plt.show()
```

Isomap Isomap is another non-linear manifold learning technique that preserves global geometric structures (Tenenbaum et al. 2000). It constructs a neighborhood graph, computes the shortest paths between all pairs of points using geodesic distances, and then applies classical multidimensional scaling (MDS) to embed the data into a lower-dimensional space. Given a dataset $\mathbf{X}$, Isomap constructs a graph G where edges connect neighboring points. The geodesic distances d_{ij} are computed using the shortest path algorithm, and MDS is applied to these distances to obtain the low-dimensional embedding.

In Python, Isomap can be implemented using the scikit-learn library (see Fig. 2.5):

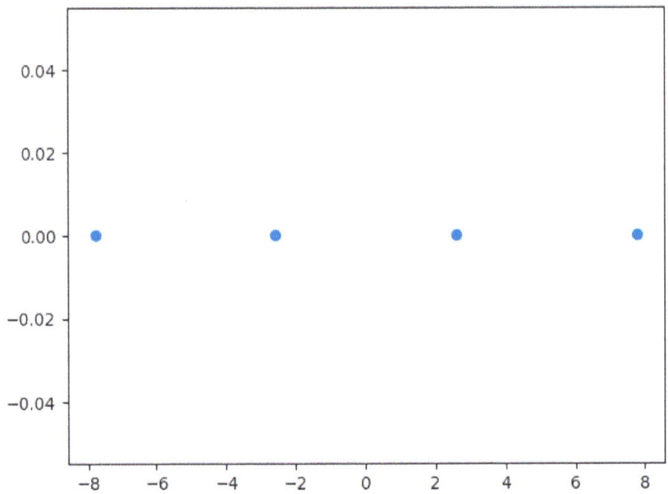

Fig. 2.5 Isomap visualization

```
from sklearn.manifold import Isomap

# Define a dataset
X = np.array([[1, 2, 3], [4, 5, 6], [7, 8, 9], [10, 11, 12]])

# Apply Isomap with n_neighbors less than or equal to the
number of samples
isomap = Isomap(n_components=2, n_neighbors=2)
X_embedded = isomap.fit_transform(X)

# Plot the results
plt.scatter(X_embedded[:, 0], X_embedded[:, 1])
plt.title('Isomap Visualization')
plt.show()
```

Embeddings and Projections

Embeddings and projections are techniques used to represent high-dimensional data in lower dimensions, often to visualize or simplify the data while preserving its essential structure and relationships.

Embeddings Embeddings map high-dimensional data to a lower-dimensional space while preserving specific properties. In deep learning, embeddings are often used to represent categorical data (e.g., words, items) as dense vectors in a continuous vector space. Word embeddings are a type of embedding used in natural language processing to represent words in a continuous vector space. Word2Vec, GloVe, and FastText are popular algorithms for generating word embeddings. These embeddings capture semantic relationships between words, allowing similar words to be close to each other in the vector space. For example, consider a vocabulary with words $\{w_1, w_2, \ldots, w_V\}$. Word2Vec learns embeddings $\mathbf{e}_i \in \mathbb{R}^d$ for each word w_i, such that the similarity between words is preserved:

$$\mathbf{e}_i = \text{Word2Vec}(w_i)$$

In Python, word embeddings can be generated using the gensim library:

```
from gensim.models import Word2Vec

# Define a corpus
sentences = [['this', 'is', 'a', 'sentence'],
['this', 'is', 'another', 'sentence']]

# Train a Word2Vec model
model = Word2Vec(sentences, vector_size=100, window=5,
min_count=1, workers=4)

# Get the embedding for a word
embedding = model.wv['sentence']
```

Projections Projections involve transforming high-dimensional data to a lower-dimensional space using linear or non-linear transformations. Principal Component Analysis (PCA) is a commonly used linear projection technique, while t-SNE and UMAP are popular non-linear projection methods. As mentioned earlier, PCA projects high-dimensional data onto the top k principal components, preserving as much variance as possible. The projection is given by:

$$\mathbf{X}_{\text{projected}} = \mathbf{X}\mathbf{W}_k$$

where $\mathbf{W}_k$ contains the top k eigenvectors of the covariance matrix.

UMAP (Uniform Manifold Approximation and Projection) UMAP is a non-linear projection technique that preserves both local and global structures of the data (McInnes and Healy 2018). It constructs a high-dimensional graph representation of the data and optimizes a low-dimensional layout that preserves the graph's structure.

Example: In Python, UMAP can be implemented using the umap-learn library (see Fig. 2.6):

```
import umap.umap_ as umap

# Define a dataset
```

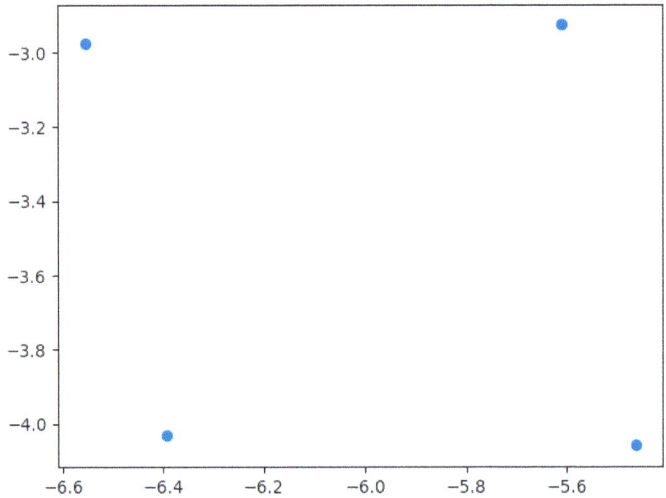

Fig. 2.6 UMAP visualization

```
X = np.array([[1, 2, 3], [4, 5, 6], [7, 8, 9], [10, 11, 12]])

# Apply UMAP
umap_model = umap.UMAP(n_components=2)
X_embedded = umap_model.fit_transform(X)

# Plot the results
plt.scatter(X_embedded[:, 0], X_embedded[:, 1])
plt.title('UMAP Visualization')
plt.show()
```

2.7 Practical Examples and Applications

2.7.1 Tensor Representation of Data

Tensors are versatile data structures that allow for the representation of various types of data in a unified manner. Understanding how to represent different kinds of data as tensors is crucial for applying deep learning techniques effectively. Two common types of data that are often represented as tensors are images and sequences.

Image Data as Tensors

Images are naturally represented as tensors due to their inherent multi-dimensional structure. An image can be described by its height, width, and the number of color channels (e.g., RGB channels). Thus, a colored image is typically represented as a 3D tensor. For example, consider an RGB image of size $H \times W$. This image can be represented as a tensor $\mathcal{I} \in \mathbb{R}^{H \times W \times 3}$, where the third dimension corresponds to the three color channels (Red, Green, and Blue). For a batch of N images, the data can be represented as a 4D tensor $\mathcal{B} \in \mathbb{R}^{N \times H \times W \times 3}$. Each slice $\mathcal{B}[i, :, :, :]$ corresponds to the i-th image in the batch. If $\mathcal{I}$ is an image tensor, the value at pixel location (i, j) for the c-th color channel is denoted as $\mathcal{I}_{i,j,c}$. For example, for a 3x3 RGB image, the tensor might look like:

$$\mathcal{I} = \left(\begin{pmatrix} I_{111} & I_{112} & I_{113} \\ I_{121} & I_{122} & I_{123} \\ I_{131} & I_{132} & I_{133} \end{pmatrix}, \begin{pmatrix} I_{211} & I_{212} & I_{213} \\ I_{221} & I_{222} & I_{223} \\ I_{231} & I_{232} & I_{233} \end{pmatrix}, \begin{pmatrix} I_{311} & I_{312} & I_{313} \\ I_{321} & I_{322} & I_{323} \\ I_{331} & I_{332} & I_{333} \end{pmatrix} \right)$$

Here, I_{ijk} represents the intensity of the color channel k at pixel (i, j).

Tensor Operations on Images:

1. Convolutional operations apply filters to images to extract features. For a convolutional kernel $\mathcal{K} \in \mathbb{R}^{kH \times kW \times C}$ applied to an image $\mathcal{I} \in \mathbb{R}^{H \times W \times C}$, the convolution operation is given by:

$$\mathcal{Y}_{i,j} = \sum_{m=1}^{kH} \sum_{n=1}^{kW} \sum_{c=1}^{C} \mathcal{I}_{i+m-1,j+n-1,c} \cdot \mathcal{K}_{m,n,c}$$

2. Pooling operations reduce the spatial dimensions of the image tensor. Max-pooling, for example, selects the maximum value within a window:

$$\mathcal{Y}_{i,j} = \max_{m,n} \mathcal{I}_{i+m-1,j+n-1}$$

Example in Python: Using a library like TensorFlow or PyTorch, image tensors can be manipulated efficiently.

```
import tensorflow as tf

# Create a batch of RGB images (batch_size, height, width,
channels)
images = tf.random.normal([32, 128, 128, 3])

# Define a convolutional layer
conv_layer = tf.keras.layers.Conv2D(filters=16, kernel_size=3,
padding='same', activation='relu')

# Apply the convolutional layer
conv_output = conv_layer(images)

print("Convolutional layer output shape:", conv_output.shape)
```

Sequence Data as Tensors

Sequence data, such as time series or text, can also be represented as tensors. A sequence can be described by its length and the number of features at each timestep. For example, a time series with T timesteps and F features per timestep is represented as a 2D tensor $\mathcal{S} \in \mathbb{R}^{T \times F}$. For a batch of sequences, the data can be represented as a 3D tensor $\mathcal{B} \in \mathbb{R}^{N \times T \times F}$, where N is the batch size, T is the sequence length, and F is the number of features. For example, consider a time series with three timesteps and two features:

$$\mathcal{S} = \begin{pmatrix} s_{11} & s_{12} \\ s_{21} & s_{22} \\ s_{31} & s_{32} \end{pmatrix}$$

In this representation, s_{ij} denotes the value of the j-th feature at the i-th timestep. Tensor Operations on Sequences:

1. RNNs and their variants, such as LSTMs and GRUs, process sequences by maintaining a hidden state that is updated at each timestep based on the input and the previous hidden state. The RNN update rule for hidden state $\mathbf{h}_t$ at timestep t is:

$$\mathbf{h}_t = \sigma(\mathbf{W}_{xh}\mathbf{x}_t + \mathbf{W}_{hh}\mathbf{h}_{t-1} + \mathbf{b}_h)$$

where $\mathbf{x}_t$ is the input at timestep t, $\mathbf{W}_{xh}$ and $\mathbf{W}_{hh}$ are weight matrices, $\mathbf{b}_h$ is the bias, and σ is an activation function.
2. Attention mechanisms enhance sequence models by allowing the model to focus on different parts of the input sequence when generating each output. The attention score α_{ij} between input i and output j is computed as:

$$\alpha_{ij} = \frac{\exp(e_{ij})}{\sum_{k=1}^{T} \exp(e_{ik})}$$

where e_{ij} is the compatibility function, often computed as a dot product between the hidden states of the input and output sequences.

Example in Python: Using TensorFlow, sequence data can be processed with RNNs or attention mechanisms:

```
import tensorflow as tf

# Create a batch of sequences (batch_size, sequence_length,
features)
sequences = tf.random.normal([32, 10, 8])

# Define an LSTM layer
lstm_layer = tf.keras.layers.LSTM(16, return_sequences=True)
```

```
# Apply the LSTM layer
lstm_output = lstm_layer(sequences)

print("LSTM layer output shape:", lstm_output.shape)
```

2.7.2 Implementing Basic Neural Networks with Tensors

Implementing neural networks with tensors involves creating the network architecture, training it on data, and evaluating its performance. Tensors facilitate the representation and manipulation of data and weights in the network, making the implementation of neural networks efficient and scalable. Here, we focus on building a simple multi-layer perceptron (MLP) and the process of training and evaluating the network.

Building a Simple MLP

An MLP is a type of feedforward neural network consisting of multiple layers of neurons, each fully connected to the neurons in the subsequent layer. The basic components of an MLP include the input layer, hidden layers, and the output layer. Each layer performs a linear transformation followed by a non-linear activation function.

Network Architecture Consider an MLP with an input layer of size d, one hidden layer of size h, and an output layer of size k. The network can be represented as:

$$\mathbf{X} \rightarrow \mathbf{W}_1 \rightarrow \sigma \rightarrow \mathbf{W}_2 \rightarrow \mathbf{Y}$$

where $\mathbf{X} \in \mathbb{R}^{N \times d}$ is the input tensor (batch size N, input features d), $\mathbf{W}_1 \in \mathbb{R}^{d \times h}$ is the weight matrix of the hidden layer, $\mathbf{W}_2 \in \mathbb{R}^{h \times k}$ is the weight matrix of the output layer, and σ is the activation function (e.g., ReLU).

Forward Pass: The forward pass involves computing the activations of each layer. For the hidden layer:

$$\mathbf{H} = \sigma(\mathbf{X}\mathbf{W}_1 + \mathbf{b}_1)$$

where $\mathbf{H} \in \mathbb{R}^{N \times h}$ is the hidden layer activations, and $\mathbf{b}_1 \in \mathbb{R}^h$ is the bias vector for the hidden layer. For the output layer:

$$\mathbf{Y} = \mathbf{H}\mathbf{W}_2 + \mathbf{b}_2$$

where $\mathbf{Y} \in \mathbb{R}^{N \times k}$ is the output tensor, and $\mathbf{b}_2 \in \mathbb{R}^k$ is the bias vector for the output layer. For example, consider an MLP with 4 input features, 3 hidden neurons, and 2 output neurons. The weight matrices and biases are:

$$\mathbf{W}_1 = \begin{pmatrix} w_{11} & w_{12} & w_{13} \\ w_{21} & w_{22} & w_{23} \\ w_{31} & w_{32} & w_{33} \\ w_{41} & w_{42} & w_{43} \end{pmatrix}, \quad \mathbf{b}_1 = \begin{pmatrix} b_{11} & b_{12} & b_{13} \end{pmatrix}$$

$$\mathbf{W}_2 = \begin{pmatrix} w_{11} & w_{12} \\ w_{21} & w_{22} \\ w_{31} & w_{32} \end{pmatrix}, \quad \mathbf{b}_2 = \begin{pmatrix} b_{21} & b_{22} \end{pmatrix}$$

The forward pass for a single input $\mathbf{X}$ is computed as:

$$\mathbf{H} = \sigma(\mathbf{X}\mathbf{W}_1 + \mathbf{b}_1)$$

$$\mathbf{Y} = \mathbf{H}\mathbf{W}_2 + \mathbf{b}_2$$

Implementation in Python: Using TensorFlow, an MLP can be implemented as follows:

```
import tensorflow as tf
from tensorflow.keras.layers import Dense

# Define the MLP model
model = tf.keras.Sequential([
    Dense(3, input_shape=(4,), activation='relu'),
    # Hidden layer with 3 neurons
    Dense(2)  # Output layer with 2 neurons
])

# Print the model summary
model.summary()
```

Training and Evaluating the Network

Training a neural network involves optimizing the weights and biases to minimize a loss function. This process typically uses gradient descent-based optimization algorithms. Evaluating the network involves measuring its performance on a validation or test set.

Loss Function: The choice of loss function depends on the task. For a regression task, mean squared error (MSE) is commonly used:

$$L = \frac{1}{N} \sum_{i=1}^{N} (\mathbf{y}_i - \hat{\mathbf{y}}_i)^2$$

For a classification task, cross-entropy loss is used:

$$L = -\frac{1}{N}\sum_{i=1}^{N}\sum_{j=1}^{k} y_{ij}\log(\hat{y}_{ij})$$

Backpropagation: Backpropagation computes the gradient of the loss function with respect to each weight and bias in the network. The gradients are used to update the parameters via gradient descent.

Example: For the MLP, the gradients are computed as follows:

1. Compute the loss gradient with respect to the output:

$$\frac{\partial L}{\partial \mathbf{Y}}$$

2. Propagate the gradient back through the output layer:

$$\frac{\partial L}{\partial \mathbf{W}_2} = \mathbf{H}^T \frac{\partial L}{\partial \mathbf{Y}}, \quad \frac{\partial L}{\partial \mathbf{b}_2} = \sum_{i=1}^{N} \frac{\partial L}{\partial \mathbf{Y}_i}$$

3. Propagate the gradient back through the hidden layer:

$$\frac{\partial L}{\partial \mathbf{H}} = \frac{\partial L}{\partial \mathbf{Y}} \mathbf{W}_2^T$$

$$\frac{\partial L}{\partial \mathbf{W}_1} = \mathbf{X}^T (\frac{\partial L}{\partial \mathbf{H}} \odot \sigma'(\mathbf{Z}_1)), \quad \frac{\partial L}{\partial \mathbf{b}_1} = \sum_{i=1}^{N} \frac{\partial L}{\partial \mathbf{H}_i} \odot \sigma'(\mathbf{Z}_1)$$

where $\odot$ denotes the element-wise multiplication, and $\sigma'(\mathbf{Z}_1)$ is the derivative of the activation function.

Optimization: The parameters are updated using an optimization algorithm such as stochastic gradient descent (SGD):

$$\mathbf{W} \leftarrow \mathbf{W} - \eta \frac{\partial L}{\partial \mathbf{W}}$$

where η is the learning rate.

Example in Python: Training the MLP using TensorFlow involves compiling the model with a loss function and optimizer, and then calling the 'fit' method:

```
# Compile the model
model.compile(optimizer='adam', loss='mse')

# Generate dummy data
```

```
import numpy as np
X_train = np.random.rand(100, 4)
y_train = np.random.rand(100, 2)

# Train the model
model.fit(X_train, y_train, epochs=10, batch_size=32)

# Evaluate the model
loss = model.evaluate(X_train, y_train)
print("Training loss:", loss)
```

2.8 Exercises

1. Given a tensor $\mathbf{A} \in \mathbb{R}^{3\times3\times3}$, write down its mode-1, mode-2, and mode-3 unfoldings. Explain the significance of each mode of unfolding.
2. Perform the following operations on tensors $\mathbf{B} \in \mathbb{R}^{2\times2}$ and $\mathbf{C} \in \mathbb{R}^{2\times2}$:

$$\mathbf{B} = \begin{pmatrix} 1 & 2 \\ 3 & 4 \end{pmatrix}, \quad \mathbf{C} = \begin{pmatrix} 5 & 6 \\ 7 & 8 \end{pmatrix}$$

   (a) Compute $\mathbf{B} + \mathbf{C}$.
   (b) Compute the outer product $\mathbf{B} \otimes \mathbf{C}$.
   (c) Compute the tensor contraction of $\mathbf{B}$ and $\mathbf{C}$ over the first mode.

3. Given two matrices $\mathbf{D} \in \mathbb{R}^{2\times2}$ and $\mathbf{E} \in \mathbb{R}^{2\times2}$, compute the Kronecker product $\mathbf{D} \otimes \mathbf{E}$:

$$\mathbf{D} = \begin{pmatrix} 0 & 1 \\ 2 & 3 \end{pmatrix}, \quad \mathbf{E} = \begin{pmatrix} 4 & 5 \\ 6 & 7 \end{pmatrix}$$

4. Given a vector $\mathbf{v} \in \mathbb{R}^3$ and a rotation matrix $\mathbf{R} \in \mathbb{R}^{3\times3}$, compute the rotated vector $\mathbf{v}' = \mathbf{R}\mathbf{v}$:

$$\mathbf{v} = \begin{pmatrix} 1 \\ 0 \\ 0 \end{pmatrix}, \quad \mathbf{R} = \begin{pmatrix} 0 & -1 & 0 \\ 1 & 0 & 0 \\ 0 & 0 & 1 \end{pmatrix}$$

5. Given a tensor field $\mathbf{H}(x, y) = \begin{pmatrix} x & y \\ y & x \end{pmatrix}$, compute the gradient and divergence of the tensor field.
6. Consider a 3D mesh represented by a set of vertices $\mathbf{V} \in \mathbb{R}^{3\times4}$:

$$\mathbf{V} = \begin{pmatrix} 1 & 0 & 0 & 1 \\ 0 & 1 & 0 & 1 \\ 0 & 0 & 1 & 1 \end{pmatrix}$$

and a transformation matrix $\mathbf{T} \in \mathbb{R}^{3\times3}$:

$$\mathbf{T} = \begin{pmatrix} 1 & 0 & 0 \\ 0 & \cos(\theta) & -\sin(\theta) \\ 0 & \sin(\theta) & \cos(\theta) \end{pmatrix}$$

where θ is a rotation angle.

(a) Perform the matrix multiplication to apply the transformation $\mathbf{T}$ to the vertices $\mathbf{V}$. Write down the transformed vertices $\mathbf{V}'$.
(b) For $\theta = \frac{\pi}{4}$, compute the explicit values of the transformed vertices.

7. Given tensors $\mathbf{A} \in \mathbb{R}^{3\times3}$ and $\mathbf{B} \in \mathbb{R}^{3\times3}$:

$$A_{ij} = i + j, \quad B_{ij} = i \cdot j$$

(a) Using Einstein summation notation, compute the tensor contraction $C = A_{ij} B_{ji}$.
(b) Verify the result by explicitly computing the summation over indices.

8. Consider the fourth-order tensor $\mathbf{C} \in \mathbb{R}^{2\times2\times2\times2}$:

$$C_{ijkl} = i + j + k + l$$

(a) Perform a tensor contraction on the first and third dimensions to obtain a matrix $\mathbf{D} \in \mathbb{R}^{2\times2}$. Write down the resulting elements.
(b) Perform another contraction on the resulting matrix $\mathbf{D}$ by summing over both dimensions to get a scalar.

9. Given vectors $\mathbf{u} \in \mathbb{R}^3$ and $\mathbf{v} \in \mathbb{R}^3$:

$$\mathbf{u} = \begin{pmatrix} 1 \\ 2 \\ 3 \end{pmatrix}, \quad \mathbf{v} = \begin{pmatrix} 4 \\ 5 \\ 6 \end{pmatrix}$$

(a) Compute the outer product $\mathbf{u} \otimes \mathbf{v}$ resulting in a matrix $\mathbf{M} \in \mathbb{R}^{3\times3}$.
(b) Verify the outer product by showing that each element $M_{ij} = u_i v_j$.

10. Given matrices $\mathbf{X} \in \mathbb{R}^{2\times2}$ and $\mathbf{Y} \in \mathbb{R}^{2\times2}$:

$$\mathbf{X} = \begin{pmatrix} 1 & 3 \\ 2 & 4 \end{pmatrix}, \quad \mathbf{Y} = \begin{pmatrix} 2 & 0 \\ 1 & 5 \end{pmatrix}$$

(a) Compute the Hadamard product $\mathbf{Z} = \mathbf{X} \cdot \mathbf{Y}$.
(b) For $\mathbf{Z}$, calculate the element-wise square root of each element to obtain a new matrix $\mathbf{W}$.

11. Consider a third-order tensor $\mathbf{H} \in \mathbb{R}^{2\times2\times2}$:

$$H_{ijk} = \begin{cases} 1 & \text{if } i = j = k = 1 \\ 0 & \text{otherwise} \end{cases}$$

(a) Perform the Higher-Order Singular Value Decomposition (HOSVD) on $\mathbf{H}$. Write down the core tensor and the factor matrices.
(b) Verify that the reconstruction of $\mathbf{H}$ using its HOSVD components yields the original tensor.

12. Consider a grayscale image represented as a matrix $\mathbf{I} \in \mathbb{R}^{4\times 4}$:

$$\mathbf{I} = \begin{pmatrix} 1 & 2 & 3 & 4 \\ 5 & 6 & 7 & 8 \\ 9 & 10 & 11 & 12 \\ 13 & 14 & 15 & 16 \end{pmatrix}$$

and a filter (kernel) $\mathbf{K} \in \mathbb{R}^{2\times 2}$:

$$\mathbf{K} = \begin{pmatrix} 0 & 1 \\ -1 & 0 \end{pmatrix}$$

(a) Perform a 2D convolution of $\mathbf{I}$ with $\mathbf{K}$ using a stride of 1 and no padding. Write down the resulting matrix.
(b) Calculate the gradient of the convolution output with respect to the input image $\mathbf{I}$.

13. Let $\mathbf{S} \in \mathbb{R}^{5\times 3}$ represent a sequence of 5 timesteps, each with 3 features. Assume:

$$\mathbf{S} = \begin{pmatrix} 0.1 & 0.2 & 0.3 \\ 0.4 & 0.5 & 0.6 \\ 0.7 & 0.8 & 0.9 \\ 1.0 & 1.1 & 1.2 \\ 1.3 & 1.4 & 1.5 \end{pmatrix}$$

(a) Apply a temporal convolution with a filter $\mathbf{F} \in \mathbb{R}^{2\times 3}$ defined as:

$$\mathbf{F} = \begin{pmatrix} 0.1 & 0.2 & 0.3 \\ 0.4 & 0.5 & 0.6 \end{pmatrix}$$

using a stride of 1 and no padding. Write down the resulting sequence matrix.
(b) Compute the element-wise activation using the ReLU function on the convolution output.

14. Consider a point cloud represented as a tensor $\mathbf{P} \in \mathbb{R}^{3\times 5}$, where each column represents a point in 3D space:

$$\mathbf{P} = \begin{pmatrix} 1 & 2 & 3 & 4 & 5 \\ 5 & 4 & 3 & 2 & 1 \\ 1 & 3 & 5 & 7 & 9 \end{pmatrix}$$

(a) Compute the pairwise Euclidean distance matrix $\mathbf{D} \in \mathbb{R}^{5\times5}$ for the points in $\mathbf{P}$.
(b) Using the distance matrix $\mathbf{D}$, perform a multidimensional scaling (MDS) to project the points into a 2D space. Write down the resulting 2D coordinates for each point.

15. Given a color image represented as a tensor $\mathbf{I} \in \mathbb{R}^{3\times4\times4}$ (3 channels, 4x4 spatial dimensions):

$$\mathbf{I}(:,:,1) = \begin{pmatrix} 1 & 2 & 3 & 4 \\ 5 & 6 & 7 & 8 \\ 9 & 10 & 11 & 12 \\ 13 & 14 & 15 & 16 \end{pmatrix}, \quad \mathbf{I}(:,:,2) = \begin{pmatrix} 17 & 18 & 19 & 20 \\ 21 & 22 & 23 & 24 \\ 25 & 26 & 27 & 28 \\ 29 & 30 & 31 & 32 \end{pmatrix}, \quad \mathbf{I}(:,:,3) = \begin{pmatrix} 33 & 34 & 35 & 36 \\ 37 & 38 & 39 & 40 \\ 41 & 42 & 43 & 44 \\ 45 & 46 & 47 & 48 \end{pmatrix}$$

Apply a 3D convolution using a filter $\mathbf{K} \in \mathbb{R}^{3\times2\times2}$:

$$\mathbf{K}(:,:,1) = \begin{pmatrix} 0 & 1 \\ -1 & 0 \end{pmatrix}, \quad \mathbf{K}(:,:,2) = \begin{pmatrix} 1 & 0 \\ 0 & -1 \end{pmatrix}, \quad \mathbf{K}(:,:,3) = \begin{pmatrix} -1 & 0 \\ 0 & 1 \end{pmatrix}$$

with a stride of 1 and no padding. Calculate the resulting feature map.

16. Consider a sequence data tensor $\mathbf{X} \in \mathbb{R}^{6\times4}$, where 6 represents the sequence length and 4 represents the feature dimension:

$$\mathbf{X} = \begin{pmatrix} 0.1 & 0.2 & 0.3 & 0.4 \\ 0.5 & 0.6 & 0.7 & 0.8 \\ 0.9 & 1.0 & 1.1 & 1.2 \\ 1.3 & 1.4 & 1.5 & 1.6 \\ 1.7 & 1.8 & 1.9 & 2.0 \\ 2.1 & 2.2 & 2.3 & 2.4 \end{pmatrix}$$

with RNN weight matrices $\mathbf{W}_h \in \mathbb{R}^{4\times4}$ and $\mathbf{W}_x \in \mathbb{R}^{4\times4}$:

$$\mathbf{W}_h = \begin{pmatrix} 0.1 & 0.2 & 0.3 & 0.4 \\ 0.5 & 0.6 & 0.7 & 0.8 \\ 0.9 & 1.0 & 1.1 & 1.2 \\ 1.3 & 1.4 & 1.5 & 1.6 \end{pmatrix}, \quad \mathbf{W}_x = \begin{pmatrix} 0.2 & 0.4 & 0.6 & 0.8 \\ 1.0 & 1.2 & 1.4 & 1.6 \\ 1.8 & 2.0 & 2.2 & 2.4 \\ 2.6 & 2.8 & 3.0 & 3.2 \end{pmatrix}$$

(a) Calculate the hidden state $\mathbf{h}_t$ at $t = 2$ assuming initial hidden state $\mathbf{h}_0 = \mathbf{0}$.
(b) Compute the output of the RNN for the entire sequence.

17. Given a point cloud represented as a tensor $\mathbf{P} \in \mathbb{R}^{3\times6}$, where each column is a point in 3D space:

$$\mathbf{P} = \begin{pmatrix} 1 & 2 & 3 & 4 & 5 & 6 \\ 6 & 5 & 4 & 3 & 2 & 1 \\ 1 & 3 & 5 & 7 & 9 & 11 \end{pmatrix}$$

(a) Apply a rotation matrix $\mathbf{R} \in \mathbb{R}^{3\times 3}$ to the point cloud, where:

$$\mathbf{R} = \begin{pmatrix} 0.866 & -0.5 & 0 \\ 0.5 & 0.866 & 0 \\ 0 & 0 & 1 \end{pmatrix}$$

Calculate the new coordinates of the points.

(b) Compute the covariance matrix of the transformed point cloud and perform Principal Component Analysis (PCA) to find the principal directions.

Chapter 3
Building Blocks

I believe that the brain uses some very unsophisticated computational principles to represent the world... by combining these principles appropriately, we can make computers do truly intelligent things.

Geoffrey Hinton, Google AI Blog

3.1 Introduction to Multi-layer Perceptrons (MLPs)

Multi-layer perceptrons (MLPs) are a class of feedforward artificial neural networks that consist of multiple layers of nodes. Each node, or neuron, in the network performs a linear transformation followed by a non-linear activation function. MLPs are the fundamental building blocks of many deep learning models and are used for a variety of tasks, including classification, regression, and function approximation. An MLP typically consists of an input layer, one or more hidden layers, and an output layer. The nodes in each layer are fully connected to the nodes in the subsequent layer, meaning every node in a given layer receives input from every node in the previous layer and sends its output to every node in the next layer. Mathematically, the operation of an MLP can be described as a series of matrix multiplications and non-linear transformations. For an input vector $\mathbf{x} \in \mathbb{R}^d$, the output of the first hidden layer is computed as:

$$\mathbf{h}^{(1)} = \sigma(\mathbf{W}^{(1)}\mathbf{x} + \mathbf{b}^{(1)})$$

where $\mathbf{W}^{(1)} \in \mathbb{R}^{h_1 \times d}$ is the weight matrix for the first hidden layer, $\mathbf{b}^{(1)} \in \mathbb{R}^{h_1}$ is the bias vector for the first hidden layer, σ is a non-linear activation function (e.g., ReLU, sigmoid, tanh). The output of the second hidden layer is:

$$\mathbf{h}^{(2)} = \sigma(\mathbf{W}^{(2)}\mathbf{h}^{(1)} + \mathbf{b}^{(2)})$$

P. Singh and B. Raman, *Deep Learning Through the Prism of Tensors*, Studies in Big Data 162, https://doi.org/10.1007/978-981-97-8019-8_3

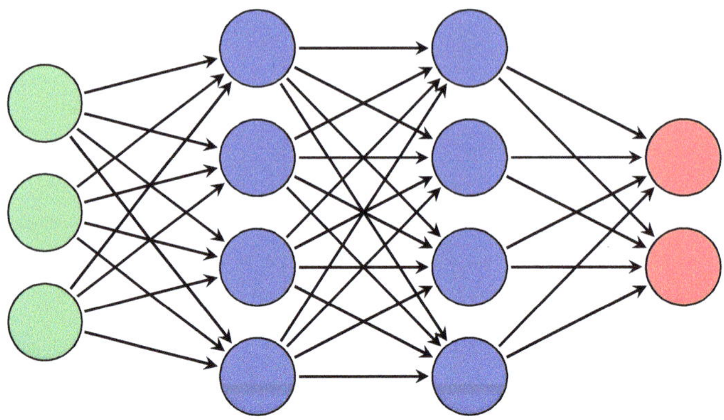

Fig. 3.1 A feed-forward neural network with 3 input neurons, 2 hidden layers each with 4 neurons, and 2 output neurons

This process continues for all subsequent hidden layers. Finally, the output layer produces the network's output:

$$\mathbf{y} = \mathbf{W}^{(L)}\mathbf{h}^{(L-1)} + \mathbf{b}^{(L)}$$

where L is the total number of layers in the network. The choice of activation function σ is crucial as it introduces non-linearity into the network, allowing it to learn complex patterns. Common activation functions include the Rectified Linear Unit (ReLU), sigmoid, and hyperbolic tangent (tanh).

Example: Consider an MLP with 2 input features, 3 hidden neurons, and 1 output neuron (see Fig. 3.1). The weight matrices and biases are:

$$\mathbf{W}^{(1)} = \begin{pmatrix} w_{11} & w_{12} \\ w_{21} & w_{22} \\ w_{31} & w_{32} \end{pmatrix}, \quad \mathbf{b}^{(1)} = \begin{pmatrix} b_{11} \\ b_{21} \\ b_{31} \end{pmatrix}$$

$$\mathbf{W}^{(2)} = \begin{pmatrix} w_{11} & w_{21} & w_{31} \end{pmatrix}, \quad \mathbf{b}^{(2)} = \begin{pmatrix} b_{12} \end{pmatrix}$$

The forward pass for an input $\mathbf{x} = (x_1, x_2)^T$ is computed as:

$$\mathbf{h}^{(1)} = \sigma(\mathbf{W}^{(1)}\mathbf{x} + \mathbf{b}^{(1)})$$

$$y = \mathbf{W}^{(2)}\mathbf{h}^{(1)} + \mathbf{b}^{(2)}$$

3.1.1 Historical Context and Development

The development of MLPs dates back to the 1950s and 1960s with the advent of the perceptron, a single-layer neural network proposed by Rosenblatt (1958). The perceptron could solve simple linear classification problems, but its limitations were highlighted by Minsky and Papert (1969) in their book "Perceptrons," which demonstrated that a single-layer perceptron could not solve non-linearly separable problems such as the XOR problem. The limitations of single-layer perceptrons motivated the development of multi-layer architectures. The breakthrough came in the 1980s with the invention of the backpropagation algorithm by Rumelhart et al. (1986). Backpropagation efficiently computes the gradient of the loss function with respect to the network's weights, enabling the training of multi-layer networks using gradient descent.

The backpropagation algorithm can be summarized as follows:

1. Perform a forward pass to compute the network's output.
2. Compute the loss between the predicted output and the true output.
3. Perform a backward pass to compute the gradients of the loss with respect to each weight and bias.
4. Update the weights and biases using the computed gradients.

Mathematically, the gradient of the loss L with respect to the weights $\mathbf{W}^{(l)}$ and biases $\mathbf{b}^{(l)}$ in layer l is given by:

$$\frac{\partial L}{\partial \mathbf{W}^{(l)}} = \delta^{(l)} \mathbf{h}^{(l-1)T}, \quad \frac{\partial L}{\partial \mathbf{b}^{(l)}} = \delta^{(l)}$$

where $\delta^{(l)}$ is the error term for layer l, computed as:

$$\delta^{(L)} = \nabla_{\mathbf{y}} L \odot \sigma'(\mathbf{z}^{(L)})$$

$$\delta^{(l)} = (\mathbf{W}^{(l+1)T} \delta^{(l+1)}) \odot \sigma'(\mathbf{z}^{(l)})$$

where $\mathbf{z}^{(l)} = \mathbf{W}^{(l)} \mathbf{h}^{(l-1)} + \mathbf{b}^{(l)}$ is the pre-activation output of layer l, and $\odot$ denotes element-wise multiplication.

The success of MLPs and backpropagation paved the way for the development of more sophisticated neural network architectures, including convolutional neural networks (CNNs) and recurrent neural networks (RNNs). Despite their simplicity, MLPs remain a powerful tool for many applications and serve as the foundation for understanding more complex models.

3.1.2 Applications of MLPs

MLPs have been successfully applied to a wide range of tasks across various domains. Some notable applications include:

1. Classification: MLPs are commonly used for classification tasks, where the goal is to assign input data to one of several predefined categories. Examples include digit recognition in the MNIST dataset, where each image of a handwritten digit is classified into one of ten classes (0–9).
2. Regression: MLPs can also be used for regression tasks, where the goal is to predict a continuous output value. An example is predicting house prices based on features such as size, location, and number of bedrooms.
3. Function Approximation: MLPs can approximate complex functions and are used in scenarios where an explicit mathematical model is not available. This capability is leveraged in control systems, robotics, and financial modeling.
4. Image Recognition: While CNNs are more commonly used for image recognition tasks, MLPs can still be applied to simpler image classification problems or as part of a hybrid model.
5. Natural Language Processing: MLPs can be used in natural language processing (NLP) tasks, such as sentiment analysis and text classification. In these applications, text data is often transformed into numerical representations, such as word embeddings, before being fed into the MLP.

Example: In a sentiment analysis task, each review can be represented as a vector of word embeddings. An MLP can be trained to classify the review as positive or negative based on these embeddings.

```
import tensorflow as tf
from tensorflow.keras.layers import Dense, Embedding, GlobalAveragePooling1D

# Define the MLP model for text classification
model = tf.keras.Sequential([
    Embedding(input_dim=5000, output_dim=50, input_length=100),
    GlobalAveragePooling1D(),
    Dense(10, activation='relu'),
    Dense(1, activation='sigmoid')
])

# Compile the model
model.compile(optimizer='adam', loss='binary_crossentropy', metrics=['accuracy'])

# Generate dummy data
X_train = np.random.randint(5000, size=(100, 100))
y_train = np.random.randint(2, size=(100,))

# Train the model
model.fit(X_train, y_train, epochs=10, batch_size=32)

# Evaluate the model
loss, accuracy = model.evaluate(X_train, y_train)
print("Training accuracy:", accuracy)
```

3.2 Architecture of Multi-layer Perceptrons

3.2.1 *Input Layer*

The input layer of a MLP is the first layer of the network, where the data is fed into the model. This layer does not perform any computation; rather, it serves as a conduit for passing the data into the subsequent layers of the network. The design and preparation of the input layer are crucial for the effective functioning of the MLP, as it determines how the data is represented and preprocessed before being processed by the hidden layers.

Data Representation and Input Features

The representation of data in the input layer depends on the type of data and the task at hand. Data can be represented in various forms, such as vectors, matrices, or higher-dimensional tensors, depending on its structure.

Vector Representation: For tasks where the input data is naturally represented as vectors, such as tabular data, each feature in the dataset corresponds to an element in the input vector. For example, consider a dataset with N samples, each having d features. The input data can be represented as a matrix $\mathbf{X} \in \mathbb{R}^{N \times d}$, where each row represents a sample and each column represents a feature.

Example: A dataset of house prices with features such as size, number of bedrooms, and location can be represented as:

$$\mathbf{X} = \begin{pmatrix} x_{11} & x_{12} & x_{13} \\ x_{21} & x_{22} & x_{23} \\ \vdots & \vdots & \vdots \\ x_{N1} & x_{N2} & x_{N3} \end{pmatrix}$$

where x_{ij} represents the j-th feature of the i-th sample.

Matrix Representation: For tasks involving images, the input data is often represented as matrices or higher-dimensional tensors. For example, a grayscale image of size $H \times W$ can be represented as a matrix $\mathcal{I} \in \mathbb{R}^{H \times W}$, while a colored image (RGB) can be represented as a 3D tensor $\mathcal{I} \in \mathbb{R}^{H \times W \times 3}$.

Example: Consider a batch of N RGB images, each of size $H \times W$. The input data can be represented as a 4D tensor:

$$\mathcal{X} = \left(\mathcal{I}_1 \; \mathcal{I}_2 \ldots \mathcal{I}_N\right), \quad \mathcal{I}_i \in \mathbb{R}^{H \times W \times 3}$$

Sequence Representation: For tasks involving sequential data, such as time series or text, the input data is represented as sequences. Each sequence can be represented as a matrix, where one dimension corresponds to the sequence length and the other dimension corresponds to the feature vector at each time step.

Example: Consider a dataset of text sequences, where each sequence is a sentence represented by a sequence of word embeddings. If each word embedding is a vector of size d and the maximum sequence length is T, the input data can be represented as a 3D tensor:

$$\mathcal{S} = \left(\mathbf{s}_1 \ \mathbf{s}_2 \ \dots \ \mathbf{s}_N\right), \quad \mathbf{s}_i \in \mathbb{R}^{T\times d}$$

Preprocessing Techniques

Preprocessing is a critical step in preparing the data for input into an MLP. Effective preprocessing techniques can improve the performance of the model by ensuring that the input data is in a suitable format and scale for the network to learn from.

Normalization: Normalization involves scaling the input features to a common range, typically [0, 1] or [– 1, 1]. This helps to stabilize and speed up the training process by ensuring that the gradients do not become too large or too small. Mathematically, normalization can be achieved by:

$$x' = \frac{x - \min(x)}{\max(x) - \min(x)}$$

where x is the original feature value, $\min(x)$ and $\max(x)$ are the minimum and maximum values of the feature, respectively, and x' is the normalized feature value.

Standardization: Standardization involves transforming the input features to have zero mean and unit variance. This ensures that the features have the same scale, which helps the model to converge more quickly during training. Mathematically, standardization is given by:

$$x' = \frac{x - \mu}{\sigma}$$

where x is the original feature value, μ is the mean of the feature, σ is the standard deviation of the feature, and x' is the standardized feature value.

One-Hot Encoding: For categorical data, one-hot encoding is a common preprocessing technique. It involves representing each category as a binary vector, where the position corresponding to the category is set to 1, and all other positions are set to 0. For example, consider a categorical feature with three categories: "Red," "Green," and "Blue." One-hot encoding transforms these categories into:

$$\text{Red} \rightarrow (1, 0, 0), \quad \text{Green} \rightarrow (0, 1, 0), \quad \text{Blue} \rightarrow (0, 0, 1)$$

Embedding: For high-cardinality categorical features, embeddings can be used to represent the categories as dense vectors in a continuous space. This technique is commonly used in natural language processing for representing words as word embeddings. For example, consider a dataset of sentences, where each word is represented by an index. An embedding layer transforms these indices into dense vectors:

```
import tensorflow as tf

# Define an embedding layer
embedding_layer = tf.keras.layers.Embedding(input_dim=5000, output_dim=50, input_length=100)

# Example input sequence (batch of sentences with word indices)
input_sequences = tf.constant([[1, 2, 3], [4, 5, 6]])

# Apply the embedding layer
embedded_sequences = embedding_layer(input_sequences)
print(embedded_sequences.shape)
```

Feature Engineering: Feature engineering involves creating new features or transforming existing features to improve the performance of the model. This can include techniques such as polynomial feature generation, interaction terms, and domain-specific transformations. For example, for a dataset with features x_1 and x_2, polynomial feature generation can create new features such as x_1^2, x_2^2, and $x_1 \cdot x_2$.

3.2.2 Hidden Layers

Hidden layers are the core computational units of a Multi-layer Perceptron (MLP). They consist of neurons that perform linear transformations followed by non-linear activation functions. The design and configuration of hidden layers significantly influence the learning capacity and performance of the network. Here, we delve into the role and design of hidden layers, and the considerations for choosing the number of hidden layers and neurons.

Role and Design of Hidden Layers

The hidden layers in an MLP are responsible for capturing and learning complex patterns in the input data. Each hidden layer applies a transformation to its input, enabling the network to learn hierarchical representations of the data. This hierarchical learning process allows MLPs to model intricate relationships and dependencies within the data.

Linear Transformation: In each hidden layer, the input is transformed by a weight matrix and a bias vector, followed by the application of an activation function. Mathematically, the output $\mathbf{h}^{(l)}$ of the l-th hidden layer is given by:

$$\mathbf{h}^{(l)} = \sigma(\mathbf{W}^{(l)}\mathbf{h}^{(l-1)} + \mathbf{b}^{(l)})$$

where $\mathbf{W}^{(l)} \in \mathbb{R}^{n_l \times n_{l-1}}$ is the weight matrix, $\mathbf{b}^{(l)} \in \mathbb{R}^{n_l}$ is the bias vector, σ is the activation function (e.g., ReLU, sigmoid, tanh), n_l is the number of neurons in the l-th layer.

Activation Function: The activation function introduces non-linearity into the network, allowing it to learn complex functions. Common activation functions include:

ReLU (Rectified Linear Unit):

$$\text{ReLU}(z) = \max(0, z)$$

Sigmoid:

$$\sigma(z) = \frac{1}{1 + e^{-z}}$$

Tanh (Hyperbolic Tangent):

$$\tanh(z) = \frac{e^z - e^{-z}}{e^z + e^{-z}}$$

The choice of activation function affects the learning dynamics and performance of the network. ReLU is particularly popular due to its simplicity and effectiveness in mitigating the vanishing gradient problem.

For example, consider a simple MLP with one hidden layer consisting of three neurons, each using the ReLU activation function. For an input vector $\mathbf{x} \in \mathbb{R}^2$, the hidden layer transformation can be represented as:

$$\mathbf{W}^{(1)} = \begin{pmatrix} w_{11} & w_{12} \\ w_{21} & w_{22} \\ w_{31} & w_{32} \end{pmatrix}, \quad \mathbf{b}^{(1)} = \begin{pmatrix} b_1 \\ b_2 \\ b_3 \end{pmatrix}$$

The output of the hidden layer is:

$$\mathbf{h}^{(1)} = \text{ReLU}(\mathbf{W}^{(1)}\mathbf{x} + \mathbf{b}^{(1)}) = \text{ReLU}\left(\begin{pmatrix} w_{11} & w_{12} \\ w_{21} & w_{22} \\ w_{31} & w_{32} \end{pmatrix}\begin{pmatrix} x_1 \\ x_2 \end{pmatrix} + \begin{pmatrix} b_1 \\ b_2 \\ b_3 \end{pmatrix}\right)$$

Choosing the Number of Hidden Layers and Neurons

Selecting the appropriate number of hidden layers and neurons is a crucial aspect of designing an MLP. This choice impacts the model's capacity to learn from data, its ability to generalize to unseen data, and its computational efficiency.

Shallow Networks: MLPs with one or two hidden layers are often sufficient for simple tasks. These networks are easier to train and less prone to overfitting, making them suitable for problems with limited data and low complexity.

Deep Networks: For more complex tasks, such as image recognition or natural language processing, deeper networks with multiple hidden layers are often required. Deep networks can capture hierarchical features and model intricate patterns in the data. However, they require more data and computational resources and are more challenging to train.

The number of neurons in each hidden layer determines the layer's capacity to represent the input data. There are several heuristics and methods for choosing the number of neurons:

Rule of Thumb: A common heuristic is to start with a number of neurons in the hidden layer that is between the size of the input layer and the size of the output layer. For example, if the input layer has 10 neurons and the output layer has 3 neurons, a hidden layer might have 6 to 8 neurons.

Grid Search and Cross-Validation: Systematically searching for the best combination of hyperparameters using techniques like grid search and cross-validation can help determine the optimal number of neurons. This approach evaluates different configurations on a validation set to identify the best-performing architecture.

Incremental Testing: Starting with a small number of neurons and gradually increasing the number while monitoring the model's performance can help identify a good configuration. This approach helps avoid overfitting and ensures that the network has enough capacity to learn from the data.

Example: Consider an MLP designed for a classification task with 100 input features and 10 output classes. A reasonable starting architecture might be:

Input layer: 100 neurons
First hidden layer: 64 neurons, ReLU activation
Second hidden layer: 32 neurons, ReLU activation
Output layer: 10 neurons, softmax activation

Mathematically, the transformations can be represented as:

$$\mathbf{h}^{(1)} = \text{ReLU}(\mathbf{W}^{(1)}\mathbf{x} + \mathbf{b}^{(1)})$$

$$\mathbf{h}^{(2)} = \text{ReLU}(\mathbf{W}^{(2)}\mathbf{h}^{(1)} + \mathbf{b}^{(2)})$$

$$\mathbf{y} = \text{softmax}(\mathbf{W}^{(3)}\mathbf{h}^{(2)} + \mathbf{b}^{(3)})$$

where $\mathbf{W}^{(1)} \in \mathbb{R}^{64\times 100}$, $\mathbf{b}^{(1)} \in \mathbb{R}^{64}$, $\mathbf{W}^{(2)} \in \mathbb{R}^{32\times 64}$, $\mathbf{b}^{(2)} \in \mathbb{R}^{32}$, and $\mathbf{W}^{(3)} \in \mathbb{R}^{10\times 32}$, $\mathbf{b}^{(3)} \in \mathbb{R}^{10}$.

When designing and training an MLP, it's important to monitor the model's performance and adjust the architecture as needed. Techniques such as dropout, early stopping, and regularization can help prevent overfitting and improve generalization.

Example in Python: Using TensorFlow, an MLP with two hidden layers can be implemented as follows:

```
import tensorflow as tf
from tensorflow.keras.layers import Dense

# Define the MLP model
model = tf.keras.Sequential([
    Dense(64, input_shape=(100,), activation='relu'),  # First hidden layer
    Dense(32, activation='relu'),  # Second hidden layer
    Dense(10, activation='softmax')  # Output layer
])
```

```
# Compile the model
model.compile(optimizer='adam', loss='sparse_categorical_crossentropy', metrics=['accuracy'])

# Generate dummy data
import numpy as np
X_train = np.random.rand(1000, 100)
y_train = np.random.randint(10, size=(1000,))

# Train the model
model.fit(X_train, y_train, epochs=10, batch_size=32)

# Evaluate the model
loss, accuracy = model.evaluate(X_train, y_train)
print("Training accuracy:", accuracy)
```

3.2.3 Output Layer

The output layer of a multi-layer perceptron (MLP) is the final layer that produces the predictions or outputs of the network. The configuration of the output layer depends on the nature of the task, such as classification, regression, or other specific requirements. This section explores different output layer configurations and the role of activation functions in determining the nature of the output.

Output Layer Configurations for Different Tasks

The design of the output layer is tailored to the specific requirements of the task at hand. Different tasks necessitate different output structures and activation functions.

Classification Tasks: For classification tasks, the output layer typically consists of a number of neurons equal to the number of classes. The output of each neuron represents the predicted probability of the corresponding class. The softmax activation function is commonly used to convert the raw outputs (logits) into probabilities that sum to one. For example, for a classification task with K classes, the output layer has K neurons. The output vector $\mathbf{y} \in \mathbb{R}^K$ is computed as:

$$y_i = \frac{e^{z_i}}{\sum_{j=1}^{K} e^{z_j}}$$

where z_i is the logit (raw output) for class i.

Regression Tasks: For regression tasks, the output layer typically has a single neuron (for univariate regression) or multiple neurons (for multivariate regression) without any activation function, allowing the network to produce continuous values. The linear activation function is used in the output layer. For example, for a univariate regression task, the output is a single continuous value:

$$y = z$$

where z is the raw output from the final layer.

Binary Classification Tasks: For binary classification tasks, the output layer usually has a single neuron with a sigmoid activation function. The sigmoid function maps the raw output to a probability value between 0 and 1. For example, for a binary classification task, the output is:

$$y = \sigma(z) = \frac{1}{1 + e^{-z}}$$

where z is the logit.

Multi-label Classification Tasks: For multi-label classification tasks, where each input can belong to multiple classes simultaneously, the output layer has multiple neurons with sigmoid activation functions. Each neuron represents the probability of a specific class. For example, for a multi-label classification task with K classes, the output vector $\mathbf{y} \in \mathbb{R}^K$ is computed as:

$$y_i = \sigma(z_i) = \frac{1}{1 + e^{-z_i}}$$

where z_i is the logit for class i.

Example in Python: Using TensorFlow, different output layer configurations can be implemented as follows:

```
import tensorflow as tf
from tensorflow.keras.layers import Dense

# Classification task with softmax activation
classification_model = tf.keras.Sequential([
    Dense(64, activation='relu', input_shape=(100,)),
    Dense(32, activation='relu'),
    Dense(10, activation='softmax')  # 10 classes
])

# Regression task with linear activation
regression_model = tf.keras.Sequential([
    Dense(64, activation='relu', input_shape=(100,)),
    Dense(32, activation='relu'),
    Dense(1)  # Single continuous output
])

# Binary classification task with sigmoid activation
binary_classification_model = tf.keras.Sequential([
    Dense(64, activation='relu', input_shape=(100,)),
    Dense(32, activation='relu'),
    Dense(1, activation='sigmoid')  # Single output for binary classification
])

# Multi-label classification task with sigmoid activation
multi_label_classification_model = tf.keras.Sequential([
    Dense(64, activation='relu', input_shape=(100,)),
    Dense(32, activation='relu'),
    Dense(10, activation='sigmoid')  # 10 classes, each with independent probability
])
```

3.3 Activation Functions

Activation functions are a critical component of neural networks, introducing non-linearity into the model, which allows the network to learn complex patterns in the data. They determine the output of each neuron and, consequently, the output of the entire network. Different activation functions have various properties that make them suitable for different tasks and network architectures. In this section, we will delve into the sigmoid function, its mathematical formulation, properties, use cases, and limitations.

3.3.1 Introduction to Activation Functions

Activation functions transform the weighted sum of inputs into an output for a neuron in a neural network. Without activation functions, a neural network would essentially perform linear transformations, regardless of the number of layers, and thus would not be able to model complex, non-linear relationships. By introducing non-linear activation functions, neural networks can approximate any continuous function, enabling them to solve a wide range of tasks. Common activation functions include the sigmoid function, hyperbolic tangent (tanh), and Rectified Linear Unit (ReLU). Each function has unique characteristics and is chosen based on the specific needs of the model and the nature of the task.

3.3.2 Sigmoid Function

The sigmoid function is one of the most commonly used activation functions, particularly in the early days of neural network research. It maps any real-valued number to a value between 0 and 1, making it useful for models where we need to predict probabilities (see Fig. 3.2). The sigmoid function, also known as the logistic function, is defined as:

$$\sigma(z) = \frac{1}{1 + e^{-z}}$$

where z is the input to the neuron.

Properties:

1. Range: The output of the sigmoid function lies between 0 and 1, making it suitable for binary classification tasks where the output needs to be interpreted as a probability.
2. Derivative: The derivative of the sigmoid function is:

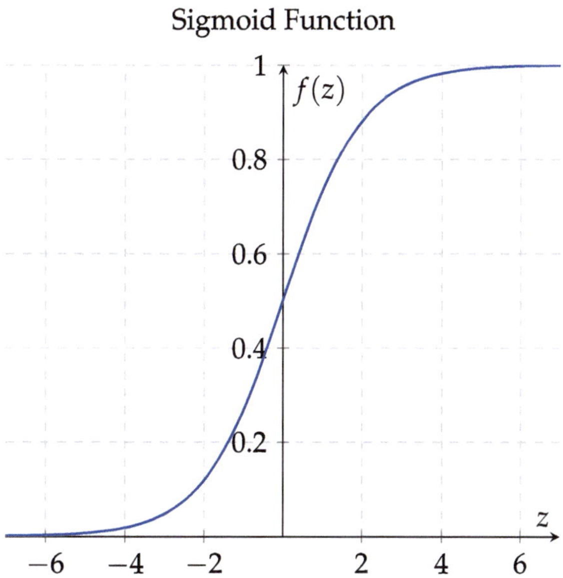

Fig. 3.2 Plot of the sigmoid function $f(z) = \frac{1}{1+e^{-z}}$

$$\sigma'(z) = \sigma(z) \cdot (1 - \sigma(z))$$

This derivative is useful for backpropagation, allowing the network to update its weights during training.

3. Saturating Non-linearity: For very large positive or negative values of z, the sigmoid function outputs values close to 1 or 0, respectively. This property can cause the gradient to vanish during training, making it difficult for the network to learn.
4. Smooth and Differentiable: The sigmoid function is smooth and differentiable, which is a desirable property for optimization during training.

Example: Consider a neuron with input $z = \mathbf{w} \cdot \mathbf{x} + b$, where $\mathbf{w}$ is the weight vector, $\mathbf{x}$ is the input vector, and b is the bias. The output of the neuron using the sigmoid activation function is:

$$y = \sigma(z) = \frac{1}{1 + e^{-(\mathbf{w} \cdot \mathbf{x} + b)}}$$

The gradient of the output with respect to the input z is:

$$\frac{\partial y}{\partial z} = \sigma(z) \cdot (1 - \sigma(z))$$

Use Cases and Limitations

Use Cases:

1. Binary Classification: The sigmoid function is widely used in the output layer of binary classification models, where the goal is to predict the probability of a binary outcome.
2. Logistic Regression: In logistic regression, the sigmoid function is used to model the probability of the dependent variable belonging to a particular class.
3. Neural Networks: In neural networks, the sigmoid function can be used in hidden layers, especially in simpler models or when dealing with certain types of data where non-linearity is essential.

Example in Python: Using TensorFlow, a simple neural network with a sigmoid activation function in the output layer for binary classification can be implemented as follows:

```
import tensorflow as tf
from tensorflow.keras.layers import Dense

# Define the model
model = tf.keras.Sequential([
    Dense(64, activation='relu', input_shape=(100,)),  # Hidden layer
    Dense(1, activation='sigmoid')  # Output layer for binary classification
])

# Compile the model
model.compile(optimizer='adam', loss='binary_crossentropy', metrics=['accuracy'])

# Generate dummy data
import numpy as np
X_train = np.random.rand(1000, 100)
y_train = np.random.randint(2, size=(1000,))

# Train the model
model.fit(X_train, y_train, epochs=10, batch_size=32)

# Evaluate the model
loss, accuracy = model.evaluate(X_train, y_train)
print("Training accuracy:", accuracy)
```

Limitations:

1. Vanishing Gradient Problem: The sigmoid function can cause the vanishing gradient problem during backpropagation. When the inputs to the sigmoid function are very large or very small, the gradients become very small, leading to slow learning or convergence issues.
2. Output Saturation: For inputs far from zero, the sigmoid function outputs values very close to 0 or 1, which can result in the saturation of neurons and limit the model's capacity to learn from data.
3. Not Zero-Centered: The sigmoid function outputs values between 0 and 1, which means the activations are not zero-centered. This can lead to inefficient gradient updates and affect the convergence speed of the model.

3.3.3 Hyperbolic Tangent (Tanh) Function

The hyperbolic tangent (tanh) function is another widely used activation function in neural networks. It is particularly popular in hidden layers due to its properties that often result in better performance compared to the sigmoid function. The tanh function is defined as:

$$\tanh(z) = \frac{e^z - e^{-z}}{e^z + e^{-z}}$$

where z is the input to the neuron.

Properties:

1. Range: The output of the tanh function lies between – 1 and 1. This symmetric range around zero makes it zero-centered, which can help in balancing the gradients during training (see Fig. 3.3).
2. Derivative: The derivative of the tanh function is:

$$\tanh'(z) = 1 - \tanh^2(z)$$

   This derivative is useful for backpropagation, allowing the network to update its weights during training effectively.
3. Non-linearity: The tanh function introduces non-linearity into the network, enabling it to learn complex relationships in the data.
4. Smooth and Differentiable: The tanh function is smooth and differentiable, which is a desirable property for optimization during training.

Example: Consider a neuron with input $z = \mathbf{w} \cdot \mathbf{x} + b$, where $\mathbf{w}$ is the weight vector, $\mathbf{x}$ is the input vector, and b is the bias. The output of the neuron using the tanh activation function is:

$$y = \tanh(z) = \tanh(\mathbf{w} \cdot \mathbf{x} + b) = \frac{e^{(\mathbf{w} \cdot \mathbf{x}+b)} - e^{-(\mathbf{w} \cdot \mathbf{x}+b)}}{e^{(\mathbf{w} \cdot \mathbf{x}+b)} + e^{-(\mathbf{w} \cdot \mathbf{x}+b)}}$$

The gradient of the output with respect to the input z is:

$$\frac{\partial y}{\partial z} = 1 - \tanh^2(z)$$

Visualization: To understand the behavior of the tanh function, it's helpful to visualize it. The function smoothly transitions from -1 to 1, with a steep slope around the origin, which helps the model to learn faster and more effectively in the regions close to zero.

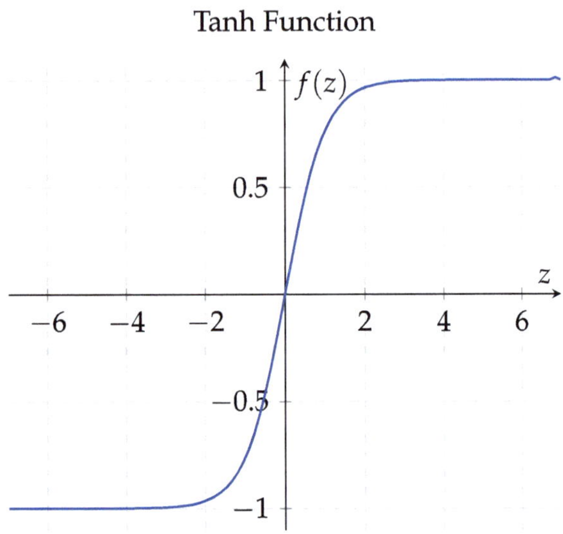

Fig. 3.3 Plot of the Tanh function $f(z) = \frac{e^z - e^{-z}}{e^z + e^{-z}}$

```
import numpy as np
import matplotlib.pyplot as plt

z = np.linspace(-5, 5, 100)
tanh_z = np.tanh(z)

plt.plot(z, tanh_z)
plt.title("Tanh Activation Function")
plt.xlabel("z")
plt.ylabel("tanh(z)")
plt.grid(True)
plt.show()
```

Use Cases and Limitations

Use Cases:

1. Hidden Layers in Neural Networks: The tanh function is widely used in hidden layers of neural networks. Its zero-centered output helps in mitigating issues related to vanishing and exploding gradients, which can improve the efficiency and speed of training.
2. Recurrent Neural Networks (RNNs): The tanh function is commonly used in RNNs and their variants (like LSTMs and GRUs) to model sequential data. It helps in capturing temporal dependencies by maintaining a balanced range of activations.
3. Binary Classification: While not as common as the sigmoid function in the output layer, the tanh function can be used in binary classification tasks where the output needs to be mapped to a range of -1 to 1.

Example in Python: Using TensorFlow, a simple neural network with the tanh activation function in the hidden layers can be implemented as follows:

```
import tensorflow as tf
from tensorflow.keras.layers import Dense

# Define the model
model = tf.keras.Sequential([
    Dense(64, activation='tanh', input_shape=(100,)),  # Hidden layer with tanh activation
    Dense(32, activation='tanh'),  # Another hidden layer with tanh activation
    Dense(1, activation='sigmoid')  # Output layer for binary classification
])

# Compile the model
model.compile(optimizer='adam', loss='binary_crossentropy', metrics=['accuracy'])

# Generate dummy data
import numpy as np
X_train = np.random.rand(1000, 100)
y_train = np.random.randint(2, size=(1000,))

# Train the model
model.fit(X_train, y_train, epochs=10, batch_size=32)

# Evaluate the model
loss, accuracy = model.evaluate(X_train, y_train)
print("Training accuracy:", accuracy)
```

Limitations:

1. Vanishing Gradient Problem: Similar to the sigmoid function, the tanh function can also suffer from the vanishing gradient problem, especially for very large positive or negative input values. This can slow down or even stall the training process.
2. Computationally Expensive: The tanh function involves exponential calculations, which can be computationally more expensive compared to simpler activation functions like ReLU.
3. Output Saturation: For inputs far from zero, the tanh function outputs values close to -1 or 1, leading to saturation. When neurons are in this saturated regime, their gradients are close to zero, which can hinder learning.

3.3.4 Rectified Linear Unit (ReLU) and Variants

The Rectified Linear Unit (ReLU) and its variants are some of the most popular activation functions in deep learning due to their simplicity and effectiveness in training deep neural networks. This section delves into the ReLU function, its extensions like Leaky ReLU, Parametric ReLU, and Exponential Linear Unit (ELU), highlighting their mathematical formulations, properties, use cases, and limitations.

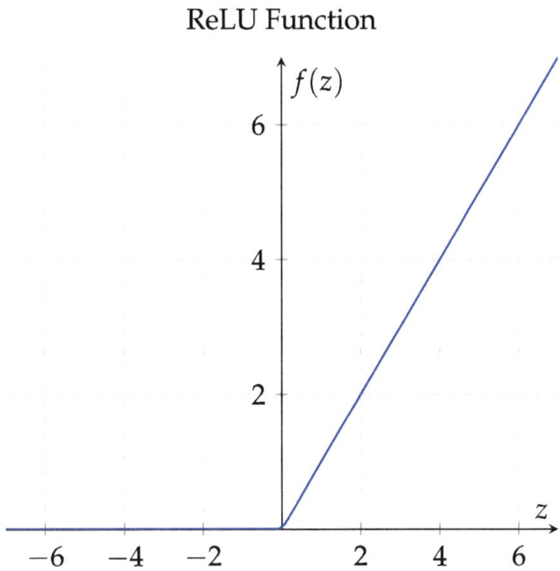

Fig. 3.4 Plot of the ReLU function $f(z) = \max(0, z)$

ReLU Function

The Rectified Linear Unit (ReLU) is defined as:

$$\text{ReLU}(z) = \max(0, z)$$

where z is the input to the neuron (see Fig. 3.4).

Properties:

1. Non-linearity: ReLU introduces non-linearity into the network, allowing it to learn complex patterns in the data.
2. Simplicity: The function is simple and computationally efficient, involving only a threshold operation.
3. Avoids Saturation: Unlike sigmoid and tanh, ReLU does not saturate in the positive domain, which helps mitigate the vanishing gradient problem.

Derivative: The derivative of ReLU is straightforward:

$$\text{ReLU}'(z) = \begin{cases} 1 & \text{if } z > 0 \\ 0 & \text{if } z \leq 0 \end{cases}$$

This derivative enables efficient backpropagation, allowing gradients to flow through the network without diminishing.

Example: Consider a neuron with input $z = \mathbf{w} \cdot \mathbf{x} + b$. The output of the neuron using the ReLU activation function is:

$$y = \text{ReLU}(z) = \max(0, \mathbf{w} \cdot \mathbf{x} + b)$$

In Python, ReLU can be implemented easily using deep learning libraries like TensorFlow:

```
import tensorflow as tf
from tensorflow.keras.layers import Dense

# Define a model with ReLU activation
model = tf.keras.Sequential([
    Dense(64, activation='relu', input_shape=(100,)),  # Hidden layer with ReLU activation
    Dense(32, activation='relu'),  # Another hidden layer with ReLU activation
    Dense(10, activation='softmax')  # Output layer for classification
])
```

Limitations:

1. Dying ReLU Problem: During training, some neurons might output zero for all inputs. This can happen when the weights are updated such that the input to ReLU is always negative, causing the neuron to "die" and stop learning.
2. Not Zero-Centered: ReLU outputs are not zero-centered, which can lead to sub-optimal gradient updates.

Leaky ReLU and Parametric ReLU

Leaky ReLU and Parametric ReLU are variants of the standard ReLU designed to address the dying ReLU problem. Leaky ReLU introduces a small slope for negative inputs to ensure that neurons continue to learn even if they receive negative inputs (see Fig. 3.5). It is defined as:

$$\text{Leaky ReLU}(z) = \begin{cases} z & \text{if } z > 0 \\ \alpha z & \text{if } z \leq 0 \end{cases}$$

where α is a small constant, typically 0.01.

Derivative: The derivative of Leaky ReLU is:

$$\text{Leaky ReLU}'(z) = \begin{cases} 1 & \text{if } z > 0 \\ \alpha & \text{if } z \leq 0 \end{cases}$$

Parametric ReLU (PReLU): PReLU is a generalization of Leaky ReLU where the parameter α is learned during training. It is defined as:

$$\text{PReLU}(z) = \begin{cases} z & \text{if } z > 0 \\ \alpha z & \text{if } z \leq 0 \end{cases}$$

where α is a learnable parameter.

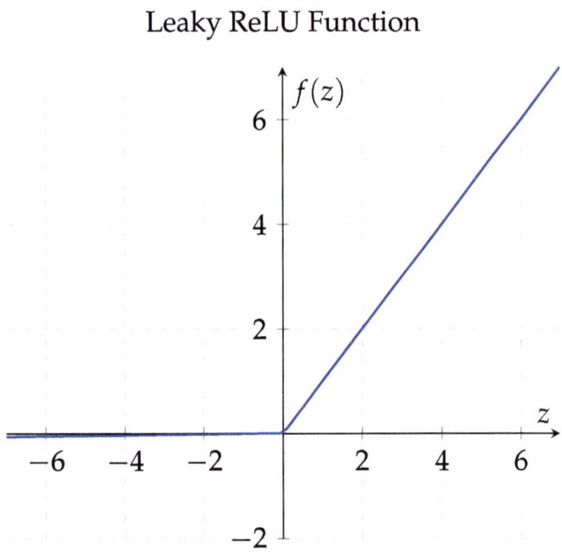

Fig. 3.5 Plot of the Leaky ReLU function

$$f(z) = \begin{cases} z, & \text{if } z \geq 0 \\ 0.01z, & \text{if } z < 0 \end{cases}$$

Derivative: The derivative of PReLU is:

$$\text{PReLU}'(z) = \begin{cases} 1 & \text{if } z > 0 \\ \alpha & \text{if } z \leq 0 \end{cases}$$

Example in Python: Using TensorFlow, Leaky ReLU and PReLU can be implemented as follows:

```
from tensorflow.keras.layers import LeakyReLU, PReLU

# Define a model with Leaky ReLU activation
leaky_relu_model = tf.keras.Sequential([
    Dense(64, input_shape=(100,)),
    LeakyReLU(alpha=0.01),
    Dense(32),
    LeakyReLU(alpha=0.01),
    Dense(10, activation='softmax')
])

# Define a model with PReLU activation
prelu_model = tf.keras.Sequential([
    Dense(64, input_shape=(100,)),
    PReLU(),
    Dense(32),
    PReLU(),
    Dense(10, activation='softmax')
])
```

Use Cases and Limitations:

Leaky ReLU: Suitable for addressing the dying ReLU problem by allowing a small gradient when the unit is not active. However, the choice of α is arbitrary and may require tuning.

PReLU: Adaptively learns the parameter α, potentially improving performance. However, it introduces additional parameters to the model, which might increase complexity.

Exponential Linear Unit (ELU)

The Exponential Linear Unit (ELU) is another variant designed to improve the learning characteristics of deep neural networks. It tends to converge faster and produce better performance compared to ReLU and its variants. ELU is defined as:

$$\text{ELU}(z) = \begin{cases} z & \text{if } z > 0 \\ \alpha(e^z - 1) & \text{if } z \leq 0 \end{cases}$$

where α is a hyperparameter that controls the value to which an ELU saturates for negative inputs (see Fig. 3.6).

Derivative: The derivative of ELU is:

$$\text{ELU}'(z) = \begin{cases} 1 & \text{if } z > 0 \\ \alpha e^z & \text{if } z \leq 0 \end{cases}$$

Properties:

1. Non-linearity: ELU introduces non-linearity and allows for negative values, which helps the network to capture a wider range of features.
2. Smoother Transition: The exponential term provides a smoother transition from negative to positive values compared to ReLU, potentially leading to faster convergence.

Example in Python: Using TensorFlow, ELU can be implemented as follows:

```
from tensorflow.keras.layers import ELU

# Define a model with ELU activation
elu_model = tf.keras.Sequential([
    Dense(64, activation='elu', input_shape=(100,)),  # Hidden layer with ELU activation
    Dense(32, activation='elu'),  # Another hidden layer with ELU activation
    Dense(10, activation='softmax')  # Output layer for classification
])
```

Use Cases and Limitations: Suitable for deep networks where faster convergence and improved performance are desired. However, it introduces additional computational complexity due to the exponential calculation.

3.3.5 Advanced Activation Functions

As the field of deep learning evolves, new activation functions are developed to address the limitations of traditional functions and to improve the performance of

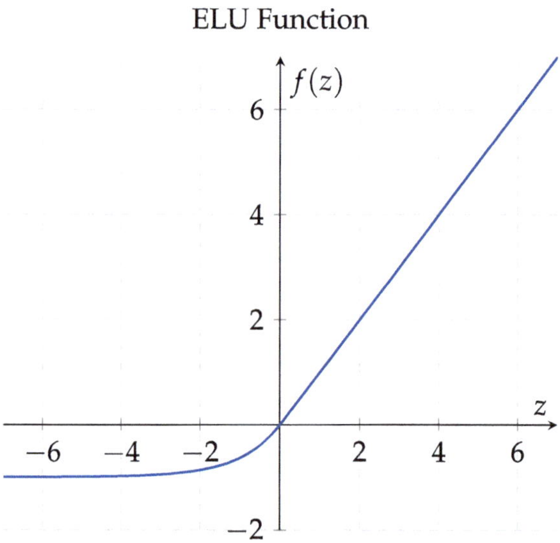

$$f(z) = \begin{cases} z, & \text{if } z \geq 0 \\ \exp(z) - 1, & \text{if } z < 0 \end{cases}$$

Fig. 3.6 Plot of the ELU function

neural networks. Two notable advanced activation functions are the Swish and Mish functions. These functions aim to combine the benefits of existing functions while mitigating their drawbacks.

Swish Function

The Swish function is a smooth, non-monotonic activation function proposed by researchers at Google. It is defined as:

$$\text{Swish}(z) = z \cdot \sigma(z) = \frac{z}{1 + e^{-z}}$$

where $\sigma(z)$ is the sigmoid function (see Fig. 3.7).

Properties:

1. Smoothness: Swish is a smooth function, which can lead to better gradient flow during training.
2. Non-monotonicity: Unlike ReLU, Swish is non-monotonic, which allows it to model more complex relationships.
3. Self-Gated: The function scales the input z by the output of the sigmoid function, which acts as a gating mechanism.

Derivative: The derivative of Swish is given by:

$$\text{Swish}'(z) = \sigma(z) + z \cdot \sigma(z) \cdot (1 - \sigma(z)) = \sigma(z) + z \cdot \sigma'(-z)$$

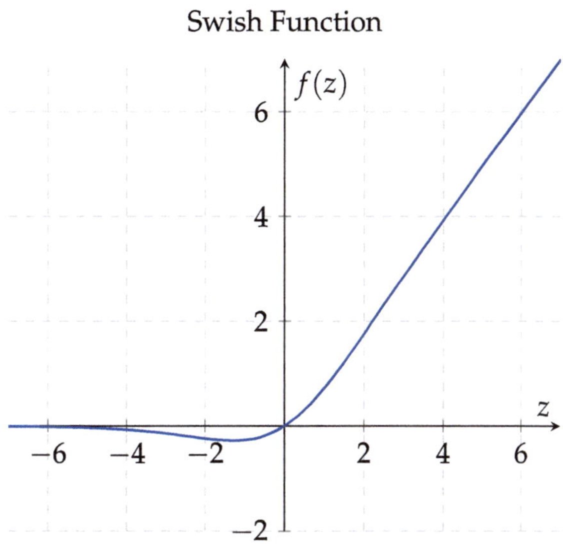

Fig. 3.7 Plot of the Swish function $f(z) = z \cdot \sigma(z)$, where $\sigma(z) = \frac{1}{1+e^{-z}}$

Example: Consider a neuron with input $z = \mathbf{w} \cdot \mathbf{x} + b$. The output of the neuron using the Swish activation function is:

$$y = \text{Swish}(z) = z \cdot \sigma(z)$$

In Python, Swish can be implemented as follows:

```
import tensorflow as tf

# Define a Swish activation function
def swish(x):
    return x * tf.nn.sigmoid(x)

# Define a model with Swish activation
model = tf.keras.Sequential([
    tf.keras.layers.Dense(64, activation=swish, input_shape=(100,)),  # Hidden layer with
            Swish activation
    tf.keras.layers.Dense(32, activation=swish),  # Another hidden layer with Swish
            activation
    tf.keras.layers.Dense(10, activation='softmax')  # Output layer for classification
])
```

Mish Functions

The Mish function is another smooth, non-monotonic activation function (see Fig. 3.8). It is defined as:

$$\text{Mish}(z) = z \cdot \tanh(\ln(1 + e^z))$$

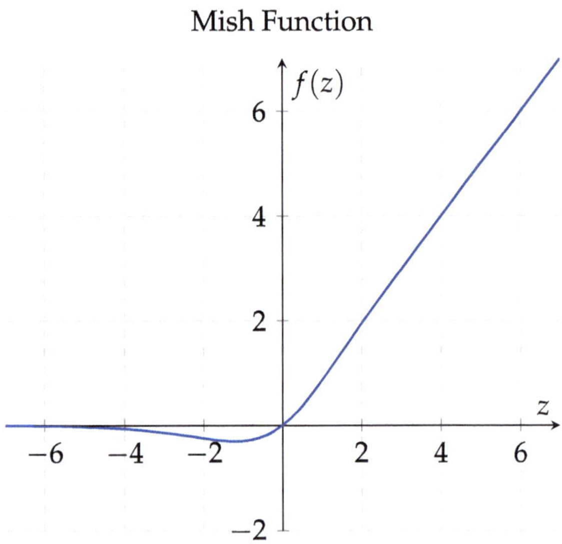

Fig. 3.8 Plot of the Mish function $f(z) = z \cdot \tanh(\ln(1 + e^z))$

Properties:

1. Smoothness: Mish is a smooth function, similar to Swish, which facilitates better gradient flow.
2. Non-monotonicity: Mish is also non-monotonic, providing flexibility in learning complex patterns.
3. Self-Regularization: The combination of tanh and ln introduces a form of self-regularization, potentially leading to better generalization.

Derivative: The derivative of Mish is more complex but can be approximated as:

$$\text{Mish}'(z) = \text{Mish}(z) + z \cdot \sigma(z) \cdot (1 - \text{Mish}(z)^2)$$

Example: For a neuron with input $z = \mathbf{w} \cdot \mathbf{x} + b$, the output using the Mish activation function is:

$$y = \text{Mish}(z) = z \cdot \tanh(\ln(1 + e^z))$$

In Python, Mish can be implemented as follows:

```
import tensorflow as tf

# Define a Mish activation function
def mish(x):
    return x * tf.math.tanh(tf.math.softplus(x))

# Define a model with Mish activation
model = tf.keras.Sequential([
    tf.keras.layers.Dense(64, activation=mish, input_shape=(100,)),  # Hidden layer
                with Mish activation
    tf.keras.layers.Dense(32, activation=mish),  # Another hidden layer with Mish activation
    tf.keras.layers.Dense(10, activation='softmax')  # Output layer for classification
])
```

Use Cases and Limitations:

Swish: Swish is suitable for a variety of tasks, including image classification and natural language processing, due to its smooth gradient and non-monotonic properties. However, it can be computationally more expensive than ReLU.

Mish: Mish is also versatile and can be used in similar tasks as Swish. Its smoothness and self-regularization properties can lead to improved performance in some cases. However, its more complex derivative can introduce additional computational overhead.

Activation Function Selection Criteria

Choosing the right activation function is crucial for the performance and training efficiency of neural networks. The selection depends on several factors, including the nature of the task, the depth of the network, and computational considerations.

Criteria for Selection:

1. Task Requirements:
   Classification: For binary classification tasks, sigmoid and tanh are commonly used in the output layer. For multi-class classification, softmax is typically used.
   Regression: Linear activation is used in the output layer for regression tasks, allowing the network to predict continuous values.
2. Network Depth:
   Shallow Networks: Functions like sigmoid and tanh can be used in shallow networks but may suffer from vanishing gradients in deeper networks.
   Deep Networks: ReLU and its variants (Leaky ReLU, PReLU, ELU) are preferred in deep networks due to their ability to mitigate the vanishing gradient problem.
3. Computational Efficiency:
   ReLU: Simple and efficient, making it suitable for most tasks.
   Swish and Mish: Although potentially more effective in some scenarios, these functions can be computationally more expensive.
4. Gradient Flow:
   Smoothness: Functions like Swish and Mish, which are smooth and have better gradient flow, can lead to improved training performance.
   Zero-Centered Output: Functions like tanh, which have zero-centered outputs, can lead to better-balanced gradient updates.
   Example: Consider a deep neural network for image classification. A suitable choice of activation functions might be:
   Hidden Layers: ReLU or its variants (e.g., Leaky ReLU, ELU) to ensure efficient gradient flow and prevent vanishing gradients.
   Output Layer: Softmax for multi-class classification to obtain probability distributions over classes.

3.4 Training Multi-layer Perceptrons

Training a MLP involves adjusting its weights and biases to minimize the error between its predictions and the actual targets. This process is typically performed using the backpropagation algorithm, combined with gradient descent optimization. In this section, we will rigorously explore the derivation of backpropagation, the process of gradient descent and weight updates, and address common pitfalls and their solutions.

3.4.1 The Backpropagation Algorithm

The backpropagation algorithm is a supervised learning method used for training neural networks. It computes the gradient of the loss function with respect to each weight by the chain rule, allowing for efficient computation of gradients in multi-layer networks.

Derivation of Backpropagation

To derive the backpropagation algorithm, we start by considering a simple MLP with one hidden layer. However, the principles apply to networks with more layers.

Network Structure:

1. Input Layer: $\mathbf{x} \in \mathbb{R}^d$
2. Hidden Layer: $\mathbf{h} = \sigma(\mathbf{W}^{(1)}\mathbf{x} + \mathbf{b}^{(1)})$, where $\mathbf{W}^{(1)} \in \mathbb{R}^{n_1 \times d}$ and $\mathbf{b}^{(1)} \in \mathbb{R}^{n_1}$
3. Output Layer: $\hat{\mathbf{y}} = f(\mathbf{W}^{(2)}\mathbf{h} + \mathbf{b}^{(2)})$, where $\mathbf{W}^{(2)} \in \mathbb{R}^{n_2 \times n_1}$ and $\mathbf{b}^{(2)} \in \mathbb{R}^{n_2}$

Loss Function: Let the loss function be $L(\hat{\mathbf{y}}, \mathbf{y})$, where $\hat{\mathbf{y}}$ is the predicted output and $\mathbf{y}$ is the true output.

Forward Pass:

1. Compute the activations of the hidden layer:

$$\mathbf{h} = \sigma(\mathbf{W}^{(1)}\mathbf{x} + \mathbf{b}^{(1)})$$

2. Compute the output:

$$\hat{\mathbf{y}} = f(\mathbf{W}^{(2)}\mathbf{h} + \mathbf{b}^{(2)})$$

Backward Pass: To compute the gradients, we apply the chain rule of calculus.

1. Output Layer:
   Compute the gradient of the loss with respect to the output layer weights $\mathbf{W}^{(2)}$:

$$\frac{\partial L}{\partial \mathbf{W}^{(2)}} = \frac{\partial L}{\partial \hat{\mathbf{y}}} \cdot \frac{\partial \hat{\mathbf{y}}}{\partial \mathbf{z}^{(2)}} \cdot \frac{\partial \mathbf{z}^{(2)}}{\partial \mathbf{W}^{(2)}}$$

where $\mathbf{z}^{(2)} = \mathbf{W}^{(2)}\mathbf{h} + \mathbf{b}^{(2)}$.

Since $\hat{\mathbf{y}} = f(\mathbf{z}^{(2)})$, we have:

$$\frac{\partial L}{\partial \mathbf{W}^{(2)}} = \delta^{(2)}\mathbf{h}^T$$

where $\delta^{(2)} = \frac{\partial L}{\partial \hat{\mathbf{y}}} \cdot f'(\mathbf{z}^{(2)})$.

2. Hidden Layer: Compute the gradient of the loss with respect to the hidden layer weights $\mathbf{W}^{(1)}$:

$$\frac{\partial L}{\partial \mathbf{W}^{(1)}} = \frac{\partial L}{\partial \mathbf{h}} \cdot \frac{\partial \mathbf{h}}{\partial \mathbf{z}^{(1)}} \cdot \frac{\partial \mathbf{z}^{(1)}}{\partial \mathbf{W}^{(1)}}$$

where $\mathbf{z}^{(1)} = \mathbf{W}^{(1)}\mathbf{x} + \mathbf{b}^{(1)}$.

Since $\mathbf{h} = \sigma(\mathbf{z}^{(1)})$, we have:

$$\frac{\partial L}{\partial \mathbf{W}^{(1)}} = \delta^{(1)}\mathbf{x}^T$$

where $\delta^{(1)} = (\mathbf{W}^{(2)})^T\delta^{(2)} \cdot \sigma'(\mathbf{z}^{(1)})$.

Bias Gradients: Similarly, the gradients with respect to the biases are:

$$\frac{\partial L}{\partial \mathbf{b}^{(2)}} = \delta^{(2)}$$

$$\frac{\partial L}{\partial \mathbf{b}^{(1)}} = \delta^{(1)}$$

Gradient Descent and Weight Updates

Once the gradients are computed, we use gradient descent to update the weights and biases. Gradient descent aims to minimize the loss function by iteratively adjusting the parameters in the direction of the negative gradient.

Gradient Descent Update Rule: For each weight $\mathbf{W}$ and bias $\mathbf{b}$:

$$\mathbf{W} \leftarrow \mathbf{W} - \eta\frac{\partial L}{\partial \mathbf{W}}$$

$$\mathbf{b} \leftarrow \mathbf{b} - \eta\frac{\partial L}{\partial \mathbf{b}}$$

where η is the learning rate, a hyperparameter that controls the step size of each update.

Example: Consider a simple MLP with weights $\mathbf{W}^{(1)}$ and $\mathbf{W}^{(2)}$. After computing the gradients $\frac{\partial L}{\partial \mathbf{W}^{(1)}}$ and $\frac{\partial L}{\partial \mathbf{W}^{(2)}}$, the weights are updated as follows:

```
# Assume gradients have been computed
dW1 = ...  # Gradient of the loss with respect to W1
dW2 = ...  # Gradient of the loss with respect to W2
db1 = ...  # Gradient of the loss with respect to b1
db2 = ...  # Gradient of the loss with respect to b2

# Learning rate
learning_rate = 0.01

# Update weights and biases
W1 -= learning_rate * dW1
W2 -= learning_rate * dW2
b1 -= learning_rate * db1
b2 -= learning_rate * db2
```

Common Pitfalls and Solutions

Training MLPs using backpropagation and gradient descent can encounter several pitfalls. Understanding these issues and their solutions is crucial for effective training.

1. Vanishing and Exploding Gradients:
   Problem: In deep networks, gradients can become very small (vanishing gradients) or very large (exploding gradients), leading to slow or unstable training.
   Solution: Use activation functions like ReLU, which mitigate the vanishing gradient problem. Implement gradient clipping to cap gradients at a maximum value. Use initialization techniques like Xavier or He initialization.
2. Overfitting:
   Problem: The model performs well on training data but poorly on validation data.
   Solution: Use regularization techniques such as L2 regularization (weight decay) or L1 regularization. Implement dropout to randomly disable neurons during training. Collect more training data or use data augmentation techniques.
3. Poor Learning Rate:
   Problem: A learning rate that is too high can cause the model to diverge, while a learning rate that is too low can result in slow convergence.
   Solution: Use learning rate schedules to adjust the learning rate during training. Implement adaptive learning rate methods like Adam, RMSprop, or Adagrad.
4. Local Minima and Saddle Points:
   Problem: The optimization process can get stuck in local minima or saddle points.
   Solution: Use stochastic gradient descent (SGD) or its variants, which incorporate randomness and can help escape local minima. Implement momentum to accelerate convergence and avoid oscillations.

Example in Python:

```
import tensorflow as tf
from tensorflow.keras.layers import Dense, Dropout

# Define the model with dropout
model = tf.keras.Sequential([
    Dense(64, activation='relu', input_shape=(100,)),
    Dropout(0.5),  # Dropout layer to prevent overfitting
    Dense(32, activation='relu'),
    Dropout(0.5),
    Dense(10, activation='softmax')
])

# Compile the model with Adam optimizer
model.compile(optimizer='adam', loss='sparse_categorical_crossentropy', metrics=['accuracy'])

# Generate dummy data
import numpy as np
X_train = np.random.rand(1000, 100)
y_train = np.random.randint(10, size=(1000,))

# Train the model
model.fit(X_train, y_train,

 epochs=10, batch_size=32, validation_split=0.2)
```

3.4.2 Tensor-Based Gradient Computation

The computation of gradients is a critical component of training neural networks using backpropagation. Modern deep learning frameworks like TensorFlow and PyTorch provide efficient and flexible tools for automatic differentiation, which greatly simplify the implementation of gradient-based optimization. In this section, we explore how gradient computation is performed using tensors in TensorFlow and PyTorch.

Gradient Computation in TensorFlow

TensorFlow uses dataflow graphs to represent computation. The core of TensorFlow's ability to compute gradients lies in its automatic differentiation engine, which constructs computational graphs and efficiently computes derivatives.

Automatic Differentiation in TensorFlow TensorFlow employs automatic differentiation (autodiff) to compute the gradients of scalar-valued functions with respect to their inputs. This is particularly useful for training neural networks, where the loss function needs to be differentiated with respect to the model parameters. For example, consider a simple neural network with a single hidden layer. We will demonstrate how to compute gradients using TensorFlow.

```
import tensorflow as tf

# Define a simple neural network model
class SimpleNN(tf.keras.Model):
    def __init__(self):
        super(SimpleNN, self).__init__()
        self.dense1 = tf.keras.layers.Dense(64, activation='relu')
        self.dense2 = tf.keras.layers.Dense(10, activation='softmax')

    def call(self, inputs):
        x = self.dense1(inputs)
        return self.dense2(x)

# Instantiate the model
model = SimpleNN()

# Generate dummy data
X_train = tf.random.normal([100, 100])
y_train = tf.random.uniform([100], maxval=10, dtype=tf.int32)

# Define the loss function
loss_fn = tf.keras.losses.SparseCategoricalCrossentropy()

# Forward pass and compute loss
with tf.GradientTape() as tape:
    logits = model(X_train)
    loss = loss_fn(y_train, logits)

# Compute gradients
gradients = tape.gradient(loss, model.trainable_variables)

# Print gradients
for var, grad in zip(model.trainable_variables, gradients):
    print(f"Variable: {var.name}, Gradient: {grad}")
```

Explanation:

1. GradientTape Context: TensorFlow uses 'tf.GradientTape' to record operations for automatic differentiation. This context manager tracks the operations on tensors within its scope.
2. Forward Pass: Inside the 'with tf.GradientTape()' block, we perform the forward pass and compute the loss.
3. Gradient Computation: After the forward pass, we call 'tape.gradient(loss, model.trainable_variables)' to compute the gradients of the loss with respect to the model parameters.
4. Output Gradients: The gradients are then printed, showing how the loss changes with respect to each parameter.

Efficient Computation: TensorFlow's automatic differentiation is highly optimized, making use of techniques such as symbolic differentiation and graph optimization to ensure efficient gradient computation.

Gradient Computation in PyTorch

PyTorch is another widely used deep learning framework, known for its dynamic computation graph and ease of use. PyTorch uses the Autograd package for automatic differentiation, which records operations in a dynamic computation graph and computes gradients on the fly.

Automatic Differentiation in PyTorch Autograd in PyTorch allows for automatic computation of gradients by dynamically building a computation graph during the forward pass. This flexibility makes it particularly suitable for tasks that require dynamic computation graphs.

Example: Let's compute gradients for a simple neural network in PyTorch.

```
import torch
import torch.nn as nn
import torch.optim as optim

# Define a simple neural network model
class SimpleNN(nn.Module):
    def __init__(self):
        super(SimpleNN, self).__init__()
        self.dense1 = nn.Linear(100, 64)
        self.dense2 = nn.Linear(64, 10)
        self.relu = nn.ReLU()

    def forward(self, x):
        x = self.relu(self.dense1(x))
        return self.dense2(x)

# Instantiate the model
model = SimpleNN()

# Generate dummy data
X_train = torch.randn(100, 100)
y_train = torch.randint(0, 10, (100,))

# Define the loss function
loss_fn = nn.CrossEntropyLoss()

# Forward pass and compute loss
logits = model(X_train)
loss = loss_fn(logits, y_train)

# Compute gradients
loss.backward()

# Print gradients
for name, param in model.named_parameters():
    if param.grad is not None:
        print(f"Variable: {name}, Gradient: {param.grad}")

# Update weights using gradient descent
optimizer = optim.SGD(model.parameters(), lr=0.01)
optimizer.step()
```

Explanation:

1. Define Model: We define a simple neural network model with one hidden layer and ReLU activation.
2. Forward Pass: We perform a forward pass to compute the logits and then compute the loss using 'nn.CrossEntropyLoss()'.
3. Backward Pass: We call 'loss.backward()' to compute the gradients of the loss with respect to the model parameters. This method accumulates the gradients in the 'grad' attribute of each parameter.
4. Output Gradients: We print the gradients to inspect how each parameter affects the loss.

5. Weight Update: Using 'torch.optim.SGD', we update the weights of the model based on the computed gradients.

Advantages of Dynamic Graphs: PyTorch's dynamic computation graph allows for more flexibility in building and modifying models, making it particularly suitable for research and development. The 'backward()' method ensures that gradients are computed efficiently and correctly, leveraging the dynamic nature of the graph.

3.4.3 *Initialization Techniques*

Proper initialization of the weights in a neural network is crucial for effective training and fast convergence. Poor initialization can lead to problems such as vanishing or exploding gradients, which can hinder the learning process. Two widely used initialization techniques that address these issues are Xavier Initialization (Glorot and Bengio 2010) and He Initialization (He et al. 2015a). This section explores these techniques and their impact on convergence.

Xavier and He Initialization

Xavier Initialization Xavier Initialization, also known as Glorot Initialization, was introduced by Xavier Glorot and Yoshua Bengio. It is designed to keep the scale of the gradients roughly the same across all layers of the network.

The Xavier Initialization for a layer with n_{in} input neurons and n_{out} output neurons is defined as:

$$W \sim \mathcal{U}\left(-\frac{\sqrt{6}}{\sqrt{n_{\text{in}} + n_{\text{out}}}}, \frac{\sqrt{6}}{\sqrt{n_{\text{in}} + n_{\text{out}}}}\right)$$

where $\mathcal{U}$ denotes a uniform distribution.

Alternatively, if a normal distribution is preferred:

$$W \sim \mathcal{N}\left(0, \frac{2}{n_{\text{in}} + n_{\text{out}}}\right)$$

Example: For a layer with 100 input neurons and 50 output neurons:

$$W \sim \mathcal{U}\left(-\frac{\sqrt{6}}{\sqrt{100 + 50}}, \frac{\sqrt{6}}{\sqrt{100 + 50}}\right) = \mathcal{U}\left(-\frac{\sqrt{6}}{\sqrt{150}}, \frac{\sqrt{6}}{\sqrt{150}}\right)$$

He Initialization He Initialization, introduced by Kaiming He et al., is particularly suited for layers with ReLU activation functions. It aims to keep the variance of the activations throughout the network consistent. The He Initialization for a layer with n_{in} input neurons is defined as:

$$W \sim \mathcal{U}\left(-\sqrt{\frac{6}{n_{\text{in}}}}, \sqrt{\frac{6}{n_{\text{in}}}}\right)$$

Alternatively, if a normal distribution is preferred:

$$W \sim \mathcal{N}\left(0, \frac{2}{n_{\text{in}}}\right)$$

Example: For a layer with 100 input neurons:

$$W \sim \mathcal{U}\left(-\sqrt{\frac{6}{100}}, \sqrt{\frac{6}{100}}\right) = \mathcal{U}\left(-\sqrt{\frac{6}{100}}, \sqrt{\frac{6}{100}}\right) = \mathcal{U}\left(-\sqrt{\frac{3}{50}}, \sqrt{\frac{3}{50}}\right)$$

Implementation in TensorFlow:

```
import tensorflow as tf

# Xavier Initialization
initializer = tf.keras.initializers.GlorotUniform()
layer = tf.keras.layers.Dense(50, input_shape=(100,), kernel_initializer=initializer)

# He Initialization
initializer = tf.keras.initializers.HeUniform()
layer = tf.keras.layers.Dense(50, input_shape=(100,), kernel_initializer=initializer)
```

In PyTorch:

```
import torch.nn as nn

# Xavier Initialization
layer = nn.Linear(100, 50)
nn.init.xavier_uniform_(layer.weight)

# He Initialization
layer = nn.Linear(100, 50)
nn.init.kaiming_uniform_(layer.weight, nonlinearity='relu')
```

Impact on Convergence

Xavier Initialization is designed to address the issue of the vanishing and exploding gradients in deep networks. By ensuring that the variance of the weights is appropriately scaled, it helps maintain the gradients' magnitude across layers, leading to more stable and faster convergence. The Xavier Initialization ensures that the variance of the activations is preserved across layers. For an activation function with zero mean and unit variance, the expected variance of the output of a layer is given by:

$$\text{Var}(Wx) = \frac{1}{n_{\text{in}}}\text{Var}(x)$$

By scaling the weights with $\frac{1}{\sqrt{n_{\text{in}}}}$, Xavier Initialization ensures that the output variance remains constant, thus preventing the gradients from vanishing or exploding.

He Initialization further refines this approach for ReLU activations by taking into account that ReLU activation functions only propagate positive values. This means that only half of the weights are effectively contributing to the output, necessitating a larger variance to compensate. For ReLU activations, the variance of the outputs can be derived as:

$$\text{Var}(Wx) = \frac{2}{n_{\text{in}}}\text{Var}(x)$$

He Initialization scales the weights with $\sqrt{\frac{2}{n_{\text{in}}}}$ to maintain this variance, ensuring that the activations do not diminish or explode as they propagate through the network.

Practical Impact:

Convergence Speed: Proper initialization can significantly speed up convergence. Xavier and He Initialization help the network start in a region of the parameter space where gradients are neither too small nor too large, leading to faster learning.

Training Stability: These initialization techniques contribute to training stability by maintaining consistent gradients throughout the network. This reduces the likelihood of encountering numerical instability issues.

Performance: By ensuring that activations and gradients have the appropriate scale, networks initialized with Xavier or He techniques often achieve better performance compared to those with naïve initialization methods.

Example:

Consider training a deep neural network with and without proper initialization:

```
# Without proper initialization
layer = tf.keras.layers.Dense(50, input_shape=(100,))
# Default initializer, which may lead to poor convergence

# With Xavier Initialization
initializer = tf.keras.initializers.GlorotUniform()
layer = tf.keras.layers.Dense(50, input_shape=(100,), kernel_initializer=initializer)

# With He Initialization for ReLU
initializer = tf.keras.initializers.HeUniform()
layer = tf.keras.layers.Dense(50, input_shape=(100,), kernel_initializer=initializer)
```

Experimental Results: Empirical studies have shown that networks initialized with Xavier or He techniques tend to converge faster and achieve higher accuracy compared to networks with random or constant initialization. This is especially true for deep networks, where the risk of vanishing or exploding gradients is higher.

3.5 Optimization Techniques

Optimization is a crucial aspect of training neural networks, involving the process of adjusting the weights to minimize the loss function. Various optimization algorithms have been developed to improve the efficiency and effectiveness of this process. In this section, we provide an introduction to optimization in neural networks, followed by a detailed discussion on stochastic gradient descent (SGD), its basic algorithm, variants, and advanced techniques like mini-batch SGD, momentum, and Nesterov Accelerated Gradient.

3.5.1 Introduction to Optimization in Neural Networks

Optimization in neural networks involves finding the set of parameters (weights and biases) that minimize the loss function, which measures the discrepancy between the predicted outputs and the actual targets. This process is guided by the gradients of the loss function with respect to the parameters, computed through backpropagation.

Loss Function: For a neural network with parameters θ, input data $\mathbf{X}$, and target labels $\mathbf{y}$, the loss function $L(\mathbf{X}, \mathbf{y}; \theta)$ quantifies the prediction error. The goal of optimization is to find θ^* that minimizes L:

$$\theta^* = \arg\min_{\theta} L(\mathbf{X}, \mathbf{y}; \theta)$$

Gradient Descent: Gradient descent is the foundational algorithm for optimization in neural networks. It iteratively updates the parameters in the direction of the negative gradient of the loss function:

$$\theta \leftarrow \theta - \eta \nabla_{\theta} L(\mathbf{X}, \mathbf{y}; \theta)$$

where η is the learning rate, controlling the step size of each update.

3.5.2 Stochastic Gradient Descent (SGD)

Stochastic gradient descent (SGD) is a variant of gradient descent that updates the parameters using a single data point or a small batch of data at each iteration, rather than the entire dataset. This introduces stochasticity into the optimization process, which can help escape local minima and improve convergence.

Basic Algorithm and Variants

Basic SGD Algorithm: In SGD, the parameter update rule is given by:

$$\theta \leftarrow \theta - \eta \nabla_{\theta} L(\mathbf{x}_i, y_i; \theta)$$

where $(\mathbf{x}_i, y_i)$ is a single training example. This makes the updates noisy but computationally efficient, especially for large datasets.

Variants of SGD:

1. Batch Gradient Descent: Uses the entire dataset for each update, resulting in stable but computationally expensive updates.
2. Mini-batch SGD: Uses small batches of data for each update, balancing the efficiency and stability of updates.

Mini-batch SGD

Mini-batch SGD combines the benefits of both batch gradient descent and basic SGD. It updates the parameters using small, randomly selected subsets of the data, called mini-batches. The update rule for mini-batch SGD is:

$$\theta \leftarrow \theta - \eta \nabla_\theta L(\mathbf{X}_{\text{mini}}, \mathbf{y}_{\text{mini}}; \theta)$$

where $(\mathbf{X}_{\text{mini}}, \mathbf{y}_{\text{mini}})$ is a mini-batch of training examples.

Advantages:

1. Efficiency: Mini-batch updates are computationally more efficient than batch updates.
2. Noise Reduction: By averaging over a mini-batch, the updates are less noisy than basic SGD.
3. Parallelism: Mini-batch SGD can leverage parallelism in hardware for faster computation.

Example in Python:

```
import tensorflow as tf

# Define a simple neural network model
model = tf.keras.Sequential([
    tf.keras.layers.Dense(64, activation='relu', input_shape=(100,)),
    tf.keras.layers.Dense(10, activation='softmax')
])

# Compile the model with SGD optimizer
model.compile(optimizer=tf.keras.optimizers.SGD(learning_rate=0.01),
              loss='sparse_categorical_crossentropy',
              metrics=['accuracy'])

# Generate dummy data
import numpy as np
X_train = np.random.rand(1000, 100)
y_train = np.random.randint(10, size=(1000,))

# Train the model using mini-batch SGD
model.fit(X_train, y_train, epochs=10, batch_size=32)
```

Momentum and Nesterov Accelerated Gradient

Momentum Momentum is an enhancement to SGD that helps accelerate convergence, particularly in the presence of high curvature, small but consistent gradients, or noisy gradients. It accumulates a velocity vector in the direction of the gradient and updates the parameters accordingly. The momentum update rule is:

$$v_t = \gamma v_{t-1} + \eta \nabla_\theta L(\mathbf{x}_i, y_i; \theta)$$

$$\theta \leftarrow \theta - v_t$$

where v_t is the velocity at iteration t, and γ is the momentum coefficient, typically between 0.9 and 0.99.

Nesterov Accelerated Gradient (NAG) NAG is an extension of momentum that improves the lookahead capability by computing the gradient with respect to the anticipated future position of the parameters. This can lead to more responsive updates (Nesterov 1983).

The NAG update rule is:

$$v_t = \gamma v_{t-1} + \eta \nabla_\theta L(\mathbf{x}_i, y_i; \theta - \gamma v_{t-1})$$

$$\theta \leftarrow \theta - v_t$$

Example in Python: Using TensorFlow, momentum and NAG can be implemented as follows:

```
# SGD with momentum
optimizer = tf.keras.optimizers.SGD(learning_rate=0.01, momentum=0.9)

# SGD with Nesterov Accelerated Gradient
optimizer = tf.keras.optimizers.SGD(learning_rate=0.01, momentum=0.9, nesterov=True)

# Compile the model with the chosen optimizer
model.compile(optimizer=optimizer,
              loss='sparse_categorical_crossentropy',
              metrics=['accuracy'])
```

Impact on Convergence:

1. Momentum: Helps in accelerating convergence and smoothing out the optimization path by dampening oscillations. It is particularly effective in navigating along the shallow, long valleys of the loss landscape.
2. NAG: Provides a more accurate adjustment by considering the anticipated future gradient, leading to faster and more stable convergence.

3.5.3 Adaptive Optimization Methods

Adaptive optimization methods adjust the learning rate based on the gradients' magnitude and history, allowing the learning rate to change dynamically during training. These methods often result in faster convergence and better performance, especially for complex neural networks. Two widely used adaptive optimization methods are Adagrad and RMSprop (Duchi et al. 2011), along with their variants, including Adam (Kingma and Ba 2014) and its variants.

Adagrad and RMSprop

Adagrad Adagrad (Adaptive Gradient Algorithm) adjusts the learning rate for each parameter individually based on the historical sum of squared gradients. This means that parameters with larger gradients receive smaller updates, and those with smaller gradients receive larger updates. The update rule for Adagrad is:

$$g_t = \nabla_\theta L(\theta_t)$$

$$G_t = G_{t-1} + g_t^2$$

$$\theta_{t+1} = \theta_t - \frac{\eta}{\sqrt{G_t + \epsilon}} g_t$$

where: g_t is the gradient at time step t, G_t is the sum of squared gradients up to time t, η is the initial learning rate, ϵ is a small constant to prevent division by zero.

Properties:

1. Adaptiveness: Adagrad adapts the learning rate based on the parameter's historical gradients, which is beneficial for sparse data.
2. Diminishing Learning Rate: The accumulation of squared gradients can lead to a continuously decaying learning rate, potentially slowing down convergence.

Example in Python:

```
import tensorflow as tf

# Compile the model with Adagrad optimizer
model.compile(optimizer=tf.keras.optimizers.Adagrad(learning_rate=0.01),
              loss='sparse_categorical_crossentropy',
              metrics=['accuracy'])
```

RMSprop RMSprop (Root Mean Square Propagation) was introduced by Geoffrey Hinton to address the diminishing learning rate issue in Adagrad. RMSprop maintains a moving average of the squared gradients and normalizes the parameter updates accordingly. The update rule for RMSprop is:

$$g_t = \nabla_\theta L(\theta_t)$$

$$E[g^2]_t = \gamma E[g^2]_{t-1} + (1 - \gamma) g_t^2$$

$$\theta_{t+1} = \theta_t - \frac{\eta}{\sqrt{E[g^2]_t + \epsilon}} g_t$$

where: $E[g^2]_t$ is the moving average of squared gradients, γ is the decay rate, η is the learning rate, ϵ is a small constant to prevent division by zero.

Properties:

1. Adaptiveness: RMSprop adjusts the learning rate for each parameter individually, similar to Adagrad.

2. Stabilized Learning Rate: The moving average of squared gradients prevents the learning rate from decaying too quickly.

Example in Python:

```
import tensorflow as tf

# Compile the model with RMSprop optimizer
model.compile(optimizer=tf.keras.optimizers.RMSprop(learning_rate=0.001),
              loss='sparse_categorical_crossentropy',
              metrics=['accuracy'])
```

Adam and its Variants

Adam Adam (Adaptive Moment Estimation) combines the benefits of both Adagrad and RMSprop, incorporating momentum by computing an exponentially decaying average of past gradients and past squared gradients (Kingma and Ba 2014). The update rule for Adam is:

$$m_t = \beta_1 m_{t-1} + (1 - \beta_1) g_t$$

$$v_t = \beta_2 v_{t-1} + (1 - \beta_2) g_t^2$$

$$\hat{m}_t = \frac{m_t}{1 - \beta_1^t}$$

$$\hat{v}_t = \frac{v_t}{1 - \beta_2^t}$$

$$\theta_{t+1} = \theta_t - \frac{\eta \hat{m}_t}{\sqrt{\hat{v}_t} + \epsilon}$$

where m_t is the first moment estimate (mean of gradients), v_t is the second moment estimate (uncentered variance), β_1 and β_2 are decay rates for the first and second moment estimates, $\hat{m}_t$ and $\hat{v}_t$ are bias-corrected estimates, η is the learning rate, ϵ is a small constant to prevent division by zero.

Properties:

1. Adaptiveness: Adam adapts the learning rate based on the first and second moments of the gradients.
2. Momentum: Incorporates momentum, which helps accelerate convergence and smooth out the optimization path.

Example in Python:

```
import tensorflow as tf

# Compile the model with Adam optimizer
model.compile(optimizer=tf.keras.optimizers.Adam(learning_rate=0.001),
              loss='sparse_categorical_crossentropy',
              metrics=['accuracy'])
```

Adam Variants:

1. AdaMax: A variant of Adam based on the infinity norm, which can be more stable in some cases.
2. Nadam: Combines Adam with Nesterov Accelerated Gradient, providing a lookahead mechanism to improve convergence.

AdaMax Update Rule:

$$m_t = \beta_1 m_{t-1} + (1 - \beta_1) g_t$$

$$u_t = \max(\beta_2 u_{t-1}, |g_t|)$$

$$\theta_{t+1} = \theta_t - \frac{\eta}{u_t + \epsilon} m_t$$

Nadam Update Rule:

$$m_t = \beta_1 m_{t-1} + (1 - \beta_1) g_t$$

$$v_t = \beta_2 v_{t-1} + (1 - \beta_2) g_t^2$$

$$\hat{m}_t = \frac{m_t}{1 - \beta_1^t}$$

$$\hat{v}_t = \frac{v_t}{1 - \beta_2^t}$$

$$\theta_{t+1} = \theta_t - \frac{\eta(\beta_1 \hat{m}_t + (1 - \beta_1) g_t)}{\sqrt{\hat{v}_t} + \epsilon}$$

Example in Python:

```
import tensorflow as tf

# Compile the model with AdaMax optimizer
model.compile(optimizer=tf.keras.optimizers.Adamax(learning_rate=0.002),
              loss='sparse_categorical_crossentropy',
              metrics=['accuracy'])

# Compile the model with Nadam optimizer
model.compile(optimizer=tf.keras.optimizers.Nadam(learning_rate=0.002),
              loss='sparse_categorical_crossentropy',
              metrics=['accuracy'])
```

Impact on Convergence:

1. Adagrad: Effective for sparse data and features, but learning rate may decay too quickly.
2. RMSprop: Stabilizes learning rate, suitable for non-stationary objectives.

3. Adam: Combines the benefits of Adagrad and RMSprop with momentum, generally leading to faster and more stable convergence.
4. Variants: AdaMax and Nadam provide further enhancements, potentially improving performance in specific scenarios.

3.5.4 *Advanced Optimization Techniques*

Advanced optimization techniques can further enhance the training process of neural networks by improving convergence speed, stability, and performance. These techniques include learning rate schedules, gradient clipping, and second-order methods. Each of these methods addresses specific challenges encountered during the optimization process.

Learning Rate Schedules

Learning rate schedules dynamically adjust the learning rate during training to improve convergence and model performance. Instead of using a fixed learning rate, these schedules decrease the learning rate over time or based on specific conditions.

Types of Learning Rate Schedules:

1. Step Decay: In step decay, the learning rate is reduced by a factor at specific intervals (epochs). This helps the model to make large updates initially and fine-tune as training progresses.

$$\eta_t = \eta_0 \cdot \text{drop}^{\left\lfloor \frac{t}{\text{epochs}} \right\rfloor}$$

where η_t is the learning rate at epoch t, η_0 is the initial learning rate, drop is the factor by which the learning rate is reduced, epochs is the number of epochs after which the learning rate is reduced.
2. Exponential Decay: Exponential decay reduces the learning rate exponentially at each step, providing a smooth reduction.

$$\eta_t = \eta_0 \cdot e^{-\lambda t}$$

where λ is the decay rate.
3. Polynomial Decay: Polynomial decay reduces the learning rate following a polynomial function of the current epoch.

$$\eta_t = \eta_0 \left(1 - \frac{t}{T}\right)^p$$

where T is the total number of epochs, p is the power of the polynomial.

4. Cosine Annealing: Cosine annealing adjusts the learning rate following a cosine function, providing periodic decreases.

$$\eta_t = \eta_0 \cdot \frac{1}{2}\left(1 + \cos\left(\frac{t\pi}{T}\right)\right)$$

Example in Python: Using TensorFlow, learning rate schedules can be implemented as follows:

```
import tensorflow as tf

# Step Decay Schedule
initial_lr = 0.1
lr_schedule = tf.keras.optimizers.schedules.ExponentialDecay(
    initial_lr, decay_steps=100000, decay_rate=0.96, staircase=True)

# Exponential Decay Schedule
lr_schedule = tf.keras.optimizers.schedules.ExponentialDecay(
    initial_lr, decay_steps=100000, decay_rate=0.96, staircase=False)

# Polynomial Decay Schedule
lr_schedule = tf.keras.optimizers.schedules.PolynomialDecay(
    initial_lr, decay_steps=100000, end_learning_rate=0.01, power=1.0)

# Cosine Annealing Schedule
lr_schedule = tf.keras.experimental.CosineDecay(
    initial_lr, decay_steps=100000)

# Compile the model with the learning rate schedule
model.compile(optimizer=tf.keras.optimizers.SGD(learning_rate=lr_schedule),
              loss='sparse_categorical_crossentropy',
              metrics=['accuracy'])
```

Gradient Clipping

Gradient clipping is a technique used to prevent the exploding gradient problem by capping the gradients to a maximum value. This ensures that the gradients do not become excessively large, which can destabilize the training process.

Methods of Gradient Clipping:

1. Clipping by Norm: Limits the norm of the gradients to a specified maximum value.

$$\mathbf{g} \leftarrow \mathbf{g} \cdot \min\left(1, \frac{\text{clip\_norm}}{\|\mathbf{g}\|_2}\right)$$

2. Clipping by Value: Limits each individual gradient value to a specified range.

$$\mathbf{g}_i \leftarrow \text{clip}(\mathbf{g}_i, \text{clip\_min}, \text{clip\_max})$$

Example in Python: Using TensorFlow, gradient clipping can be implemented as follows:

```
import tensorflow as tf

# Define the optimizer with gradient clipping by norm
optimizer = tf.keras.optimizers.SGD(learning_rate=0.01, clipnorm=1.0)

# Define the optimizer with gradient clipping by value
optimizer = tf.keras.optimizers.SGD(learning_rate=0.01, clipvalue=0.5)

# Compile the model with the optimizer
model.compile(optimizer=optimizer,
              loss='sparse_categorical_crossentropy',
              metrics=['accuracy'])
```

Second-order Methods (e.g., L-BFGS)

Second-order optimization methods, such as the Limited-memory Broyden-Fletcher-Goldfarb-Shanno (L-BFGS) algorithm, leverage second-order derivative information (the Hessian matrix) to make more informed updates to the parameters. These methods can converge faster and more reliably for certain types of problems (Liu and Nocedal 1989).

L-BFGS Algorithm L-BFGS is an iterative method for solving large-scale optimization problems. It approximates the inverse Hessian matrix to compute the search direction.

$$\theta_{t+1} = \theta_t - \alpha_t \mathbf{H}_t^{-1} \nabla_\theta L(\theta_t)$$

where α_t is the step size, $\mathbf{H}_t^{-1}$ is the approximate inverse Hessian matrix.
Properties:

1. Efficiency: L-BFGS is more efficient than full BFGS for large-scale problems because it uses limited memory to approximate the Hessian.
2. Accuracy: By incorporating second-order information, L-BFGS can achieve faster and more accurate convergence for certain optimization problems.

Example in Python: Using PyTorch, L-BFGS can be implemented as follows:

```
import torch
import torch.nn as nn
import torch.optim as optim

# Define a simple neural network model
class SimpleNN(nn.Module):
    def __init__(self):
        super(SimpleNN, self).__init__()
        self.dense1 = nn.Linear(100, 64)
        self.dense2 = nn.Linear(64, 10)
        self.relu = nn.ReLU()

    def forward(self, x):
        x = self.relu(self.dense1(x))
        return self.dense2(x)
```

```
# Instantiate the model
model = SimpleNN()

# Define the loss function and optimizer
criterion = nn.CrossEntropyLoss()
optimizer = optim.LBFGS(model.parameters(), lr=0.01)

# Generate dummy data
X_train = torch.randn(100, 100)
y_train = torch.randint(0, 10, (100,))

# Training step
def closure():
    optimizer.zero_grad()
    outputs = model(X_train)
    loss = criterion(outputs, y_train)
    loss.backward()
    return loss

# Perform optimization
optimizer.step(closure)
```

3.6 Regularization Techniques

Regularization techniques are essential in training neural networks to prevent overfitting, which occurs when a model performs well on training data but poorly on unseen test data. Regularization introduces additional constraints or penalties on the model parameters, helping to improve generalization. Common regularization methods include L1 and L2 regularization, dropout, and data augmentation.

3.6.1 Dropout

Dropout is a widely used regularization technique that prevents overfitting by randomly dropping units (along with their connections) during training. This introduces noise and forces the network to learn redundant representations, improving robustness and generalization. Dropout can be viewed as sampling from an exponential number of different "thinned" networks, and at test time, it approximates the effect of averaging these networks. During training, each unit in a neural network is retained with a probability p and dropped with a probability $1 - p$. Formally, for a given layer with activations $\mathbf{h}$:

$$\mathbf{h}' = \mathbf{h} \cdot \mathbf{r}$$

$$\mathbf{r} \sim \text{Bernoulli}(p)$$

where $\mathbf{r}$ is a binary mask vector sampled from a Bernoulli distribution with parameter p.

Forward Pass with Dropout: During the forward pass, the dropout mask $\mathbf{r}$ is applied to the activations:

$$\mathbf{h}_i' = \mathbf{h}_i \cdot \mathbf{r}_i$$

Scaling at Test Time: At test time, to maintain the expected value of the activations, the weights are scaled by p:

$$\mathbf{h}_{\text{test}} = p \cdot \mathbf{h}$$

Implementation in TensorFlow: Dropout can be easily implemented using TensorFlow:

```
import tensorflow as tf

# Define a simple neural network model with dropout
model = tf.keras.Sequential([
    tf.keras.layers.Dense(64, activation='relu', input_shape=(100,)),
    tf.keras.layers.Dropout(0.5),  # Dropout layer with p = 0.5
    tf.keras.layers.Dense(32, activation='relu'),
    tf.keras.layers.Dropout(0.5),  # Dropout layer with p = 0.5
    tf.keras.layers.Dense(10, activation='softmax')
])

# Compile the model
model.compile(optimizer='adam', loss='sparse_categorical_crossentropy', metrics=['accuracy'])
```

Implementation in PyTorch: Similarly, dropout can be implemented in PyTorch:

```
import torch
import torch.nn as nn

# Define a simple neural network model with dropout
class SimpleNN(nn.Module):
    def __init__(self):
        super(SimpleNN, self).__init__()
        self.dense1 = nn.Linear(100, 64)
        self.dropout1 = nn.Dropout(0.5)  # Dropout layer with p = 0.5
        self.dense2 = nn.Linear(64, 32)
        self.dropout2 = nn.Dropout(0.5)  # Dropout layer with p = 0.5
        self.dense3 = nn.Linear(32, 10)

    def forward(self, x):
        x = torch.relu(self.dense1(x))
        x = self.dropout1(x)
        x = torch.relu(self.dense2(x))
        x = self.dropout2(x)
        return torch.softmax(self.dense3(x), dim=1)

# Instantiate the model
model = SimpleNN()

# Define the loss function and optimizer
criterion = nn.CrossEntropyLoss()
optimizer = torch.optim.Adam(model.parameters(), lr=0.001)

# Generate dummy data
X_train = torch.randn(1000, 100)
y_train = torch.randint(0, 10, (1000,))

# Training loop
```

```
model.train()
for epoch in range(10):
    optimizer.zero_grad()
    outputs = model(X_train)
    loss = criterion(outputs, y_train)
    loss.backward()
    optimizer.step()
```

Impact on Overfitting

Dropout significantly reduces overfitting by introducing randomness into the training process. This has several beneficial effects:

1. Prevents Co-adaptation: By randomly dropping units, dropout forces the network to learn redundant representations, reducing the reliance on specific neurons and promoting a more robust model.
2. Acts as Model Averaging: Dropout can be seen as an efficient form of model averaging. Each forward pass uses a different "thinned" network, and at test time, the full network is used with scaled weights, approximating the effect of averaging many different models.
3. Improves Generalization: The noise introduced by dropout acts as a form of regularization, making the model less sensitive to small changes in the input data and improving generalization to unseen data.

Experimental Evidence: Empirical studies have shown that dropout can significantly improve the performance of neural networks on various tasks, especially when the network is prone to overfitting. For example, in image classification tasks, dropout has been shown to improve accuracy and robustness.

Practical Considerations:

Dropout Rate p: The choice of the dropout rate p is crucial. A common practice is to use $p = 0.5$ for hidden layers and a lower rate for input layers.

Training Time: Dropout can increase training time because the effective number of units is reduced, requiring more epochs to converge.

Example Results: Without dropout, a neural network might achieve high accuracy on training data but lower accuracy on validation data due to overfitting. Introducing dropout typically reduces the training accuracy but improves the validation accuracy, indicating better generalization.

3.6.2 Batch Normalization

Batch normalization is a regularization technique that helps stabilize and accelerate the training of deep neural networks by normalizing the inputs of each layer. It addresses issues related to internal covariate shift, where the distribution of each layer's inputs changes during training, making optimization difficult. Batch normalization normalizes the activations of the previous layer at each batch, maintaining the mean and variance close to 0 and 1, respectively.

Normalization during Training and Inference

During training, batch normalization standardizes the inputs of each layer by adjusting and scaling the activations. For a given mini-batch, the normalization is performed as follows:

1. Compute the Mean and Variance:
   For a mini-batch $\mathbf{X} = \{\mathbf{x}_1, \mathbf{x}_2, \ldots, \mathbf{x}_m\}$:

$$\mu_{\text{batch}} = \frac{1}{m}\sum_{i=1}^{m} \mathbf{x}_i$$

$$\sigma_{\text{batch}}^2 = \frac{1}{m}\sum_{i=1}^{m} (\mathbf{x}_i - \mu_{\text{batch}})^2$$

2. Normalize the Batch:

$$\hat{\mathbf{x}}_i = \frac{\mathbf{x}_i - \mu_{\text{batch}}}{\sqrt{\sigma_{\text{batch}}^2 + \epsilon}}$$

   where ϵ is a small constant to prevent division by zero.
3. Scale and Shift:
   Each normalized activation is then scaled and shifted using learned parameters γ and β:

$$\mathbf{y}_i = \gamma \hat{\mathbf{x}}_i + \beta$$

Here, γ and β are learned during training and allow the network to recover the original activations if necessary.

During inference, the mean and variance of the entire training set are used instead of the batch statistics. The population mean μ_{pop} and variance σ_{pop}^2 are computed during training as running averages:

$$\mu_{\text{pop}} = \frac{1}{N}\sum_{k=1}^{N} \mu_{\text{batch}}^k$$

$$\sigma_{\text{pop}}^2 = \frac{1}{N}\sum_{k=1}^{N} \sigma_{\text{batch}}^2 + \frac{1}{N}\sum_{k=1}^{N} (\mu_{\text{batch}}^k - \mu_{\text{pop}})^2$$

During inference, the normalization uses these population statistics:

$$\hat{\mathbf{x}} = \frac{\mathbf{x} - \mu_{\text{pop}}}{\sqrt{\sigma_{\text{pop}}^2 + \epsilon}}$$

$$\mathbf{y} = \gamma \hat{\mathbf{x}} + \beta$$

Implementation in TensorFlow:

```
import tensorflow as tf

# Define a simple neural network model with batch normalization
model = tf.keras.Sequential([
    tf.keras.layers.Dense(64, activation='relu', input_shape=(100,)),
    tf.keras.layers.BatchNormalization(),
    tf.keras.layers.Dense(32, activation='relu'),
    tf.keras.layers.BatchNormalization(),
    tf.keras.layers.Dense(10, activation='softmax')
])

# Compile the model
model.compile(optimizer='adam', loss='sparse_categorical_crossentropy', metrics=['accuracy'])
```

Implementation in PyTorch:

```
import torch
import torch.nn as nn

# Define a simple neural network model with batch normalization
class SimpleNN(nn.Module):
    def __init__(self):
        super(SimpleNN, self).__init__()
        self.dense1 = nn.Linear(100, 64)
        self.bn1 = nn.BatchNorm1d(64)
        self.dense2 = nn.Linear(64, 32)
        self.bn2 = nn.BatchNorm1d(32)
        self.dense3 = nn.Linear(32, 10)

    def forward(self, x):
        x = torch.relu(self.bn1(self.dense1(x)))
        x = torch.relu(self.bn2(self.dense2(x)))
        return torch.softmax(self.dense3(x), dim=1)

# Instantiate the model
model = SimpleNN()

# Define the loss function and optimizer
criterion = nn.CrossEntropyLoss()
optimizer = torch.optim.Adam(model.parameters(), lr=0.001)

# Generate dummy data
X_train = torch.randn(1000, 100)
y_train = torch.randint(0, 10, (1000,))

# Training loop
model.train()
for epoch in range(10):
    optimizer.zero_grad()
    outputs = model(X_train)
    loss = criterion(outputs, y_train)
    loss.backward()
    optimizer.step()
```

Benefits and Challenges

Benefits:

1. Accelerated Training: By normalizing the inputs of each layer, batch normalization allows the use of higher learning rates, which accelerates training.
2. Improved Stability: Normalizing the activations reduces internal covariate shift, leading to more stable and faster convergence.
3. Regularization Effect: The stochastic nature of batch normalization introduces noise, similar to dropout, which helps in preventing overfitting and improving generalization.

Challenges:

1. Dependence on Batch Size: The effectiveness of batch normalization can be sensitive to the batch size. Very small batch sizes may not provide stable estimates of the mean and variance.
2. Additional Computation: Batch normalization introduces additional computation and memory overhead due to the calculation of batch statistics and normalization during training.
3. Interaction with Dropout: Using dropout and batch normalization together can sometimes lead to suboptimal performance. It is important to carefully tune their parameters and understand their combined effect on the network.

Empirical Evidence: Empirical studies have shown that batch normalization significantly improves the performance and training speed of deep neural networks. For instance, networks trained with batch normalization often converge faster and achieve higher accuracy on test data compared to those without it.

Example Results: In practice, models with batch normalization tend to achieve higher validation accuracy and exhibit smoother learning curves. This demonstrates the regularization effect and the stability provided by normalizing the inputs of each layer.

3.6.3 Weight Regularization

Weight regularization is a critical technique to prevent overfitting in neural networks. It involves adding a penalty to the loss function to constrain the magnitude of the weights. The most common methods are L1 and L2 regularization. These techniques help control the complexity of the model by discouraging large weights, thereby promoting simpler models that generalize better to unseen data.

L1 and L2 Regularization

L1 regularization, also known as Lasso (Least Absolute Shrinkage and Selection Operator), adds a penalty equal to the absolute value of the weights to the loss function. This encourages sparsity, meaning many weights will be driven to zero, effectively performing feature selection. Given a loss function $L(\theta)$ for the model parameters θ, the L1 regularized loss is:

$$L_{\text{L1}}(\theta) = L(\theta) + \lambda\|\theta\|_1$$

where $\|\theta\|_1 = \sum_i |\theta_i|$ is the L1 norm of the weights, and λ is the regularization strength. The gradient of the L1 regularized loss with respect to the weights is:

$$\nabla_\theta L_{\text{L1}}(\theta) = \nabla_\theta L(\theta) + \lambda \text{sign}(\theta)$$

where $\text{sign}(\theta)$ is the element-wise sign function.

L2 regularization, also known as Ridge regression or Tikhonov regularization, adds a penalty equal to the square of the magnitude of the weights to the loss function. This encourages smaller weights but does not necessarily drive them to zero. The L2 regularized loss is:

$$L_{\text{L2}}(\theta) = L(\theta) + \lambda\|\theta\|_2^2$$

where $\|\theta\|_2^2 = \sum_i \theta_i^2$ is the L2 norm of the weights. The gradient of the L2 regularized loss with respect to the weights is:

$$\nabla_\theta L_{\text{L2}}(\theta) = \nabla_\theta L(\theta) + 2\lambda\theta$$

Implementation in TensorFlow: In TensorFlow, L1 and L2 regularization can be applied using the 'kernel_regularizer' parameter in the 'Dense' layer:

```
import tensorflow as tf

# Define a simple neural network model with L1 and L2 regularization
model = tf.keras.Sequential([
    tf.keras.layers.Dense(64, activation='relu', input_shape=(100,),
                          kernel_regularizer=tf.keras.regularizers.l1(0.01)),
    tf.keras.layers.Dense(32, activation='relu',
                          kernel_regularizer=tf.keras.regularizers.l2(0.01)),
    tf.keras.layers.Dense(10, activation='softmax')
])

# Compile the model
model.compile(optimizer='adam', loss='sparse_categorical_crossentropy', metrics=['accuracy'])
```

Implementation in PyTorch: In PyTorch, L1 and L2 regularization can be applied using the 'weight_decay' parameter in the optimizer for L2 regularization, and manually for L1 regularization:

```
import torch
import torch.nn as nn
import torch.optim as optim

# Define a simple neural network model
class SimpleNN(nn.Module):
    def __init__(self):
        super(SimpleNN, self).__init__()
        self.dense1 = nn.Linear(100, 64)
        self.dense2 = nn.Linear(64, 32)
        self.dense3 = nn.Linear(32, 10)

    def forward(self, x):
        x = torch.relu(self.dense1(x))
        x = torch.relu(self.dense2(x))
        return torch.softmax(self.dense3(x), dim=1)

# Instantiate the model
model = SimpleNN()

# Define the loss function and optimizer
criterion = nn.CrossEntropyLoss()
optimizer = optim.Adam(model.parameters(), lr=0.001, weight_decay=0.01)  # L2 regularization

# Training loop with L1 regularization
for epoch in range(10):
    optimizer.zero_grad()
    outputs = model(X_train)
    loss = criterion(outputs, y_train)

    # L1 regularization
    l1_lambda = 0.01
    l1_norm = sum(p.abs().sum() for p in model.parameters())
    loss += l1_lambda * l1_norm

    loss.backward()
    optimizer.step()
```

Impact on Model Complexity

L1 Regularization:

1. Sparsity: L1 regularization encourages sparsity in the model parameters. Many weights will be driven to zero, effectively performing feature selection. This can lead to simpler and more interpretable models.
2. Feature Selection: By driving certain weights to zero, L1 regularization effectively removes some input features, which can be beneficial in high-dimensional spaces with irrelevant features.

L2 Regularization:

1. Weight Shrinkage: L2 regularization penalizes large weights by adding their squared values to the loss function. This leads to weight shrinkage, where the weights are generally smaller but not necessarily zero.
2. Smooth Solutions: L2 regularization tends to produce smoother solutions by distributing the weight more evenly across all features, which can prevent the model from becoming too complex and overfitting the training data.

Combination of L1 and L2 (Elastic Net): A combination of L1 and L2 regularization, known as Elastic Net, can leverage the benefits of both methods:

$$L_{\text{Elastic Net}}(\theta) = L(\theta) + \lambda_1 \|\theta\|_1 + \lambda_2 \|\theta\|_2^2$$

This approach encourages sparsity while also controlling the overall complexity of the model.

Example in Python (TensorFlow):

```
# Define a model with Elastic Net regularization
model = tf.keras.Sequential([
    tf.keras.layers.Dense(64, activation='relu', input_shape=(100,),
                          kernel_regularizer=tf.keras.regularizers.l1_l2(l1=0.01, l2=0.01)),
    tf.keras.layers.Dense(32, activation='relu',
                          kernel_regularizer=tf.keras.regularizers.l1_l2(l1=0.01, l2=0.01)),
    tf.keras.layers.Dense(10, activation='softmax')
])

# Compile the model
model.compile(optimizer='adam', loss='sparse_categorical_crossentropy', metrics=['accuracy'])
```

Practical Impact:

Training Dynamics: Regularization techniques like L1 and L2 can significantly affect the training dynamics. They help in preventing overfitting by adding a penalty to large weights, thereby promoting simpler models.

Generalization: Regularized models tend to generalize better to unseen data, as they avoid overfitting the noise in the training data.

Hyperparameter Tuning: The regularization strength (λ) is a hyperparameter that needs careful tuning. Too high a value can lead to underfitting, while too low a value may not effectively prevent overfitting.

3.7 Practical Examples and Exercises

3.7.1 Building MLPs with TensorFlow

Building MLPs with TensorFlow involves several key steps, including defining the model architecture, compiling the model, training it on data, and evaluating its performance. Additionally, using TensorBoard for visualization can greatly aid in understanding the training process and model performance. This section provides a detailed, step-by-step implementation, training and evaluation process, and TensorBoard visualization.

Step-by-step Implementation

1. Define the Model Architecture: An MLP consists of multiple layers, typically including an input layer, one or more hidden layers, and an output layer. Each layer is defined using TensorFlow's 'Dense' layer.

```
import tensorflow as tf

# Define the model
model = tf.keras.Sequential([
    tf.keras.layers.Dense(128, activation='relu', input_shape=(784,)),  # Input layer with
            128 units
    tf.keras.layers.Dense(64, activation='relu'),  # Hidden layer with 64 units
    tf.keras.layers.Dense(10, activation='softmax')  # Output layer with 10 units
            (for classification)
])
```

In this example, the input shape is (784,) which is suitable for a flattened 28 × 28 pixel image (such as MNIST dataset). The hidden layers use ReLU activation, and the output layer uses softmax activation for classification.

2. Compile the Model: Compiling the model involves specifying the optimizer, loss function, and evaluation metrics.

```
model.compile(optimizer='adam',
              loss='sparse_categorical_crossentropy',
              metrics=['accuracy'])
```

3. Generate Dummy Data: For demonstration purposes, we can use the MNIST dataset, a standard dataset for image classification tasks.

```
from tensorflow.keras.datasets import mnist

# Load the data
(X_train, y_train), (X_test, y_test) = mnist.load_data()

# Normalize the data
X_train = X_train / 255.0
X_test = X_test / 255.0

# Flatten the images
X_train = X_train.reshape(-1, 784)
X_test = X_test.reshape(-1, 784)
```

4. Train the Model: Training the model involves feeding the data into the model and iterating over the dataset for a specified number of epochs.

```
model.fit(X_train, y_train, epochs=10, batch_size=32,
```

Training and Evaluation

1. Training Process: The training process involves optimizing the model parameters to minimize the loss function. TensorFlow handles the backpropagation and weight updates during the training phase.

```
history = model.fit(X_train, y_train, epochs=10, batch_size=32, validation_split=0.2)
```

'epochs': Number of times the model will pass through the entire training dataset.
'batch_size': Number of samples per gradient update.
'validation_split': Fraction of the training data to be used as validation data.

2. Evaluate the Model: Evaluating the model on the test dataset helps in assessing its generalization performance.

```
test_loss, test_accuracy = model.evaluate(X_test, y_test)
print(f"Test accuracy: {test_accuracy}")
```

4. Visualizing Training with TensorBoard: TensorBoard is a powerful visualization tool that helps monitor and visualize the training process. It can be used to track metrics such as loss and accuracy, and to visualize the computational graph.

Setup TensorBoard Callback:

```
import datetime

log_dir = "logs/fit/" + datetime.datetime.now().strftime("%Y%m%d-%H%M%S")
tensorboard_callback = tf.keras.callbacks.TensorBoard(log_dir=log_dir, histogram_freq=1)
```

Train the Model with TensorBoard:

```
model.fit(X_train, y_train, epochs=10, batch_size=32, validation_split=0.2,
          callbacks=[tensorboard_callback])
```

Launching TensorBoard: To launch TensorBoard, use the following command in your terminal:

```
tensorboard --logdir logs/fit
```

Open the provided URL in your browser to visualize the training process. TensorBoard provides various visualizations, including:
Scalars: Track metrics such as loss and accuracy.
Graphs: Visualize the computational graph.
Distributions and Histograms: Inspect the distribution of weights and biases.

Example: Training an MLP on the MNIST dataset typically yields high accuracy, demonstrating the effectiveness of neural networks for image classification tasks. The validation accuracy and loss plots in TensorBoard can help identify potential issues such as overfitting or underfitting, allowing for adjustments to the model architecture or training parameters.

3.7.2 Building MLPs with PyTorch

Building MLPs with PyTorch involves defining the model architecture, training it on data, and evaluating its performance. PyTorch provides a dynamic computation graph and intuitive API, making it a popular choice for deep learning research and development. This section covers a step-by-step implementation, training and evaluation process, and visualization using TorchVision.

Step-by-step Implementation

1. Define the Model Architecture: In PyTorch, we define the model architecture by subclassing 'nn.Module' and specifying the layers in the '__init__' method and the forward pass in the 'forward' method.

```
import torch
import torch.nn as nn

# Define the MLP model
class MLP(nn.Module):
    def __init__(self):
        super(MLP, self).__init__()
        self.dense1 = nn.Linear(784, 128)  # Input layer with 128 units
        self.dense2 = nn.Linear(128, 64)   # Hidden layer with 64 units
        self.dense3 = nn.Linear(64, 10)    # Output layer with 10 units (for classification)
        self.relu = nn.ReLU()
        self.softmax = nn.Softmax(dim=1)

    def forward(self, x):
        x = self.relu(self.dense1(x))
        x = self.relu(self.dense2(x))
        x = self.softmax(self.dense3(x))
        return x

# Instantiate the model
model = MLP()
```

2. Define the Loss Function and Optimizer: Next, we define the loss function and optimizer. For a classification task, we use cross-entropy loss and an optimizer like Adam.

```
criterion = nn.CrossEntropyLoss()
optimizer = torch.optim.Adam(model.parameters(), lr=0.001)
```

3. Load and Preprocess the Data: For demonstration purposes, we use the MNIST dataset, which is available through 'torchvision'.

```
from torchvision import datasets, transforms
from torch.utils.data import DataLoader

# Define data transformations
transform = transforms.Compose([
    transforms.ToTensor(),
    transforms.Normalize((0.1307,), (0.3081,))
])

# Load the MNIST dataset
train_dataset = datasets.MNIST(root='./data', train=True, transform=transform, download=True)
test_dataset = datasets.MNIST(root='./data', train=False, transform=transform)

# Create data loaders
train_loader = DataLoader(train_dataset, batch_size=32, shuffle=True)
test_loader = DataLoader(test_dataset, batch_size=32, shuffle=False)
```

Training and Evaluation

1. Training the Model:

```
# Training loop
num_epochs = 10
for epoch in range(num_epochs):
    model.train()
    for images, labels in train_loader:
        # Flatten the images
        images = images.view(-1, 784)

        # Forward pass
        outputs = model(images)
        loss = criterion(outputs, labels)

        # Backward pass and optimization
        optimizer.zero_grad()
        loss.backward()
        optimizer.step()

    print(f"Epoch [{epoch+1}/{num_epochs}], Loss: {loss.item():.4f}")
```

2. Evaluating the Model:

```
# Evaluation loop
model.eval()
with torch.no_grad():
    correct = 0
    total = 0
    for images, labels in test_loader:
        images = images.view(-1, 784)
        outputs = model(images)
        _, predicted = torch.max(outputs.data, 1)
        total += labels.size(0)
        correct += (predicted == labels).sum().item()

    accuracy = correct / total
    print(f"Test Accuracy: {accuracy * 100:.2f}%")
```

Visualizing with TorchVision

1. Visualizing Data Samples: Using TorchVision, we can visualize sample images from the dataset.

```
import matplotlib.pyplot as plt

# Get a batch of training data
dataiter = iter(train_loader)
images, labels = dataiter.next()

# Show images
fig, axes = plt.subplots(1, 6, figsize=(12, 2))
for i in range(6):
    ax = axes[i]
    ax.imshow(images[i].numpy().squeeze(), cmap='gray')
    ax.set_title(f"Label: {labels[i].item()}")
    ax.axis('off')
plt.show()
```

2. Visualizing Training Progress: While PyTorch itself does not include an integrated visualization tool like TensorBoard, it can be used with TensorBoard by integrating the 'torch.utils.tensorboard' module.

Setup TensorBoard:

```
from torch.utils.tensorboard import SummaryWriter

# Initialize the TensorBoard writer
writer = SummaryWriter(log_dir='./logs')

# Training loop with TensorBoard logging
for epoch in range(num_epochs):
    model.train()
    running_loss = 0.0
    for i, (images, labels) in enumerate(train_loader):
        images = images.view(-1, 784)
        outputs = model(images)
        loss = criterion(outputs, labels)

        optimizer.zero_grad()
        loss.backward()
        optimizer.step()

        running_loss += loss.item()
        if (i+1) % 100 == 0:
            writer.add_scalar('training_loss', running_loss / 100, epoch * len(train_loader) + i)
            running_loss = 0.0

# Close the writer
writer.close()
```

Launching TensorBoard: To launch TensorBoard, use the following command in your terminal:

```
tensorboard --logdir ./logs
```

Open the provided URL in your browser to visualize the training process.

3.7.3 Case Studies

Case studies are essential for understanding the practical application as they provide hands-on experience in tackling real-world problems and help solidify the theoretical concepts discussed in earlier sections. This section presents detailed case studies for classification and regression tasks, as well as exercises for hyperparameter tuning and custom MLP design.

Classification Task Case Study

Problem Statement: Consider a classification task where we need to classify handwritten digits from the MNIST dataset. The goal is to build an MLP that can accurately predict the digit (0-9) in a given image.
Data Preprocessing: Load and preprocess the MNIST dataset, ensuring the data is normalized and reshaped appropriately.

```
from torchvision import datasets, transforms
from torch.utils.data import DataLoader

# Define data transformations
transform = transforms.Compose([
    transforms.ToTensor(),
    transforms.Normalize((0.1307,), (0.3081,))
])

# Load the MNIST dataset
train_dataset = datasets.MNIST(root='./data', train=True, transform=transform, download=True)
test_dataset = datasets.MNIST(root='./data', train=False, transform=transform)

# Create data loaders
train_loader = DataLoader(train_dataset, batch_size=32, shuffle=True)
test_loader = DataLoader(test_dataset, batch_size=32, shuffle=False)
```

Model Architecture: Define an MLP with an input layer, two hidden layers, and an output layer. Use ReLU activations for the hidden layers and softmax for the output layer.

```
import torch
import torch.nn as nn

# Define the MLP model
class MLP(nn.Module):
    def __init__(self):
        super(MLP, self).__init__()
        self.dense1 = nn.Linear(784, 128)
        self.dense2 = nn.Linear(128, 64)
        self.dense3 = nn.Linear(64, 10)
        self.relu = nn.ReLU()
        self.softmax = nn.Softmax(dim=1)

    def forward(self, x):
        x = self.relu(self.dense1(x))
        x = self.relu(self.dense2(x))
        x = self.softmax(self.dense3(x))
        return x

# Instantiate the model
model = MLP()
```

Training and Evaluation: Train the model using cross-entropy loss and the Adam optimizer. Evaluate its performance on the test set.

```
criterion = nn.CrossEntropyLoss()
optimizer = torch.optim.Adam(model.parameters(), lr=0.001)

# Training loop
num_epochs = 10
for epoch in range(num_epochs):
    model.train()
    for images, labels in train_loader:
        images = images.view(-1, 784)
        outputs = model(images)
        loss = criterion(outputs, labels)

        optimizer.zero_grad()
        loss.backward()
        optimizer.step()

    print(f"Epoch [{epoch+1}/{num_epochs}], Loss: {loss.item():.4f}")
```

```
# Evaluation loop
model.eval()
with torch.no_grad():
    correct = 0
    total = 0
    for images, labels in test_loader:
        images = images.view(-1, 784)
        outputs = model(images)
        _, predicted = torch.max(outputs.data, 1)
        total += labels.size(0)
        correct += (predicted == labels).sum().item()

    accuracy = correct / total
    print(f"Test Accuracy: {accuracy * 100:.2f}%")
```

Regression Task Case Study

Problem Statement: Consider a regression task where we need to predict house prices based on various features such as the number of rooms, square footage, and location. Data Preprocessing: Load and preprocess the housing dataset, normalizing the features for better convergence.

```
import pandas as pd
from sklearn.model_selection import train_test_split
from sklearn.preprocessing import StandardScaler

# Load the dataset
data = pd.read_csv('housing.csv')
X = data.drop('price', axis=1).values
y = data['price'].values

# Split the dataset
X_train, X_test, y_train, y_test = train_test_split(X, y, test_size=0.2, random_state=42)

# Standardize the features
scaler = StandardScaler()
X_train = scaler.fit_transform(X_train)
X_test = scaler.transform(X_test)
```

Model Architecture: Define an MLP for regression with an input layer, two hidden layers, and an output layer. Use ReLU activations for the hidden layers and a linear activation for the output layer.

```
import torch
import torch.nn as nn

# Define the MLP model for regression
class MLPRegression(nn.Module):
    def __init__(self):
        super(MLPRegression, self).__init__()
        self.dense1 = nn.Linear(X_train.shape[1], 128)
        self.dense2 = nn.Linear(128, 64)
        self.dense3 = nn.Linear(64, 1)
        self.relu = nn.ReLU()

    def forward(self, x):
        x = self.relu(self.dense1(x))
        x = self.relu(self.dense2(x))
        x = self.dense3(x)
        return x

# Instantiate the model
model = MLPRegression()
```

Training and Evaluation: Train the model using mean squared error (MSE) loss and the Adam optimizer. Evaluate its performance on the test set.

```
criterion = nn.MSELoss()
optimizer = torch.optim.Adam(model.parameters(), lr=0.001)

# Training loop
num_epochs = 50
for epoch in range(num_epochs):
    model.train()
    for i in range(0, len(X_train), 32):
        inputs = torch.tensor(X_train[i:i+32], dtype=torch.float32)
        targets = torch.tensor(y_train[i:i+32], dtype=torch.float32).view(-1, 1)

        outputs = model(inputs)
        loss = criterion(outputs, targets)

        optimizer.zero_grad()
        loss.backward()
        optimizer.step()

    print(f"Epoch [{epoch+1}/{num_epochs}], Loss: {loss.item():.4f}")

# Evaluation loop
model.eval()
with torch.no_grad():
    inputs = torch.tensor(X_test, dtype=torch.float32)
    targets = torch.tensor(y_test, dtype=torch.float32).view(-1, 1)
    outputs = model(inputs)
    mse_loss = criterion(outputs, targets)
    print(f"Test MSE Loss: {mse_loss.item():.4f}")
```

Hyperparameter Tuning Exercise

Objective: Optimize the hyperparameters of an MLP model using grid search or random search. Hyperparameters to tune may include the learning rate, batch size, number of layers, number of neurons per layer, and activation functions.
Implementation: Use a library like 'scikit-learn' for hyperparameter tuning.

```
from sklearn.model_selection import GridSearchCV
from skorch import NeuralNetClassifier

# Define the model class
class MLPHyperparam(nn.Module):
    def __init__(self, num_units=128, dropout=0.5):
        super(MLPHyperparam, self).__init__()
        self.dense1 = nn.Linear(784, num_units)
        self.dropout = nn.Dropout(dropout)
        self.dense2 = nn.Linear(num_units, 64)
        self.dense3 = nn.Linear(64, 10)
        self.relu = nn.ReLU()
        self.softmax = nn.Softmax(dim=1)

    def forward(self, x):
        x = self.relu(self.dense1(x))
        x = self.dropout(x)
        x = self.relu(self.dense2(x))
        x = self.softmax(self.dense3(x))
        return x

# Wrap the model with skorch
net = NeuralNetClassifier(
    MLPHyperparam,
```

```
        max_epochs=10,
        lr=0.1,
        iterator_train__shuffle=True
)

# Define the hyperparameter grid
params = {
        'module__num_units': [128, 256],
        'module__dropout': [0.3, 0.5, 0.7],
        'lr': [0.01, 0.1, 0.2],
}

# Perform grid search
gs = GridSearchCV(net, params, refit=False, cv=3, scoring='accuracy')
gs.fit(X_train.reshape(-1, 784), y_train)

print("Best parameters found:", gs.best_params_)
```

3.8 Exercises

1. Given an input vector $\mathbf{x} \in \mathbb{R}^4$, describe how the input layer processes this vector in an MLP with 5 neurons in the first hidden layer. Write down the mathematical formulation for the input transformation.
2. For an MLP with 3 hidden layers where each layer has 10 neurons, compute the total number of parameters (weights and biases) if the input dimension is 5 and the output dimension is 1.
3. Design an output layer for a classification task with 3 classes. Explain the choice of activation function and the interpretation of the output.
4. Compute the output of the sigmoid activation function for the input $\mathbf{z} = \begin{pmatrix} -1 \\ 0 \\ 1 \end{pmatrix}$.
5. Derive the gradient of the tanh function. Verify the gradient computation for the input $\mathbf{z} = \begin{pmatrix} -1 \\ 0 \\ 1 \end{pmatrix}$.
6. For the input vector $\mathbf{z} = \begin{pmatrix} -1 \\ 0 \\ 1 \end{pmatrix}$, compute the output of the ReLU, Leaky ReLU (with $\alpha = 0.01$), and ELU (with $\alpha = 1.0$) activation functions.
7. Implement the backpropagation algorithm for a simple MLP with one hidden layer. Use tensor operations to compute the gradients and update the weights.
8. Consider a simple MLP with one hidden layer. The input $\mathbf{x} \in \mathbb{R}^3$ is given by:

$$\mathbf{x} = \begin{pmatrix} 1 \\ 2 \\ 3 \end{pmatrix}$$

The weights of the hidden layer are $\mathbf{W}_1 \in \mathbb{R}^{4\times3}$ and the weights of the output layer are $\mathbf{W}_2 \in \mathbb{R}^{2\times4}$:

$$\mathbf{W}_1 = \begin{pmatrix} 0.2 & -0.1 & 0.4 \\ 0.5 & 0.3 & -0.2 \\ -0.3 & 0.6 & 0.1 \\ 0.1 & -0.4 & 0.2 \end{pmatrix}, \quad \mathbf{W}_2 = \begin{pmatrix} 0.3 & -0.6 & 0.1 & 0.5 \\ -0.2 & 0.4 & 0.7 & -0.3 \end{pmatrix}$$

(a) Compute the activations of the hidden layer using ReLU activation function.
(b) Compute the output of the MLP using the softmax activation function.

9. Consider a categorical feature with 4 possible values: $\{A, B, C, D\}$. The input data is a sequence $[A, C, D, B, A]$.

(a) Represent this sequence using one-hot encoding.
(b) Given an embedding matrix $\mathbf{E} \in \mathbb{R}^{4\times3}$:

$$\mathbf{E} = \begin{pmatrix} 0.1 & 0.2 & 0.3 \\ 0.4 & 0.5 & 0.6 \\ 0.7 & 0.8 & 0.9 \\ 1.0 & 1.1 & 1.2 \end{pmatrix}$$

Compute the embedding representations for the input sequence.

10. Given an embedding matrix $\mathbf{E} \in \mathbb{R}^{5\times2}$ initialized using a normal distribution with mean 0 and standard deviation 0.1, and an input sequence of word indices $[2, 3, 1, 4, 0]$.

(a) Generate a sample embedding matrix $\mathbf{E}$ using the given initialization method.
(b) Compute the embedding lookups for the input sequence.

11. Consider an MLP with an input layer, one hidden layer, and an output layer. Let the input $\mathbf{x} \in \mathbb{R}^2$ be:

$$\mathbf{x} = \begin{pmatrix} 0.5 \\ -0.5 \end{pmatrix}$$

The weights for the hidden layer $\mathbf{W}_1 \in \mathbb{R}^{3\times2}$ and the weights for the output layer $\mathbf{W}_2 \in \mathbb{R}^{2\times3}$ are:

$$\mathbf{W}_1 = \begin{pmatrix} 0.1 & -0.3 \\ 0.2 & 0.4 \\ -0.5 & 0.1 \end{pmatrix}, \quad \mathbf{W}_2 = \begin{pmatrix} 0.3 & -0.2 & 0.5 \\ -0.1 & 0.4 & -0.3 \end{pmatrix}$$

(a) Compute the forward pass for the given input.
(b) Assume the true label for this input is $[1, 0]$. Calculate the cross-entropy loss for the predicted output.

Chapter 4
Journey into Convolutions

Our intelligence is what makes us human, and AI is an extension of that quality.
– Yann LeCun, *Gradient-Based Learning Applied to Document Recognition*

Abstract This chapter introduces convolutional neural networks (CNNs), highlighting their significance in image processing. It explains convolution operations, implementing convolutions with tensors, and the structure of convolutional, pooling, and fully connected layers. The chapter explores advanced CNN architectures, techniques to improve performance, and practical applications such as image classification, object detection, and image segmentation. Implementation examples with TensorFlow and PyTorch, along with exercises, are provided to solidify understanding.

Keywords Convolutional neural networks (CNNs) · Convolution operations · CNN layers

4.1 Introduction to Convolutional Neural Networks

Convolutional neural networks (CNNs) have revolutionized the field of image processing and computer vision, providing state-of-the-art performance in a wide range of applications. They are a class of deep learning models specifically designed to process and analyze visual data. Unlike traditional neural networks, CNNs use convolutional layers to automatically and adaptively learn spatial hierarchies of features from input images. The key components of a CNN include convolutional layers, pooling layers, and fully connected layers.

Convolutional Layers The core building block of a CNN is the convolutional layer, which performs the convolution operation. A convolution operation involves sliding a filter (or kernel) over the input image and computing the dot product between the filter and local regions of the input. Mathematically, the convolution operation for a 2D input image **I** and a filter **K** is defined as:

P. Singh and B. Raman, *Deep Learning Through the Prism of Tensors*, Studies in Big Data 162, https://doi.org/10.1007/978-981-97-8019-8_4

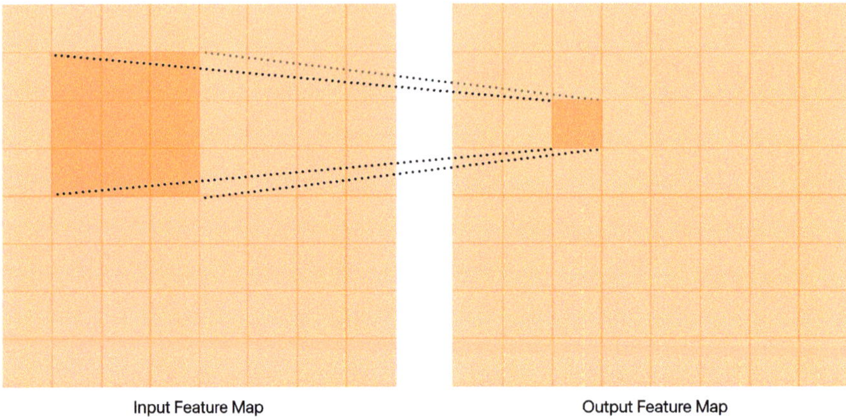

Fig. 4.1 Illustration of the convolution operation

$$(\mathbf{I} * \mathbf{K})(i, j) = \sum_m \sum_n \mathbf{I}(i + m, j + n)\mathbf{K}(m, n)$$

where $\mathbf{I}(i, j)$ denotes the pixel value at position (i, j), and $\mathbf{K}(m, n)$ represents the filter weights (see Fig. 4.1).

Pooling Layers Pooling layers are used to downsample the spatial dimensions of the input, reducing the number of parameters and computational load. The most common pooling operation is max-pooling, which selects the maximum value within a local region. For an input region $\mathbf{R}$, the max-pooling operation is defined as:

$$\text{Max-Pool}(\mathbf{R}) = \max_{i,j} \mathbf{R}(i, j)$$

Fully Connected Layers After several convolutional and pooling layers, the output is usually flattened and passed through one or more fully connected layers. These layers are similar to those in traditional neural networks, where each neuron is connected to every neuron in the previous layer.

4.1.1 Historical Context and Development

The development of CNNs can be traced back to the 1980s and 1990s, with significant contributions from researchers like Yann LeCun. The pioneering work on CNNs was driven by the need to improve the performance of models on image recognition tasks.

Early Milestones:

1. **Neocognitron (1980)**: Proposed by Fukushima (1980), the neocognitron is considered a precursor to modern CNNs. It introduced the concept of hierarchical feature extraction using a series of convolutional and pooling layers.
2. **LeNet-5 (1998)**: Developed by LeCun et al. (1998), LeNet-5 was one of the first successful applications of CNNs for handwritten digit recognition. It consisted of two convolutional layers, followed by subsampling layers (pooling) and fully connected layers.

Modern Advancements:

1. **AlexNet (2012)**: AlexNet Krizhevsky et al. (2012) demonstrated the potential of deep CNNs by winning the ImageNet Large Scale Visual Recognition Challenge (ILSVRC). It introduced the use of ReLU activations, dropout, and data augmentation.
2. **VGGNet (2014)**: The Visual Geometry Group (VGG) network Simonyan and Zisserman (2014) showed that increasing the depth of CNNs could improve performance. VGGNet used small (3×3) convolution filters and demonstrated the importance of network depth.
3. **ResNet (2015)**: Residual Networks (ResNet) He et al. (2015b) introduced the concept of residual learning with skip connections. This allowed the training of extremely deep networks, addressing the vanishing gradient problem.

4.1.2 Advantages of CNNs in Image Processing

CNNs have several advantages over traditional neural networks and other machine learning methods in the context of image processing:

1. **Automatic Feature Extraction**: CNNs automatically learn hierarchical representations of the input data, capturing low-level features (e.g., edges, textures) in the initial layers and higher-level features (e.g., objects, shapes) in deeper layers. This eliminates the need for manual feature engineering.
2. **Parameter Sharing**: The use of convolutional layers with shared weights (filters) reduces the number of parameters compared to fully connected networks. This parameter sharing leads to more efficient models with lower memory requirements.
3. **Spatial Invariance**: Convolutional and pooling operations provide spatial invariance, allowing CNNs to recognize patterns regardless of their position in the input image. This property is particularly useful for tasks like object detection and image classification.
4. **Scalability**: CNNs can be scaled to handle large and complex datasets. Techniques such as transfer learning allow pre-trained CNNs to be fine-tuned for specific tasks, leveraging the knowledge learned from large datasets.

4.1.3 *Challenges and Limitations of CNNs*

Despite their success, CNNs also face several challenges and limitations:

1. **Data Requirements**: CNNs require large amounts of labeled data for training to achieve high performance. Acquiring and annotating such datasets can be expensive and time-consuming.
2. **Computational Resources**: Training deep CNNs is computationally intensive and requires significant hardware resources, such as GPUs. This can limit their accessibility for researchers and practitioners with limited resources.
3. **Interpretability**: CNNs are often considered black-box models, making it challenging to interpret and understand the learned features and decision-making process. Techniques for visualizing and interpreting CNNs are an active area of research.
4. **Vulnerability to Adversarial Attacks**: CNNs can be susceptible to adversarial attacks, where small perturbations to the input image can lead to incorrect predictions. Ensuring robustness and security against such attacks is a critical challenge.

Example Consider the task of classifying images of handwritten digits. A traditional machine learning approach might require manual feature extraction, such as edge detection or texture analysis. In contrast, a CNN automatically learns these features through convolutional layers. The initial layers might learn edges and textures, while deeper layers recognize complex patterns and shapes, leading to accurate digit classification (see Fig. 4.2 for an overview of the architecture). Despite requiring large datasets and computational resources, the automatic feature extraction and hierarchical learning make CNNs a powerful tool for image processing tasks.

4.2 Basics of Convolution Operations

Convolution operations are at the heart of CNNs, enabling them to learn spatial hierarchies of features from input images. This section delves into the fundamental aspects of convolution operations, including kernels and filters, strides, padding, dilated convolutions, and transposed convolutions.

4.2.1 *Convolution Operations*

Convolution operations involve sliding a filter (kernel) over the input image to produce feature maps. These operations are essential for detecting patterns and structures within the image data.

Kernels and Filters Kernels (or filters) are small matrices used to perform the convolution operation. Each element in the kernel is a learnable parameter. The kernel slides over the input image, computing the dot product between the kernel elements and the corresponding input elements. This operation produces a single output value, which forms part of the feature map. For a 2D input image **I** of size $H \times W$ and a 2D kernel **K** of size $k \times k$, the convolution operation is defined as:

$$(\mathbf{I} * \mathbf{K})(i, j) = \sum_{m=0}^{k-1} \sum_{n=0}^{k-1} \mathbf{I}(i+m, j+n)\mathbf{K}(m, n)$$

where $\mathbf{I}(i+m, j+n)$ denotes the pixel value at position $(i+m, j+n)$ in the input image, and $\mathbf{K}(m, n)$ represents the kernel weight at position (m, n). For example, consider a 3×3 kernel applied to a 5×5 input image. The kernel moves across the image, performing the dot product at each position to produce a 3×3 feature map.

Strides Strides determine the step size by which the kernel moves across the input image. A stride of 1 means the kernel moves one pixel at a time, while a stride of 2 means the kernel moves two pixels at a time. Larger strides reduce the spatial dimensions of the output feature map. For a stride s, the output feature map dimension (H_o, W_o) is given by:

$$H_o = \left\lfloor \frac{H-k}{s} + 1 \right\rfloor$$

$$W_o = \left\lfloor \frac{W-k}{s} + 1 \right\rfloor$$

where H and W are the height and width of the input image, k is the kernel size, and s is the stride. For example, using a 3×3 kernel with a stride of 2 on a 5×5 input image results in a 2×2 feature map.

Padding Padding involves adding a border of zeros around the input image to control the spatial dimensions of the output feature map. Padding ensures that the edges of the input image are adequately processed and helps preserve the spatial dimensions.

Types of Padding:

1. Valid Padding (No Padding): No padding is added, resulting in smaller output dimensions.

2. Same Padding (Zero Padding): Padding is added such that the output dimensions are the same as the input dimensions.

For a padding p, the output feature map dimension (H_o, W_o) is given by:

$$H_o = \left\lfloor \frac{H+2p-k}{s} + 1 \right\rfloor$$

$$W_o = \left\lfloor \frac{W+2p-k}{s} + 1 \right\rfloor$$

where p is the padding size. For example, using a 3×3 kernel with a stride of 1 and padding of 1 on a 5×5 input image results in a 5×5 feature map.

Dilated Convolutions

Dilated convolutions introduce a spacing between the kernel elements, allowing the network to have a larger receptive field without increasing the number of parameters. This is particularly useful for capturing long-range dependencies in the input. For a dilation rate d, the dilated convolution is defined as:

$$(\mathbf{I} * \mathbf{K})(i, j) = \sum_{m=0}^{k-1} \sum_{n=0}^{k-1} \mathbf{I}(i + md, j + nd)\mathbf{K}(m, n)$$

where d is the dilation rate, k is the kernel size, and $\mathbf{I}$ and $\mathbf{K}$ are the input image and kernel, respectively. For example, a 3×3 kernel with a dilation rate of 2 applied to a 5×5 input image results in a larger receptive field compared to a standard convolution. Let the input image $\mathbf{I}$ and the kernel $\mathbf{K}$ be:

$$\mathbf{I} = \begin{bmatrix} i_{00} & i_{01} & i_{02} & i_{03} & i_{04} \\ i_{10} & i_{11} & i_{12} & i_{13} & i_{14} \\ i_{20} & i_{21} & i_{22} & i_{23} & i_{24} \\ i_{30} & i_{31} & i_{32} & i_{33} & i_{34} \\ i_{40} & i_{41} & i_{42} & i_{43} & i_{44} \end{bmatrix}$$

$$\mathbf{K} = \begin{bmatrix} k_{00} & k_{01} & k_{02} \\ k_{10} & k_{11} & k_{12} \\ k_{20} & k_{21} & k_{22} \end{bmatrix}$$

With a dilation rate $d = 2$, the positions in the input image that the kernel elements will cover are spaced out by 2. Specifically, each element of the kernel will cover every second element of the input image. To simplify, we'll look at just the center element where the output is computed at position (2, 2):

$$(\mathbf{I} * \mathbf{K})(2, 2) = i_{00} \cdot k_{00} + i_{02} \cdot k_{01} + i_{04} \cdot k_{02} + i_{20} \cdot k_{10} + i_{22} \cdot k_{11} + i_{24} \cdot k_{12} + i_{40} \cdot k_{20} + i_{42} \cdot k_{21} + i_{44} \cdot k_{22}$$

When this process is applied to the entire input image, the resulting output feature map will have a larger receptive field compared to a standard convolution. In this case, the 3×3 kernel with a dilation rate of 2 effectively covers an area of 5×5 in the input image due to the spacing introduced by the dilation. Thus, the dilated convolution allows the kernel to capture more context from the input image without increasing the number of parameters, making it an efficient way to expand the receptive field.

Transposed Convolutions

Transposed convolutions, also known as deconvolutions or upsampling convolutions, are used to increase the spatial dimensions of the input feature map. They are commonly used in tasks such as image generation and semantic segmentation. They perform the reverse operation of standard convolutions by inserting zeros between the elements of the input feature map before applying the convolution.

$$(\mathbf{I} * \mathbf{K})_{\text{transposed}}(i, j) = \sum_{m=0}^{k-1} \sum_{n=0}^{k-1} \mathbf{I}_{\text{upsampled}}(i - m, j - n)\mathbf{K}(m, n)$$

Example: Applying a 3×3 transposed convolution to a 2×2 input feature map can result in a 4×4 output feature map, effectively upsampling the input. Assume the input feature map $\mathbf{I}$ and the kernel $\mathbf{K}$ are given by:

$$\mathbf{I} = \begin{bmatrix} a & b \\ c & d \end{bmatrix}$$

$$\mathbf{K} = \begin{bmatrix} k_{00} & k_{01} & k_{02} \\ k_{10} & k_{11} & k_{12} \\ k_{20} & k_{21} & k_{22} \end{bmatrix}$$

The first step in transposed convolution is to upsample the input by inserting zeros between the original values. This increases the dimensions of the input from 2×2 to 4×4, resulting in the upsampled input feature map $\mathbf{I}_{\text{upsampled}}$:

$$\mathbf{I}_{\text{upsampled}} = \begin{bmatrix} a & 0 & b & 0 \\ 0 & 0 & 0 & 0 \\ c & 0 & d & 0 \\ 0 & 0 & 0 & 0 \end{bmatrix}$$

Next, we apply the 3×3 kernel $\mathbf{K}$ to the upsampled input. This involves sliding the kernel over the upsampled input and performing element-wise multiplication followed by summation. The computation for each element in the resulting output feature map $\mathbf{Y}$ is as follows: $y(0, 0) = a \cdot k_{22}$, $y(0, 1) = a \cdot k_{21} + b \cdot k_{22}$, $y(0, 2) = a \cdot k_{20} + b \cdot k_{21}$, $y(0, 3) = b \cdot k_{20}$, $y(1, 0) = a \cdot k_{12} + c \cdot k_{22}$, $y(1, 1) = a \cdot k_{11} + b \cdot k_{12} + c \cdot k_{21} + d \cdot k_{22}$, $y(1, 2) = a \cdot k_{10} + b \cdot k_{11} + c \cdot k_{20} + d \cdot k_{21}$, $y(1, 3) = b \cdot k_{10} + d \cdot k_{20}$, $y(2, 0) = c \cdot k_{12}$, $y(2, 1) = c \cdot k_{11} + d \cdot k_{12}$, $y(2, 2) = c \cdot k_{10} + d \cdot k_{11}$, and $y(2, 3) = d \cdot k_{10}$.

Combining these elements, the resulting 4×4 output feature map $\mathbf{Y}$ is:

$$\mathbf{Y} = \begin{bmatrix} a \cdot k_{22} & a \cdot k_{21} + b \cdot k_{22} & a \cdot k_{20} + b \cdot k_{21} & b \cdot k_{20} \\ a \cdot k_{12} + c \cdot k_{22} & a \cdot k_{11} + b \cdot k_{12} + c \cdot k_{21} + d \cdot k_{22} & a \cdot k_{10} + b \cdot k_{11} + c \cdot k_{20} + d \cdot k_{21} & b \cdot k_{10} + d \cdot k_{20} \\ c \cdot k_{12} & c \cdot k_{11} + d \cdot k_{12} & c \cdot k_{10} + d \cdot k_{11} & d \cdot k_{10} \\ 0 & 0 & 0 & 0 \end{bmatrix}$$

This example demonstrates how transposed convolution works by upsampling the input feature map and applying a convolutional kernel, effectively spreading the influence of the original input values across a larger output space.

3D Convolutions

3D convolutions extend the concept of 2D convolutions to three dimensions, making them suitable for processing volumetric data such as video sequences or medical imaging scans (e.g., MRI, CT scans). The input data in 3D convolutions is a 3D array, and the kernel is also a 3D filter. Given a 3D input volume $\mathbf{I}$ of size $D \times H \times W$ and a 3D kernel $\mathbf{K}$ of size $d \times k \times k$, the output feature map $\mathbf{O}$ of size $D_o \times H_o \times W_o$ is computed as follows:

$$\mathbf{O}(i, j, k) = (\mathbf{I} * \mathbf{K})(i, j, k) = \sum_{p=0}^{d-1} \sum_{q=0}^{k-1} \sum_{r=0}^{k-1} \mathbf{I}(i + p, j + q, k + r)\mathbf{K}(p, q, r)$$

where $\mathbf{I}(i, j, k)$ denotes the voxel value at position (i, j, k) in the input volume, $\mathbf{K}(p, q, r)$ represents the kernel weight at position (p, q, r), and $\mathbf{O}(i, j, k)$ is the output value at position (i, j, k) in the feature map.

4.2.2 *Implementing Convolutions with Tensors*

Implementing convolutions with tensors in deep learning frameworks like TensorFlow and PyTorch is crucial for developing CNNs. This section provides detailed implementations of convolutions in both TensorFlow and PyTorch.

Convolutions in TensorFlow

Implementing convolutions in TensorFlow involves defining convolutional layers, specifying kernel sizes, strides, and padding.

Defining Convolutional Layers In TensorFlow, convolutional layers can be defined using the 'Conv2D' class from the 'tf.keras.layers' module.

Example:

```
import tensorflow as tf

# Define a convolutional layer
conv_layer = tf.keras.layers.Conv2D(
    filters=32,             # Number of filters (output feature maps)
    kernel_size=(3, 3),     # Size of the kernel
    strides=(1, 1),         # Stride of the convolution
    padding='same',         # Padding type ('valid' or 'same')
    activation='relu'       # Activation function
)

# Print the configuration of the convolutional layer
print(conv_layer)
```

Applying the Convolutional Layer To apply the convolutional layer to an input tensor, the input tensor must have the shape '(batch_size, height, width, channels)'.

Example:

```
# Define an input tensor with shape (batch_size, height, width, channels)
input_tensor = tf.random.normal([1, 28, 28, 1])

# Example for a grayscale image of size 28x28

# Apply the convolutional layer
output_tensor = conv_layer(input_tensor)

# Print the shape of the output tensor
print(output_tensor.shape)
```

Given an input tensor $\mathbf{I}$ of shape (H, W, C) and a kernel $\mathbf{K}$ of shape (k, k, C, F), where H and W are the height and width of the input, C is the number of input channels, k is the kernel size, and F is the number of filters, the convolution operation can be expressed as:

$$\mathbf{O}(i, j, f) = \sum_{m=0}^{k-1} \sum_{n=0}^{k-1} \sum_{c=0}^{C-1} \mathbf{I}(i+m, j+n, c)\mathbf{K}(m, n, c, f)$$

where $\mathbf{O}$ is the output tensor of shape (H_o, W_o, F).

Practical Example Let's create a simple CNN model using TensorFlow and apply it to the MNIST dataset.

```
import tensorflow as tf
from tensorflow.keras.datasets import mnist
from tensorflow.keras.models import Sequential
from tensorflow.keras.layers import Conv2D, MaxPooling2D, Flatten, Dense

# Load and preprocess the MNIST dataset
(X_train, y_train), (X_test, y_test) = mnist.load_data()
X_train = X_train.reshape(-1, 28, 28, 1) / 255.0
X_test = X_test.reshape(-1, 28, 28, 1) / 255.0

# Define a simple CNN model
model = Sequential([
    Conv2D(32, (3, 3), activation='relu', input_shape=(28, 28, 1)),
    MaxPooling2D((2, 2)),
    Flatten(),
    Dense(128, activation='relu'),
```

```
    Dense(10, activation='softmax')
])

# Compile the model
model.compile(optimizer='adam', loss='sparse_categorical_crossentropy', metrics=['accuracy'])

# Train the model
model.fit(X_train, y_train, epochs=5, batch_size=32, validation_split=0.2)

# Evaluate the model
test_loss, test_accuracy = model.evaluate(X_test, y_test)
print(f"Test accuracy: {test_accuracy:.4f}")
```

Convolutions in PyTorch

Defining Convolutional Layers In PyTorch, convolutional layers can be defined using the‘nn.Conv2d’ class.

Example:

```
import torch
import torch.nn as nn

# Define a convolutional layer
conv_layer = nn.Conv2d(
    in_channels=1,          # Number of input channels
    out_channels=32,        # Number of filters (output feature maps)
    kernel_size=3,          # Size of the kernel
    stride=1,               # Stride of the convolution
    padding=1               # Padding size
)

# Print the configuration of the convolutional layer
print(conv_layer)
```

Applying the Convolutional Layer To apply the convolutional layer to an input tensor, the input tensor must have the shape ‘(batch_size, channels, height, width)’.

Example:

```
# Define an input tensor with shape (batch_size, channels, height, width)
input_tensor = torch.randn(1, 1, 28, 28)  # Example for a grayscale image of size 28x28

# Apply the convolutional layer
output_tensor = conv_layer(input_tensor)

# Print the shape of the output tensor
print(output_tensor.shape)
```

Practical Example Let’s create a simple CNN model using PyTorch and apply it to the MNIST dataset.

```
import torch
import torch.nn as nn
import torch.optim as optim
from torchvision import datasets, transforms
from torch.utils.data import DataLoader

# Define data transformations
transform = transforms.Compose([
    transforms.ToTensor(),
    transforms.Normalize((0.1307,), (0.3081,))
])
```

```
# Load the MNIST dataset
train_dataset = datasets.MNIST(root='./data', train=True, transform=transform, download=True)
test_dataset = datasets.MNIST(root='./data', train=False, transform=transform)

# Create data loaders
train_loader = DataLoader(train_dataset, batch_size=32, shuffle=True)
test_loader = DataLoader(test_dataset, batch_size=32, shuffle=False)

# Define a simple CNN model
class CNN(nn.Module):
    def __init__(self):
        super(CNN, self).__init__()
        self.conv1 = nn.Conv2d(1, 32, kernel_size=3, stride=1, padding=1)
        self.pool = nn.MaxPool2d(kernel_size=2, stride=2, padding=0)
        self.fc1 = nn.Linear(32 * 14 * 14, 128)
        self.fc2 = nn.Linear(128, 10)
        self.relu = nn.ReLU()
        self.softmax = nn.Softmax(dim=1)

    def forward(self, x):
        x = self.pool(self.relu(self.conv1(x)))
        x = x.view(-1, 32 * 14 * 14)
        x = self.relu(self.fc1(x))
        x = self.softmax(self.fc2(x))
        return x

# Instantiate the model
model = CNN()

# Define the loss function and optimizer
criterion = nn.CrossEntropyLoss()
optimizer = optim.Adam(model.parameters(), lr=0.001)

# Training loop
num_epochs = 5
for epoch in range(num_epochs):
    model.train()
    for images, labels in train_loader:
        outputs = model(images)
        loss = criterion(outputs, labels)

        optimizer.zero_grad()
        loss.backward()
        optimizer.step()

    print(f"Epoch [{epoch+1}/{num_epochs}], Loss: {loss.item():.4f}")

# Evaluation loop
model.eval()
with torch.no_grad():
    correct = 0
    total = 0
    for images, labels in test_loader:
        outputs = model(images)
        _, predicted = torch.max(outputs.data, 1)
        total += labels.size(0)
        correct += (predicted == labels).sum().item()

    accuracy = correct / total
    print(f"Test Accuracy: {accuracy * 100:.2f}%")
```

4.3 Layers in Convolutional Neural Networks

Layers in CNNs play a crucial role in extracting features and learning hierarchical representations from input data. Each layer type contributes uniquely to the network's ability to process and analyze images. This section focuses on convolutional layers, detailing their role, functionality, and the tensor operations involved.

4.3.1 Convolutional Layers

Convolutional layers are the cornerstone of CNNs. They are responsible for detecting local patterns and features in the input data through the convolution operation.

Role and Functionality
Convolutional layers apply a set of learnable filters to the input data to produce feature maps. These filters are designed to detect specific features such as edges, textures, and patterns in the input images. By stacking multiple convolutional layers, CNNs can learn complex and abstract features, enabling them to perform tasks like image classification, object detection, and segmentation.

Functionality of Convolutional Layers:

1. **Feature Detection**: Each filter in a convolutional layer detects a specific feature. For example, early layers might detect simple features like edges and corners, while deeper layers might detect more complex features like shapes and objects.
2. **Parameter Sharing**: Convolutional layers use parameter sharing, meaning the same filter (kernel) is applied across the entire input image. This reduces the number of parameters and computational complexity compared to fully connected layers.
3. **Spatial Invariance**: The convolution operation provides spatial invariance, allowing the network to recognize features regardless of their position in the input image.

Tensor Operations in Convolutional Layers
Convolutional layers perform several tensor operations, including convolution, activation, and normalization.

Convolution Operation The convolution operation involves sliding a kernel over the input tensor and computing the dot product between the kernel and the local regions of the input. For a 4D input tensor $\mathbf{I}$ of shape (N, C, H, W) and a 4D kernel tensor $\mathbf{K}$ of shape (F, C, k, k), where N is the batch size, C is the number of channels, H and W are the height and width of the input, F is the number of filters, and k is the kernel size, the output tensor $\mathbf{O}$ is computed as:

$$\mathbf{O}(n, f, i, j) = \sum_{c=0}^{C-1} \sum_{m=0}^{k-1} \sum_{n=0}^{k-1} \mathbf{I}(n, c, i+m, j+n)\mathbf{K}(f, c, m, n)$$

where $\mathbf{I}(n, c, i+m, j+n)$ is the input value at position $(n, c, i+m, j+n)$, $\mathbf{K}(f, c, m, n)$ is the kernel weight at position (f, c, m, n), and $\mathbf{O}(n, f, i, j)$ is the output value at position (n, f, i, j).

Activation Function After the convolution operation, an activation function is applied element-wise to introduce non-linearity into the network. Common activation functions include ReLU, sigmoid, and tanh.

Example in TensorFlow

```
import tensorflow as tf
from tensorflow.keras.layers import Conv2D, ReLU, BatchNormalization

# Define the convolutional layer
conv_layer = Conv2D(filters=32, kernel_size=(3, 3), strides=(1, 1), padding='same')

# Define the activation and batch normalization layers
activation_layer = ReLU()
bn_layer = BatchNormalization()

# Apply the layers to an input tensor
input_tensor = tf.random.normal([1, 28, 28, 1])
conv_output = conv_layer(input_tensor)
activated_output = activation_layer(conv_output)
normalized_output = bn_layer(activated_output)

# Print the shapes of the outputs
print(conv_output.shape)
print(activated_output.shape)
print(normalized_output.shape)
```

Example in PyTorch

```
import torch
import torch.nn as nn

# Define the convolutional layer
conv_layer = nn.Conv2d(in_channels=1, out_channels=32, kernel_size=3, stride=1, padding=1)

# Define the activation and batch normalization layers
activation_layer = nn.ReLU()
bn_layer = nn.BatchNorm2d(num_features=32)

# Apply the layers to an input tensor
input_tensor = torch.randn(1, 1, 28, 28)
conv_output = conv_layer(input_tensor)
activated_output = activation_layer(conv_output)
normalized_output = bn_layer(activated_output)

# Print the shapes of the outputs
print(conv_output.shape)
print(activated_output.shape)
print(normalized_output.shape)
```

4.3.2 Pooling Layers

Pooling layers are essential components of CNNs that reduce the spatial dimensions of the input feature maps. This reduction helps decrease the computational load, memory usage, and the number of parameters in the network. Pooling layers summarize the presence of features in patches of the feature map, making the representations more invariant to small translations and distortions in the input. This section covers various types of pooling operations: max-pooling, average pooling, global pooling, and fractional pooling.

Max-Pooling

Max-pooling is a pooling operation that selects the maximum value within a local region (window) of the feature map. This operation helps retain the most salient features while reducing the spatial dimensions. Given an input feature map $\mathbf{I}$ of size $H \times W$ and a pooling window of size $k \times k$, the max-pooling operation is defined as:

$$\mathbf{O}(i, j) = \max_{0 \leq m < k, 0 \leq n < k} \mathbf{I}(i \cdot k + m, j \cdot k + n)$$

where $\mathbf{O}$ is the output feature map, (i, j) is the position in the output feature map, and max denotes the maximum value within the window. For example, consider a 4×4 input feature map and a 2×2 pooling window. The max-pooling operation would result in a 2×2 output feature map, where each value is the maximum of the corresponding 2×2 window in the input.

Example in TensorFlow

```
import tensorflow as tf

# Define a max pooling layer
max_pool_layer = tf.keras.layers.MaxPooling2D(pool_size=(2, 2), strides=(2, 2))

# Apply the max pooling layer to an input tensor
input_tensor = tf.random.normal([1, 4, 4, 1])  # Example for a 4x4 input feature map
output_tensor = max_pool_layer(input_tensor)

# Print the shapes of the input and output tensors
print("Input shape:", input_tensor.shape)
print("Output shape:", output_tensor.shape)
```

Example in PyTorch

```
import torch
import torch.nn as nn

# Define a max pooling layer
max_pool_layer = nn.MaxPool2d(kernel_size=2, stride=2)

# Apply the max pooling layer to an input tensor
input_tensor = torch.randn(1, 1, 4, 4)  # Example for a 4x4 input feature map
output_tensor = max_pool_layer(input_tensor)

# Print the shapes of the input and output tensors
print("Input shape:", input_tensor.shape)
print("Output shape:", output_tensor.shape)
```

Average Pooling

Average pooling is a pooling operation that computes the average value within a local region (window) of the feature map. This operation reduces the spatial dimensions while preserving the average presence of features. Given an input feature map $\mathbf{I}$ of size $H \times W$ and a pooling window of size $k \times k$, the average pooling operation is defined as:

$$\mathbf{O}(i, j) = \frac{1}{k^2} \sum_{m=0}^{k-1} \sum_{n=0}^{k-1} \mathbf{I}(i \cdot k + m, j \cdot k + n)$$

where $\mathbf{O}$ is the output feature map, (i, j) is the position in the output feature map, and $\sum$ denotes the summation over the window. For example, consider a 4×4 input feature map and a 2×2 pooling window. The average pooling operation would result in a 2×2 output feature map, where each value is the average of the corresponding 2×2 window in the input.

Example in TensorFlow

```
import tensorflow as tf

# Define an average pooling layer
avg_pool_layer = tf.keras.layers.AveragePooling2D(pool_size=(2, 2), strides=(2, 2))

# Apply the average pooling layer to an input tensor
input_tensor = tf.random.normal([1, 4, 4, 1])  # Example for a 4x4 input feature map
output_tensor = avg_pool_layer(input_tensor)

# Print the shapes of the input and output tensors
print("Input shape:", input_tensor.shape)
print("Output shape:", output_tensor.shape)
```

Example in PyTorch

```
import torch
import torch.nn as nn

# Define an average pooling layer
avg_pool_layer = nn.AvgPool2d(kernel_size=2, stride=2)

# Apply the average pooling layer to an input tensor
input_tensor = torch.randn(1, 1, 4, 4)  # Example for a 4x4 input feature map
output_tensor = avg_pool_layer(input_tensor)

# Print the shapes of the input and output tensors
print("Input shape:", input_tensor.shape)
print("Output shape:", output_tensor.shape)
```

Global Pooling

Global pooling operations reduce each feature map to a single value by taking the maximum or average value across the entire feature map. This operation is often used before fully connected layers to reduce the dimensionality of the data.

Types of Global Pooling:

1. **Global Max-Pooling**: Takes the maximum value across the entire feature map.
2. **Global Average Pooling**: Takes the average value across the entire feature map.

For global average pooling, given an input feature map $\mathbf{I}$ of size $H \times W$:

$$\mathbf{O} = \frac{1}{H \times W} \sum_{i=0}^{H-1} \sum_{j=0}^{W-1} \mathbf{I}(i, j)$$

where $\mathbf{O}$ is the output scalar value.

Example in TensorFlow

```
import tensorflow as tf

# Define a global average pooling layer
global_avg_pool_layer = tf.keras.layers.GlobalAveragePooling2D()

# Apply the global average pooling layer to an input tensor
input_tensor = tf.random.normal([1, 4, 4, 1])  # Example for a 4x4 input feature map
output_tensor = global_avg_pool_layer(input_tensor)

# Print the shapes of the input and output tensors
print("Input shape:", input_tensor.shape)
print("Output shape:", output_tensor.shape)
```

Example in PyTorch

```
import torch
import torch.nn as nn

# Define a global average pooling layer
global_avg_pool_layer = nn.AdaptiveAvgPool2d(1)

# Apply the global average pooling layer to an input tensor
input_tensor = torch.randn(1, 1, 4, 4)  # Example for a 4x4 input feature map
output_tensor = global_avg_pool_layer(input_tensor)

# Print the shapes of the input and output tensors
print("Input shape:", input_tensor.shape)
print("Output shape:", output_tensor.shape)
```

Fractional Pooling
Fractional pooling is a pooling operation that reduces the spatial dimensions of the input feature maps by a non-integer factor. This allows for more flexible downsampling compared to standard pooling operations. It can be implemented using various methods, such as random pooling or linear interpolation. The key idea is to reduce the dimensions by a fraction, rather than an integer factor. For example, if the desired output size is a fraction of the input size, fractional pooling can be achieved through interpolation methods.

Example in TensorFlow TensorFlow does not have a direct implementation of fractional pooling, but it can be achieved using interpolation:

```
import tensorflow as tf

# Define an input tensor
input_tensor = tf.random.normal([1, 4, 4, 1])  # Example for a 4x4 input feature map

# Apply fractional pooling using bilinear interpolation
output_tensor = tf.image.resize(input_tensor, [3, 3], method='bilinear')
```

```
# Print the shapes of the input and output tensors
print("Input shape:", input_tensor.shape)
print("Output shape:", output_tensor.shape)
```

Example in PyTorch Similarly, in PyTorch, fractional pooling can be achieved using interpolation:

```
import torch
import torch.nn.functional as F

# Define an input tensor
input_tensor = torch.randn(1, 1, 4, 4)  # Example for a 4x4 input feature map

# Apply fractional pooling using bilinear interpolation
output_tensor = F.interpolate(input_tensor, size=(3, 3),
mode='bilinear', align_corners=False)

# Print the shapes of the input and output tensors
print("Input shape:", input_tensor.shape)
print("Output shape:", output_tensor.shape)
```

4.3.3 Fully Connected Layers

Fully connected layers, also known as dense layers, are a fundamental component of CNNs, especially toward the end of the network architecture. These layers are responsible for integrating the features extracted by previous convolutional and pooling layers and performing the final classification or regression tasks. Here we discuss the role and functionality of fully connected layers, as well as the operations involved in flattening the input tensors to make them compatible with these layers.

Role and Functionality

Fully connected layers serve as the final stage in a CNN, where high-level reasoning about the input data is performed. Each neuron in a fully connected layer is connected to every neuron in the preceding layer, allowing the network to synthesize information from the entire feature map and make predictions based on the learned representations.

Functionality of Fully Connected Layers:

1. **Integration of Features**: Fully connected layers combine the features extracted by convolutional layers into a cohesive representation. This allows the network to learn complex relationships between high-level features and the target labels.
2. **Decision Making**: In classification tasks, the output of the fully connected layers is typically passed through a softmax activation function, which converts the raw scores into probabilities for each class. In regression tasks, a linear activation function might be used to predict continuous values.
3. **Parameter Learning**: Fully connected layers contain a large number of parameters (weights and biases), which are learned during the training process. These parameters contribute significantly to the network's capacity to model complex functions but also make the layers prone to overfitting if not regularized properly.

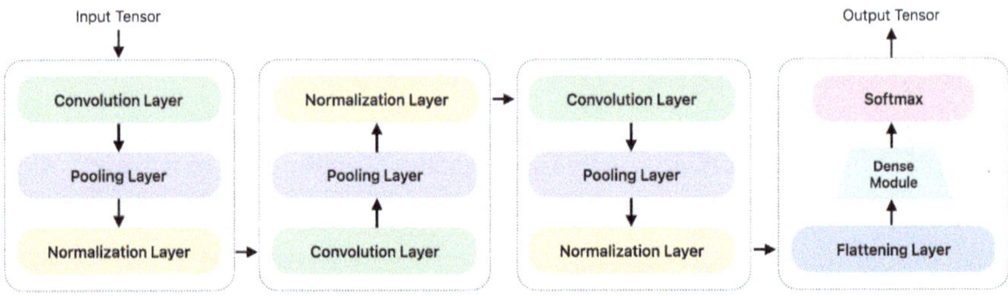

Fig. 4.2 Illustration of a typical pipeline of a CNN

The operation of a fully connected layer can be described mathematically as follows:

$$\mathbf{y} = \mathbf{W}\mathbf{x} + \mathbf{b}$$

where $\mathbf{x}$ is the input vector, $\mathbf{W}$ is the weight matrix, $\mathbf{b}$ is the bias vector, and $\mathbf{y}$ is the output vector. Each element of the output vector $\mathbf{y}$ is computed as:

$$y_i = \sum_j W_{ij} x_j + b_i$$

For example, consider a fully connected layer with an input vector of size 4 and an output vector of size 3. The weight matrix $\mathbf{W}$ would have dimensions 3×4, and the bias vector $\mathbf{b}$ would have dimensions 3×1. The output vector $\mathbf{y}$ is computed as:

$$\mathbf{y} = \begin{bmatrix} W_{11} & W_{12} & W_{13} & W_{14} \\ W_{21} & W_{22} & W_{23} & W_{24} \\ W_{31} & W_{32} & W_{33} & W_{34} \end{bmatrix} \begin{bmatrix} x_1 \\ x_2 \\ x_3 \\ x_4 \end{bmatrix} + \begin{bmatrix} b_1 \\ b_2 \\ b_3 \end{bmatrix}$$

Flattening and Tensor Operations

Before passing data to a fully connected layer, the multi-dimensional feature maps produced by convolutional layers must be flattened into a 1D vector. This process involves rearranging the elements of the feature maps into a single vector while preserving their order.

Flattening is the process of converting a multi-dimensional tensor into a 1D tensor (vector). For example, a 3D tensor with shape (H, W, C) is converted into a 1D tensor with shape $(H \cdot W \cdot C)$. Given a 3D tensor $\mathbf{T}$ of shape (H, W, C), the flattened tensor $\mathbf{f}$ is defined as:

$$\mathbf{f} = \text{flatten}(\mathbf{T})$$

where:

$$f(i) = T\left(\left\lfloor \frac{i}{WC} \right\rfloor, \left\lfloor \frac{i \mod WC}{C} \right\rfloor, i \mod C\right)$$

for $i = 0, 1, \ldots, HWC - 1$. For example, consider a tensor $\mathbf{T}$ with shape (2, 2, 3):

$$\mathbf{T} = \begin{bmatrix} \begin{bmatrix} 1 & 2 & 3 \\ 4 & 5 & 6 \end{bmatrix} \\ \begin{bmatrix} 7 & 8 & 9 \\ 10 & 11 & 12 \end{bmatrix} \end{bmatrix}$$

Flattening $\mathbf{T}$ results in:

$$\mathbf{f} = [1, 2, 3, 4, 5, 6, 7, 8, 9, 10, 11, 12]$$

Example in TensorFlow

```
import tensorflow as tf
from tensorflow.keras.layers import Flatten, Dense

# Define a flatten layer and a dense layer
flatten_layer = Flatten()
dense_layer = Dense(units=10, activation='softmax')

# Apply the flatten layer to an input tensor
input_tensor = tf.random.normal([1, 4, 4, 3])  # Example for a 4x4x3 input feature map
flattened_tensor = flatten_layer(input_tensor)

# Apply the dense layer to the flattened tensor
output_tensor = dense_layer(flattened_tensor)

# Print the shapes of the input, flattened, and output tensors
print("Input shape:", input_tensor.shape)
print("Flattened shape:", flattened_tensor.shape)
print("Output shape:", output_tensor.shape)
```

Example in PyTorch

```
import torch
import torch.nn as nn

# Define a flatten layer and a dense layer
flatten_layer = nn.Flatten()
dense_layer = nn.Linear(in_features=4*4*3, out_features=10)

# Apply the flatten layer to an input tensor
input_tensor = torch.randn(1, 3, 4, 4)  # Example for a 4x4x3 input feature map
flattened_tensor = flatten_layer(input_tensor)

# Apply the dense layer to the flattened tensor
output_tensor = dense_layer(flattened_tensor)

# Print the shapes of the input, flattened, and output tensors
print("Input shape:", input_tensor.shape)
print("Flattened shape:", flattened_tensor.shape)
print("Output shape:", output_tensor.shape)
```

4.3.4 Normalization Layers

Normalization layers are essential components in CNNs that help stabilize and accelerate the training process by normalizing the inputs of each layer. They mitigate issues like internal covariate shift, where the distribution of layer inputs changes during training, and improve the convergence of gradient-based optimization. This section covers three widely used normalization techniques: Batch Normalization, Layer Normalization, and Group Normalization.

Batch Normalization

Batch normalization (BN) normalizes the input of a layer across the mini-batch, ensuring that each input feature has zero mean and unit variance. This helps in stabilizing the learning process and allows the use of higher learning rates, potentially leading to faster convergence. Given an input mini-batch $\mathbf{X} = \{x_1, x_2, \ldots, x_m\}$ of size m with each input $x_i \in \mathbb{R}^d$, batch normalization transforms the inputs as follows:

1. Compute the mean and variance of the mini-batch:

$$\mu_{\text{batch}} = \frac{1}{m}\sum_{i=1}^{m} x_i$$

$$\sigma_{\text{batch}}^2 = \frac{1}{m}\sum_{i=1}^{m}(x_i - \mu_{\text{batch}})^2$$

2. Normalize the inputs:

$$\hat{x}_i = \frac{x_i - \mu_{\text{batch}}}{\sqrt{\sigma_{\text{batch}}^2 + \epsilon}}$$

3. Scale and shift the normalized inputs:

$$y_i = \gamma \hat{x}_i + \beta$$

where γ (scale) and β (shift) are learnable parameters, and ϵ is a small constant added for numerical stability.

Example: Consider a mini-batch with four inputs: $\mathbf{X} = \{1, 2, 3, 4\}$. The mean and variance are computed as:

$$\mu_{\text{batch}} = \frac{1+2+3+4}{4} = 2.5$$

$$\sigma_{\text{batch}}^2 = \frac{(1-2.5)^2 + (2-2.5)^2 + (3-2.5)^2 + (4-2.5)^2}{4} = 1.25$$

The normalized inputs are:

$$\hat{x}_i = \frac{x_i - 2.5}{\sqrt{1.25 + \epsilon}}$$

After normalization, the inputs are scaled and shifted using learnable parameters γ and β.

Example in TensorFlow

```
import tensorflow as tf
from tensorflow.keras.layers import BatchNormalization

# Define a batch normalization layer
bn_layer = BatchNormalization()

# Apply the batch normalization layer to an input tensor
input_tensor = tf.random.normal([10, 4, 4, 3])  # Example for a 4x4x3 input feature map
output_tensor = bn_layer(input_tensor)

# Print the shapes of the input and output tensors
print("Input shape:", input_tensor.shape)
print("Output shape:", output_tensor.shape)
```

Example in PyTorch

```
import torch
import torch.nn as nn

# Define a batch normalization layer
bn_layer = nn.BatchNorm2d(num_features=3)

# Apply the batch normalization layer to an input tensor
input_tensor = torch.randn(10, 3, 4, 4)  # Example for a 4x4x3 input feature map
output_tensor = bn_layer(input_tensor)

# Print the shapes of the input and output tensors
print("Input shape:", input_tensor.shape)
print("Output shape:", output_tensor.shape)
```

Layer Normalization

Layer normalization (LN) normalizes the inputs across the features rather than the mini-batch. This technique is particularly useful for recurrent neural networks (RNNs) and other architectures where batch normalization may be less effective due to varying sequence lengths or batch sizes. Given an input $x \in \mathbb{R}^d$ for a single sample, layer normalization transforms the inputs as follows:

1. Compute the mean and variance for each layer:

$$\mu = \frac{1}{d} \sum_{j=1}^{d} x_j$$

$$\sigma^2 = \frac{1}{d} \sum_{j=1}^{d} (x_j - \mu)^2$$

2. Normalize the inputs:

$$\hat{x}_j = \frac{x_j - \mu}{\sqrt{\sigma^2 + \epsilon}}$$

3. Scale and shift the normalized inputs:

$$y_j = \gamma \hat{x}_j + \beta$$

where γ and β are learnable parameters, and ϵ is a small constant added for numerical stability.

Example: Consider an input tensor $x = \{1, 2, 3, 4\}$. The mean and variance are similarly computed as:

$$\mu = \frac{1 + 2 + 3 + 4}{4} = 2.5$$

$$\sigma^2 = \frac{(1 - 2.5)^2 + (2 - 2.5)^2 + (3 - 2.5)^2 + (4 - 2.5)^2}{4} = 1.25$$

The normalized inputs are:

$$\hat{x}_j = \frac{x_j - 2.5}{\sqrt{1.25 + \epsilon}}$$

After normalization, the inputs are scaled and shifted using learnable parameters γ and β.

Example in TensorFlow

```
import tensorflow as tf
from tensorflow.keras.layers import LayerNormalization

# Define a layer normalization layer
ln_layer = LayerNormalization()

# Apply the layer normalization layer to an input tensor
input_tensor = tf.random.normal([1, 4, 4, 3])  # Example for a 4x4x3 input feature map
output_tensor = ln_layer(input_tensor)

# Print the shapes of the input and output tensors
print("Input shape:", input_tensor.shape)
print("Output shape:", output_tensor.shape)
```

Example in PyTorch

```
import torch
import torch.nn as nn

# Define a layer normalization layer
ln_layer = nn.LayerNorm(normalized_shape=[4, 4, 3])

# Apply the layer normalization layer to an input tensor
input_tensor = torch.randn(1, 4, 4, 3)  # Example for a 4x4x3 input feature map
output_tensor = ln_layer(input_tensor)
```

```
# Print the shapes of the input and output tensors
print("Input shape:", input_tensor.shape)
print("Output shape:", output_tensor.shape)
```

Group Normalization

Group normalization (GN) divides the channels into groups and normalizes the features within each group. This technique combines the strengths of batch and layer normalization and is particularly effective for tasks where batch sizes are small or vary significantly. Given an input $x \in \mathbb{R}^{N \times C \times H \times W}$ with N samples, C channels, and spatial dimensions $H \times W$, group normalization divides the channels into G groups, each containing C/G channels. The normalization is applied within each group.

1. Compute the mean and variance for each group:

$$\mu_g = \frac{1}{|g|} \sum_{k \in g} x_k$$

$$\sigma_g^2 = \frac{1}{|g|} \sum_{k \in g} (x_k - \mu_g)^2$$

where $|g| = \frac{C}{G} \times H \times W$ is the number of elements in each group.

2. Normalize the inputs:

$$\hat{x}_k = \frac{x_k - \mu_g}{\sqrt{\sigma_g^2 + \epsilon}}$$

3. Scale and shift the normalized inputs:

$$y_k = \gamma \hat{x}_k + \beta$$

where γ and β are learnable parameters, and ϵ is a small constant added for numerical stability.

Example: Consider an input tensor with 4 channels and 2 groups. The input channels are divided into two groups, and normalization is applied within each group.

Example in TensorFlow TensorFlow does not have a built-in layer for group normalization, but it can be implemented using custom layers:

```
import tensorflow as tf

class GroupNormalization(tf.keras.layers.Layer):
    def __init__(self, groups=2, epsilon=1e-5):
        super(GroupNormalization, self).__init__()
        self.groups = groups
        self.epsilon = epsilon

    def build

(self, input_shape):
        self.gamma = self.add_weight(shape=(input_shape[-1],), initializer='ones',
        trainable=True)
```

```
        self.beta = self.add_weight(shape=(input_shape[-1],), initializer='zeros',
        trainable=True)

    def call(self, inputs):
        input_shape = tf.shape(inputs)
        N, H, W, C = input_shape[0], input_shape[1], input_shape[2], input_shape[3]
        G = self.groups
        x = tf.reshape(inputs, [N, H, W, G, C // G])
        mean, variance = tf.nn.moments(x, [1, 2, 4], keepdims=True)
        x = (x - mean) / tf.sqrt(variance + self.epsilon)
        x = tf.reshape(x, [N, H, W, C])
        return self.gamma * x + self.beta

# Define a group normalization layer
gn_layer = GroupNormalization(groups=2)

# Apply the group normalization layer to an input tensor
input_tensor = tf.random.normal([1, 4, 4, 4])  # Example for a 4x4x4 input feature map
output_tensor = gn_layer(input_tensor)

# Print the shapes of the input and output tensors
print("Input shape:", input_tensor.shape)
print("Output shape:", output_tensor.shape)
```

Example in PyTorch

```
import torch
import torch.nn as nn

# Define a group normalization layer
gn_layer = nn.GroupNorm(num_groups=2, num_channels=4)

# Apply the group normalization layer to an input tensor
input_tensor = torch.randn(1, 4, 4, 4)  # Example for a 4x4x4 input feature map
output_tensor = gn_layer(input_tensor)

# Print the shapes of the input and output tensors
print("Input shape:", input_tensor.shape)
print("Output shape:", output_tensor.shape)
```

4.3.5 *Advanced Layers*

Advanced layers in CNNs introduce sophisticated mechanisms that enhance the model's capacity to learn complex representations, improve gradient flow, and capture dependencies in the data more effectively. This section explores three key advanced layers: Residual Connections, Inception Modules, and Attention Mechanisms.

Residual Connections
Residual connections, introduced in Residual Networks (ResNet), address the problem of vanishing gradients in deep networks. By adding shortcut connections that skip one or more layers, residual connections allow gradients to flow directly through the network, improving training efficiency and enabling the training of much deeper networks. A residual block consists of a series of layers followed by a shortcut connection that adds the input of the block to its output. Formally, given an input $\mathbf{x}$, a residual block computes:

$$\mathbf{y} = \mathcal{F}(\mathbf{x}, \{\mathbf{W}_i\}) + \mathbf{x}$$

where $\mathbf{y}$ is the output, $\mathcal{F}(\mathbf{x}, \{\mathbf{W}_i\})$ represents the operations performed by the layers within the block (e.g., convolution, batch normalization, ReLU), and $\mathbf{W}_i$ are the learnable weights. If the dimensions of $\mathbf{x}$ and $\mathcal{F}(\mathbf{x})$ do not match, a linear projection (e.g., 1×1 convolution) is used to match dimensions. In such a case, a 1×1 convolution with C_{out} filters, each of size $1 \times 1 \times C_{\text{in}}$, is applied to $\mathbf{x}$. This operation performs a weighted sum across the input channels for each spatial location, transforming the input $\mathbf{x}$ from dimensions $H \times W \times C_{\text{in}}$ to $H \times W \times C_{\text{out}}$, thus allowing the addition operation to be valid. For example, if the input feature map has dimensions $3 \times 3 \times 2$, and we need to transform it to $3 \times 3 \times 3$, the 1×1 convolution will adjust the channels appropriately while preserving the spatial dimensions. For example, consider a residual block with two convolutional layers. The operations can be expressed as:

$$\mathbf{y} = \text{ReLU}(\text{BN}(\mathbf{W}_2 * (\text{ReLU}(\text{BN}(\mathbf{W}_1 * \mathbf{x})))) + \mathbf{x})$$

where $*$ denotes convolution, BN denotes batch normalization, and ReLU is the activation function.

Example in TensorFlow

```
import tensorflow as tf
from tensorflow.keras.layers import Conv2D, BatchNormalization, ReLU, Add

# Define a residual block
def residual_block(input_tensor, filters, kernel_size=3):
    x = Conv2D(filters, kernel_size, padding='same')(input_tensor)
    x = BatchNormalization()(x)
    x = ReLU()(x)
    x = Conv2D(filters, kernel_size, padding='same')(x)
    x = BatchNormalization()(x)
    x = Add()([x, input_tensor])
    return ReLU()(x)

# Apply the residual block to an input tensor
input_tensor = tf.random.normal([1, 32, 32, 64])  # Example for a 32x32x64 input feature map
output_tensor = residual_block(input_tensor, filters=64)

# Print the shapes of the input and output tensors
print("Input shape:", input_tensor.shape)
print("Output shape:", output_tensor.shape)
```

Example in PyTorch

```
import torch
import torch.nn as nn

# Define a residual block
class ResidualBlock(nn.Module):
    def __init__(self, in_channels, out_channels, kernel_size=3):
        super(ResidualBlock, self).__init__()
        self.conv1 = nn.Conv2d(in_channels, out_channels, kernel_size, padding=1)
        self.bn1 = nn.BatchNorm2d(out_channels)
        self.relu = nn.ReLU()
        self.conv2 = nn.Conv2d(out_channels, out_channels, kernel_size, padding=1)
```

```
        self.bn2 = nn.BatchNorm2d(out_channels)

    def forward(self, x):
        identity = x
        out = self.conv1(x)
        out = self.bn1(out)
        out = self.relu(out)
        out = self.conv2(out)
        out = self.bn2(out)
        out += identity
        return self.relu(out)

# Apply the residual block to an input tensor
input_tensor = torch.randn(1, 64, 32, 32)  # Example for a 32x32x64 input feature map
model = ResidualBlock(64, 64)
output_tensor = model(input_tensor)

# Print the shapes of the input and output tensors
print("Input shape:", input_tensor.shape)
print("Output shape:", output_tensor.shape)
```

Inception Modules

Inception modules, introduced in GoogLeNet (Inception v1) (Szegedy et al. 2015), aim to improve the utilization of computational resources inside the network by performing convolutions with multiple filter sizes in parallel, allowing the network to capture features at different scales simultaneously. Given an input tensor $\mathbf{x} \in \mathbb{R}^{H \times W \times C}$, where H and W are the height and width of the input feature map, and C is the number of channels, an inception module computes several parallel operations. These operations include a 1×1 convolution ($\mathcal{F}_1$), a 3×3 convolution ($\mathcal{F}_3$), a 5×5 convolution ($\mathcal{F}_5$), and a pooling operation followed by a 1×1 convolution ($\mathcal{F}_{\text{pool}}$). The output tensors from these operations are $\mathbf{y}_1 \in \mathbb{R}^{H \times W \times C_1}$, $\mathbf{y}_3 \in \mathbb{R}^{H \times W \times C_3}$, $\mathbf{y}_5 \in \mathbb{R}^{H \times W \times C_5}$, and $\mathbf{y}_{\text{pool}} \in \mathbb{R}^{H \times W \times C_{\text{pool}}}$, respectively. The final output tensor of the inception module, $\mathbf{y}$, is obtained by concatenating these output tensors along the channel dimension, resulting in

$$\mathbf{y} = \bigoplus([\mathbf{y}_1, \mathbf{y}_3, \mathbf{y}_5, \mathbf{y}_{\text{pool}}]),$$

where $\bigoplus$ denotes the concatenation operation. This yields a tensor $\mathbf{y}$ with the shape $\mathbb{R}^{H \times W \times (C_1 + C_3 + C_5 + C_{\text{pool}})}$, effectively integrating features from different scales into a single, rich representation.

Example in TensorFlow

```
import tensorflow as tf
from tensorflow.keras.layers import Conv2D, MaxPooling2D, concatenate

# Define an inception module
def inception_module(input_tensor, filters):
    branch1x1 = Conv2D(filters, (1, 1), padding='same', activation='relu')(input_tensor)
    branch3x3 = Conv2D(filters, (3, 3), padding='same', activation='relu')(input_tensor)
    branch5x5 = Conv2D(filters, (5, 5), padding='same', activation='relu')(input_tensor)
    branch_pool = MaxPooling2D((3, 3), strides=(1, 1), padding='same')(input_tensor)
    branch_pool = Conv2D(filters, (1, 1), padding='same', activation='relu')(branch_pool)

    output = concatenate([branch1x1, branch3x3, branch5x5, branch_pool], axis=-1)
    return output
```

```
# Apply the inception module to an input tensor
input_tensor = tf.random.normal([1, 32, 32, 64])  # Example for a 32x32x64 input feature map
output_tensor = inception_module(input_tensor, filters=32)

# Print the shapes of the input and output tensors
print("Input shape:", input_tensor.shape)
print("Output shape:", output_tensor.shape)
```

Example in PyTorch

```
import torch
import torch.nn as nn

# Define an inception module
class InceptionModule(nn.Module):
    def __init__(self, in_channels, out_channels):
        super(InceptionModule, self).__init__()
        self.branch1x1 = nn.Conv2d(in_channels, out_channels, kernel_size=1)

        self.branch3x3 = nn.Conv2d(in_channels, out_channels, kernel_size=3, padding=1)

        self.branch5x5 = nn.Conv2d(in_channels, out_channels, kernel_size=5, padding=2)

        self.branch_pool = nn.Sequential(
            nn.MaxPool2d(kernel_size=3, stride=1, padding=1),
            nn.Conv2d(in_channels, out_channels, kernel_size=1)
        )

    def forward(self, x):
        branch1x1 = self.branch1x1(x)
        branch3x3 = self.branch3x3(x)
        branch5x5 = self.branch5x5(x)
        branch_pool = self.branch_pool(x)

        outputs = [branch1x1, branch3x3, branch5x5, branch_pool]
        return torch.cat(outputs, 1)

# Apply the inception module to an input tensor
input_tensor = torch.randn(1, 64, 32, 32)  # Example for a 32x32x64 input feature map
model = InceptionModule(64, 32)
output_tensor = model(input_tensor)

# Print the shapes of the input and output tensors
print("Input shape:", input_tensor.shape)
print("Output shape:", output_tensor.shape)
```

Attention Mechanisms

Attention mechanisms, particularly effective in sequence-based tasks, allow the model to focus on relevant parts of the input data while processing it. In CNNs, attention mechanisms can enhance the feature extraction process by weighting the importance of different spatial regions. An attention mechanism computes a weighted sum of the input features, where the weights are determined by an attention function. Given an input tensor $\mathbf{x} \in \mathbb{R}^{H \times W \times C}$, attention weights $\mathbf{a} \in \mathbb{R}^{H \times W \times C}$ are computed as:

$$\mathbf{a} = \text{softmax}(\mathbf{W}_a \mathbf{x} + \mathbf{b}_a)$$

where $\mathbf{W}_a \in \mathbb{R}^{1 \times 1 \times C}$ and $\mathbf{b}_a \in \mathbb{R}^{1 \times 1 \times C}$ are learnable parameters. The softmax function is applied element-wise over the spatial dimensions (height and width).

The output tensor $\mathbf{y} \in \mathbb{R}^{H \times W \times C}$ is then computed as:

$$\mathbf{y} = \mathbf{a} \odot \mathbf{x}$$

where $\odot$ denotes element-wise multiplication.

For example, consider a simple attention mechanism applied to a feature map to highlight important regions. Suppose the input tensor $\mathbf{x}$ has the dimensions $H = 4$, $W = 4$, and $C = 3$, i.e., $\mathbf{x} \in \mathbb{R}^{4 \times 4 \times 3}$. The attention weights are computed using the learnable parameters $\mathbf{W}_a \in \mathbb{R}^{1 \times 1 \times 3}$ and $\mathbf{b}_a \in \mathbb{R}^{1 \times 1 \times 3}$. For simplicity, assume $\mathbf{W}_a = \begin{bmatrix} w_1 & w_2 & w_3 \end{bmatrix}$ and $\mathbf{b}_a = \begin{bmatrix} b_1 & b_2 & b_3 \end{bmatrix}$. Then, the attention weights tensor $\mathbf{a}$ is calculated as $\mathbf{a} = \text{softmax}(\mathbf{W}_a \cdot \mathbf{x} + \mathbf{b}_a)$, resulting in $\mathbf{a} \in \mathbb{R}^{4 \times 4 \times 3}$. The output tensor $\mathbf{y}$ is obtained by element-wise multiplying the input tensor $\mathbf{x}$ with the attention weights $\mathbf{a}$, i.e., $\mathbf{y} = \mathbf{a} \odot \mathbf{x}$. Each element y_{ijk} of the output tensor $\mathbf{y}$ is computed as $y_{ijk} = a_{ijk} \cdot x_{ijk}$ for $i = 1, \ldots, 4$, $j = 1, \ldots, 4$, and $k = 1, \ldots, 3$. By applying this attention mechanism, the resulting tensor $\mathbf{y} \in \mathbb{R}^{4 \times 4 \times 3}$ highlights the important regions in the feature map, with the attention weights $\mathbf{a}$ determining the relevance of each spatial location in the input tensor $\mathbf{x}$.

Example in TensorFlow

```
import tensorflow as tf
from tensorflow.keras.layers import Conv2D, Softmax, Multiply

# Define an attention mechanism
def attention_module(input_tensor):
    attention_weights = Conv2D(1, (1, 1), activation='sigmoid')(input_tensor)
    output = Multiply()([input_tensor, attention_weights])
    return output

# Apply the attention mechanism to an input tensor
input_tensor = tf.random.normal([1, 32, 32, 64])  # Example for a 32x32x64 input feature map
output_tensor = attention_module(input_tensor)

# Print the shapes of the input and output tensors
print("Input shape:", input_tensor.shape)
print("Output shape:", output_tensor.shape)
```

Example in PyTorch

```
import torch
import torch.nn as nn

# Define an attention mechanism
class AttentionModule(nn.Module):
    def __init__(self, in_channels):
        super(AttentionModule, self).__init__()
        self.attention = nn.Conv2d(in_channels, 1, kernel_size=1)
        self.sigmoid = nn.Sigmoid()

    def forward(self, x):
        attention_weights = self.sigmoid(self.attention(x))
        return x * attention_weights

# Apply the attention mechanism to an input tensor
input_tensor = torch.randn(1, 64, 32, 32)  # Example for a 32x32x64 input feature map
model = AttentionModule(64)
output_tensor = model(input_tensor)
```

```
# Print the shapes of the input and output tensors
print("Input shape:", input_tensor.shape)
print("Output shape:", output_tensor.shape)
```

4.4 Advanced CNN Architectures

CNNs have seen remarkable advancements since their inception, with various architectures pushing the boundaries of performance and efficiency in image recognition tasks. One of the pioneering architectures that had a profound impact on the development of CNNs is AlexNet.

4.4.1 AlexNet

AlexNet, introduced by Krizhevsky et al. (2012), marked a significant breakthrough in the field of computer vision by winning the ImageNet Large Scale Visual Recognition Challenge (ILSVRC) (Deng et al. 2009) with a substantial margin. The architecture and design choices of AlexNet laid the foundation for many subsequent deep learning models.

Architecture and Design Choices
1. **Layer Configuration**: AlexNet consists of eight learned layers: five convolutional layers followed by three fully connected layers. The architecture also includes max-pooling layers and dropout layers to enhance its performance. The first two convolutional layers are followed by max-pooling layers. These layers use relatively large kernels (11×11 in the first layer, 5×5 in the second) and a stride of 4 for the first layer to aggressively reduce the spatial dimensions. The third, fourth, and fifth convolutional layers are directly connected without pooling in between. These layers use smaller kernels (3×3) and are designed to extract increasingly abstract features from the input images. AlexNet uses the Rectified Linear Unit (ReLU) activation function, which significantly accelerates the convergence of the training process compared to traditional sigmoid or tanh functions. Max-pooling layers follow the first and second convolutional layers to downsample the feature maps and reduce the computational load. The three fully connected layers at the end of the network perform high-level reasoning based on the extracted features. The final fully connected layer outputs a probability distribution over the classes using the softmax function.

2. **Regularization**: AlexNet employs dropout regularization in the fully connected layers to mitigate overfitting. Dropout randomly sets a fraction of the neurons to zero during training, preventing the network from becoming too reliant on any single neuron.

$$\text{Dropout}(h_i) = \begin{cases} 0 & \text{with probability } p \\ h_i & \text{with probability } 1 - p \end{cases}$$

where h_i is the activation of a neuron, and p is the dropout rate.

3. **Data Augmentation**: To further prevent overfitting and improve generalization, AlexNet utilizes data augmentation techniques such as random cropping, horizontal flipping, and color jittering. These augmentations artificially increase the diversity of the training data.

4. **Overlapping Pooling**: Instead of non-overlapping pooling, AlexNet uses overlapping pooling, where the stride is smaller than the pooling window size. This approach helps retain more information and provides better generalization.

The first convolutional layer applies 96 kernels of size $11 \times 11 \times 3$ with a stride of 4 and padding of 2. The second convolutional layer applies 256 kernels of size $5 \times 5x48$ with a stride of 1 and padding of 2. The third, fourth, and fifth convolutional layers use 384, 384, and 256 kernels of size 3×3 with a stride of 1 and padding of 1, respectively. The fully connected layers consist of 4096 neurons each, followed by a final layer with 1000 neurons for classification.

Example: Consider an input image of size $227 \times 227 \times 3$. The first convolutional layer with 96 kernels of size 11×11 and stride 4 reduces the spatial dimensions to $55 \times 55 \times 96$. After the first max-pooling layer (size 3×3, stride 2), the dimensions are further reduced to $27 \times 27 \times 96$. The process continues through the subsequent layers, progressively reducing the spatial dimensions and increasing the depth of the feature maps.

Example in TensorFlow

```
import tensorflow as tf
from tensorflow.keras.layers import Conv2D, MaxPooling2D, Flatten, Dense, Dropout
from tensorflow.keras.models import Sequential

# Define the AlexNet model
model = Sequential([
    Conv2D(96, (11, 11), strides=4, activation='relu', input_shape=(227, 227, 3),
         padding='same'),
    MaxPooling2D((3, 3), strides=2),
    Conv2D(256, (5, 5), activation='relu', padding='same'),
    MaxPooling2D((3, 3), strides=2),
    Conv2D(384, (3, 3), activation='relu', padding='same'),
    Conv2D(384, (3, 3), activation='relu', padding='same'),
    Conv2D(256, (3, 3), activation='relu', padding='same'),
    MaxPooling2D((3, 3), strides=2),
    Flatten(),
    Dense(4096, activation='relu'),
    Dropout(0.5),
    Dense(4096, activation='relu'),
    Dropout(0.5),
    Dense(1000, activation='softmax')
])

# Print the model summary
model.summary()
```

Example in PyTorch

```
import torch
import torch.nn as nn
import torch.nn.functional as F

class AlexNet(nn.Module):
    def __init__(self):
        super(AlexNet, self).__init__()
        self.conv1 = nn.Conv2d(3, 96, kernel_size=11, stride=4, padding=2)
        self.pool = nn.MaxPool2d(kernel_size=3, stride=2)
        self.conv2 = nn.Conv2d(96, 256, kernel_size=5, padding=2)
        self.conv3 = nn.Conv2d(256, 384, kernel_size=3, padding=1)
        self.conv4 = nn.Conv2d(384, 384, kernel_size=3, padding=1)
        self.conv5 = nn.Conv2d(384, 256, kernel_size=3, padding=1)
        self.fc1 = nn.Linear(256 * 6 * 6, 4096)
        self.fc2 = nn.Linear(4096, 4096)
        self.fc3 = nn.Linear(4096, 1000)
        self.dropout = nn.Dropout(0.5)

    def forward(self, x):
        x = F.relu(self.conv1(x))
        x = self.pool(x)
        x = F.relu(self.conv2(x))
        x = self.pool(x)
        x = F.relu(self.conv3(x))
        x = F.relu(self.conv4(x))
        x = F.relu(self.conv5(x))
        x = self.pool(x)
        x = torch.flatten(x, 1)
        x = F.relu(self.fc1(x))
        x = self.dropout(x)
        x = F.relu(self.fc2(x))
        x = self.dropout(x)
        x = self.fc3(x)
        return x

# Instantiate the model
model = AlexNet()

# Print the model summary
print(model)
```

Impact and Contributions

1. Breakthrough in ImageNet Challenge: AlexNet's victory in the 2012 ImageNet Challenge demonstrated the power of deep learning and CNNs, outperforming the previous state-of-the-art by a significant margin. This achievement brought widespread attention to deep learning and spurred a wave of research and development in the field.

2. ReLU Activation: The use of the ReLU activation function in AlexNet significantly improved training speed and performance, establishing ReLU as a standard activation function in deep learning.

3. GPU Utilization: AlexNet effectively utilized GPUs for training, highlighting the importance of hardware acceleration in deep learning. This led to the widespread adoption of GPUs and other specialized hardware for training deep neural networks.

4. Data Augmentation and Regularization: The innovative use of data augmentation techniques and dropout regularization in AlexNet set new standards for improving the generalization of deep learning models. These techniques are now commonly used in many deep learning architectures.

5. Influence on Subsequent Architectures: The design principles of AlexNet influenced the development of numerous subsequent architectures, including VGGNet, GoogLeNet, and ResNet. These architectures built upon the ideas introduced in AlexNet, leading to further advancements in performance and efficiency.

4.4.2 VGG Networks

VGG Networks, introduced by the Visual Geometry Group (VGG) at the University of Oxford (Simonyan and Zisserman 2014), are renowned for their simplicity and effectiveness in image recognition tasks. These networks use a straightforward architecture based on small (3×3) convolutional filters and deep networks with many layers. The VGG models demonstrated that deeper networks could lead to improved performance in image classification tasks.

Architecture
The primary architecture of VGG networks is characterized by its simplicity and depth. The VGG models consist of a series of convolutional layers with small receptive fields (3×3) and a fixed stride and padding. These layers are followed by max-pooling layers and a set of fully connected layers at the end. The key architectural choices include:

1. **Small Receptive Fields**: All convolutional layers use 3×3 filters with a stride of 1 and padding of 1. This choice ensures that the receptive field is small, but the depth of the network compensates for the limited size, allowing the network to learn complex features.

2. **Increasing Depth**: The VGG models increase the depth of the network by stacking more convolutional layers. The most common variants are VGG16 and VGG19, which have 16 and 19 weight layers, respectively.

3. **Max-Pooling**: Max-pooling layers with a 2×2 filter and a stride of 2 are used to reduce the spatial dimensions of the feature maps and control the computational load.

4. **Fully Connected Layers**: The convolutional and pooling layers are followed by three fully connected layers, with the first two layers containing 4096 neurons each and the third layer containing 1000 neurons for classification (for ImageNet).

Variants

1. **VGG16**: VGG16 has 16 weight layers: 13 convolutional layers and 3 fully connected layers. The configuration is as follows: Conv3-64, Conv3-64, Max-Pool; Conv3-128, Conv3-128, Max-Pool; Conv3-256, Conv3-256, Conv3-256, Max-Pool; Conv3-512, Conv3-512, Conv3-512, Max-Pool; Conv3-512, Conv3-512, Conv3-512, Max-Pool; 3 Fully Connected Layers.

2. **VGG19**: VGG19 extends VGG16 by adding more convolutional layers, resulting in 19 weight layers: 16 convolutional layers and 3 fully connected layers. The configuration is as follows: Conv3-64, Conv3-64, Max-Pool; Conv3-128, Conv3-128, Max-Pool; Conv3-256, Conv3-256, Conv3-256, Conv3-256, Max-Pool; Conv3-512,

Conv3-512, Conv3-512, Conv3-512, Max-Pool; Conv3-512, Conv3-512, Conv3-512, Conv3-512, Max-Pool; 3 Fully Connected Layers.

Example in TensorFlow

```
import tensorflow as tf
from tensorflow.keras.layers import Conv2D, MaxPooling2D, Flatten, Dense
from tensorflow.keras.models import Sequential

# Define the VGG16 model
model = Sequential([
    Conv2D(64, (3, 3), activation='relu', padding='same', input_shape=(224, 224, 3)),
    Conv2D(64, (3, 3), activation='relu', padding='same'),
    MaxPooling2D((2, 2)),
    Conv2D(128, (3, 3), activation='relu', padding='same'),
    Conv2D(128, (3, 3), activation='relu', padding='same'),
    MaxPooling2D((2, 2)),
    Conv2D(256, (3, 3), activation='relu', padding='same'),
    Conv2D(256, (3, 3), activation='relu', padding='same'),
    Conv2D(256, (3, 3), activation='relu', padding='same'),
    MaxPooling2D((2, 2)),
    Conv2D(512, (3, 3), activation='relu', padding='same'),
    Conv2D(512, (3, 3), activation='relu', padding='same'),
    Conv2D(512, (3, 3), activation='relu', padding='same'),
    MaxPooling2D((2, 2)),
    Conv2D(512, (3, 3), activation='relu', padding='same'),
    Conv2D(512, (3, 3), activation='relu', padding='same'),
    Conv2D(512, (3, 3), activation='relu', padding='same'),
    MaxPooling2D((2, 2)),
    Flatten(),
    Dense(4096, activation='relu'),
    Dense(4096, activation='relu'),
    Dense(1000, activation='softmax')
])

# Print the model summary
model.summary()
```

Example in PyTorch

```
import torch
import torch.nn as nn
import torch.nn.functional as F

class VGG16(nn.Module):
    def __init__(self):
        super(VGG16, self).__init__()
        self.conv1_1 = nn.Conv2d(3, 64, kernel_size=3, padding=1)
        self.conv1_2 = nn.Conv2d(64, 64, kernel_size=3, padding=1)
        self.pool1 = nn.MaxPool2d(kernel_size=2, stride=2)

        self.conv2_1 = nn.Conv2d(64, 128, kernel_size=3, padding=1)
        self.conv2_2 = nn.Conv2d(128, 128, kernel_size=3, padding=1)
        self.pool2 = nn.MaxPool2d(kernel_size=2, stride=2)

        self.conv3_1 = nn.Conv2d(128, 256, kernel_size=3, padding=1)
        self.conv3_2 = nn.Conv2d(256, 256, kernel_size=3, padding=1)
        self.conv3_3 = nn.Conv2d(256, 256, kernel_size=3, padding=1)
        self.pool3 = nn.MaxPool2d(kernel_size=2, stride=2)

        self.conv4_1 = nn.Conv2d(256, 512, kernel_size=3, padding=1)
        self.conv4_2 = nn.Conv2d(512, 512, kernel_size=3, padding=1)
        self.conv4_3 = nn.Conv2d(512, 512, kernel_size=3, padding=1)
        self.pool4 = nn.MaxPool2d(kernel_size=2, stride=2)

        self.conv5_1 = nn.Conv2d(512, 512, kernel_size=3, padding=1)
```

```
        self.conv5_2 = nn.Conv2d(512, 512, kernel_size=3, padding=1)
        self.conv5_3 = nn.Conv2d(512, 512, kernel_size=3, padding=1)
        self.pool5 = nn.MaxPool2d(kernel_size=2, stride=2)

        self.fc1 = nn.Linear(512 * 7 * 7, 4096)
        self.fc2 = nn.Linear(4096, 4096)
        self.fc3 = nn.Linear(4096, 1000)

    def forward(self, x):
        x = F.relu(self.conv1_1(x))
        x = F.relu(self.conv1_2(x))
        x = self.pool1(x)

        x = F.relu(self.conv2_1(x))
        x = F.relu(self.conv2_2(x))
        x = self.pool2(x)

        x = F.relu(self.conv3_1(x))
        x = F.relu(self.conv3_2(x))
        x = F.relu(self.conv3_3(x))
        x = self.pool3(x)

        x = F.relu(self.conv4_1(x))
        x = F.relu(self.conv4_2(x))
        x = F.relu(self.conv4_3(x))
        x = self.pool4(x)

        x = F.relu(self.conv5_1(x))
        x = F.relu(self.conv5_2(x))
        x = F.relu(self.conv5_3(x))
        x = self.pool5(x)

        x = x.view(-1, 512 * 7 * 7)
        x = F.relu(self.fc1(x))
        x = F.dropout(x, 0.5, training=self.training)
        x = F.relu(self.fc2(x))
        x = F.dropout(x, 0.5, training=self.training)
        x = self.fc3(x)
        return x

# Instantiate the model
model = VGG16()

# Print the model summary
print(model)
```

Performance and Use Cases

VGG networks, particularly VGG16 and VGG19, have demonstrated excellent performance on various image classification tasks. They achieved top-5 error rates of 7.3% and 6.8%, respectively, in the ILSVRC-2014 competition. The key strengths of VGG networks include their simplicity and the use of small convolutional filters, which make them easy to implement and understand.

Advantages:

1. Simplicity and Modularity: The consistent use of 3×3 convolutional filters makes VGG networks highly modular and easy to replicate. This simplicity has contributed to their widespread adoption and adaptation in various applications.

2. Effective Feature Extraction: The deep architecture allows VGG networks to extract rich hierarchical features from input images, which can be beneficial for various computer vision tasks.

Disadvantages: VGG networks have a large number of parameters, leading to high computational and memory demands. This can be a limitation when deploying the model on resource-constrained devices.

Use Cases:

1. Image Classification: VGG networks are widely used for image classification tasks, where they have demonstrated state-of-the-art performance.

2. Feature Extraction for Transfer Learning: The learned features from pre-trained VGG models can be used as feature extractors in other tasks, such as object detection, segmentation, and image generation.

3. Medical Imaging: VGG networks have been applied to various medical imaging tasks, including the classification of medical images, detection of anomalies, and segmentation of medical scans.

4.4.3 ResNet

ResNet, or Residual Networks, introduced by He et al. (2015b), addressed the challenge of training very deep neural networks. ResNet's key innovation is the use of residual blocks, which allow the network to learn residual functions with reference to the layer inputs, thereby mitigating the vanishing gradient problem and enabling the training of much deeper networks.

Residual Blocks and Identity Mapping

The fundamental building block of ResNet is the residual block, which consists of two or more convolutional layers and an identity shortcut connection that bypasses these layers. The residual block explicitly allows the network to learn an identity mapping by skipping layers, making it easier for the optimizer to maintain the learned information as the network depth increases. Given an input $\mathbf{x}$, a residual block computes:

$$\mathbf{y} = \mathcal{F}(\mathbf{x}, \{\mathbf{W}_i\}) + \mathbf{x}$$

where $\mathbf{y}$ is the output, $\mathcal{F}(\mathbf{x}, \{\mathbf{W}_i\})$ represents the operations performed by the layers within the block (e.g., convolution, batch normalization, ReLU), $\mathbf{W}_i$ are the learnable weights. In a basic residual block with two convolutional layers, the function $\mathcal{F}$ can be expressed as:

$$\mathcal{F}(\mathbf{x}) = \mathbf{W}_2 * \text{ReLU}(\text{BN}(\mathbf{W}_1 * \mathbf{x}))$$

where $*$ denotes convolution, BN denotes batch normalization, and ReLU is the activation function. If the dimensions of $\mathbf{x}$ and $\mathcal{F}(\mathbf{x})$ do not match, a linear projection (e.g., 1×1 convolution) is used to match dimensions:

$$\mathbf{y} = \mathcal{F}(\mathbf{x}, \{\mathbf{W}_i\}) + \mathbf{W}_s\mathbf{x}$$

where $\mathbf{W}_s$ is the projection matrix. The identity mapping in residual blocks allows gradients to flow directly through the network, which helps in mitigating the vanishing gradient problem. This is crucial for training very deep networks, as it ensures that the information can be propagated backward without degradation. For example, consider a residual block with two convolutional layers. The operations can be expressed as:

$$\mathbf{y} = \text{ReLU}(\text{BN}(\mathbf{W}_2 * (\text{ReLU}(\text{BN}(\mathbf{W}_1 * \mathbf{x})))) + \mathbf{x})$$

where $*$ denotes convolution, BN denotes batch normalization, and ReLU is the activation function.

Example in TensorFlow

```
import tensorflow as tf
from tensorflow.keras.layers import Conv2D, BatchNormalization, ReLU, Add

# Define a residual block
def residual_block(input_tensor, filters, kernel_size=3, stride=1,
 projection_shortcut=False):
    x = Conv2D(filters, kernel_size, strides=stride, padding='same')(input_tensor)
    x = BatchNormalization()(x)
    x = ReLU()(x)
    x = Conv2D(filters, kernel_size, strides=1, padding='same')(x)
    x = BatchNormalization()(x)

    if projection_shortcut:
        shortcut = Conv2D(filters, kernel_size=1, strides=stride)(input_tensor)
        shortcut = BatchNormalization()(shortcut)
    else:
        shortcut = input_tensor

    x = Add()([x, shortcut])
    return ReLU()(x)

# Apply the residual block to an input tensor
input_tensor = tf.random.normal([1, 32, 32, 64])  # Example for a 32x32x64 input feature map
output_tensor = residual_block(input_tensor, filters=64, projection_shortcut=True)

# Print the shapes of the input and output tensors
print("Input shape:", input_tensor.shape)
print("Output shape:", output_tensor.shape)
```

Example in PyTorch

```
import torch
import torch.nn as nn

# Define a residual block
class ResidualBlock(nn.Module):
    def __init__(self, in_channels, out_channels, stride=1, projection_shortcut=False):
        super(ResidualBlock, self).__init__()
        self.conv1 = nn.Conv2d(in_channels, out_channels, kernel_size=3, stride=stride, padding=1)
        self.bn1 = nn.BatchNorm2d(out_channels)
        self.relu = nn.ReLU()
        self.conv2 = nn.Conv2d(out_channels, out_channels, kernel_size=3, padding=1)
        self.bn2 = nn.BatchNorm2d(out_channels)
        self.projection_shortcut = projection_shortcut
        if self.projection_shortcut:
            self.shortcut = nn.Sequential(
                nn.Conv2d(in_channels, out_channels, kernel_size=1, stride=stride),
```

```
                nn.BatchNorm2d(out_channels)
            )
        else:
            self.shortcut = nn.Identity()

    def forward(self, x):
        identity = x
        out = self.conv1(x)
        out = self.bn1(out)
        out = self.relu(out)
        out = self.conv2(out)
        out = self.bn2(out)
        if self.projection_shortcut:
            identity = self.shortcut(x)
        out += identity
        return self.relu(out)

# Apply the residual block to an input tensor
input_tensor = torch.randn(1, 64, 32, 32)  # Example for a 32x32x64 input feature map
model = ResidualBlock(64, 64, projection_shortcut=True)
output_tensor = model(input_tensor)

# Print the shapes of the input and output tensors
print("Input shape:", input_tensor.shape)
print("Output shape:", output_tensor.shape)
```

Performance and Use Cases

ResNet models, particularly ResNet50, ResNet101, and ResNet152, have achieved remarkable performance on various image recognition tasks, including winning the ILSVRC 2015 challenge. The key strengths of ResNet include its ability to train very deep networks, robust feature learning, and improved gradient flow.

Advantages:

1. Training Deep Networks: Residual connections mitigate the vanishing gradient problem, enabling the training of networks with hundreds or even thousands of layers.
2. Improved Gradient Flow: The identity mappings in residual blocks ensure that gradients can flow through the network without degradation, leading to more stable and efficient training.
3. Robust Feature Learning: The deep architecture of ResNet models allows them to learn rich hierarchical features, which can be beneficial for various computer vision tasks.

Disadvantages: The deep architecture of ResNet models can lead to high computational and memory demands, especially when training very deep networks.

Use Cases:

1. Image Classification: ResNet models are widely used for image classification tasks, achieving state-of-the-art performance on datasets like ImageNet.
2. Object Detection and Segmentation: The robust feature learning capabilities of ResNet models make them suitable for object detection and segmentation tasks, where accurate localization and classification of objects are required.
3. Feature Extraction for Transfer Learning: Pre-trained ResNet models can be used as feature extractors for transfer learning in various applications, such as medical imaging, autonomous driving, and more.

Motivating Example: Consider the task of classifying images from the CIFAR-10 dataset, which contains 60,000 images across 10 classes. Traditional deep learning models struggle with this task due to the limited size of the dataset and the complexity of the images. ResNet models, with their deep architecture and residual connections, can learn rich hierarchical features and achieve high classification accuracy. The identity mappings in residual blocks ensure that gradients flow smoothly through the network, enabling efficient training even with limited data.

4.4.4 Inception Networks

Inception Networks, introduced by Szegedy et al. (2015) in the GoogLeNet architecture, brought a novel approach to network design by incorporating multi-scale processing within the same layer. These networks leverage inception modules to capture features at different scales, improving the network's ability to handle various types of image data and reducing the computational burden compared to traditional deep networks.

Inception Modules and Design

The inception module is the core building block of Inception Networks. It allows the network to process information at multiple scales simultaneously by performing several convolutions and pooling operations in parallel, followed by concatenation of the resulting feature maps. This design helps the network capture both fine and coarse features, enhancing its representational capacity. An inception module typically consists of the following branches:

1. 1×1 Convolution: This branch performs a convolution with a 1×1 kernel, which serves to reduce the dimensionality (number of channels) and introduces nonlinearity.

$$\mathbf{O}_1 = \text{ReLU}(\mathbf{W}_{1\times 1} * \mathbf{x})$$

2. 1×1 Convolution followed by 3×3 Convolution: This branch first reduces the dimensionality with a 1×1 convolution, followed by a 3×3 convolution to capture spatial features.

$$\mathbf{O}_2 = \text{ReLU}(\mathbf{W}_{3\times 3} * \text{ReLU}(\mathbf{W}_{1\times 1} * \mathbf{x}))$$

3. 1×1 Convolution followed by 5×5 Convolution: Similar to the previous branch, but uses a larger 5×5 convolution for a broader spatial context.

$$\mathbf{O}_3 = \text{ReLU}(\mathbf{W}_{5\times 5} * \text{ReLU}(\mathbf{W}_{1\times 1} * \mathbf{x}))$$

4. 3×3 Max-Pooling followed by 1×1 Convolution: This branch performs max-pooling to capture dominant features and reduces dimensionality using a 1×1 convolution.

$$\mathbf{O}_4 = \text{ReLU}(\mathbf{W}_{1\times1} * \text{Max-Pool}_{3\times3}(\mathbf{x}))$$

The outputs of these branches are concatenated along the channel dimension to form the final output of the inception module:

$$\mathbf{O} = \bigoplus([\mathbf{O}_1, \mathbf{O}_2, \mathbf{O}_3, \mathbf{O}_4], \text{axis} = -1)$$

Design Choices:

1. Dimensionality Reduction: The use of 1×1 convolutions helps reduce the number of parameters and computational cost, making the network more efficient without compromising on the capacity to learn complex representations.
2. Multi-scale Feature Extraction: By combining convolutions of different sizes and pooling operations, the inception module enables the network to extract features at multiple scales, which is beneficial for recognizing objects of varying sizes and shapes in images.
3. Parallel Processing: The parallel structure of inception modules allows the network to perform multiple operations simultaneously, which can be efficiently executed on modern hardware, leveraging the parallel processing capabilities of GPUs.

Example in TensorFlow

```
import tensorflow as tf
from tensorflow.keras.layers import Conv2D, MaxPooling2D, concatenate

# Define an inception module
def inception_module(input_tensor, filters):
    branch1x1 = Conv2D(filters, (1, 1), padding='same', activation='relu')(input_tensor)

    branch3x3 = Conv2D(filters, (1, 1), padding='same', activation='relu')(input_tensor)
    branch3x3 = Conv2D(filters, (3, 3), padding='same', activation='relu')(branch3x3)

    branch5x5 = Conv2D(filters, (1, 1), padding='same', activation='relu')(input_tensor)
    branch5x5 = Conv2D(filters, (5, 5), padding='same', activation='relu')(branch5x5)

    branch_pool = MaxPooling2D((3, 3), strides=(1, 1), padding='same')(input_tensor)
    branch_pool = Conv2D(filters, (1, 1), padding='same', activation='relu')(branch_pool)

    output = concatenate([branch1x1, branch3x3, branch5x5, branch_pool], axis=-1)
    return output

# Apply the inception module to an input tensor
input_tensor = tf.random.normal([1, 28, 28, 256])  # Example for a 28x28x256 input feature map
output_tensor = inception_module(input_tensor, filters=64)

# Print the shapes of the input and output tensors
print("Input shape:", input_tensor.shape)
print("Output shape:", output_tensor.shape)
```

Example in PyTorch

```
import torch
import torch.nn as nn

# Define an inception module
class InceptionModule(nn.Module):
    def __init__(self, in_channels, out_channels):
        super(InceptionModule, self).__init__()
        self.branch1x1 = nn.Conv2d(in_channels, out_channels, kernel_size=1)

        self.branch3x3_1 = nn.Conv2d(in_channels, out_channels, kernel_size=1)
        self.branch3x3_2 = nn.Conv2d(out_channels, out_channels, kernel_size=3, padding=1)

        self.branch5x5_1 = nn.Conv2d(in_channels, out_channels, kernel_size=1)
        self.branch5x5_2 = nn.Conv2d(out_channels, out_channels, kernel_size=5, padding=2)

        self.branch_pool = nn.Sequential(
            nn.MaxPool2d(kernel_size=3, stride=1, padding=1),
            nn.Conv2d(in_channels, out_channels, kernel_size=1)
        )

    def forward(self, x):
        branch1x1 = self.branch1x1(x)

        branch3x3 = self.branch3x3_1(x)
        branch3x3 = self.branch3x3_2(branch3x3)

        branch5x5 = self.branch5x5_1(x)
        branch5x5 = self.branch5x5_2(branch5x5)

        branch_pool = self.branch_pool(x)

        outputs = [branch1x1, branch3x3, branch5x5, branch_pool]
        return torch.cat(outputs, 1)

# Apply the inception module to an input tensor
input_tensor = torch.randn(1, 256, 28, 28)  # Example for a 28x28x256 input feature map
model = InceptionModule(256, 64)
output_tensor = model(input_tensor)

# Print the shapes of the input and output tensors
print("Input shape:", input_tensor.shape)
print("Output shape:", output_tensor.shape)
```

Performance and Use Cases

Inception Networks, particularly GoogLeNet (Inception v1), Inception v2, and Inception v3, have demonstrated outstanding performance in various image recognition tasks. GoogLeNet achieved a top-5 error rate of 6.67% in the ILSVRC 2014 competition, significantly improving upon previous architectures.

Advantages:

1. Efficient Use of Parameters: Inception modules allow the network to be deeper and wider without a significant increase in computational cost. The use of 1×1 convolutions for dimensionality reduction is particularly effective in controlling the number of parameters.
2. Multi-scale Feature Learning: By processing information at multiple scales, inception modules enhance the network's ability to capture diverse features, making it more robust to variations in the input data.

Disadvantages: The intricate structure of inception modules can make the design and implementation of Inception Networks more complex compared to traditional CNNs.

Use Cases:

1. Image Classification: Inception Networks are widely used for image classification tasks, achieving state-of-the-art performance on datasets like ImageNet.
2. Object Detection and Segmentation: The multi-scale processing capabilities of inception modules make them suitable for object detection and segmentation tasks, where it is essential to capture features at different resolutions.
3. Feature Extraction for Transfer Learning: Pre-trained Inception Networks can be used as feature extractors for transfer learning in various applications, such as medical imaging, autonomous driving, and more.

4.4.5 MobileNet and EfficientNet

MobileNet and EfficientNet are two prominent architectures designed to achieve high performance with lightweight and efficient models, making them suitable for deployment on mobile and edge devices. These architectures emphasize the trade-off between accuracy and computational cost, leveraging innovative techniques to optimize model size and inference speed.

MobileNet
MobileNet, introduced by Howard et al. (2017), is designed specifically for mobile and embedded vision applications. The core idea behind MobileNet is to use depthwise separable convolutions to build lightweight deep neural networks. Depthwise separable convolutions factorize a standard convolution into two simpler operations: a depthwise convolution and a pointwise convolution. This reduces the computational complexity and the number of parameters significantly.

Depthwise Convolution In a depthwise convolution, each input channel is filtered independently using a single convolutional filter. Given an input tensor $\mathbf{X}$ with dimensions $H \times W \times C$ and depthwise filters $\mathbf{K}$ of size $k \times k \times C$, the depthwise convolution output is:

$$\mathbf{Y}_{\text{depthwise}}(i, j, c) = \sum_{m=0}^{k-1}\sum_{n=0}^{k-1}\mathbf{X}(i+m, j+n, c)\mathbf{K}(m, n, c)$$

Here, H and W are the height and width of the input tensor, C is the number of channels, and k is the kernel size. Each channel c is convolved independently with its corresponding kernel.

Pointwise Convolution A pointwise convolution applies a 1×1 convolution to combine the output of the depthwise convolution across the channels. Given a pointwise filter $\mathbf{W}$ of size $1 \times 1 \times C \times N$, the output is:

$$\mathbf{Y}_{\text{pointwise}}(i, j, n) = \sum_{c=0}^{C-1} \mathbf{Y}_{\text{depthwise}}(i, j, c)\mathbf{W}(1, 1, c, n)$$

Here, N is the number of output channels.

The computational complexity of a standard convolution is:

$$D_{\text{standard}} = H \times W \times C \times N \times k^2$$

For depthwise separable convolutions, the complexity is:

$$D_{\text{depthwise}} = H \times W \times C \times k^2$$

$$D_{\text{pointwise}} = H \times W \times C \times N$$

$$D_{\text{total}} = D_{\text{depthwise}} + D_{\text{pointwise}}$$

This results in a significant reduction in computational cost, especially when k is small.

MobileNetV2 introduced the concept of inverted residuals and linear bottlenecks, further improving the efficiency of the architecture.

Inverted Residuals: Inverted residuals use a shortcut connection between the input and output of a sequence of layers, but with an expanded representation in between. This helps preserve information and gradient flow.

Linear Bottlenecks: Linear bottlenecks use linear activation functions at the end of each residual block to avoid non-linear transformations that can destroy information.

Example: Consider an input tensor of size $224 \times 224 \times 3$. A MobileNet architecture can include: Depthwise convolution with 32 filters, Pointwise convolution with 64 filters, Inverted residual block with 128 filters, and a stride of 2.

Example in TensorFlow

```
import tensorflow as tf
from tensorflow.keras.layers import DepthwiseConv2D, Conv2D, BatchNormalization, ReLU, Add

# Define a MobileNet block
def mobilenet_block(input_tensor, filters, strides=1):
    x = DepthwiseConv2D(kernel_size=3, strides=strides, padding='same')(input_tensor)
    x = BatchNormalization()(x)
    x = ReLU()(x)
    x = Conv2D(filters, kernel_size=1, padding='same')(x)
    x = BatchNormalization()(x)
    return ReLU()(x)

# Apply the MobileNet block to an input tensor
input_tensor = tf.random.normal([1, 224, 224, 3])  # Example for a 224x224x3 input image
output_tensor = mobilenet_block(input_tensor, filters=64, strides=2)

# Print the shapes of the input and output tensors
print("Input shape:", input_tensor.shape)
print("Output shape:", output_tensor.shape)
```

Example in PyTorch

```
import torch
import torch.nn as nn

# Define a MobileNet block
class MobileNetBlock(nn.Module):
    def __init__(self, in_channels, out_channels, stride=1):
        super(MobileNetBlock, self).__init__()
        self.depthwise = nn.Conv2d(in_channels, in_channels, kernel_size=3, stride=stride,
                padding=1, groups=in_channels)
        self.pointwise = nn.Conv2d(in_channels, out_channels, kernel_size=1)
        self.bn = nn.BatchNorm2d(out_channels)
        self.relu = nn.ReLU()

    def forward(self, x):
        x = self.relu(self.bn(self.depthwise(x)))
        x = self.relu(self.bn(self.pointwise(x)))
        return x

# Apply the MobileNet block to an input tensor
input_tensor = torch.randn(1, 3, 224, 224)  # Example for a 224x224x3 input image
model = MobileNetBlock(3, 64, stride=2)
output_tensor = model(input_tensor)

# Print the shapes of the input and output tensors
print("Input shape:", input_tensor.shape)
print("Output shape:", output_tensor.shape)
```

EfficientNet

EfficientNet, introduced by Tan and Le (2019), uses a compound scaling method to systematically scale up the network's width, depth, and resolution. This approach achieves better accuracy and efficiency by optimizing all three dimensions simultaneously.

Compound Scaling The compound scaling method uses a set of scaling coefficients to determine how to scale each dimension. Given a baseline network, the scaled version can be described as:

$$d = \alpha^n, \quad w = \beta^n, \quad r = \gamma^n$$

where d is the network depth, w is the width, r is the input resolution, and α, β, and γ are the scaling coefficients. n is a user-defined compound coefficient. For example, consider EfficientNet-B0 as the baseline network. EfficientNet-B1 to B7 are scaled versions with increasing depth, width, and resolution. For instance, EfficientNet-B1 might use: Depth multiplier $\alpha = 1.2$, Width multiplier $\beta = 1.1$, and Resolution multiplier $\gamma = 1.15$.

Example in TensorFlow

```
import tensorflow as tf
from tensorflow.keras.applications import EfficientNetB0

# Load the EfficientNetB0 model
model = EfficientNetB0(weights='imagenet')

# Print the model summary
model.summary()
```

Example in PyTorch

```
import torch
import torchvision.models as models

# Load the EfficientNetB0 model
model = models.efficientnet_b0(pretrained=True)

# Print the model summary
print(model)
```

Performance and Applications

Both MobileNet and EfficientNet achieve high accuracy while maintaining low computational complexity and memory usage, making them ideal for mobile and embedded applications.

Advantages:

1. Efficiency: MobileNet and EfficientNet are designed to be efficient, with fewer parameters and reduced computational cost compared to traditional CNNs.
2. Scalability: EfficientNet's compound scaling method allows for systematic scaling of network dimensions, enabling the creation of models that balance accuracy and efficiency.
3. Flexibility: These models can be easily adapted for various applications, from image classification to object detection and segmentation.

Disadvantages: While efficient, these models may have lower capacity compared to larger, more complex architectures, potentially limiting their performance on very large or complex datasets.

Use Cases:

1. Mobile and Edge Devices: MobileNet and EfficientNet are widely used in applications requiring real-time processing on mobile and edge devices, such as facial recognition, augmented reality, and mobile health diagnostics.
2. Embedded Systems: These models are suitable for deployment on embedded systems, such as smart cameras, IoT devices, and autonomous drones, where computational resources are limited.
3. Transfer Learning: Pre-trained MobileNet and EfficientNet models are commonly used for transfer learning in various applications, providing a strong starting point for custom tasks with limited data.

Motivating Example: Consider the task of real-time object detection on a smartphone. Traditional deep learning models are too resource-intensive for such applications. MobileNet, with its depthwise separable convolutions, offers a lightweight alternative that can run efficiently on mobile hardware without sacrificing too much accuracy. Similarly, EfficientNet provides a scalable solution that can be tailored to the specific requirements of the application, achieving a balance between performance and resource usage.

4.4.6 *Other Notable Architectures*

In addition to MobileNet and EfficientNet, several other innovative architectures have made significant contributions to the field of deep learning. This section explores DenseNet, SqueezeNet, and Capsule Networks, highlighting their unique design principles and their impact on deep learning.

DenseNet
DenseNet, introduced by Huang et al. (2016), addresses the issue of vanishing gradients and enhances feature propagation by connecting each layer to every other layer in a feed-forward fashion. This dense connectivity pattern encourages feature reuse and reduces the number of parameters.

Dense Connectivity In DenseNet, each layer receives inputs from all preceding layers and passes its own feature maps to all subsequent layers. Formally, the output of the lth layer is defined as:

$$\mathbf{x}_l = H_l([\mathbf{x}_0, \mathbf{x}_1, \ldots, \mathbf{x}_{l-1}])$$

where $[\mathbf{x}_0, \mathbf{x}_1, \ldots, \mathbf{x}_{l-1}]$ denotes the concatenation of feature maps from layers 0 to $l-1$, and H_l is a composite function of operations such as Batch Normalization (BN), ReLU, and convolution.

Bottleneck Layers and Compression To improve computational efficiency, DenseNet employs bottleneck layers (1×1 convolutions) before the 3×3 convolutions to reduce the number of input feature maps. Additionally, transition layers between dense blocks use 1×1 convolutions followed by 2×2 average pooling to control the complexity and size of the model.

Given an input tensor $\mathbf{X}$ and filters $\mathbf{K}$ for a dense layer, the operations can be expressed as:

$$\mathbf{Y} = \text{BN}(\mathbf{X})$$

$$\mathbf{Y} = \text{ReLU}(\mathbf{Y})$$

$$\mathbf{Y} = \mathbf{K}_{1\times1} * \mathbf{Y}$$

$$\mathbf{Y} = \text{BN}(\mathbf{Y})$$

$$\mathbf{Y} = \text{ReLU}(\mathbf{Y})$$

$$\mathbf{Y} = \mathbf{K}_{3\times3} * \mathbf{Y}$$

where $*$ denotes convolution.

Example Consider a DenseNet architecture with dense blocks and transition layers. Each dense block comprises multiple dense layers, and the transition layers reduce the number of feature maps and spatial dimensions.

Example in TensorFlow

```
import tensorflow as tf
from tensorflow.keras.layers import Conv2D, BatchNormalization, ReLU, Concatenate,
 AveragePooling2D, Input
from tensorflow.keras.models import Model

# Define a dense block
def dense_block(x, num_layers, growth_rate):
    for i in range(num_layers):
        y = BatchNormalization()(x)
        y = ReLU()(y)
        y = Conv2D(4 * growth_rate, (1, 1), padding='same')(y)
        y = BatchNormalization()(y)
        y = ReLU()(y)
        y = Conv2D(growth_rate, (3, 3), padding='same')(y)
        x = Concatenate()([x, y])
    return x

# Define a transition layer
def transition_layer(x, reduction):
    x = BatchNormalization()(x)
    x = Conv2D(int(x.shape[-1] * reduction), (1, 1), padding='same')(x)
    x = AveragePooling2D((2, 2), strides=2)(x)
    return x

# Apply the dense block and transition layer to an input tensor
input_tensor = Input(shape=(32, 32, 64))  # Example input shape
x = dense_block(input_tensor, num_layers=4, growth_rate=32)
output_tensor = transition_layer(x, reduction=0.5)

# Define the model
model = Model(inputs=input_tensor, outputs=output_tensor)

# Print the model summary
model.summary()
```

Example in PyTorch

```
import torch
import torch.nn as nn

# Define a dense layer
class DenseLayer(nn.Module):
    def __init__(self, in_channels, growth_rate):
        super(DenseLayer, self).__init__()
        self.bn1 = nn.BatchNorm2d(in_channels)
        self.relu = nn.ReLU()
        self.conv1 = nn.Conv2d(in_channels, 4 * growth_rate, kernel_size=1, bias=False)
        self.bn2 = nn.BatchNorm2d(4 * growth_rate)
        self.conv2 = nn.Conv2d(4 * growth_rate, growth_rate, kernel_size=3, padding=1,
                bias=False)

    def forward(self, x):
        out = self.conv1(self.relu(self.bn1(x)))
        out = self.conv2(self.relu(self.bn2(out)))
        out = torch.cat([x, out], 1)
        return out

# Define a dense block
class DenseBlock(nn.Module):
```

```
    def __init__(self, num_layers, in_channels, growth_rate):
        super(DenseBlock, self).__init__()
        layers = []
        for i in range(num_layers):
            layers.append(DenseLayer(in_channels + i * growth_rate, growth_rate))
        self.denseblock = nn.Sequential(*layers)

    def forward(self, x):
        return self.denseblock(x)

# Define a transition layer
class TransitionLayer(nn.Module):
    def __init__(self, in_channels, out_channels):
        super(TransitionLayer, self).__init__()
        self.bn = nn.BatchNorm2d(in_channels)
        self.conv = nn.Conv2d(in_channels, out_channels, kernel_size=1, bias=False)
        self.pool = nn.AvgPool2d(kernel_size=2, stride=2)

    def forward(self, x):
        x = self.conv(self.bn(x))
        x = self.pool(x)
        return x

# Apply the dense block and transition layer to an input tensor
input_tensor = torch.randn(1, 64, 32, 32)  # Example input shape
model = nn.Sequential(
    DenseBlock(num_layers=4, in_channels=64, growth_rate=32),
    TransitionLayer(in_channels=192, out_channels=96)
)
output_tensor = model(input_tensor)

# Print the shapes of the input and output tensors
print("Input shape:", input_tensor.shape)
print("Output shape:", output_tensor.shape)
```

SqueezeNet

SqueezeNet, introduced by Iandola et al. (2016), aims to achieve AlexNet-level accuracy with significantly fewer parameters. This is accomplished through the use of Fire modules and aggressive parameter reduction techniques.

Fire Modules Fire modules are the key building blocks of SqueezeNet. Each Fire module consists of a squeeze layer and an expand layer.

The squeeze layer uses 1×1 convolutions to reduce the number of input channels, significantly decreasing the number of parameters.

$$\mathbf{Y}_{\text{squeeze}} = \text{ReLU}(\mathbf{W}_{1\times1} * \mathbf{X})$$

The expand layer uses both 1×1 and 3×3 convolutions to process the reduced input from the squeeze layer. This dual-path approach allows for a rich combination of features.

$$\mathbf{Y}_{\text{expand1}} = \text{ReLU}(\mathbf{W}_{1\times1} * \mathbf{Y}_{\text{squeeze}})$$

$$\mathbf{Y}_{\text{expand3}} = \text{ReLU}(\mathbf{W}_{3\times3} * \mathbf{Y}_{\text{squeeze}})$$

The outputs of the 1×1 and 3×3 convolutions are concatenated to form the final output of the Fire module.

$$\mathbf{Y} = \bigoplus([\mathbf{Y}_{\text{expand1}}, \mathbf{Y}_{\text{expand3}}], \text{axis} = -1)$$

Parameter Reduction SqueezeNet reduces the number of parameters through the extensive use of 1×1 convolutions and by decreasing the number of input channels to 3×3 convolutions.

Example Consider a Fire module with an input tensor of size $224 \times 224x64$. The squeeze layer might use 16 filters, while the expand layer uses 32 filters for both 1×1 and 3×3 convolutions.

Example in TensorFlow

```
import tensorflow as tf
from tensorflow.keras.layers import Conv2D, Concatenate, ReLU

# Define a Fire module
def fire_module(input_tensor, squeeze_filters, expand_filters):
    squeeze = Conv2D(squeeze_filters, (1, 1), activation='relu', padding='same')
        (input_tensor)
    expand1x1 = Conv2D(expand_filters, (1, 1), activation='relu', padding='same')(squeeze)
    expand3x3 = Conv2D(expand_filters, (3, 3), activation='relu', padding='same')(squeeze)
    output = Concatenate(axis=-1)([expand1x1, expand3x3])
    return output

# Apply the Fire module to an input tensor
input_tensor = tf.random.normal([1, 224, 224, 64])  # Example for a 224x224x64 input image
output_tensor = fire_module(input_tensor, squeeze_filters=16, expand_filters=32)

# Print the shapes of the input and output tensors
print("Input shape:", input_tensor.shape)
print("Output shape:", output_tensor.shape)
```

Example in PyTorch

```
import torch
import torch.nn as nn

# Define a Fire module
class FireModule(nn.Module):
    def __init__(self, in_channels, squeeze_channels, expand_channels):
        super(FireModule, self).__init__()
        self.squeeze = nn.Conv2d(in_channels, squeeze_channels, kernel_size=1)
        self.expand1x1 = nn.Conv2d(squeeze_channels, expand_channels, kernel_size=1)
        self.expand3x3 = nn.Conv2d(squeeze_channels, expand_channels, kernel_size=3,
                padding=1)
        self.relu = nn.ReLU()

    def forward(self, x):
        x = self.relu(self.squeeze(x))
        return torch.cat([self.relu(self.expand1x1(x)), self.relu(self.expand3x3(x))], 1)

# Apply the Fire module to an input tensor
input_tensor = torch.randn(1, 64, 224, 224)  # Example for a 224x224x64 input image
model = FireModule(in_channels=64, squeeze_channels=16, expand_channels=32)
output_tensor = model(input_tensor)

# Print the shapes of the input and output tensors
print("Input shape:", input_tensor.shape)
print("Output shape:", output_tensor.shape)
```

Capsule Networks

Capsule Networks, introduced by Sabour et al. (2017), address the limitations of traditional CNNs in capturing spatial hierarchies and relationships between features. Capsule Networks use capsules, which are groups of neurons that capture different properties of objects.

Capsules Capsules are groups of neurons that represent various properties of an object, such as pose, texture, and deformation. Each capsule outputs a vector rather than a scalar, allowing it to encode more information about the object's attributes.

Dynamic Routing Capsule Networks use dynamic routing to determine the connection strength between capsules in adjacent layers. This process ensures that capsules in higher layers receive inputs from the most relevant lower-layer capsules.

Given an input capsule $\mathbf{u}_i$ and output capsule $\mathbf{v}_j$, the coupling coefficient c_{ij} is determined through dynamic routing. The output capsule $\mathbf{v}_j$ is computed as:

$$\mathbf{s}_j = \sum_i c_{ij} \mathbf{u}_i$$

$$\mathbf{v}_j = \frac{||\mathbf{s}_j||^2}{1 + ||\mathbf{s}_j||^2} \frac{\mathbf{s}_j}{||\mathbf{s}_j||}$$

where $\mathbf{s}_j$ is the total input to capsule j, and c_{ij} is the coupling coefficient that is dynamically updated. For example, consider a capsule layer with input capsules of size 8 and output capsules of size 16. The dynamic routing algorithm updates the coupling coefficients based on the agreement between the capsules.

Example in TensorFlow

```
import tensorflow as tf
from tensorflow.keras.layers import Layer

# Define a capsule layer
class CapsuleLayer(Layer):
    def __init__(self, num_capsules, dim_capsules, num_routing):
        super(CapsuleLayer, self).__init__()
        self.num_capsules = num_capsules
        self.dim_capsules = dim_capsules
        self.num_routing = num_routing

    def build(self, input_shape):
        self.W = self.add_weight(shape=[self.num_capsules, input_shape[-1], self.dim_capsules],
                                 initializer='glorot_uniform', trainable=True)

    def call(self, inputs):
        inputs_expand = tf.expand_dims(inputs, 1)
        inputs_tiled = tf.tile(inputs_expand, [1, self.num_capsules, 1])
        u_hat = tf.matmul(inputs_tiled, self.W)
        b = tf.zeros(shape=[tf.shape(inputs)[0], self.num_capsules, 1])
        for i in range(self.num_routing):
            c = tf.nn.softmax(b, axis=1)
            s = tf.reduce_sum(c * u_hat, axis=1, keepdims=True)
            v = self.squash(s)
            if i < self.num_routing - 1:
                b += tf.reduce_sum(u_hat * v, axis=-1, keepdims=True)
        return tf.squeeze(v, axis=1)
```

```
    def squash(self, s):
        s_norm = tf.reduce_sum(tf.square(s), axis=-1, keepdims=True)
        scale = s_norm / (1 + s_norm) / tf.sqrt(s_norm + 1e-7)
        return scale * s

# Apply the capsule layer to an input tensor
input_tensor = tf.random.normal([1, 1152, 8])  # Example for a 1152x8 input capsule
capsule_layer = CapsuleLayer(num_capsules=10, dim_capsules=16, num_routing=3)
output_tensor = capsule_layer(input_tensor)

# Print the shapes of the input and output tensors
print("Input shape:", input_tensor.shape)
print("Output shape:", output_tensor.shape)
```

Example in PyTorch

```
import torch
import torch.nn as nn
import torch.nn.functional as F

# Define a capsule layer
class CapsuleLayer(nn.Module):
    def __init__(self, num_capsules, dim_capsules, num_routing):
        super(CapsuleLayer, self).__init__()
        self.num_capsules = num_capsules
        self.dim_capsules = dim_capsules
        self.num_routing = num_routing
        self.W = nn.Parameter(torch.randn(num_capsules, dim_capsules, dim_capsules))

    def forward(self, x):
        x = x.unsqueeze(1).expand(-1, self.num_capsules, -1, -1)
        u_hat = torch.matmul(x, self.W)
        b = torch.zeros(x.size(0), self.num_capsules, 1, device=x.device)
        for i in range(self.num_routing):
            c = F.softmax(b, dim=1)
            s = (c * u_hat).sum(dim=2, keepdim=True)
            v = self.squash(s)
            if i < self.num_routing - 1:
                b = b + (u_hat * v).sum(dim=-1, keepdim=True)
        return v.squeeze(2)

    def squash(self, s):
        s_norm = (s  2).sum(dim=-1, keepdim=True)
        return s_norm / (1 + s_norm) * s / torch.sqrt(s_norm + 1e-8)

# Apply the capsule layer to an input tensor
input_tensor = torch.randn(1, 1152, 8)  # Example for a 1152x8 input capsule
capsule_layer = CapsuleLayer(num_capsules=10, dim_capsules=16, num_routing=3)
output_tensor = capsule_layer(input_tensor)

# Print the shapes of the input and output tensors
print("Input shape:", input_tensor.shape)
print("Output shape:", output_tensor.shape)
```

4.5 Techniques to Improve CNN Performance

CNNs have revolutionized the field of computer vision, achieving remarkable performance across various tasks. However, their performance can be further enhanced through various techniques. One such crucial technique is data augmentation. Data augmentation involves artificially increasing the size and diversity of the training dataset by applying various transformations to the existing data. This technique helps improve the generalization ability of CNNs, making them more robust to variations in the input data.

4.5.1 Data Augmentation

Data augmentation plays a vital role in enhancing the performance of CNNs by providing more diverse training samples. It helps prevent overfitting and improves the model's ability to generalize to new, unseen data. By applying various transformations to the input data, we can create new training examples that capture the variability present in real-world scenarios.

Types of Augmentation Techniques

Geometric Transformations Geometric transformations alter the spatial properties of the input images, simulating different perspectives and viewpoints. Common geometric transformations include:

Rotation: Rotating the image by a certain angle, typically within a specified range.

$$\mathbf{I}'(x, y) = \mathbf{I}(x \cos\theta - y \sin\theta, x \sin\theta + y \cos\theta)$$

where θ is the rotation angle.

Translation: Shifting the image horizontally or vertically.

$$\mathbf{I}'(x, y) = \mathbf{I}(x + \Delta x, y + \Delta y)$$

where Δx and Δy are the translation offsets.

Scaling: Resizing the image by a certain factor.

$$\mathbf{I}'(x, y) = \mathbf{I}(\alpha x, \alpha y)$$

where α is the scaling factor.

Shearing: Applying a shear transformation to the image.

$$\mathbf{I}'(x, y) = \mathbf{I}(x + ky, y)$$

where k is the shear factor.

Photometric Transformations Photometric transformations modify the pixel values of the input images, simulating different lighting conditions and color variations. Common photometric transformations include:

Brightness Adjustment: Altering the brightness of the image.

$$\mathbf{I}'(x, y) = \mathbf{I}(x, y) + \beta$$

where β is the brightness adjustment factor.

Contrast Adjustment: Modifying the contrast of the image.

$$\mathbf{I}'(x, y) = \alpha(\mathbf{I}(x, y) - \mu) + \mu$$

where α is the contrast adjustment factor and μ is the mean pixel value.

Color Jittering: Randomly changing the hue, saturation, and value (HSV) of the image.

$$\mathbf{I}' = \text{HSV}(\mathbf{I}) + (\Delta h, \Delta s, \Delta v)$$

where Δh, Δs, and Δv are the changes in hue, saturation, and value, respectively.

Gaussian Noise: Adding random noise to the image.

$$\mathbf{I}'(x, y) = \mathbf{I}(x, y) + \mathcal{N}(0, \sigma^2)$$

where $\mathcal{N}(0, \sigma^2)$ is Gaussian noise with mean 0 and variance σ^2.

Random Erasing Random erasing involves randomly selecting a rectangular region in the image and erasing its content by setting it to a constant value or random noise. This technique helps the model become invariant to occlusions.

$$\mathbf{I}'(x, y) = \begin{cases} \mathbf{I}(x, y) & \text{if } (x, y) \notin \text{erased region} \\ \text{random value} & \text{if } (x, y) \in \text{erased region} \end{cases}$$

Mixup Mixup generates new training samples by linearly interpolating between pairs of images and their corresponding labels.

$$\mathbf{I}' = \lambda \mathbf{I}_1 + (1 - \lambda)\mathbf{I}_2$$

$$\mathbf{y}' = \lambda \mathbf{y}_1 + (1 - \lambda)\mathbf{y}_2$$

where $\mathbf{I}_1$ and $\mathbf{I}_2$ are two images, $\mathbf{y}_1$ and $\mathbf{y}_2$ are their corresponding labels, and λ is a mixing parameter sampled from a Beta distribution.

Cutout Cutout involves randomly masking out square regions of the input image. This technique forces the model to focus on the surrounding context, improving its robustness.

$$\mathbf{I}'(x, y) = \begin{cases} \mathbf{I}(x, y) & \text{if } (x, y) \notin \text{cutout region} \\ 0 & \text{if } (x, y) \in \text{cutout region} \end{cases}$$

Implementation and Benefits

Data augmentation can be easily implemented using popular deep learning frameworks such as TensorFlow and PyTorch. Both frameworks provide built-in functions and libraries to apply various augmentation techniques.

Example in TensorFlow

```
import tensorflow as tf
from tensorflow.keras.preprocessing.image import ImageDataGenerator

# Define an image data generator with augmentation
datagen = ImageDataGenerator(
    rotation_range=20,
    width_shift_range=0.2,
    height_shift_range=0.2,
    shear_range=0.2,
    zoom_range=0.2,
    horizontal_flip=True,
    fill_mode='nearest'
)

# Load and augment an image
image = tf.keras.preprocessing.image.load_img('path/to/image.jpg')
x = tf.keras.preprocessing.image.img_to_array(image)
x = x.reshape((1,) + x.shape)

# Generate augmented images
i = 0
for batch in datagen.flow(x, batch_size=1):
    plt.figure(i)
    imgplot = plt.imshow(tf.keras.preprocessing.image.array_to_img(batch[0]))
    i += 1
    if i % 4 == 0:
        break
plt.show()
```

Example in PyTorch

```
import torchvision.transforms as transforms
from PIL import Image

# Define a set of transformations
transform = transforms.Compose([
    transforms.RandomRotation(20),
    transforms.RandomHorizontalFlip(),
    transforms.RandomResizedCrop(224),
    transforms.ColorJitter(brightness=0.2, contrast=0.2, saturation=0.2, hue=0.2),
    transforms.ToTensor()
])

# Load and augment an image
image = Image.open('path/to/image.jpg')
augmented_image = transform(image)

# Visualize the augmented image
plt.imshow(transforms.ToPILImage()(augmented_image))
plt.show()
```

Benefits:

1. Improved Generalization: Data augmentation increases the diversity of the training dataset, helping the model generalize better to new, unseen data.
2. Reduced Overfitting: By providing more varied training examples, data augmentation reduces the risk of overfitting, where the model performs well on the training data but poorly on the test data.
3. Enhanced Robustness: Augmented data simulates different real-world conditions, making the model more robust to variations in lighting, orientation, and occlusions.

4. Efficient Use of Limited Data: Data augmentation is particularly beneficial when the available training data is limited, as it effectively multiplies the dataset size without the need for additional data collection.

Motivating Example Consider a scenario where a CNN is trained to recognize handwritten digits using the MNIST dataset. The dataset contains 60,000 training images, which might not capture all possible variations of handwritten digits. By applying data augmentation techniques such as rotation, scaling, and translation, we can create new training examples that mimic the variability found in real-world handwriting. This enhanced training dataset helps the CNN learn more robust features, improving its accuracy and generalization on the test set.

4.5.2 Regularization Methods

Regularization techniques are crucial for improving the generalization ability of CNNs. They help mitigate overfitting by adding constraints or modifications to the learning process. This section delves into three widely used regularization methods: Dropout, Batch Normalization, and L2 Regularization.

Dropout

Dropout is a regularization technique that prevents overfitting by randomly setting a fraction of the neurons to zero during each training iteration. This forces the network to learn redundant representations, making it more robust and less likely to overfit to the training data.

During each training iteration, each neuron i is retained with a probability p and dropped with a probability $1 - p$. Formally, for a given layer with activations $\mathbf{h}$, the dropout operation can be described as:

$$\mathbf{h}' = \mathbf{h} \odot \mathbf{r}$$

where $\odot$ denotes element-wise multiplication and $\mathbf{r}$ is a binary mask vector with each element r_i sampled from a Bernoulli distribution:

$$r_i \sim \text{Bernoulli}(p)$$

During inference, the activations are scaled by the retention probability p to account for the reduced number of neurons:

$$\mathbf{h}_{\text{inference}} = p\mathbf{h}$$

Example: Consider a neural network layer with activations $\mathbf{h} = [0.5, 0.2, 0.8]$ and a dropout probability $p = 0.5$. A possible dropout mask $\mathbf{r}$ might be $[1, 0, 1]$, resulting in the dropped activations $\mathbf{h}' = [0.5, 0, 0.8]$.

Example in TensorFlow

```
import tensorflow as tf
from tensorflow.keras.layers import Dropout

# Define a simple model with Dropout
model = tf.keras.Sequential([
    tf.keras.layers.Dense(128, activation='relu', input_shape=(784,)),
    Dropout(0.5),
    tf.keras.layers.Dense(10, activation='softmax')
])

# Print the model summary
model.summary()
```

Example in PyTorch

```
import torch
import torch.nn as nn

# Define a simple model with Dropout
class SimpleModel(nn.Module):
    def __init__(self):
        super(SimpleModel, self).__init__()
        self.fc1 = nn.Linear(784, 128)
        self.dropout = nn.Dropout(p=0.5)
        self.fc2 = nn.Linear(128, 10)

    def forward(self, x):
        x = torch.relu(self.fc1(x))
        x = self.dropout(x)
        x = self.fc2(x)
        return x

# Instantiate the model
model = SimpleModel()

# Print the model summary
print(model)
```

Batch Normalization

Batch Normalization (BN) is a technique that normalizes the activations of a layer for each mini-batch, stabilizing the learning process and enabling the use of higher learning rates. It reduces the internal covariate shift, making the optimization landscape smoother. For a given mini-batch of activations $\mathbf{h}$, Batch Normalization transforms the activations to have zero mean and unit variance:

$$\mu_B = \frac{1}{m}\sum_{i=1}^{m} h_i$$

$$\sigma_B^2 = \frac{1}{m}\sum_{i=1}^{m}(h_i - \mu_B)^2$$

$$\hat{h}_i = \frac{h_i - \mu_B}{\sqrt{\sigma_B^2 + \epsilon}}$$

where μ_B and σ_B^2 are the mean and variance of the mini-batch, and ϵ is a small constant added for numerical stability. The normalized activations $\hat{h}_i$ are then scaled and shifted using learned parameters γ and β:

$$h_i' = \gamma \hat{h}_i + \beta$$

Example: Consider a mini-batch of activations $\mathbf{h} = [0.5, 0.2, 0.8, 0.4]$. The mean and variance are $\mu_B = 0.475$ and $\sigma_B^2 = 0.045$. Normalizing and scaling these activations with $\gamma = 1.0$ and $\beta = 0.0$ gives $\hat{h}_i = [-0.078, -1.214, 1.536, -0.243]$.

Example in TensorFlow

```
import tensorflow as tf
from tensorflow.keras.layers import BatchNormalization

# Define a simple model with Batch Normalization
model = tf.keras.Sequential([
    tf.keras.layers.Conv2D(32, (3, 3), activation='relu', input_shape=(28, 28, 1)),
    BatchNormalization(),
    tf.keras.layers.Conv2D(64, (3, 3), activation='relu'),
    BatchNormalization(),
    tf.keras.layers.Flatten(),
    tf.keras.layers.Dense(128, activation='relu'),
    BatchNormalization(),
    tf.keras.layers.Dense(10, activation='softmax')
])

# Print the model summary
model.summary()
```

Example in PyTorch

```
import torch
import torch.nn as nn

# Define a simple model with Batch Normalization
class SimpleModel(nn.Module):
    def __init__(self):
        super(SimpleModel, self).__init__()
        self.conv1 = nn.Conv2d(1, 32, kernel_size=3)
        self.bn1 = nn.BatchNorm2d(32)
        self.conv2 = nn.Conv2d(32, 64, kernel_size=3)
        self.bn2 = nn.BatchNorm2d(64)
        self.fc1 = nn.Linear(9216, 128)
        self.bn3 = nn.BatchNorm1d(128)
        self.fc2 = nn.Linear(128, 10)

    def forward(self, x):
        x = torch.relu(self.bn1(self.conv1(x)))
        x = torch.relu(self.bn2(self.conv2(x)))
        x = x.view(-1, 9216)
        x = torch.relu(self.bn3(self.fc1(x)))
        x = self.fc2(x)
        return x

# Instantiate the model
model = SimpleModel()

# Print the model summary
print(model)
```

L2 Regularization

L2 Regularization, also known as weight decay, adds a penalty to the loss function proportional to the sum of the squared weights. This encourages the model to keep the weights small, preventing overfitting by discouraging complex models with large weights. The regularized loss function L_{reg} is given by:

$$L_{\text{reg}} = L + \lambda \sum_{j=1}^{n} w_j^2$$

where L is the original loss function, λ is the regularization parameter, and w_j are the weights of the model. The gradient update rule with L2 regularization modifies the weight update to include the penalty term:

$$w_j \leftarrow w_j - \eta \left(\frac{\partial L}{\partial w_j} + 2\lambda w_j \right)$$

where η is the learning rate. For example, consider a linear regression model with weights $\mathbf{w} = [0.5, -0.3, 0.8]$ and a regularization parameter $\lambda = 0.01$. The L2 penalty term is $0.01 \times (0.5^2 + (-0.3)^2 + 0.8^2) = 0.01 \times 0.98 = 0.0098$.

Example in TensorFlow

```
import tensorflow as tf
from tensorflow.keras.regularizers import l2

# Define a simple model with L2 Regularization
model = tf.keras.Sequential([
    tf.keras.layers.Dense(128, activation='relu', kernel_regularizer=l2(0.01),
         input_shape=(784,)),
    tf.keras.layers.Dense(10, activation='softmax', kernel_regularizer=l2(0.01))
])

# Print the model summary
model.summary()
```

Example in PyTorch

```
import torch
import torch.nn as nn
import torch.optim as optim

# Define a simple model with L2 Regularization
class SimpleModel(nn.Module):
    def __init__(self):
        super(SimpleModel, self).__init__()
        self.fc1 = nn.Linear(784, 128)
        self.fc2 = nn.Linear(128, 10)

    def forward(self, x):
        x = torch.relu(self.fc1(x))
        x = self.fc2(x)
        return x

# Instantiate the model
model = Simple
```

```
Model()

# Define the optimizer with L2 regularization (weight decay)
optimizer = optim.SGD(model.parameters(), lr=0.01, weight_decay=0.01)

# Print the model summary
print(model)
```

4.5.3 Transfer Learning

Transfer learning is a powerful technique in deep learning that leverages pre-trained models on large datasets to solve similar tasks with smaller datasets. By utilizing the knowledge acquired from the pre-trained model, transfer learning enables faster convergence, improved performance, and reduced computational cost, especially in scenarios where labeled data is limited.

Pre-trained Models and Fine-Tuning
Pre-trained models are neural networks that have been previously trained on extensive datasets, such as ImageNet, which contains millions of labeled images across thousands of categories. These models have learned rich hierarchical features that can be transferred to new tasks.

Popular pre-trained models include:

1. VGG: Known for its simplicity and depth, VGG models (e.g., VGG16, VGG19) consist of many convolutional layers followed by fully connected layers. The small (3×3) convolutional filters enable effective feature extraction.
2. ResNet: Residual Networks (e.g., ResNet50, ResNet101) use residual blocks with identity mappings to enable the training of very deep networks. These models are highly effective in learning complex features and have demonstrated state-of-the-art performance on various benchmarks.
3. Inception: Inception models (e.g., Inception v3) incorporate multi-scale processing within the same layer using inception modules, capturing features at different scales and improving performance.
4. EfficientNet: EfficientNet models scale the network's width, depth, and resolution systematically using a compound scaling method, achieving a balance between accuracy and efficiency.

Fine-Tuning: Fine-tuning is the process of adapting a pre-trained model to a new task by continuing the training on a new dataset. This process typically involves the following steps:

1. Freezing Layers: Initially, most of the layers from the pre-trained model are frozen, meaning their weights are not updated during training. This allows the model to retain the learned features from the original dataset.
2. Adding Custom Layers: New layers specific to the target task are added on top of the pre-trained model. These layers are usually unfrozen and trained from scratch.

3. Unfreezing Layers: Optionally, some of the pre-trained layers can be unfrozen and fine-tuned with a lower learning rate, allowing the model to adjust the learned features to the new dataset.

Let $\mathbf{W}_{\text{pre}}$ represent the weights of the pre-trained model and $\mathbf{W}_{\text{new}}$ represent the weights of the new layers added for the specific task. The objective is to minimize the loss function L:

$$L(\mathbf{W}_{\text{pre}}, \mathbf{W}_{\text{new}}) = L_{\text{task}}(\mathbf{W}_{\text{new}}) + \lambda \sum_{i=1}^{n} ||\mathbf{W}_{\text{pre},i}||^2$$

where L_{task} is the task-specific loss function, and λ is the regularization parameter. For example, consider fine-tuning a pre-trained ResNet50 model on a new dataset of flower images. The steps involve freezing the ResNet50 layers, adding new fully connected layers for flower classification, and fine-tuning the model.

Example in TensorFlow

```
import tensorflow as tf
from tensorflow.keras.applications import ResNet50
from tensorflow.keras.layers import Dense, GlobalAveragePooling2D
from tensorflow.keras.models import Model

# Load the pre-trained ResNet50 model
base_model = ResNet50(weights='imagenet', include_top=False, input_shape=(224, 224, 3))

# Freeze the base model layers
for layer in base_model.layers:
    layer.trainable = False

# Add custom layers on top of the base model
x = base_model.output
x = GlobalAveragePooling2D()(x)
x = Dense(1024, activation='relu')(x)
predictions = Dense(5, activation='softmax')(x)  # Assuming 5 classes for flower
 classification

# Create the new model
model = Model(inputs=base_model.input, outputs=predictions)

# Compile the model
model.compile(optimizer='adam', loss='categorical_crossentropy', metrics=['accuracy'])

# Print the model summary
model.summary()
```

Example in PyTorch

```
import torch
import torch.nn as nn
import torchvision.models as models

# Load the pre-trained ResNet50 model
model = models.resnet50(pretrained=True)

# Freeze the base model layers
for param in model.parameters():
    param.requires_grad = False
```

```
# Modify the final layer for flower classification (assuming 5 classes)
num_ftrs = model.fc.in_features
model.fc = nn.Sequential(
    nn.Linear(num_ftrs, 1024),
    nn.ReLU(),
    nn.Linear(1024, 5)
)

# Print the modified model
print(model)

# Define the loss function and optimizer
criterion = nn.CrossEntropyLoss()
optimizer = torch.optim.Adam(model.fc.parameters(), lr=0.001)

# Training loop
# for epoch in range(num_epochs):
#     for inputs, labels in dataloader:
#         outputs = model(inputs)
#         loss = criterion(outputs, labels)
#         optimizer.zero_grad()
#         loss.backward()
#         optimizer.step()
```

Applications and Benefits
Applications:

1. Image Classification: Transfer learning is widely used in image classification tasks where pre-trained models are fine-tuned on specific datasets such as medical imaging, wildlife classification, and product identification.
2. Object Detection: Pre-trained models serve as feature extractors for object detection frameworks like Faster R-CNN, YOLO, and SSD, which are then fine-tuned on target datasets to detect specific objects.
3. Segmentation: In semantic and instance segmentation tasks, pre-trained models provide a strong backbone for extracting features, which are then used to segment images into meaningful regions.
4. Natural Language Processing: Transfer learning extends beyond computer vision to natural language processing (NLP) tasks. Pre-trained language models like BERT, GPT, and RoBERTa are fine-tuned for tasks such as text classification, named entity recognition, and question answering.

Benefits:

1. Improved Performance: Transfer learning leverages the rich feature representations learned by pre-trained models, leading to improved performance on the target task, even with limited data.
2. Faster Convergence: Starting with a pre-trained model accelerates the training process, as the model has already learned general features. Fine-tuning requires fewer epochs compared to training from scratch.
3. Reduced Data Requirements: Transfer learning is particularly beneficial when labeled data is scarce. The knowledge transferred from the pre-trained model reduces the need for large annotated datasets.

4. Resource Efficiency: Using pre-trained models reduces the computational cost and time required for training, making it feasible to deploy deep learning models on resource-constrained devices.

Motivating Example: Consider a medical imaging application where the goal is to classify different types of skin lesions. Collecting and labeling a large dataset of medical images is challenging and expensive. By leveraging a pre-trained ResNet model on ImageNet, we can fine-tune it on a smaller dataset of skin lesion images. The pre-trained model provides a strong foundation, capturing generic features such as edges and textures, which can be fine-tuned to recognize specific patterns in skin lesions. This approach significantly improves classification accuracy while reducing the need for extensive data collection and labeling.

4.5.4 Hyperparameter Optimization

Hyperparameter optimization is a critical step in the training of convolutional neural networks (CNNs). Proper selection of hyperparameters can significantly influence the performance and efficiency of the model. This section explores two widely used methods for hyperparameter optimization: Grid Search and Random Search, followed by Bayesian Optimization, a more sophisticated approach.

Grid Search
Grid Search is a brute-force approach that exhaustively searches through a specified subset of the hyperparameter space. It evaluates all possible combinations of hyperparameters by training the model for each combination and selecting the one that yields the best performance based on a validation metric. Given a set of hyperparameters $\{\theta_1, \theta_2, \ldots, \theta_k\}$ and their corresponding ranges $\{\Theta_1, \Theta_2, \ldots, \Theta_k\}$, Grid Search evaluates the performance $f(\theta_1, \theta_2, \ldots, \theta_k)$ for all combinations:

$$(\theta_1^*, \theta_2^*, \ldots, \theta_k^*) = \arg \max_{\theta_1 \in \Theta_1, \theta_2 \in \Theta_2, \ldots, \theta_k \in \Theta_k} f(\theta_1, \theta_2, \ldots, \theta_k)$$

where f is the performance metric (e.g., accuracy, loss).

Advantages:

1. Exhaustive Search: Grid Search guarantees that all possible combinations within the specified ranges are evaluated, ensuring that the global optimum is found if it exists within the grid.
2. Simple Implementation: The method is straightforward to implement and parallelize, making it suitable for distributed computing environments.

Disadvantages:

1. Computationally Expensive: The exhaustive nature of Grid Search can lead to a combinatorial explosion in the number of evaluations, making it impractical for large hyperparameter spaces.
2. Inefficiency: Grid Search often evaluates many hyperparameter combinations that yield similar performance, leading to wasted computational resources.

Example: Consider optimizing the learning rate η and the batch size B for a CNN. The ranges are defined as $\eta \in \{0.001, 0.01, 0.1\}$ and $B \in \{16, 32, 64\}$. Grid Search evaluates all 9 combinations to find the optimal pair.

Example in TensorFlow

```
import tensorflow as tf
from tensorflow.keras.models import Sequential
from tensorflow.keras.layers import Conv2D, Flatten, Dense

# Define a simple CNN model
def create_model(learning_rate):
    model = Sequential([
        Conv2D(32, (3, 3), activation='relu', input_shape=(28, 28, 1)),
        Flatten(),
        Dense(10, activation='softmax')
    ])
    model.compile(optimizer=tf.keras.optimizers.Adam(learning_rate=learning_rate),
                  loss='sparse_categorical_crossentropy',
                  metrics=['accuracy'])
    return model

# Define the grid search parameters
learning_rates = [0.001, 0.01, 0.1]
batch_sizes = [16, 32, 64]

# Perform grid search
for lr in learning_rates:
    for bs in batch_sizes:
        model = create_model(learning_rate=lr)
        model.fit(train_images, train_labels, epochs=5, batch_size=bs, validation_data
              =(test_images, test_labels))
        _, accuracy = model.evaluate(test_images, test_labels)
        print(f'Learning rate: {lr}, Batch size: {bs}, Accuracy: {accuracy}')
```

Example in PyTorch

```
import torch
import torch.nn as nn
import torch.optim as optim
from torchvision import datasets, transforms

# Define a simple CNN model
class SimpleCNN(nn.Module):
    def __init__(self):
        super(SimpleCNN, self).__init__()
        self.conv1 = nn.Conv2d(1, 32, kernel_size=3)
        self.fc1 = nn.Linear(26*26*32, 10)

    def forward(self, x):
        x = torch.relu(self.conv1(x))
        x = x.view(-1, 26*26*32)
        x = self.fc1(x)
```

```
        return x

# Define the grid search parameters
learning_rates = [0.001, 0.01, 0.1]
batch_sizes = [16, 32, 64]

# Define the dataset and data loaders
transform = transforms.Compose([transforms.ToTensor()])
train_dataset = datasets.MNIST(root='./data', train=True, download=True, transform=transform)
test_dataset = datasets.MNIST(root='./data', train=False, download=True, transform=transform)

# Perform grid search
for lr in learning_rates:
    for bs in batch_sizes:
        train_loader = torch.utils.data.DataLoader(train_dataset, batch_size=bs, shuffle=True)
        test_loader = torch.utils.data.DataLoader(test_dataset, batch_size=bs, shuffle=False)
        model = SimpleCNN()
        criterion = nn.CrossEntropyLoss()
        optimizer = optim.Adam(model.parameters(), lr=lr)
        # Training loop (simplified)
        for epoch in range(5):
            model.train()
            for data, target in train_loader:
                optimizer.zero_grad()
                output = model(data)
                loss = criterion(output, target)
                loss.backward()
                optimizer.step()
        # Evaluation
        model.eval()
        correct = 0
        with torch.no_grad():
            for data, target in test_loader:
                output = model(data)
                pred = output.argmax(dim=1, keepdim=True)
                correct += pred.eq(target.view_as(pred)).sum().item()
        accuracy = correct / len(test_loader.dataset)
        print(f'Learning rate: {lr}, Batch size: {bs}, Accuracy: {accuracy}')
```

Random Search

Random Search, proposed by Bergstra and Bengio, addresses some limitations of Grid Search by sampling hyperparameters from specified distributions. This method randomly selects combinations of hyperparameters, allowing for a more efficient exploration of the hyperparameter space. Given a set of hyperparameters $\{\theta_1, \theta_2, \ldots, \theta_k\}$ and their corresponding distributions $\{\Theta_1, \Theta_2, \ldots, \Theta_k\}$, Random Search samples n random combinations:

$$(\theta_1^*, \theta_2^*, \ldots, \theta_k^*) = \arg \max_{(\theta_1, \theta_2, \ldots, \theta_k) \sim \Theta} f(\theta_1, \theta_2, \ldots, \theta_k)$$

where f is the performance metric (e.g., accuracy, loss).

Advantages:

1. Efficiency: Random Search is more efficient than Grid Search, especially when only a few hyperparameters significantly impact the model's performance.
2. Scalability: The method can explore larger hyperparameter spaces without a combinatorial explosion in the number of evaluations.

Disadvantages: Random Search does not guarantee that the global optimum will be found, as it relies on random sampling.

Consider optimizing the learning rate η sampled from a log-uniform distribution $\eta \sim \text{log-uniform}(10^{-4}, 10^{-2})$ and the batch size B sampled from a uniform distribution $B \sim \{16, 32, 64\}$. Random Search evaluates a fixed number of random combinations to find the optimal pair.

Example in TensorFlow

```
import tensorflow as tf
import numpy as np

# Define the random search parameters
learning_rates = np.random.uniform(1e-4, 1e-2, size=10)
batch_sizes = np.random.choice([16, 32, 64], size=10)

# Perform random search
for lr, bs in zip(learning_rates, batch_sizes):
    model = create_model(learning_rate=lr)
    model.fit(train_images, train_labels, epochs=5, batch_size=bs, validation_data
        =(test_images, test_labels))
    _, accuracy = model.evaluate(test_images, test_labels)
    print(f'Learning rate: {lr}, Batch size: {bs}, Accuracy: {accuracy}')
```

Example in PyTorch

```
import numpy as np

# Define the random search parameters
learning_rates = np.random.uniform(1e-4, 1e-2, size=10)
batch_sizes = np.random.choice([16, 32, 64], size=10)

# Perform random search
for lr, bs in zip(learning_rates, batch_sizes):
    train_loader = torch.utils.data.DataLoader(train_dataset, batch_size=bs, shuffle=True)
    test_loader = torch.utils.data.DataLoader(test_dataset, batch_size=bs, shuffle=False)
    model = SimpleCNN()
    criterion = nn.CrossEntropyLoss()
    optimizer = optim.Adam(model.parameters(), lr=lr)
    # Training loop (simplified)
    for epoch in range(5):
        model.train()
        for data, target in train_loader:
            optimizer.zero_grad()
            output = model(data)
            loss = criterion(output, target)
            loss.backward()
            optimizer.step()
    # Evaluation
    model.eval()
    correct = 0
    with torch.no_grad():
        for data, target in test_loader:
            output = model(data)
            pred = output.argmax(dim=1, keepdim=True)
            correct += pred.eq(target.view_as(pred)).sum().item()
    accuracy = correct / len(test_loader.dataset)
    print(f'Learning rate: {lr}, Batch size: {bs}, Accuracy: {accuracy}')
```

Bayesian Optimization

Bayesian Optimization is a more sophisticated approach that models the hyperparameter space as a probabilistic function. It uses a surrogate model to predict the performance of different hyperparameter combinations and iteratively refines the model based on observed results. This method balances exploration and exploitation, efficiently navigating the hyperparameter space. Bayesian Optimization models the objective function $f(\theta)$ using a surrogate model, typically a Gaussian Process (GP). The GP provides a distribution over functions, capturing the uncertainty in the objective function. Given the observed data $D = \{(\theta_i, f(\theta_i))\}_{i=1}^n$, the posterior distribution of the objective function is updated as new observations are made. The next hyperparameter combination is chosen by maximizing an acquisition function $a(\theta)$, which balances exploration (uncertainty) and exploitation (mean prediction):

$$\theta_{\text{next}} = \arg\max_{\theta} a(\theta|D)$$

Advantages:

1. Efficiency: Bayesian Optimization is more sample-efficient than Grid Search and Random Search, requiring fewer evaluations to find optimal hyperparameters.
2. Incorporates Uncertainty: The surrogate model accounts for uncertainty, enabling a more informed exploration of the hyperparameter space.

Disadvantages: The method is more complex to implement and requires careful tuning of the surrogate model and acquisition function.

Example: Consider optimizing the learning rate η and batch size B using Bayesian Optimization. The GP models the performance as a function of η and B, iteratively refining its predictions based on observed results.

Example in Python Using Scikit-Optimize

```
from skopt import gp_minimize
from skopt.space import Real, Integer

# Define the objective function
def objective(params):
    lr, bs = params
    model = create_model(learning_rate=lr)
    model.fit(train_images, train_labels, epochs=5, batch_size=bs, validation_data
        =(test_images, test_labels))
    _, accuracy = model.evaluate(test_images, test_labels)
    return -accuracy  # Minimize negative accuracy

# Define the search space
search_space = [Real(1e-4, 1e-2, prior='log-uniform', name='learning_rate'),
                Integer(16, 64, name='batch_size')]

# Perform Bayesian Optimization
result = gp_minimize(objective, search_space, n_calls=20, random_state=0)

# Print the best hyperparameters
print(f'Best learning rate: {result.x[0]}, Best batch size: {result.x[1]}')
```

4.5.5 Ensemble Methods

Ensemble methods are powerful techniques that combine the predictions of multiple models to improve the overall performance, robustness, and generalization of machine learning systems. By leveraging the diversity of different models, ensemble methods can achieve better accuracy and mitigate the risk of overfitting. This section explores two primary types of ensemble methods: Bagging and Boosting, and Stacking and Blending.

Bagging

Bagging (Bootstrap Aggregating) is an ensemble technique that aims to reduce the variance of the model by training multiple models on different subsets of the data and averaging their predictions. Each model in the ensemble is trained on a bootstrap sample (a random subset of the training data with replacement). Given a training set $D = \{(x_i, y_i)\}_{i=1}^N$, bagging generates M bootstrap samples D_m and trains a model f_m on each sample. The final prediction is obtained by averaging the predictions of the individual models:

$$\hat{y} = \frac{1}{M} \sum_{m=1}^{M} f_m(x)$$

Advantages:

1. Variance Reduction: By averaging the predictions, bagging reduces the variance of the model, leading to more stable and robust predictions.
2. Parallelization: Each model in the ensemble can be trained independently, making bagging highly parallelizable and efficient.

Disadvantages: Training multiple models can be computationally expensive and requires more memory.

Example: Random Forests, a popular ensemble method, use bagging with decision trees as base learners. Each tree is trained on a bootstrap sample, and the final prediction is the average (or majority vote) of the predictions from all trees.

Example in Python Using Scikit-Learn

```
from sklearn.ensemble import RandomForestClassifier
from sklearn.datasets import load_iris
from sklearn.model_selection import train_test_split
from sklearn.metrics import accuracy_score

# Load the Iris dataset
iris = load_iris()
X_train, X_test, y_train, y_test = train_test_split(iris.data, iris.target, test_size=0.3,
 random_state=42)

# Train a Random Forest classifier
model = RandomForestClassifier(n_estimators=100, random_state=42)
model.fit(X_train, y_train)

# Make predictions and evaluate accuracy
y_pred = model.predict(X_test)
accuracy = accuracy_score(y_test, y_pred)
print(f'Accuracy: {accuracy}')
```

Boosting
Boosting is an ensemble technique that aims to reduce bias and variance by sequentially training models, where each model attempts to correct the errors of its predecessor. The models are trained in a sequential manner, with each model focusing on the mistakes made by the previous models. Given a training set $D = \{(x_i, y_i)\}_{i=1}^N$, boosting adjusts the weights of the training samples based on the errors of the previous model. The final prediction is a weighted sum of the predictions from all models:

$$\hat{y} = \sum_{m=1}^{M} \alpha_m f_m(x)$$

where α_m is the weight assigned to the mth model based on its accuracy.

Advantages:

1. Bias and Variance Reduction: Boosting reduces both bias and variance, leading to more accurate and robust models.
2. Adaptability: Boosting can adapt to the complexity of the data by focusing on difficult-to-predict instances.

Disadvantages: Boosting can be prone to overfitting if not properly regularized, especially when using complex base learners.

Example: AdaBoost (Adaptive Boosting) is a popular boosting algorithm that combines weak learners (e.g., decision stumps) to create a strong classifier. Each weak learner is trained to correct the errors of the previous learners.

Example in Python Using Scikit-Learn

```
from sklearn.ensemble import AdaBoostClassifier

# Train an AdaBoost classifier
model = AdaBoostClassifier(n_estimators=100, random_state=42)
model.fit(X_train, y_train)

# Make predictions and evaluate accuracy
y_pred = model.predict(X_test)
accuracy = accuracy_score(y_test, y_pred)
print(f'Accuracy: {accuracy}')
```

Stacking
Stacking, or stacked generalization, is an ensemble method that combines multiple base models using a meta-model. The base models are trained on the original training data, and their predictions are used as input features for the meta-model, which is trained to make the final prediction. Let $\{f_1, f_2, \ldots, f_M\}$ be the base models, and g be the meta-model. The base models are first trained on the training data D, and their predictions are combined to form a new dataset D':

$$D' = \{(f_1(x_i), f_2(x_i), \ldots, f_M(x_i), y_i)\}_{i=1}^N$$

The meta-model g is then trained on D' to make the final prediction:

$$\hat{y} = g(f_1(x), f_2(x), \ldots, f_M(x))$$

Advantages:

1. Leverage Diverse Models: Stacking can leverage the strengths of different models by combining their predictions in a non-linear fashion.
2. Improved Performance: The meta-model can learn to correct the biases and errors of the base models, leading to improved overall performance.

Disadvantages: Stacking involves training multiple models and a meta-model, increasing the complexity and computational cost.

Example: Consider stacking logistic regression, decision trees, and k-nearest neighbors as base models, with a random forest as the meta-model.

Example in Python Using Scikit-Learn

```
from sklearn.ensemble import StackingClassifier
from sklearn.linear_model import LogisticRegression
from sklearn.tree import DecisionTreeClassifier
from sklearn.neighbors import KNeighborsClassifier
from sklearn.ensemble import RandomForestClassifier

# Define the base models
base_models = [
    ('lr', LogisticRegression()),
    ('dt', DecisionTreeClassifier()),
    ('knn', KNeighborsClassifier())
]

# Define the meta-model
meta_model = RandomForestClassifier(n_estimators=100, random_state=42)

# Create the stacking classifier
model = StackingClassifier(estimators=base_models, final_estimator=meta_model)

# Train the model
model.fit(X_train, y_train)

# Make predictions and evaluate accuracy
y_pred = model.predict(X_test)
accuracy = accuracy_score(y_test, y_pred)
print(f'Accuracy: {accuracy}')
```

Blending

Blending is similar to stacking but uses a holdout validation set to train the meta-model. The base models are trained on the training set, and their predictions on the holdout set are used to train the meta-model. Let D_{train} and D_{val} be the training and validation sets, respectively. The base models are trained on D_{train}, and their predictions on D_{val} are used to form a new dataset D':

$$D' = \{(f_1(x_i), f_2(x_i), \ldots, f_M(x_i), y_i)\}_{i=1}^{|D_{\text{val}}|}$$

The meta-model g is then trained on D' to make the final prediction.

Advantages:

1. Simplicity: Blending is simpler than stacking as it requires only a holdout set for training the meta-model.
2. Efficiency: By using a holdout set, blending reduces the risk of overfitting compared to using the same data for training both base and meta-models.

Disadvantages: Blending requires a sufficient amount of data to create a holdout set without compromising the training of the base models.

Example: Consider blending logistic regression, decision trees, and k-nearest neighbors as base models, with a random forest as the meta-model, using a holdout set.

Example in Python

```
from sklearn.model_selection import train_test_split

# Split the training data into a new training set and holdout set
X_train_new, X_val, y_train_new, y_val = train_test_split(X_train, y_train, test_size=0.2,
 random_state=42)

# Train the base models on the new training set
base_model_lr = LogisticRegression().fit(X_train_new, y_train_new)
base_model_dt = DecisionTreeClassifier().fit(X_train_new, y_train_new)
base_model_knn = KNeighborsClassifier().fit(X_train_new, y_train_new)

# Get predictions from the base models on the holdout set
val_preds_lr = base_model_lr.predict(X_val)
val_preds_dt = base_model_dt.predict(X_val)
val_preds_knn = base_model_knn.predict(X_val)

# Combine the predictions to form the new dataset for the meta-model
val_preds_combined = np.column_stack((val_preds_lr, val_preds_dt, val_preds_knn))

# Train the meta-model on the holdout set predictions
meta_model = RandomForestClassifier(n_estimators=100, random_state=42)
meta_model.fit(val_preds_combined, y_val)

# Make final predictions using the meta-model
test_preds_lr = base_model_lr.predict(X_test)
test_preds_dt = base_model_dt.predict(X_test)
test_preds_knn = base_model_knn.predict(X_test)
test_preds_combined = np.column_stack((test_preds_lr, test_preds_dt, test_preds_knn))
final_preds = meta_model.predict(test_preds_combined)

# Evaluate accuracy
accuracy = accuracy_score(y_test, final_preds)
print(f'Accuracy: {accuracy}')
```

4.6 Practical Applications of CNNs

CNNs have proven to be highly effective in various computer vision tasks. One of the most prominent applications of CNNs is image classification. This section delves into the practical aspects of applying CNNs to image classification, covering dataset preparation and preprocessing, model building and training, and evaluation metrics.

4.6.1 *Image Classification*

Image classification involves assigning a label to an image from a predefined set of categories. The process requires careful preparation of the dataset, designing and training an effective CNN model, and evaluating its performance using appropriate metrics.

Dataset Preparation and Preprocessing
Dataset Collection: The first step in image classification is to gather a representative dataset. Commonly used datasets include CIFAR-10, CIFAR-100, ImageNet, and MNIST. These datasets provide a diverse set of images across various categories, enabling robust training and evaluation of CNN models.

Data Augmentation: To enhance the diversity and robustness of the training data, data augmentation techniques are applied. These techniques include random rotations, translations, scaling, flipping, and color jittering. Data augmentation helps the model generalize better by exposing it to different variations of the input images.

Consider an image $\mathbf{I}$ of size $H \times W \times C$, where H is the height, W is the width, and C is the number of channels. Data augmentation involves applying transformations T to generate augmented images $\mathbf{I}'$:

$$\mathbf{I}' = T(\mathbf{I})$$

For example, a rotation transformation T_r by an angle θ can be represented as:

$$\mathbf{I}'(x, y) = \mathbf{I}(x \cos\theta - y \sin\theta, x \sin\theta + y \cos\theta)$$

Normalization: Normalization is crucial to ensure that the input images have a consistent scale, which facilitates faster convergence during training. Typically, the pixel values of the images are normalized to a range of [0, 1] or standardized to have zero mean and unit variance.

$$\mathbf{I}' = \frac{\mathbf{I} - \mu}{\sigma}$$

where μ and σ are the mean and standard deviation of the pixel values, respectively.

Example: Consider the CIFAR-10 dataset, which consists of 60,000 32×32 color images in 10 classes. Data augmentation and normalization can be implemented using popular deep learning frameworks such as TensorFlow and PyTorch.

Example in TensorFlow

```
import tensorflow as tf
from tensorflow.keras.preprocessing.image import ImageDataGenerator

# Load the CIFAR-10 dataset
(x_train, y_train), (x_test, y_test) = tf.keras.datasets.cifar10.load_data()

# Define an image data generator with data augmentation
datagen = ImageDataGenerator(
    rotation_range=15,
    width_shift_range=0.1,
    height_shift_range=0.1,
    horizontal_flip=True,
    rescale=1.0/255.0
)

# Fit the data generator on the training data
datagen.fit(x_train)

# Normalize the test data
x_test = x_test.astype('float32') / 255.0
```

Example in PyTorch

```
import torch
import torchvision.transforms as transforms
from torchvision.datasets import CIFAR10
from torch.utils.data import DataLoader

# Define transformations including data augmentation and normalization
transform = transforms.Compose([
    transforms.RandomHorizontalFlip(),
    transforms.RandomRotation(15),
    transforms.ToTensor(),
    transforms.Normalize((0.4914, 0.4822, 0.4465), (0.2470, 0.2435, 0.2616))
])

# Load the CIFAR-10 dataset with the defined transformations
train_dataset = CIFAR10(root='./data', train=True, download=True, transform=transform)
test_dataset = CIFAR10(root='./data', train=False, download=True, transform=transforms.
    Compose([transforms.ToTensor(),
    transforms.Normalize((0.4914, 0.4822, 0.4465), (0.2470, 0.2435, 0.2616))
]))

# Create data loaders
train_loader = DataLoader(train_dataset, batch_size=64, shuffle=True)
test_loader = DataLoader(test_dataset, batch_size=64, shuffle=False)
```

Model Building and Training

Model Architecture: Designing an effective CNN model architecture is crucial for achieving high performance in image classification tasks. We will recall the main components below.

Convolutional Layers: Convolutional layers apply convolutional filters to the input images to extract spatial features. The output feature maps are obtained by convolving the filters with the input image.

$$\mathbf{O}(x, y, k) = \sum_{i=0}^{H_k-1} \sum_{j=0}^{W_k-1} \sum_{c=0}^{C-1} \mathbf{I}(x+i, y+j, c)\mathbf{K}(i, j, c, k) + b_k$$

where $\mathbf{O}(x, y, k)$ is the output feature map, $\mathbf{I}(x, y, c)$ is the input image, $\mathbf{K}(i, j, c, k)$ is the filter, and b_k is the bias term.

Pooling Layers: Pooling layers reduce the spatial dimensions of the feature maps, retaining the most important features while reducing computational complexity.

$$\mathbf{P}(x, y, k) = \max_{i,j\in\{0,\ldots,s-1\}} \mathbf{O}(sx + i, sy + j, k)$$

where $\mathbf{P}(x, y, k)$ is the pooled feature map and s is the pooling size.

Fully Connected Layers: Fully connected layers combine the features extracted by the convolutional and pooling layers to make the final classification decision.

$$\mathbf{z} = \mathbf{W}\mathbf{h} + \mathbf{b}$$

where $\mathbf{z}$ is the output logits, $\mathbf{W}$ is the weight matrix, $\mathbf{h}$ is the flattened feature vector, and $\mathbf{b}$ is the bias term.

Training: Training a CNN involves optimizing the model parameters using a suitable loss function and optimization algorithm. The cross-entropy loss is commonly used for classification tasks, and optimization algorithms like stochastic gradient descent (SGD) or Adam are used to update the model parameters.

$$L(\mathbf{y}, \hat{\mathbf{y}}) = -\sum_{i=1}^{N} y_i \log(\hat{y}_i)$$

where $\mathbf{y}$ is the true label and $\hat{\mathbf{y}}$ is the predicted probability.

Example in TensorFlow

```
from tensorflow.keras.models import Sequential
from tensorflow.keras.layers import Conv2D, MaxPooling2D, Flatten, Dense

# Define the CNN model
model = Sequential([
    Conv2D(32, (3, 3), activation='relu', input_shape=(32, 32, 3)),
    MaxPooling2D((2, 2)),
    Conv2D(64, (3, 3), activation='relu'),
    MaxPooling2D((2, 2)),
    Flatten(),
    Dense(128, activation='relu'),
    Dense(10, activation='softmax')
])

# Compile the model
model.compile(optimizer='adam', loss='sparse_categorical_crossentropy', metrics=['accuracy'])

# Train the model
model.fit(datagen.flow(x_train, y_train, batch_size=64), epochs=10, validation_data=(x_test,
        y_test))
```

Example in PyTorch

```
import torch.nn as nn
import torch.optim as optim

# Define the CNN model
class SimpleCNN(nn.Module):
    def __init__(self):
        super(SimpleCNN, self).__init__()
        self.conv1 = nn.Conv2d(3, 32, kernel_size=3, activation='relu')
        self.pool = nn.MaxPool2d(kernel_size=2, stride=2)
        self.conv2 = nn.Conv2d(32, 64, kernel_size=3, activation='relu')
        self.fc1 = nn.Linear(64 * 6 * 6, 128)
        self.fc2 = nn.Linear(128, 10)

    def forward(self, x):
        x = self.pool(torch.relu(self.conv1(x)))
        x = self.pool(torch.relu(self.conv2(x)))
        x = x.view(-1, 64 * 6 * 6)
        x = torch.relu(self.fc1(x))
        x = self.fc2(x)
        return x

# Instantiate the model
model = SimpleCNN()

# Define the loss function and optimizer
criterion = nn.CrossEntropyLoss()
optimizer = optim.Adam(model.parameters(), lr=0.001)

# Training loop
for epoch in range(10):
    for inputs, labels in train_loader:
        optimizer.zero_grad()
        outputs = model(inputs)
        loss = criterion(outputs, labels)
        loss.backward()
        optimizer.step()

    # Evaluate on test set
    correct = 0
    total = 0
    with torch.no_grad():
        for inputs, labels in test_loader:
            outputs = model(inputs)
            _, predicted = torch.max(outputs, 1)
            total += labels.size(0)
            correct += (predicted == labels).sum().item()
    print(f'Epoch {epoch+1}, Accuracy: {100 * correct / total}%')
```

Evaluation Metrics

Accuracy: Accuracy is the most straightforward evaluation metric, representing the percentage of correctly classified samples out of the total samples.

$$\text{Accuracy} = \frac{\text{Number of Correct Predictions}}{\text{Total Number of Predictions}}$$

Confusion Matrix: The confusion matrix provides a detailed breakdown of the classification performance by showing the true positives, true negatives, false positives, and false negatives for each class.

Precision, Recall, and F1-Score: Precision, recall, and F1-score are important metrics for imbalanced datasets. Precision measures the proportion of true positives among the predicted positives, recall measures the proportion of true positives among the actual positives, and F1-score is the harmonic mean of precision and recall.

$$\text{Precision} = \frac{\text{True Positives}}{\text{True Positives} + \text{False Positives}}$$

$$\text{Recall} = \frac{\text{True Positives}}{\text{True Positives} + \text{False Negatives}}$$

$$\text{F1-Score} = 2 \times \frac{\text{Precision} \times \text{Recall}}{\text{Precision} + \text{Recall}}$$

Example in Python

```
from sklearn.metrics import accuracy_score, precision_score, recall_score, f1_score

# Predict on the test set
y_pred = model.predict(x_test)

# Calculate evaluation metrics
accuracy = accuracy_score(y_test, y_pred)
precision = precision_score(y_test, y_pred, average='weighted')
recall = recall_score(y_test, y_pred, average='weighted')
f1 = f1_score(y_test, y_pred, average='weighted')

print(f'Accuracy: {accuracy}')
print(f'Precision: {precision}')
print(f'Recall: {recall}')
print(f'F1-Score: {f1}')
```

4.6.2 Object Detection

Object detection is a crucial task in computer vision that involves identifying and localizing objects within an image. Unlike image classification, which assigns a single label to an image, object detection provides bounding boxes and labels for multiple objects within the image. This section explores key object detection frameworks, including R-CNN and its variants, YOLO, and SSD, along with the evaluation metrics used to assess their performance.

R-CNN and Its Variants

Regions with Convolutional Neural Networks (R-CNN) is one of the pioneering frameworks for object detection. It combines region proposal methods with CNNs to detect objects within an image.

1. Region Proposal: The algorithm generates around 2000 region proposals using selective search, a technique that merges superpixels to form object proposals.
2. Feature Extraction: Each region proposal is warped to a fixed size and passed through a pre-trained CNN (e.g., AlexNet) to extract features.

$$\mathbf{f}_i = \text{CNN}(\mathbf{R}_i)$$

where $\mathbf{R}_i$ is the ith region proposal and $\mathbf{f}_i$ is the extracted feature vector.

3. Classification and Regression: The extracted features are fed into a set of SVM classifiers to predict object classes and a linear regressor to refine the bounding box coordinates.

$$\hat{\mathbf{b}}_i = \mathbf{W}_{\text{reg}}\mathbf{f}_i + \mathbf{b}_{\text{reg}}$$

where $\hat{\mathbf{b}}_i$ is the refined bounding box for the ith region proposal.

Limitations: Computationally expensive due to the need to run CNNs on thousands of region proposals. Slow inference speed.

Fast R-CNN Fast R-CNN improves upon R-CNN by processing the entire image with a CNN and then extracting region proposals from the feature map, rather than the input image.

1. Feature Map Extraction: The entire image is processed through a CNN to obtain a feature map.

$$\mathbf{F} = \text{CNN}(\mathbf{I})$$

where $\mathbf{I}$ is the input image and $\mathbf{F}$ is the feature map.

2. Region of Interest (RoI) Pooling: Region proposals are projected onto the feature map, and RoI pooling is used to extract fixed-size feature vectors for each proposal.

$$\mathbf{f}_i = \text{RoI Pooling}(\mathbf{F}, \mathbf{R}_i)$$

3. Classification and Regression: The feature vectors are fed into fully connected layers for classification and bounding box regression.

Faster R-CNN Faster R-CNN further improves the efficiency by introducing a Region Proposal Network (RPN) that shares convolutional layers with the detection network.

1. RPN: The RPN generates region proposals directly from the feature map.

$$\mathbf{R}_i = \text{RPN}(\mathbf{F})$$

2. RoI Pooling and Classification: The rest of the pipeline follows the Fast R-CNN approach.

Example in TensorFlow

```
import tensorflow as tf
from tensorflow.keras.applications import ResNet50
from tensorflow.keras.layers import Dense, Flatten, Conv2D
from tensorflow.keras.models import Model

# Define the base model (feature extractor)
base_model = ResNet50(weights='imagenet', include_top=False, input_shape=(224, 224, 3))
```

```
# Define the RPN model
x = base_model.output
rpn_conv = Conv2D(512, (3, 3), activation='relu', padding='same')(x)
rpn_cls = Conv2D(9 * 2, (1, 1), activation='softmax')(rpn_conv)  # 9 anchors, 2 classes
        (object/not-object)
rpn_reg = Conv2D(9 * 4, (1, 1))(rpn_conv)  # 9 anchors, 4 coordinates (dx, dy, dw, dh)

# Create the model
rpn_model = Model(inputs=base_model.input, outputs=[rpn_cls, rpn_reg])

# Compile the model
rpn_model.compile(optimizer='adam', loss=['binary_crossentropy', 'mse'])

# Print the model summary
rpn_model.summary()
```

Example in PyTorch

```
import torch
import torch.nn as nn
import torchvision.models as models

# Define the base model (feature extractor)
base_model = models.resnet50(pretrained=True)
modules = list(base_model.children())[:-2]  # Remove the last two layers
base_model = nn.Sequential(*modules)

# Define the RPN model
class RPN(nn.Module):
    def __init__(self, in_channels, num_anchors):
        super(RPN, self).__init__()
        self.conv = nn.Conv2d(in_channels, 512, kernel_size=3, stride=1, padding=1)
        self.cls = nn.Conv2d(512, num_anchors * 2, kernel_size=1, stride=1)  # 2 classes
        self.reg = nn.Conv2d(512, num_anchors * 4, kernel_size=1, stride=1)  # 4 coordinates

    def forward(self, x):
        x = torch.relu(self.conv(x))
        cls = self.cls(x)
        reg = self.reg(x)
        return cls, reg

# Instantiate the RPN model
rpn_model = RPN(in_channels=2048, num_anchors=9)  # 2048 channels from ResNet50

# Print the model summary
print(rpn_model)
```

YOLO and SSD Frameworks

You Only Look Once (YOLO) is an object detection framework that treats object detection as a single regression problem, predicting bounding boxes and class probabilities directly from the input image in one pass.

1. Grid Division: The input image is divided into an $S \times S$ grid. Each grid cell predicts B bounding boxes, confidence scores, and class probabilities.
2. Bounding Box Prediction: Each bounding box is represented by four coordinates (x, y, w, h), where (x, y) is the center, and (w, h) are the width and height.
3. Confidence Score: The confidence score reflects the Intersection over Union (IoU) between the predicted box and the ground truth box.

$$\text{Confidence} = P(\text{object}) \times \text{IoU}$$

4. Class Probability: Each grid cell predicts class probabilities for the object contained within it.

$$P(C_i|\text{object})$$

Advantages: Real-time object detection, Simplicity and efficiency.

Disadvantages: Struggles with small objects and objects in close proximity.

Single Shot MultiBox Detector (SSD) is another efficient object detection framework that makes predictions at multiple scales using feature maps from different layers of the network.

1. Multi-scale Predictions: SSD generates predictions from multiple feature maps of different resolutions, capturing objects of various sizes.
2. Default Boxes: Each feature map cell predicts a set of default boxes with different aspect ratios and scales.
3. Prediction: Each default box is associated with a confidence score and a set of class probabilities.

Advantages: High accuracy and speed, Better handling of objects of different sizes. Disadvantages: Complexity in training due to multi-scale predictions.

Example in TensorFlow (YOLO)

```
import tensorflow as tf
from tensorflow.keras.layers import Conv2D, Input, Reshape
from tensorflow.keras.models import Model

# Define the YOLO model
inputs = Input(shape=(416, 416, 3))
x = Conv2D(16, (3, 3), padding='same', activation='relu')(inputs)
x = Conv2D(32, (3, 3), padding='same', activation='relu')(x)
x = Conv2D(5 * (4 + 1 + 20), (1, 1), padding='same')(x)  # 5 boxes, 4 coords, 1 conf,
        20 classes
outputs = Reshape((13, 13, 5, 25))(x)  # Reshape to (grid, grid, boxes, box attributes)

# Create the model
model = Model(inputs, outputs)

# Print the model summary
model.summary()
```

Example in PyTorch (SSD)

```
import torch.nn as nn
import torchvision.models as models

# Define the SSD model
class SSD(nn.Module):
    def __init__(self, num_classes):
        super(SSD, self).__init__()
        base_model = models.vgg16(pretrained=True).features
        self.feature_extractor = nn.Sequential(*list(base_model)[:-1])
        self.loc_layers = nn.ModuleList

([
            nn.Conv2d(512, 4 * 4, kernel_size=3, padding=1),
            nn.Conv2d(1024, 6 * 4, kernel_size=3, padding=1)
        ])
        self.cls_layers = nn.ModuleList([
```

```
            nn.Conv2d(512, 4 * num_classes, kernel_size=3, padding=1),
            nn.Conv2d(1024, 6 * num_classes, kernel_size=3, padding=1)
        ])

    def forward(self, x):
        x = self.feature_extractor(x)
        locs = []
        confs = []
        for loc_layer, cls_layer in zip(self.loc_layers, self.cls_layers):
            locs.append(loc_layer(x))
            confs.append(cls_layer(x))
        return locs, confs

# Instantiate the SSD model
ssd_model = SSD(num_classes=21)

# Print the model summary
print(ssd_model)
```

Evaluation Metrics

Precision and recall are essential metrics for evaluating object detection performance. Precision measures the proportion of true positive detections among all detections, while recall measures the proportion of true positive detections among all actual objects.

Intersection over Union (IoU) is a metric that measures the overlap between the predicted bounding box and the ground truth bounding box. It is calculated as the ratio of the area of intersection to the area of union.

$$\text{IoU} = \frac{\text{Area of Intersection}}{\text{Area of Union}}$$

Average Precision (AP) and Mean Average Precision (mAP): AP summarizes the precision-recall curve into a single value by calculating the area under the curve (AUC). mAP is the mean of the AP values across all classes. AP is calculated as the average precision at different recall levels:

$$\text{AP} = \int_0^1 P(R)\,\mathrm{d}R$$

where $P(R)$ is the precision as a function of recall R.

mAP is the mean of the AP values across all classes C:

$$\text{mAP} = \frac{1}{C}\sum_{i=1}^{C} \text{AP}_i$$

4.6.3 *Image Segmentation*

Image segmentation is a critical task in computer vision that involves partitioning an image into multiple segments or regions, each corresponding to different objects or parts of objects. Unlike object detection, which provides bounding boxes around objects, image segmentation provides pixel-level annotations. There are two main types of image segmentation: semantic segmentation and instance segmentation.

Semantic Segmentation
Semantic segmentation assigns a class label to each pixel in an image, effectively classifying each pixel into a predefined category. The goal is to label each pixel with the corresponding object class, such as "car," "tree," or "building." Given an image $\mathbf{I}$ of size $H \times W$, the output of semantic segmentation is a label map $\mathbf{L}$ of the same size, where each pixel (i, j) is assigned a class label c:

$$\mathbf{L}(i, j) = \arg \max_{c \in \{1,...,C\}} P(c|\mathbf{I}(i, j))$$

where C is the total number of classes, and $P(c|\mathbf{I}(i, j))$ is the probability of class c given the pixel value at (i, j).

Example Architecture: Fully Convolutional Networks (FCNs) are commonly used for semantic segmentation. An FCN replaces the fully connected layers of a standard CNN with convolutional layers, enabling it to produce spatially dense predictions.

Example in TensorFlow

```
import tensorflow as tf
from tensorflow.keras.layers import Conv2D, UpSampling2D, Input
from tensorflow.keras.models import Model

# Define the FCN model for semantic segmentation
inputs = Input(shape=(256, 256, 3))
x = Conv2D(64, (3, 3), activation='relu', padding='same')(inputs)
x = Conv2D(128, (3, 3), activation='relu', padding='same')(x)
x = Conv2D(256, (3, 3), activation='relu', padding='same')(x)
x = UpSampling2D((2, 2))(x)
x = Conv2D(128, (3, 3), activation='relu', padding='same')(x)
x = UpSampling2D((2, 2))(x)
x = Conv2D(64, (3, 3), activation='relu', padding='same')(x)
x = Conv2D(21, (1, 1), activation='softmax')(x)  # 21 classes
model = Model(inputs, x)

# Compile the model
model.compile(optimizer='adam', loss='categorical_crossentropy', metrics=['accuracy'])

# Print the model summary
model.summary()
```

Example in PyTorch

```
import torch
import torch.nn as nn

# Define the FCN model for semantic segmentation
class FCN(nn.Module):
    def __init__(self, num_classes):
        super(FCN, self).__init__()
        self.conv1 = nn.Conv2d(3, 64, kernel_size=3, padding=1)
        self.conv2 = nn.Conv2d(64, 128, kernel_size=3, padding=1)
        self.conv3 = nn.Conv2d(128, 256, kernel_size=3, padding=1)
        self.upsample1 = nn.Upsample(scale_factor=2, mode='bilinear', align_corners=True)
        self.conv4 = nn.Conv2d(256, 128, kernel_size=3, padding=1)
        self.upsample2 = nn.Upsample(scale_factor=2, mode='bilinear', align_corners=True)
        self.conv5 = nn.Conv2d(128, 64, kernel_size=3, padding=1)
        self.conv6 = nn.Conv2d(64, num_classes, kernel_size=1)

    def forward(self, x):
        x = torch.relu(self.conv1(x))
        x = torch.relu(self.conv2(x))
        x = torch.relu(self.conv3(x))
        x = self.upsample1(x)
        x = torch.relu(self.conv4(x))
        x = self.upsample2(x)
        x = torch.relu(self.conv5(x))
        x = self.conv6(x)
        return x

# Instantiate the model
model = FCN(num_classes=21)

# Print the model summary
print(model)
```

Instance Segmentation

Instance segmentation goes beyond semantic segmentation by not only classifying each pixel but also distinguishing between different instances of the same class. It provides a separate label for each object instance, allowing the model to segment individual objects separately. Given an image $\mathbf{I}$ of size $H \times W$, the output of instance segmentation is a set of masks $\{\mathbf{M}_k\}$, where each mask $\mathbf{M}_k$ corresponds to a different object instance:

$$\mathbf{M}_k(i, j) = \begin{cases} 1 & \text{if pixel } (i, j) \text{ belongs to instance } k \\ 0 & \text{otherwise} \end{cases}$$

Each mask $\mathbf{M}_k$ is associated with a class label c_k and bounding box $\mathbf{B}_k$.

Example Architecture: Mask R-CNN extends Faster R-CNN by adding a branch for predicting segmentation masks on each Region of Interest (RoI) in parallel with the existing branches for classification and bounding box regression.

Practical Example in TensorFlow Using Matterport's Mask R-CNN

```
import mrcnn.config
import mrcnn.model as modellib

# Define the configuration for the model
class InferenceConfig(mrcnn.config.Config):
    NAME = "coco"
    IMAGES_PER_GPU = 1
    NUM_CLASSES = 1 + 80  # COCO has 80 classes

# Instantiate the model
config = InferenceConfig()
model = modellib.MaskRCNN(mode="inference", model_dir="./", config=config)

# Load pre-trained weights
model.load_weights("mask_rcnn_coco.h5", by_name=True)

# Print the model summary
model.keras_model.summary()
```

Example in PyTorch Using Detectron2

```
from detectron2.engine import DefaultPredictor
from detectron2.config import get_cfg
from detectron2 import model_zoo

# Configure the model
cfg = get_cfg()
cfg.merge_from_file(model_zoo.get_config_file
  ("COCO-InstanceSegmentation/mask_rcnn_R_50_FPN_3x.yaml"))
cfg.MODEL.ROI_HEADS.SCORE_THRESH_TEST = 0.5  # Set threshold for this model
cfg.MODEL.WEIGHTS = model_zoo.get_checkpoint_url
  ("COCO-InstanceSegmentation/mask_rcnn_R_50_FPN_3x.yaml")

# Instantiate the predictor
predictor = DefaultPredictor(cfg)

# Print the model configuration
print(cfg)
```

Evaluation Metrics

Intersection over Union (IoU) is used to evaluate the accuracy of object segmentation. It measures the overlap between the predicted segmentation mask and the ground truth mask.

$$\text{IoU} = \frac{\text{Area of Intersection}}{\text{Area of Union}}$$

Mean IoU (mIoU) is the mean IoU across all classes. It provides an overall performance metric for the segmentation model.

$$\text{mIoU} = \frac{1}{C}\sum_{c=1}^{C} \text{IoU}_c$$

where C is the number of classes and IoU_c is the IoU for class c.

Pixel accuracy measures the proportion of correctly classified pixels.

$$\text{Pixel Accuracy} = \frac{\sum_{i=1}^{H} \sum_{j=1}^{W} \mathbf{1}(\mathbf{L}(i, j) = \hat{\mathbf{L}}(i, j))}{H \times W}$$

where **1** is the indicator function that is 1 if the prediction is correct and 0 otherwise, $\mathbf{L}(i, j)$ is the true label, and $\hat{\mathbf{L}}(i, j)$ is the predicted label.

Example in Python

```
import numpy as np
from sklearn.metrics import jaccard_score, accuracy_score

# True and predicted labels (flattened for simplicity)
y_true = np.array([0, 1, 1, 0, 1])
y_pred = np.array([0, 1, 0, 0, 1])

# Calculate IoU
iou = jaccard_score(y_true, y_pred, average='macro')
print(f'IoU: {iou}')

# Calculate pixel accuracy
accuracy = accuracy_score(y_true, y_pred)
print(f'Pixel Accuracy: {accuracy}')
```

4.6.4 *Other Applications*

CNNs have revolutionized various fields in computer vision and beyond. In this section, we explore three additional applications of CNNs: face recognition, style transfer, and super-resolution. Each application leverages the powerful feature extraction capabilities of CNNs to address unique challenges and provide innovative solutions.

Face Recognition
Face recognition involves identifying or verifying a person from a digital image or video frame. It encompasses both face identification (recognizing who the person is) and face verification (confirming if the person matches a given identity). Given an input image **I**, the goal is to extract a feature vector **f** that uniquely represents the face. This feature vector is then compared with the feature vectors of known faces using a distance metric, such as Euclidean distance.

$$\mathbf{f} = \text{CNN}(\mathbf{I})$$

The similarity score between two feature vectors $\mathbf{f}_1$ and $\mathbf{f}_2$ is calculated as:

$$\text{Similarity}(\mathbf{f}_1, \mathbf{f}_2) = \frac{\mathbf{f}_1 \cdot \mathbf{f}_2}{\|\mathbf{f}_1\| \|\mathbf{f}_2\|}$$

One of the most popular architectures for face recognition is the FaceNet model, which uses a triplet loss function to learn an embedding space where faces of the same person are close together, and faces of different people are far apart.

Example in TensorFlow

```
import tensorflow as tf
from tensorflow.keras.layers import Input, Conv2D, MaxPooling2D, Flatten, Dense, Lambda
from tensorflow.keras.models import Model
import tensorflow.keras.backend as K

# Define the base CNN model for feature extraction
def create_base_model(input_shape):
    inputs = Input(shape=input_shape)
    x = Conv2D(64, (3, 3), activation='relu', padding='same')(inputs)
    x = MaxPooling2D((2, 2))(x)
    x = Conv2D(128, (3, 3), activation='relu', padding='same')(x)
    x = MaxPooling2D((2, 2))(x)
    x = Flatten()(x)
    x = Dense(128, activation='relu')(x)
    model = Model(inputs, x)
    return model

# Create the triplet loss function
def triplet_loss(y_true, y_pred, alpha=0.2):
    anchor, positive, negative = y_pred[0], y_pred[1], y_pred[2]
    pos_dist = K.sum(K.square(anchor - positive), axis=-1)
    neg_dist = K.sum(K.square(anchor - negative), axis=-1)
    basic_loss = pos_dist - neg_dist + alpha
    loss = K.maximum(basic_loss, 0.0)
    return loss

# Define the model
input_shape = (96, 96, 3)
base_model = create_base_model(input_shape)
anchor_input = Input(shape=input_shape, name='anchor_input')
positive_input = Input(shape=input_shape, name='positive_input')
negative_input = Input(shape=input_shape, name='negative_input')
anchor_embedding = base_model(anchor_input)
positive_embedding = base_model(positive_input)
negative_embedding = base_model(negative_input)
outputs = Lambda(lambda x: x)([anchor_embedding, positive_embedding, negative_embedding])
model = Model([anchor_input, positive_input, negative_input], outputs)

# Compile the model
model.compile(optimizer='adam', loss=triplet_loss)

# Print the model summary
model.summary()
```

Example in PyTorch

```
import torch
import torch.nn as nn
import torch.nn.functional as F

# Define the base CNN model for feature extraction
class BaseModel(nn.Module):
    def __init__(self):
        super(BaseModel, self).__init__()
        self.conv1 = nn.Conv2d(3, 64, kernel_size=3, padding=1)
        self.conv2 = nn.Conv2d(64, 128, kernel_size=3, padding=1)
        self.fc1 = nn.Linear(128 * 24 * 24, 128)

    def forward(self, x):
        x = F.relu(self.conv1(x))
        x = F.max_pool2d(x, 2)
        x = F.relu(self.conv2(x))
        x = F.max_pool2d(x, 2)
        x = x.view(-1, 128 * 24 * 24)
```

```
        x = F.relu(self.fc1(x))
        return x

# Define the triplet loss function
def triplet_loss(anchor, positive, negative, alpha=0.2):
    pos_dist = F.pairwise_distance(anchor, positive)
    neg_dist = F.pairwise_distance(anchor, negative)
    loss = F.relu(pos_dist - neg_dist + alpha)
    return loss.mean()

# Instantiate the model
model = BaseModel()

# Print the model summary
print(model)
```

Style Transfer

Style transfer involves modifying an image's style while preserving its content. This technique combines the content of one image with the style of another to create a new, stylized image. Given a content image $\mathbf{I}_c$ and a style image $\mathbf{I}_s$, the goal is to generate a new image $\mathbf{I}_g$ that has the content of $\mathbf{I}_c$ and the style of $\mathbf{I}_s$. This is achieved by minimizing a loss function that combines content and style losses.

Content Loss: The content loss measures the difference between the content representations of the generated image and the content image. Let $\mathbf{F}_c$ and $\mathbf{F}_g$ be the feature maps of $\mathbf{I}_c$ and $\mathbf{I}_g$ at a certain layer l:

$$\mathcal{L}_{\text{content}} = \frac{1}{2} \sum_{i,j} (\mathbf{F}_g^{(l)}(i, j) - \mathbf{F}_c^{(l)}(i, j))^2$$

Style Loss: The style loss measures the difference between the style representations (Gram matrices) of the generated image and the style image. Let $\mathbf{G}_s$ and $\mathbf{G}_g$ be the Gram matrices of $\mathbf{I}_s$ and $\mathbf{I}_g$:

$$\mathcal{L}_{\text{style}} = \frac{1}{4N^2M^2} \sum_{i,j} (\mathbf{G}_g(i, j) - \mathbf{G}_s(i, j))^2$$

where N is the number of feature maps and M is the number of elements in each feature map.

Total Loss: The total loss is a weighted sum of the content and style losses:

$$\mathcal{L}_{\text{total}} = \alpha \mathcal{L}_{\text{content}} + \beta \mathcal{L}_{\text{style}}$$

Example Architecture: The VGG network, pre-trained on ImageNet, is commonly used for style transfer due to its rich feature representations.

Example in TensorFlow

```
import tensorflow as tf
from tensorflow.keras.applications import VGG19
from tensorflow.keras.models import Model
from tensorflow.keras.preprocessing.image import load_img, img_to_array
import numpy as np

# Load VGG19 model pre-trained on ImageNet
vgg = VGG19(include_top=False, weights='imagenet')
vgg.trainable = False

# Define content and style layers
content_layer = 'block5_conv2'
style_layers = ['block1_conv1', 'block2_conv1', 'block3_conv1', 'block4_conv1',
 'block5_conv1']

# Create model outputs for style and content layers
content_model = Model(inputs=vgg.input, outputs=vgg.get_layer(content_layer).output)
style_models = [Model(inputs=vgg.input, outputs=vgg.get_layer(layer).output)
 for layer in style_layers]

# Function to compute content loss
def content_loss(content, generated):
    return tf.reduce_mean(tf.square(generated - content))

# Function to compute style loss
def gram_matrix(tensor):
    channels = int(tensor.shape[-1])
    a = tf.reshape(tensor, [-1, channels])
    n = tf.shape(a)[0]
    gram = tf.matmul(a, a, transpose_a=True)
    return gram / tf.cast(n, tf.float32)

def style_loss(style, generated):
    S = gram_matrix(style)
    G = gram_matrix(generated)
    return tf.reduce_mean(tf.square(S - G))

# Load and preprocess images
def load_and_process_image(image_path):
    img = load_img(image_path, target_size=(224, 224))
    img = img_to_array(img)
    img = np.expand_dims(img, axis=0)
    img = tf.keras.applications.vgg19.preprocess_input(img)
    return img

# Example
content_image = load_and_process_image('path_to_content_image.jpg')
style_image = load_and_process_image('path_to_style_image.jpg')

# Extract content and style features
content_features = content_model(content_image)
style_features = [model(style_image) for model in style_models]

# Print content and style features shapes
print(content_features.shape)
for feature in style_features:
    print(feature.shape)
```

Example in PyTorch

```
import torch
import torch.nn as nn
import torch.optim as optim
from torchvision import models, transforms
from PIL import Image

# Load VGG19 model pre-trained on ImageNet
vgg = models.vgg19(pretrained=True).features
for param in vgg.parameters():
    param.requires_grad = False

# Define content and style layers
content_layer = '21'  # Corresponds to block5_conv2
style_layers = ['0', '5', '10', '19', '28']  # Corresponds to block1_conv1, block2_conv1,
 block3_conv1, block4_conv1, block5_conv1

# Function to load and preprocess image
def load_image(image_path, max_size=400, shape=None):
    image = Image.open(image_path).convert('RGB')
    size = max_size if max(image.size) > max_size else max(image.size)
    if shape is not None:
        size = shape
    transform = transforms.Compose([
        transforms.Resize(size),
        transforms.ToTensor(),
        transforms.Normalize(mean=[0.485, 0.456, 0.406], std=[0.229, 0.224, 0.225]),
    ])
    image = transform(image)[:3, :, :].unsqueeze(0)
    return image

# Function to compute content loss
class ContentLoss(nn.Module):
    def __init__(self, target):
        super(ContentLoss, self).__init__()
        self.target = target.detach()

    def forward(self, x):
        self.loss = F.mse_loss(x, self.target)
        return x

# Function to compute style loss
class StyleLoss(nn.Module):
    def __init__(self, target_feature):
        super(StyleLoss, self).__init__()
        self.target = self.gram_matrix(target_feature).detach()

    def gram_matrix(self, input):
        a, b, c, d = input.size()
        features = input.view(a * b, c * d)
        G = torch.mm(features, features.t())
        return G.div(a * b * c * d)

    def forward(self, x):
        G = self.gram_matrix(x)
        self.loss = F.mse_loss(G, self.target)
        return x

# Example
content_image = load_image('path_to_content_image.jpg')
style_image = load_image('path_to_style_image.jpg', shape=content_image.shape[-2:])

# Print image shapes
print(content_image.shape)
print(style_image.shape)
```

Super-Resolution

Super-resolution involves enhancing the resolution of an image, increasing its size while preserving and enhancing its details. This technique is widely used in applications requiring high-quality images from low-resolution sources, such as medical imaging and satellite imagery. Given a low-resolution image $\mathbf{I}_{\text{LR}}$, the goal is to generate a high-resolution image $\mathbf{I}_{\text{HR}}$ that is visually close to the ground truth high-resolution image. This is often achieved by training a CNN to minimize the loss between the generated high-resolution image and the ground truth.

Example Architecture: The Super-Resolution Convolutional Neural Network (SRCNN) (Dong et al. 2014) is a popular architecture for super-resolution. It consists of three convolutional layers that progressively refine the resolution of the input image.

$$\mathbf{I}_{\text{HR}} = \text{SRCNN}(\mathbf{I}_{\text{LR}})$$

Example in TensorFlow

```
import tensorflow as tf
from tensorflow.keras.layers import Conv2D, Input
from tensorflow.keras.models import Model

# Define the SRCNN model for super-resolution
inputs = Input(shape=(None, None, 3))
x = Conv2D(64, (9, 9), activation='relu', padding='same')(inputs)
x = Conv2D(32, (1, 1), activation='relu', padding='same')(x)
x = Conv2D(3, (5, 5), activation='linear', padding='same')(x)
model = Model(inputs, x)

# Compile the model
model.compile(optimizer='adam', loss='mean_squared_error')

# Print the model summary
model.summary()
```

Example in PyTorch

```
import torch.nn as nn

# Define the SRCNN model for super-resolution
class SRCNN(nn.Module):
    def __init__(self):
        super(SRCNN, self).__init__()
        self.conv1 = nn.Conv2d(3, 64, kernel_size=9, padding=4)
        self.conv2 = nn.Conv2d(64, 32, kernel_size=1)
        self.conv3 = nn.Conv2d(32, 3, kernel_size=5, padding=2)

    def forward(self, x):
        x = F.relu(self.conv1(x))
        x = F.relu(self.conv2(x))
        x = self.conv3(x)
        return x

# Instantiate the model
model = SRCNN()

# Print the model summary
print(model)
```

4.7 Implementing CNNs with TensorFlow and PyTorch

4.7.1 Building CNNs with TensorFlow

Building a CNN in TensorFlow involves defining the model architecture, compiling the model, and preparing the data. We will demonstrate these steps with a simple example of building a CNN for image classification on the CIFAR-10 dataset.

```
import tensorflow as tf
from tensorflow.keras.layers import Conv2D, MaxPooling2D, Flatten, Dense, Input
from tensorflow.keras.models import Model

# Define the CNN model
def create_cnn_model(input_shape, num_classes):
    inputs = Input(shape=input_shape)
    x = Conv2D(32, (3, 3), activation='relu', padding='same')(inputs)
    x = MaxPooling2D((2, 2))(x)
    x = Conv2D(64, (3, 3), activation='relu', padding='same')(x)
    x = MaxPooling2D((2, 2))(x)
    x = Flatten()(x)
    x = Dense(128, activation='relu')(x)
    outputs = Dense(num_classes, activation='softmax')(x)
    model = Model(inputs, outputs)
    return model

# Create the model
input_shape = (32, 32, 3)
num_classes = 10
model = create_cnn_model(input_shape, num_classes)

# Compile the model
model.compile(optimizer='adam', loss='sparse_categorical_crossentropy', metrics=['accuracy'])

# Print the model summary
model.summary()
```

In this example, we defined a simple CNN model with two convolutional layers, each followed by a max-pooling layer, and two fully connected layers. The model is compiled using the Adam optimizer and the sparse categorical cross-entropy loss function, which is suitable for multi-class classification tasks.

Dataset Preparation: We will use the CIFAR-10 dataset, which consists of 60,000 32×32 color images in 10 classes, with 50,000 training images and 10,000 test images.

Data Preprocessing:

```
from tensorflow.keras.datasets import cifar10
from tensorflow.keras.utils import to_categorical

# Load and preprocess the CIFAR-10 dataset
(x_train, y_train), (x_test, y_test) = cifar10.load_data()
x_train = x_train.astype('float32') / 255.0
x_test = x_test.astype('float32') / 255.0
y_train = y_train.squeeze()
y_test = y_test.squeeze()
```

Training the Model:

```
# Train the model
history = model.fit(x_train, y_train, epochs=10, batch_size=64, validation_data=(x_test,
  y_test))
```

Evaluation:

```
# Evaluate the model
test_loss, test_accuracy = model.evaluate(x_test, y_test)
print(f'Test accuracy: {test_accuracy}')
```

In this example, we trained the model for 10 epochs with a batch size of 64 and evaluated its performance on the test set. The 'history' object contains training metrics, such as loss and accuracy, for each epoch.

TensorBoard for Visualization

TensorBoard is a powerful visualization tool that allows us to monitor various aspects of our model's training process, such as loss and accuracy over epochs, histograms of weights, and visualizations of the computational graph.

Enabling TensorBoard: To use TensorBoard, we need to set up a callback that logs training metrics during the training process.

```
from tensorflow.keras.callbacks import TensorBoard
import datetime

# Define TensorBoard callback
log_dir = "logs/fit/" + datetime.datetime.now().strftime("%Y%m%d-%H%M%S")
tensorboard_callback = TensorBoard(log_dir=log_dir, histogram_freq=1)

# Train the model with TensorBoard callback
history = model.fit(x_train, y_train, epochs=10, batch_size=64, validation_data=(x_test,
 y_test), callbacks=[tensorboard_callback])
```

Launching TensorBoard: After training the model with TensorBoard logging enabled, we can launch TensorBoard to visualize the training process.

```
tensorboard --logdir logs/fit
```

This command starts the TensorBoard server, which can be accessed through a web browser at 'http://localhost:6006'.

Visualizing Metrics: TensorBoard provides various tabs for visualizing different aspects of the training process:

Scalars: Plots of metrics such as loss and accuracy over epochs.

Graphs: Visual representation of the model's computational graph.

Histograms: Distribution of weights and biases in the model.

By using TensorBoard, we can gain valuable insights into the training process and identify potential issues, such as overfitting or vanishing gradients, early in the training phase.

4.7.2 Building CNNs with PyTorch

```
import torch
import torch.nn as nn
import torch.nn.functional as F

# Define the CNN model
class SimpleCNN(nn.Module):
    def __init__(self):
        super(SimpleCNN, self).__init__()
        self.conv1 = nn.Conv2d(3, 32, kernel_size=3, padding=1)
        self.conv2 = nn.Conv2d(32, 64, kernel_size=3, padding=1)
        self.pool = nn.MaxPool2d(kernel_size=2, stride=2, padding=0)
        self.fc1 = nn.Linear(64 * 8 * 8, 128)
        self.fc2 = nn.Linear(128, 10)

    def forward(self, x):
        x = F.relu(self.conv1(x))
        x = self.pool(x)
        x = F.relu(self.conv2(x))
        x = self.pool(x)
        x = x.view(-1, 64 * 8 * 8)
        x = F.relu(self.fc1(x))
        x = self.fc2(x)
        return x

# Instantiate the model
model = SimpleCNN()

# Print the model summary
print(model)
```

In this example, we defined a simple CNN model with two convolutional layers, each followed by a max-pooling layer, and two fully connected layers.

Training and Evaluation

Training a CNN in PyTorch involves defining the loss function, optimizer, and the training loop. We will also evaluate the model's performance on a validation set.

Dataset Preparation: We will use the CIFAR-10 dataset, which consists of 60,000 32×32 color images in 10 classes, with 50,000 training images and 10,000 test images.

Data Preprocessing:

```
import torchvision
import torchvision.transforms as transforms
from torch.utils.data import DataLoader

# Define data transformations
transform = transforms.Compose([
    transforms.ToTensor(),
    transforms.Normalize((0.4914, 0.4822, 0.4465), (0.2470, 0.2435, 0.2616))
])

# Load and preprocess the CIFAR-10 dataset
train_dataset = torchvision.datasets.CIFAR10(root='./data', train=True, download=True,
        transform=transform)
test_dataset = torchvision.datasets.CIFAR10(root='./data', train=False, download=True,
        transform=transform)

# Create data loaders
train_loader = DataLoader(train_dataset, batch_size=64, shuffle=True)
test_loader = DataLoader(test_dataset, batch_size=64, shuffle=False)
```

Training the Model:

```
import torch.optim as optim

# Define loss function and optimizer
criterion = nn.CrossEntropyLoss()
optimizer = optim.Adam(model.parameters(), lr=0.001)

# Training loop
for epoch in range(10):  # Number of epochs
    model.train()
    running_loss = 0.0
    for inputs, labels in train_loader:
        optimizer.zero_grad()  # Zero the parameter gradients
        outputs = model(inputs)
        loss = criterion(outputs, labels)
        loss.backward()
        optimizer.step()
        running_loss += loss.item()

    print(f'Epoch {epoch + 1}, Loss: {running_loss / len(train_loader)}')

print('Finished Training')
```

In this example, we trained the model for 10 epochs using the Adam optimizer and the cross-entropy loss function.

Evaluation:

```
# Evaluate the model
model.eval()
correct = 0
total = 0
with torch.no_grad():
    for inputs, labels in test_loader:
        outputs = model(inputs)
        _, predicted = torch.max(outputs.data, 1)
        total += labels.size(0)
        correct += (predicted == labels).sum().item()

accuracy = 100 * correct / total
print(f'Test accuracy: {accuracy}%')
```

In this example, we evaluated the model's performance on the test set and calculated the accuracy.

Visualizing with TorchVision
TorchVision is a package in the PyTorch ecosystem that provides tools for handling vision-related tasks. It includes datasets, model architectures, and utilities for transforming and visualizing images.

Visualizing Training Data: Visualizing the training data can provide insights into the dataset and help in identifying potential issues with data quality or preprocessing.

```
import matplotlib.pyplot as plt
import numpy as np

# Function to show an image
def imshow(img):
    img = img / 2 + 0.5  # Unnormalize
    npimg = img.numpy()
    plt.imshow(np.transpose(npimg, (1, 2, 0)))
    plt.show()
```

```
# Get a batch of training data
dataiter = iter(train_loader)
images, labels = dataiter.next()

# Show images
imshow(torchvision.utils.make_grid(images))
```

Visualizing Model Predictions: Visualizing the model's predictions can help in understanding how well the model performs and in identifying misclassified examples.

```
# Get a batch of test data
dataiter = iter(test_loader)
images, labels = dataiter.next()

# Make predictions
outputs = model(images)
_, predicted = torch.max(outputs, 1)

# Show images and predictions
imshow(torchvision.utils.make_grid(images))
print('GroundTruth: ', ' '.join(f'{classes[labels[j]]}' for j in range(4)))
print('Predicted: ', ' '.join(f'{classes[predicted[j]]}' for j in range(4)))
```

4.7.3 Case Studies

To solidify the understanding of CNNs, it is essential to apply the concepts in real-world scenarios. This section provides detailed case studies covering image classification, object detection, and image segmentation. These practical examples illustrate the application of CNNs to diverse tasks, demonstrating their versatility and effectiveness.

Object Detection Case Study
Model Architecture: We will use the Faster R-CNN model for object detection, which consists of a backbone CNN for feature extraction, a Region Proposal Network (RPN) for generating region proposals, and a detection head for classifying and refining bounding boxes.

1. Region Proposal Network (RPN):

$$\mathbf{R}_i = \text{RPN}(\mathbf{F})$$

2. RoI Pooling:

$$\mathbf{f}_i = \text{RoI Pooling}(\mathbf{F}, \mathbf{R}_i)$$

3. Classification and Bounding Box Regression:

$$\hat{\mathbf{b}}_i = \mathbf{W}_{\text{reg}}\mathbf{f}_i + \mathbf{b}_{\text{reg}}$$

$$\hat{\mathbf{c}}_i = \mathbf{W}_{\text{cls}}\mathbf{f}_i + \mathbf{b}_{\text{cls}}$$

Implementation in PyTorch using torchvision

```
import torch
from torchvision.models.detection import fasterrcnn_resnet50_fpn
from torchvision.datasets import VOCDetection
import torchvision.transforms as transforms
from torch.utils.data import DataLoader

# Load and preprocess the Pascal VOC dataset
transform = transforms.Compose([transforms.ToTensor()])
train_dataset = VOCDetection(root='./data', year='2012', image_set='train', download=True,
 transform=transform)
test_dataset = VOCDetection(root='./data', year='2012', image_set='val', download=True,
 transform=transform)
train_loader = DataLoader(train_dataset, batch_size=2, shuffle=True, collate_fn=lambda x:
 tuple(zip(*x)))
test_loader = DataLoader(test_dataset, batch_size=2, shuffle=False, collate_fn=lambda x:
 tuple(zip(*x)))

# Define the Faster R-CNN model
model = fasterrcnn_resnet50_fpn(pretrained=True)
model.train()

# Define optimizer
optimizer = torch.optim.SGD(model.parameters(), lr=0.005, momentum=0.9, weight_decay=0.0005)

# Training loop
for epoch in range(10):
    running_loss = 0.0
    for images, targets in train_loader:
        images = list(image for image in images)
        targets = [{k: v for k, v in t.items()} for t in targets]
        loss_dict = model(images, targets)
        losses = sum(loss for loss in loss_dict.values())
        optimizer.zero_grad()
        losses.backward()
        optimizer.step()
        running_loss += losses.item()
    print(f'Epoch {epoch + 1}, Loss: {running_loss / len(train_loader)}')

print('Finished Training')
```

In this example, we used the torchvision library to load the Pascal VOC dataset and train a Faster R-CNN model. The model is trained using SGD with momentum, and the training loop iterates over the dataset for 10 epochs.

Image Segmentation Case Study

Problem Definition: Image segmentation involves partitioning an image into multiple segments or regions, each corresponding to different objects or parts of objects. We will use the U-Net architecture for semantic segmentation on the Medical Decathlon dataset, specifically the brain tumor segmentation task.

Model Architecture: U-Net is a popular architecture for image segmentation, consisting of an encoder–decoder structure with skip connections.

1. Encoder (Down-sampling):

$$\mathbf{F}_i = \text{Conv}(\mathbf{F}_{i-1})$$

2. Decoder (Up-sampling):

$$\mathbf{F}'_i = \text{UpConv}(\mathbf{F}_{i+1})$$

3. Skip Connections:

$$\mathbf{F}_i'' = \mathbf{F}_i \oplus \mathbf{F}_i'$$

Implementation in TensorFlow

```
import tensorflow as tf
from tensorflow.keras.layers import Conv2D, MaxPooling2D, Conv2DTranspose, concatenate, Input
from tensorflow.keras.models import Model

# Define the U-Net model
def unet_model(input_shape):
    inputs = Input(shape=input_shape)
    c1 = Conv2D(64, (3, 3), activation='relu', padding='same')(inputs)
    c1 = Conv2D(64, (3, 3), activation='relu', padding='same')(c1)
    p1 = MaxPooling2D((2, 2))(c1)

    c2 = Conv2D(128, (3, 3), activation='relu', padding='same')(p1)
    c2 = Conv2D(128, (3, 3), activation='relu', padding='same')(c2)
    p2 = MaxPooling2D((2, 2))(c2)

    c3 = Conv2D(256, (3, 3), activation='relu', padding='same')(p2)
    c3 = Conv2D(256, (3, 3), activation='relu', padding='same')(c3)
    p3 = MaxPooling2D((2, 2))(c3)

    c4 = Conv2D(512, (3, 3), activation='relu', padding='same')(p3)
    c4 = Conv2D(512, (3, 3), activation='relu', padding='same')(c4)
    p4 = MaxPooling2D((2, 2))(c4)

    c5 = Conv2D(1024, (3, 3), activation='relu', padding='same')(p4)
    c5 = Conv2D(1024, (3, 3), activation='relu', padding='same')(c5)

    u6 = Conv2DTranspose(512, (2, 2), strides=(2, 2), padding='same')(c5)
    u6 = concatenate([u6, c4])
    c6 = Conv2D(512, (3, 3), activation='relu', padding='same')(u6)
    c6 = Conv2D(512, (3, 3), activation='relu', padding='same')(c6)

    u7 = Conv2DTranspose(256, (2, 2), strides=(2, 2), padding='same')(c6)
    u7 = concatenate([u7, c3])
    c7 = Conv2D(256, (3, 3), activation='relu', padding='same')(u7)
    c7 = Conv2D(256, (3, 3), activation='relu', padding='same')(c7)

    u8 = Conv2DTranspose(128, (2, 2), strides=(2, 2), padding='same')(c7)
    u8 = concatenate([u8, c2])
    c8 = Conv2D(128, (3, 3), activation='relu', padding='same')(u8)
    c8 = Conv2D(128, (3, 3), activation='relu', padding='same')(c8)

    u9 = Conv2DTranspose(64, (2, 2), strides=(2, 2), padding='same')(c8)
    u9 = concatenate([u9, c1])
    c9 = Conv2D(64, (3, 3), activation='relu', padding='same')(u9)
    c9 = Conv2D(64, (3, 3), activation='relu', padding='same')(c9)

    outputs = Conv2D(1, (1, 1), activation='sigmoid')(c9)
    model = Model(inputs, outputs)
    return model

# Create the model
input_shape = (128, 128, 1)
model = unet_model(input_shape)

# Compile the model
model.compile(optimizer='adam', loss='binary_crossentropy', metrics=['accuracy'])

# Print the model summary
model.summary()
```

Implementation in PyTorch

```
import torch
import torch.nn as nn
import torch.nn.functional as F

# Define the U-Net model
class UNet(nn.Module):
    def __init__(self):
        super(UNet, self).__init__()
        self.enc1 = self.conv_block(1, 64)
        self.enc2 = self.conv_block(64, 128)
        self.enc3 = self.conv_block(128, 256)
        self.enc4 = self.conv_block(256, 512)
        self.center = self.conv_block(512, 1024)
        self.dec4 = self.conv_block(1024 + 512, 512)
        self.dec3 = self.conv_block(512 + 256, 256)
        self.dec2 = self.conv_block(256 + 128, 128)
        self.dec1 = self.conv_block(128 + 64, 64)
        self.final = nn.Conv2d(64, 1, kernel_size=1)

    def conv_block(self, in_channels, out_channels):
        return nn.Sequential(
            nn.Conv2d(in_channels, out_channels, kernel_size=3, padding=1),
            nn.ReLU(inplace=True),
            nn.Conv2d(out_channels, out_channels, kernel_size=3, padding=1),
            nn.ReLU(inplace=True)
        )

    def forward(self, x):
        enc1 = self.enc1(x)
        enc2 = self.enc2(F.max_pool2d(enc1, 2))
        enc3 = self.enc3(F.max_pool2d(enc2, 2))
        enc4 = self.enc4(F.max_pool2d(enc3, 2))
        center = self.center(F.max_pool2d(enc4, 2))
        dec4 = self.dec4(torch.cat([F.interpolate(center, scale_factor=2, mode='bilinear',
                align_corners=True), enc4], 1))
        dec3 = self.dec3(torch.cat([F.interpolate(dec4, scale_factor=2, mode='bilinear',
                align_corners=True), enc3], 1))
        dec2 = self.dec2(torch.cat([F.interpolate(dec3, scale_factor=2, mode='bilinear',
                align_corners=True), enc2], 1))
        dec1 = self.dec1(torch.cat([F.interpolate(dec2, scale_factor=2, mode='bilinear',
                align_corners=True), enc1], 1))
        final = torch.sigmoid(self.final(dec1))
        return final

# Instantiate the model
model = UNet()

# Print the model summary
print(model)
```

4.8 Exercises

1. Given an input image $\mathbf{I} \in \mathbb{R}^{5\times5}$ and a filter $\mathbf{K} \in \mathbb{R}^{3\times3}$:

$$\mathbf{I} = \begin{pmatrix} 1 & 2 & 3 & 4 & 5 \\ 6 & 7 & 8 & 9 & 10 \\ 11 & 12 & 13 & 14 & 15 \\ 16 & 17 & 18 & 19 & 20 \\ 21 & 22 & 23 & 24 & 25 \end{pmatrix}, \quad \mathbf{K} = \begin{pmatrix} 1 & 0 & -1 \\ 1 & 0 & -1 \\ 1 & 0 & -1 \end{pmatrix}$$

(a) Perform a valid convolution operation on **I** using **K**.
(b) Perform the same convolution operation using padding to maintain the original image size.

2. Given a convolutional layer with 32 filters of size 3×3 applied to an input image of size $28 \times 28 \times 3$, compute the output dimensions:

   (a) Without padding and a stride of 1.
   (b) With padding to maintain the input dimensions and a stride of 1.

3. Compute the output of a 2×2 max-pooling operation applied to the following matrix:

$$\mathbf{M} = \begin{pmatrix} 1 & 2 & 3 & 4 \\ 5 & 6 & 7 & 8 \\ 9 & 10 & 11 & 12 \\ 13 & 14 & 15 & 16 \end{pmatrix}$$

4. Consider a 2D input tensor (image) $\mathbf{I} \in \mathbb{R}^{4\times4}$:

$$\mathbf{I} = \begin{pmatrix} 1 & 2 & 3 & 4 \\ 5 & 6 & 7 & 8 \\ 9 & 10 & 11 & 12 \\ 13 & 14 & 15 & 16 \end{pmatrix}$$

   and a filter (kernel) $\mathbf{K} \in \mathbb{R}^{2\times2}$:

$$\mathbf{K} = \begin{pmatrix} 1 & 0 \\ -1 & 1 \end{pmatrix}$$

   (a) Perform a 2D convolution of **I** with **K** using a stride of 1 and no padding. Write down the resulting matrix.
   (b) Compute the same convolution using a stride of 2 and no padding. Write down the resulting matrix.

5. Given a 3D input tensor $\mathbf{X} \in \mathbb{R}^{3\times3\times3}$ representing a multi-channel image:

$$\mathbf{X} = \left(\begin{pmatrix} 1 & 2 & 3 \\ 4 & 5 & 6 \\ 7 & 8 & 9 \end{pmatrix}, \begin{pmatrix} 9 & 8 & 7 \\ 6 & 5 & 4 \\ 3 & 2 & 1 \end{pmatrix}, \begin{pmatrix} 2 & 4 & 6 \\ 8 & 10 & 12 \\ 14 & 16 & 18 \end{pmatrix} \right)$$

   and a filter tensor $\mathbf{F} \in \mathbb{R}^{2\times2\times3}$:

$$\mathbf{F} = \left(\begin{pmatrix} 0 & 1 \\ -1 & 0 \end{pmatrix}, \begin{pmatrix} 1 & 0 \\ 0 & -1 \end{pmatrix}, \begin{pmatrix} -1 & 1 \\ 1 & -1 \end{pmatrix} \right)$$

   Perform a 3D convolution of **X** with **F** using a stride of 1 and no padding. Write down the resulting feature map.

6. Consider a 2D input tensor $\mathbf{A} \in \mathbb{R}^{5\times5}$:

$$\mathbf{A} = \begin{pmatrix} 1 & 2 & 3 & 4 & 5 \\ 6 & 7 & 8 & 9 & 10 \\ 11 & 12 & 13 & 14 & 15 \\ 16 & 17 & 18 & 19 & 20 \\ 21 & 22 & 23 & 24 & 25 \end{pmatrix}$$

and a filter (kernel) $\mathbf{B} \in \mathbb{R}^{3\times3}$:

$$\mathbf{B} = \begin{pmatrix} 1 & 0 & -1 \\ 2 & 0 & -2 \\ 1 & 0 & -1 \end{pmatrix}$$

Perform a 2D dilated convolution of **A** with **B** using a dilation rate of 2, stride of 1, and no padding. Write down the resulting matrix.

7. Given a 2D input tensor $\mathbf{C} \in \mathbb{R}^{2\times2}$:

$$\mathbf{C} = \begin{pmatrix} 1 & 2 \\ 3 & 4 \end{pmatrix}$$

and a filter (kernel) $\mathbf{D} \in \mathbb{R}^{3\times3}$:

$$\mathbf{D} = \begin{pmatrix} 1 & 0 & 1 \\ 0 & 1 & 0 \\ 1 & 0 & 1 \end{pmatrix}$$

Perform a 2D transposed convolution of **C** with **D** using a stride of 2 and no padding. Write down the resulting matrix.

8. Consider a CNN layer with a single input channel, one convolutional layer with 2 filters, and the following input tensor $\mathbf{E} \in \mathbb{R}^{4\times4}$:

$$\mathbf{E} = \begin{pmatrix} 1 & 2 & 3 & 4 \\ 5 & 6 & 7 & 8 \\ 9 & 10 & 11 & 12 \\ 13 & 14 & 15 & 16 \end{pmatrix}$$

and filters $\mathbf{F}_1$ and $\mathbf{F}_2$ both of shape $\mathbb{R}^{2\times2}$:

$$\mathbf{F}_1 = \begin{pmatrix} 1 & -1 \\ 1 & -1 \end{pmatrix}, \quad \mathbf{F}_2 = \begin{pmatrix} 0 & 1 \\ -1 & 0 \end{pmatrix}$$

(a) Perform the convolution of **E** with $\mathbf{F}_1$ and $\mathbf{F}_2$ using a stride of 1 and no padding. Write down the resulting feature maps.

(b) Apply a ReLU activation function to the resulting feature maps. Write down the activated feature maps.

9. Given a 4D input tensor $\mathbf{X} \in \mathbb{R}^{2\times4\times3\times3}$ representing a batch of 2 images with 4 channels each, and spatial dimensions of 3×3:

$$\mathbf{X}_{1,:,:,:} = \left(\begin{pmatrix} 1 & 2 & 3 \\ 4 & 5 & 6 \\ 7 & 8 & 9 \end{pmatrix}, \begin{pmatrix} 9 & 8 & 7 \\ 6 & 5 & 4 \\ 3 & 2 & 1 \end{pmatrix}, \begin{pmatrix} 2 & 4 & 6 \\ 8 & 10 & 12 \\ 14 & 16 & 18 \end{pmatrix}, \begin{pmatrix} 18 & 16 & 14 \\ 12 & 10 & 8 \\ 6 & 4 & 2 \end{pmatrix} \right)$$

Perform group normalization with 2 groups. Calculate the normalized tensor for the first image in the batch. Provide the resulting values.

10. Consider a residual block where the input tensor $\mathbf{Y} \in \mathbb{R}^{3\times3}$ is:

$$\mathbf{Y} = \begin{pmatrix} 1 & 2 & 3 \\ 4 & 5 & 6 \\ 7 & 8 & 9 \end{pmatrix}$$

and the convolution operation within the block is represented by the filter $\mathbf{K} \in \mathbb{R}^{3\times3}$:

$$\mathbf{K} = \begin{pmatrix} 0 & 1 & 0 \\ 1 & -4 & 1 \\ 0 & 1 & 0 \end{pmatrix}$$

(a) Compute the output of the convolution operation on $\mathbf{Y}$ using $\mathbf{K}$ with padding to keep the output dimensions the same.
(b) Add the residual connection by summing the input tensor $\mathbf{Y}$ with the output of the convolution. Write down the resulting tensor.

11. Let $\mathbf{F} \in \mathbb{R}^{3\times3\times4}$ be a feature map with spatial dimensions 3×3 and 4 channels:

$$\mathbf{F} = \left(\begin{pmatrix} 1 & 0 & 2 & 3 \\ 4 & 1 & 0 & 2 \\ 3 & 2 & 1 & 4 \end{pmatrix}, \begin{pmatrix} 2 & 3 & 1 & 0 \\ 0 & 1 & 3 & 2 \\ 4 & 2 & 1 & 0 \end{pmatrix}, \begin{pmatrix} 3 & 2 & 0 & 1 \\ 1 & 0 & 3 & 2 \\ 2 & 4 & 1 & 0 \end{pmatrix} \right)$$

(a) Compute the attention weights using a self-attention mechanism for each spatial location, considering only the first channel. Provide the attention map.
(b) Apply the computed attention weights to the feature map $\mathbf{F}$. Write down the resulting attended feature map.

12. Given a 2D input image tensor $\mathbf{I} \in \mathbb{R}^{4\times4}$:

$$\mathbf{I} = \begin{pmatrix} 1 & 2 & 3 & 4 \\ 5 & 6 & 7 & 8 \\ 9 & 10 & 11 & 12 \\ 13 & 14 & 15 & 16 \end{pmatrix}$$

(a) Perform a horizontal flip augmentation. Write down the resulting image tensor.
(b) Perform a 90-degree rotation clockwise. Write down the resulting image tensor.
(c) Apply a Gaussian noise augmentation with a mean of 0 and standard deviation of 1. Write down the resulting image tensor.

13. Consider a 3D input tensor $\mathbf{G} \in \mathbb{R}^{2\times3\times3}$ with 2 channels and spatial dimensions 3×3:

$$\mathbf{G}_{:,:,:} = \left(\begin{pmatrix} 1 & 2 & 3 \\ 4 & 5 & 6 \\ 7 & 8 & 9 \end{pmatrix}, \begin{pmatrix} 9 & 8 & 7 \\ 6 & 5 & 4 \\ 3 & 2 & 1 \end{pmatrix} \right)$$

and a filter (kernel) $\mathbf{H} \in \mathbb{R}^{2\times2\times2}$:

$$\mathbf{H}_{:,:,:} = \left(\begin{pmatrix} 1 & 0 \\ -1 & 1 \end{pmatrix}, \begin{pmatrix} 0 & 1 \\ 1 & -1 \end{pmatrix} \right)$$

(a) Perform a 2D dilated convolution on $\mathbf{G}$ with $\mathbf{H}$ using a dilation rate of 2 and stride of 1. Write down the resulting tensor.
(b) Apply group normalization with 2 groups to the resulting tensor. Write down the normalized tensor.

Chapter 5
Modeling Temporal Data

If you can't beat it, join it.

—Elon Musk, Artificial Intelligence, CO_2, and the Future of Energy

5.1 Introduction to Recurrent Neural Networks

Recurrent neural networks (RNNs) are a class of neural networks designed to handle sequential data, making them particularly well-suited for modeling temporal dependencies. Unlike traditional feed-forward neural networks, which assume that inputs are independent of each other, RNNs are explicitly designed to recognize patterns across sequences of data, such as time series, text, and audio. This section provides an overview of RNNs, their applications in temporal data, and their historical development. RNNs are distinguished by their ability to maintain a hidden state that captures information about previous inputs in the sequence. This hidden state is updated at each time step, allowing the network to retain memory of past events. Mathematically, an RNN can be described as follows:

Given an input sequence $\{\mathbf{x}_t\}_{t=1}^{T}$, where $\mathbf{x}_t$ is the input at time step t, the hidden state $\mathbf{h}_t$ at time step t is computed using:

$$\mathbf{h}_t = f(\mathbf{W}_{hx}\mathbf{x}_t + \mathbf{W}_{hh}\mathbf{h}_{t-1} + \mathbf{b}_h)$$

where $\mathbf{W}_{hx}$ is the weight matrix connecting the input to the hidden state, $\mathbf{W}_{hh}$ is the weight matrix connecting the hidden state at the previous time step to the hidden state at the current time step, $\mathbf{b}_h$ is the bias vector for the hidden state, f is an activation function, typically a non-linear function such as tanh or ReLU. The output $\mathbf{y}_t$ at time step t is then computed as:

$$\mathbf{y}_t = g(\mathbf{W}_{hy}\mathbf{h}_t + \mathbf{b}_y)$$

P. Singh and B. Raman, *Deep Learning Through the Prism of Tensors*, Studies in Big Data 162, https://doi.org/10.1007/978-981-97-8019-8_5

where $\mathbf{W}_{hy}$ is the weight matrix connecting the hidden state to the output, $\mathbf{b}_y$ is the bias vector for the output, and g is an activation function or softmax function for classification tasks.

One of the key challenges with basic RNNs is the vanishing gradient problem, which occurs when gradients propagated through many layers tend to diminish, making it difficult for the network to learn long-range dependencies. This issue led to the development of more advanced architectures, such as Long Short-Term Memory (LSTM) networks and Gated Recurrent Units (GRUs).

5.1.1 Applications of RNNs in Temporal Data

RNNs have been successfully applied to a wide range of temporal data tasks due to their ability to capture temporal dependencies and patterns. Some of the notable applications include:

1. Time Series Forecasting: RNNs can predict future values in a time series based on past observations. This is useful in domains such as finance (stock prices), weather forecasting, and demand prediction.
2. Natural Language Processing (NLP): RNNs are used for tasks like language modeling, text generation, machine translation, and sentiment analysis. They can process sequences of words or characters and generate coherent text outputs.
3. Speech Recognition: RNNs are employed to convert spoken language into written text by modeling the temporal dependencies in audio signals.
4. Video Analysis: RNNs can analyze sequences of video frames to recognize activities, track objects, and generate video descriptions.
5. Anomaly Detection: RNNs can identify unusual patterns or outliers in sequential data, which is useful for monitoring systems and detecting faults or fraud.

5.1.2 Historical Context and Development

The development of RNNs has a rich history, marked by significant milestones and advancements. The concept of RNNs was first introduced in the 1980s, but early models were limited by the vanishing gradient problem, which hindered their ability to learn long-term dependencies. Over time, various solutions and enhancements have been proposed, leading to the modern RNN architectures we use today.

1. Early Development: Hopfield (1982) introduced the Hopfield network, which laid the groundwork for recurrent structures in neural networks. Rumelhart et al. (1986) and colleagues described the concept of Backpropagation Through Time (BPTT), an extension of the backpropagation algorithm for training RNNs.
2. Addressing the Vanishing Gradient Problem: Hochreiter and Schmidhuber (1997) proposed the Long Short-Term Memory (LSTM) network, which introduced gating mechanisms to maintain gradients over long sequences and effectively learn long-term dependencies. Cho et al. (2014) introduced the Gated Recurrent Unit (GRU), a simpler variant of the LSTM with fewer parameters but similar performance.
3. Advancements and Applications: The advent of powerful GPUs and large datasets has enabled the training of deep RNNs, leading to breakthroughs in various fields, including natural language processing, speech recognition, and time series forecasting. The development of sequence-to-sequence (Seq2Seq) models by Sutskever et al. (2014) revolutionized machine translation and text generation.
4. Integration with Other Architectures: Recent research has explored the integration of RNNs with convolutional neural networks (CNNs) for tasks such as video analysis and image captioning. The rise of attention mechanisms, such as the transformer model introduced by Vaswani et al. (2017), has provided alternative approaches to sequence modeling, often outperforming traditional RNNs in certain tasks.

5.2 Basic Architecture of RNNs

RNNs are designed to handle sequential data, making them ideal for tasks involving time series, language, and other temporally dependent data. This section delves into the fundamental structure of RNNs, focusing on their recurrent units and hidden states, and explaining the concept of unfolding RNNs in time.

5.2.1 Structure of Recurrent Neural Networks

The basic architecture of an RNN involves an input layer, hidden layers with recurrent units, and an output layer (see Fig. 5.1 for an overview of the architecture). What sets RNNs apart from traditional feed-forward neural networks is their ability to maintain a hidden state that captures information from previous time steps. This hidden state is updated at each time step, allowing the network to learn dependencies across time.

Recurrent Units and Hidden States

At the core of an RNN is the recurrent unit, which processes each element of the input sequence and updates the hidden state. The hidden state acts as a memory, storing information about the sequence processed so far. Mathematically, the hidden state $\mathbf{h}_t$ at time step t is computed using:

$$\mathbf{h}_t = f(\mathbf{W}_{hx}\mathbf{x}_t + \mathbf{W}_{hh}\mathbf{h}_{t-1} + \mathbf{b}_h)$$

where $\mathbf{W}_{hx}$ is the weight matrix connecting the input $\mathbf{x}_t$ to the hidden state, $\mathbf{W}_{hh}$ is the weight matrix connecting the hidden state at the previous time step $\mathbf{h}_{t-1}$ to the hidden state at the current time step, $\mathbf{b}_h$ is the bias vector for the hidden state, and f is an activation function, typically tanh or ReLU. The output $\mathbf{y}_t$ at time step t is then computed as:

$$\mathbf{y}_t = g(\mathbf{W}_{hy}\mathbf{h}_t + \mathbf{b}_y)$$

where $\mathbf{W}_{hy}$ is the weight matrix connecting the hidden state to the output, $\mathbf{b}_y$ is the bias vector for the output, g is an activation function or softmax function for classification tasks.

The recurrence relation in RNNs allows them to propagate information from one time step to the next, enabling the network to learn temporal dependencies.

Unfolding RNNs in Time

To understand how RNNs process sequences, it is helpful to visualize them as being "unfolded" in time. Unfolding an RNN means representing the network at each time step as a separate layer in a deep network, with shared weights across these layers. This unfolding illustrates how the hidden state is updated over time and how it influences the outputs at each time step.

Consider a sequence of inputs $\{\mathbf{x}_1, \mathbf{x}_2, \ldots, \mathbf{x}_T\}$. When the RNN is unfolded, it can be visualized as follows:

1. Time Step 1:

$$\mathbf{h}_1 = f(\mathbf{W}_{hx}\mathbf{x}_1 + \mathbf{b}_h)$$
$$\mathbf{y}_1 = g(\mathbf{W}_{hy}\mathbf{h}_1 + \mathbf{b}_y)$$

2. Time Step 2:

$$\mathbf{h}_2 = f(\mathbf{W}_{hx}\mathbf{x}_2 + \mathbf{W}_{hh}\mathbf{h}_1 + \mathbf{b}_h)$$
$$\mathbf{y}_2 = g(\mathbf{W}_{hy}\mathbf{h}_2 + \mathbf{b}_y)$$

3. Time Step 3:

$$\mathbf{h}_3 = f(\mathbf{W}_{hx}\mathbf{x}_3 + \mathbf{W}_{hh}\mathbf{h}_2 + \mathbf{b}_h)$$
$$\mathbf{y}_3 = g(\mathbf{W}_{hy}\mathbf{h}_3 + \mathbf{b}_y)$$

And so on, until the final time step T:

$$\mathbf{h}_T = f(\mathbf{W}_{hx}\mathbf{x}_T + \mathbf{W}_{hh}\mathbf{h}_{T-1} + \mathbf{b}_h)$$
$$\mathbf{y}_T = g(\mathbf{W}_{hy}\mathbf{h}_T + \mathbf{b}_y)$$

This unfolding shows how the hidden state $\mathbf{h}_t$ at each time step t depends on the current input $\mathbf{x}_t$ and the hidden state from the previous time step $\mathbf{h}_{t-1}$. The same set of weights $\mathbf{W}_{hx}$, $\mathbf{W}_{hh}$, and $\mathbf{W}_{hy}$ are used across all time steps, which allows the network to generalize patterns learned across different parts of the sequence. The training of RNNs involves Backpropagation Through Time (BPTT), where the gradients are computed for each time step and propagated backward through the unfolded network. This process updates the weights to minimize the loss function, accounting for dependencies across the entire sequence.

Example: Consider a simple example of an RNN for predicting the next value in a sequence. Given a sequence of past values, the RNN can be trained to predict the next value at each time step.

```
import tensorflow as tf
from tensorflow.keras.models import Sequential
from tensorflow.keras.layers import SimpleRNN, Dense

# Define the RNN model
model = Sequential()
model.add(SimpleRNN(50, activation='relu', input_shape=(10, 1)))  # 10 time steps, 1 feature
model.add(Dense(1))

# Compile the model
model.compile(optimizer='adam', loss='mean_squared_error')

# Generate synthetic data for demonstration
import numpy as np
x_train = np.random.rand(1000, 10, 1)  # 1000 samples, 10 time steps, 1 feature
y_train = np.random.rand(1000, 1)      # 1000 target values

# Train the model
model.fit(x_train, y_train, epochs=20, batch_size=32)

# Evaluate the model
x_test = np.random.rand(100, 10, 1)
y_test = np.random.rand(100, 1)
loss = model.evaluate(x_test, y_test)
print(f'Test Loss: {loss}')
```

Implementation in PyTorch:

```
import torch
import torch.nn as nn
import torch.optim as optim

# Define the RNN model
class SimpleRNN(nn.Module):
    def __init__(self):
        super(SimpleRNN, self).__init__()
        self.rnn = nn.RNN(input_size=1, hidden_size=50, num_layers=1, batch_first=True)
        self.fc = nn.Linear(50, 1)

    def forward(self, x):
        h0 = torch.zeros(1, x.size(0), 50)  # Initial hidden state
        out, _ = self.rnn(x, h0)
        out = self.fc(out[:, -1, :])
        return out

# Instantiate the model
model = SimpleRNN()

# Define the loss function and optimizer
criterion = nn.MSELoss()
optimizer = optim.Adam(model.parameters(), lr=0.001)

# Generate synthetic data for demonstration
x_train = torch.rand(1000, 10, 1)
y_train = torch.rand(1000, 1)

# Train the model
num_epochs = 20
for epoch in range(num_epochs):
    model.train()
    outputs = model(x_train)
    loss = criterion(outputs, y_train)

    optimizer.zero_grad()
    loss.backward()
    optimizer.step()

    if (epoch+1) % 5 == 0:
        print(f'Epoch [{epoch+1}/{num_epochs}], Loss: {loss.item():.4f}')

# Evaluate the model
model.eval()
x_test = torch.rand(100, 10, 1)
y_test = torch.rand(100, 1)
with torch.no_grad():
    predictions = model(x_test)
    test_loss = criterion(predictions, y_test)
print(f'Test Loss: {test_loss.item():.4f}')
```

5.2.2 Mathematical Formulation

This section provides a detailed look at the forward pass in RNNs and the backward pass, specifically focusing on BPTT.

Forward Pass in RNNs
The forward pass in an RNN involves computing the hidden states and outputs for each time step in the input sequence. The hidden state at each time step t captures the information from the current input $\mathbf{x}_t$ and the previous hidden state $\mathbf{h}_{t-1}$. For a sequence of inputs $\{\mathbf{x}_1, \mathbf{x}_2, \ldots, \mathbf{x}_T\}$, the forward pass is computed as follows:

1. Initialization: The initial hidden state $\mathbf{h}_0$ is often initialized to a vector of zeros.

$$\mathbf{h}_0 = \mathbf{0}.$$

2. Hidden State Computation: At each time step t, the hidden state $\mathbf{h}_t$ is updated using the input $\mathbf{x}_t$ and the previous hidden state $\mathbf{h}_{t-1}$:

$$\mathbf{h}_t = f(\mathbf{W}_{hx}\mathbf{x}_t + \mathbf{W}_{hh}\mathbf{h}_{t-1} + \mathbf{b}_h).$$

   Here, f is an activation function such as tanh or ReLU.
3. Output Computation: The output $\mathbf{y}_t$ at time step t is computed using the current hidden state $\mathbf{h}_t$:

$$\mathbf{y}_t = g(\mathbf{W}_{hy}\mathbf{h}_t + \mathbf{b}_y),$$

   where g can be a linear function or a softmax function, depending on the task.

To summarize, the forward pass equations are:

$$\begin{aligned}\mathbf{h}_t &= f(\mathbf{W}_{hx}\mathbf{x}_t + \mathbf{W}_{hh}\mathbf{h}_{t-1} + \mathbf{b}_h)\\ \mathbf{y}_t &= g(\mathbf{W}_{hy}\mathbf{h}_t + \mathbf{b}_y)\end{aligned}$$

These equations illustrate how the RNN processes each element of the input sequence and maintains a hidden state that carries information forward through time.

Example: Consider an RNN that processes a sequence of three input vectors $\mathbf{x}_1, \mathbf{x}_2, \mathbf{x}_3$. The hidden states and outputs are computed as follows:
At $t = 1$:

$$\begin{aligned}\mathbf{h}_1 &= f(\mathbf{W}_{hx}\mathbf{x}_1 + \mathbf{b}_h)\\ \mathbf{y}_1 &= g(\mathbf{W}_{hy}\mathbf{h}_1 + \mathbf{b}_y)\end{aligned}$$

At $t = 2$:

$$\mathbf{h}_2 = f(\mathbf{W}_{hx}\mathbf{x}_2 + \mathbf{W}_{hh}\mathbf{h}_1 + \mathbf{b}_h)$$
$$\mathbf{y}_2 = g(\mathbf{W}_{hy}\mathbf{h}_2 + \mathbf{b}_y)$$

At $t = 3$:

$$\mathbf{h}_3 = f(\mathbf{W}_{hx}\mathbf{x}_3 + \mathbf{W}_{hh}\mathbf{h}_2 + \mathbf{b}_h)$$
$$\mathbf{y}_3 = g(\mathbf{W}_{hy}\mathbf{h}_3 + \mathbf{b}_y)$$

This example shows the sequential computation of hidden states and outputs for a simple RNN.

Backward Pass and Backpropagation Through Time (BPTT)
The backward pass in RNNs involves computing the gradients of the loss function with respect to the network parameters. This process, known as Backpropagation Through Time (BPTT), extends the standard backpropagation algorithm to handle the temporal dependencies in RNNs.

1. Loss Function: Suppose we have a loss function $\mathcal{L}$ that depends on the outputs $\{\mathbf{y}_t\}_{t=1}^{T}$ and the target values $\{\mathbf{y}_t^*\}_{t=1}^{T}$. The total loss is given by:

$$\mathcal{L} = \sum_{t=1}^{T} \mathcal{L}_t(\mathbf{y}_t, \mathbf{y}_t^*)$$

2. Gradients with Respect to Outputs: Compute the gradient of the loss with respect to the outputs at each time step:

$$\frac{\partial \mathcal{L}}{\partial \mathbf{y}_t} = \frac{\partial \mathcal{L}_t}{\partial \mathbf{y}_t}$$

3. Gradients with Respect to Hidden States: Use the chain rule to propagate the gradients back through the hidden states. The gradient of the loss with respect to the hidden state $\mathbf{h}_t$ is given by:

$$\frac{\partial \mathcal{L}}{\partial \mathbf{h}_t} = \frac{\partial \mathcal{L}}{\partial \mathbf{h}_{t+1}} \frac{\partial \mathbf{h}_{t+1}}{\partial \mathbf{h}_t} + \frac{\partial \mathcal{L}_t}{\partial \mathbf{h}_t}$$

Here, $\frac{\partial \mathbf{h}_{t+1}}{\partial \mathbf{h}_t}$ is the gradient of the next hidden state with respect to the current hidden state.

4. Gradients with Respect to Weights: Finally, compute the gradients of the loss with respect to the weight matrices:

$$\frac{\partial \mathcal{L}}{\partial \mathbf{W}_{hx}} = \sum_{t=1}^{T} \frac{\partial \mathcal{L}_t}{\partial \mathbf{h}_t} \frac{\partial \mathbf{h}_t}{\partial \mathbf{W}_{hx}}$$

$$\frac{\partial \mathcal{L}}{\partial \mathbf{W}_{hh}} = \sum_{t=1}^{T} \frac{\partial \mathcal{L}_t}{\partial \mathbf{h}_t} \frac{\partial \mathbf{h}_t}{\partial \mathbf{W}_{hh}}$$

$$\frac{\partial \mathcal{L}}{\partial \mathbf{W}_{hy}} = \sum_{t=1}^{T} \frac{\partial \mathcal{L}_t}{\partial \mathbf{y}_t} \frac{\partial \mathbf{y}_t}{\partial \mathbf{W}_{hy}}$$

5.2.3 *Challenges with Standard RNNs*

While RNNs are powerful tools for modeling sequential data, they come with significant challenges. Two major issues are the vanishing and exploding gradient problems. This section explores these challenges and discusses various solutions and mitigation techniques.

Vanishing and Exploding Gradient Problems

The vanishing and exploding gradient problems arise during the training of RNNs when backpropagating errors through many time steps. The vanishing gradient problem occurs when the gradients of the loss function with respect to the network parameters become exceedingly small. This issue is particularly pronounced in RNNs because the gradients are multiplied by the same weight matrices repeatedly as they are propagated back through time. Mathematically, this can be explained by considering the partial derivative of the loss $\mathcal{L}$ with respect to a weight $\mathbf{W}$ at time step t:

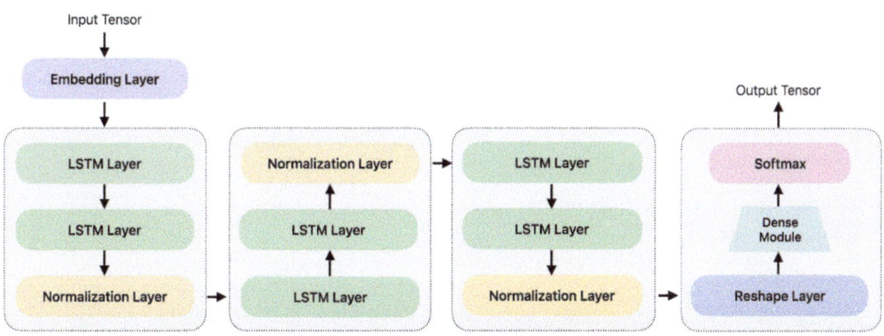

Fig. 5.1 Illustration of a typical pipeline of an RNN

$$\frac{\partial \mathcal{L}}{\partial \mathbf{W}} = \sum_{k=1}^{t} \left(\frac{\partial \mathcal{L}}{\partial \mathbf{h}_t} \prod_{j=k}^{t-1} \frac{\partial \mathbf{h}_{j+1}}{\partial \mathbf{h}_j} \right)$$

If the eigenvalues of the Jacobian matrix $\frac{\partial \mathbf{h}_{j+1}}{\partial \mathbf{h}_j}$ are less than one, the product of many such Jacobians will tend to zero exponentially fast as t increases. Consequently, the gradients become very small, causing the network to learn very slowly or not at all for long-range dependencies.

Conversely, the exploding gradient problem occurs when the gradients grow exponentially during backpropagation. This happens if the eigenvalues of the Jacobian matrix $\frac{\partial \mathbf{h}_{j+1}}{\partial \mathbf{h}_j}$ are greater than one. As the gradients are propagated back through time, they can become excessively large, leading to numerical instability and making the training process highly erratic.

Example: Consider a simple RNN with a single weight matrix $\mathbf{W}_{hh}$ and an activation function f. If the magnitude of the eigenvalues of $\mathbf{W}_{hh}$ is greater than one, the gradients will explode as they are propagated back through many time steps. Conversely, if the eigenvalues are less than one, the gradients will vanish. Let $\mathbf{W}_{hh}$ be the recurrent weight matrix, and consider the gradient of the loss with respect to $\mathbf{W}_{hh}$:

$$\frac{\partial \mathcal{L}}{\partial \mathbf{W}_{hh}} = \sum_{t=1}^{T} \frac{\partial \mathcal{L}}{\partial \mathbf{h}_t} \sum_{k=1}^{t} \left(\prod_{j=k}^{t-1} \mathbf{W}_{hh} \right) \mathbf{h}_{k-1}$$

If λ is an eigenvalue of $\mathbf{W}_{hh}$, then λ^t can grow or shrink exponentially as t increases, leading to exploding or vanishing gradients, respectively.

Solutions and Mitigation Techniques

Several techniques have been proposed to mitigate the vanishing and exploding gradient problems in RNNs:

1. Gradient Clipping: Gradient clipping is a technique used to prevent exploding gradients by scaling the gradients when they exceed a certain threshold. This ensures that the gradients remain within a manageable range during backpropagation. Given a gradient $\mathbf{g}$, if $\|\mathbf{g}\| > \tau$ (where τ is a predefined threshold), then scale $\mathbf{g}$ as follows:

$$\mathbf{g} \leftarrow \tau \frac{\mathbf{g}}{\|\mathbf{g}\|}$$

   This technique helps maintain numerical stability and prevents the gradients from growing too large.
   Example:

```
optimizer = torch.optim.Adam(model.parameters(), lr=0.001)

for epoch in range(num_epochs):
    for inputs, labels in train_loader:
```

```
optimizer.zero_grad()
outputs = model(inputs)
loss = criterion(outputs, labels)
loss.backward()
torch.nn.utils.clip_grad_norm_(model.parameters(), max_norm=1.0)
optimizer.step()
```

2. Long Short-Term Memory (LSTM) Networks: LSTM networks introduce memory cells and gating mechanisms to maintain gradients over long sequences. LSTMs effectively mitigate the vanishing gradient problem by maintaining a nearly constant error flow through the network.
3. Gated Recurrent Units (GRUs): GRUs are a simplified variant of LSTMs with fewer parameters but similar performance. They combine the forget and input gates into a single update gate and use a reset gate to control the flow of information. GRUs simplify the architecture while retaining the benefits of gating mechanisms.
4. Layer Normalization and Batch Normalization: Normalization techniques can also help stabilize the training process. Layer normalization normalizes the inputs across the features, while batch normalization normalizes the inputs across the batch.

Layer Normalization: Normalize the activations within each layer independently.

$$\hat{\mathbf{h}}_t = \frac{\mathbf{h}_t - \mu}{\sigma}\gamma + \beta$$

Batch Normalization: Normalize the activations across the batch.

$$\hat{\mathbf{h}}_t = \frac{\mathbf{h}_t - \mathbb{E}[\mathbf{h}_t]}{\sqrt{\text{Var}[\mathbf{h}_t] + \epsilon}}\gamma + \beta$$

where γ and β are learnable parameters, μ and σ are the mean and standard deviation, and ϵ is a small constant for numerical stability.

5.3 Advanced RNN Architectures

5.3.1 Long Short-Term Memory (LSTM) Networks

Long Short-Term Memory (LSTM) networks are a special kind of RNN capable of learning long-term dependencies. Introduced by Hochreiter and Schmidhuber in 1997, LSTMs were explicitly designed to address the vanishing gradient problem encountered in standard RNNs, making them highly effective for tasks that require remembering information over long sequences.

LSTM Architecture and Components

The architecture of an LSTM network consists of a chain-like structure of repeating modules, similar to standard RNNs. However, the LSTM module, or cell, has a more complex structure than that of a standard RNN. Each LSTM cell contains several interacting layers that regulate the flow of information.

Key components of an LSTM cell include:

1. Cell State ($\mathbf{C}_t$): The cell state runs through the entire chain, with only some minor linear interactions. It acts as a conveyor belt, with the ability to maintain its values relatively unchanged across time steps.
2. Forget Gate ($\mathbf{f}_t$): The forget gate decides what information to discard from the cell state. It takes the hidden state from the previous time step ($\mathbf{h}_{t-1}$) and the current input ($\mathbf{x}_t$), and passes them through a sigmoid function, producing a value between 0 and 1.

$$\mathbf{f}_t = \sigma(\mathbf{W}_f\mathbf{x}_t + \mathbf{U}_f\mathbf{h}_{t-1} + \mathbf{b}_f)$$

3. Input Gate ($\mathbf{i}_t$): The input gate decides what new information to store in the cell state. It consists of a sigmoid layer and a tanh layer. The sigmoid layer determines which values to update, while the tanh layer creates a vector of new candidate values ($\tilde{\mathbf{C}}_t$) that could be added to the cell state.

$$\begin{aligned}\mathbf{i}_t &= \sigma(\mathbf{W}_i\mathbf{x}_t + \mathbf{U}_i\mathbf{h}_{t-1} + \mathbf{b}_i)\\ \tilde{\mathbf{C}}_t &= \tanh(\mathbf{W}_c\mathbf{x}_t + \mathbf{U}_c\mathbf{h}_{t-1} + \mathbf{b}_c)\end{aligned}$$

4. Cell State Update ($\mathbf{C}_t$): The new cell state is updated by combining the old cell state and the new candidate values, modulated by the forget and input gates.

$$\mathbf{C}_t = \mathbf{f}_t \odot \mathbf{C}_{t-1} + \mathbf{i}_t \odot \tilde{\mathbf{C}}_t$$

5. Output Gate ($\mathbf{o}_t$): The output gate decides what information from the cell state to output. It involves a sigmoid layer determining which part of the cell state to output, and a tanh layer that scales the cell state values to the range [− 1, 1].

$$\begin{aligned}\mathbf{o}_t &= \sigma(\mathbf{W}_o\mathbf{x}_t + \mathbf{U}_o\mathbf{h}_{t-1} + \mathbf{b}_o)\\ \mathbf{h}_t &= \mathbf{o}_t \odot \tanh(\mathbf{C}_t)\end{aligned}$$

Mathematical Formulation of LSTMs

The LSTM cell involves several steps for updating the cell state and the hidden state at each time step:

1. Forget Gate Activation:

$$\mathbf{f}_t = \sigma(\mathbf{W}_f\mathbf{x}_t + \mathbf{U}_f\mathbf{h}_{t-1} + \mathbf{b}_f)$$

2. Input Gate Activation:

$$\mathbf{i}_t = \sigma(\mathbf{W}_i\mathbf{x}_t + \mathbf{U}_i\mathbf{h}_{t-1} + \mathbf{b}_i)$$

$$\tilde{\mathbf{C}}_t = \tanh(\mathbf{W}_c\mathbf{x}_t + \mathbf{U}_c\mathbf{h}_{t-1} + \mathbf{b}_c)$$

3. Cell State Update:

$$\mathbf{C}_t = \mathbf{f}_t \odot \mathbf{C}_{t-1} + \mathbf{i}_t \odot \tilde{\mathbf{C}}_t$$

4. Output Gate Activation:

$$\mathbf{o}_t = \sigma(\mathbf{W}_o\mathbf{x}_t + \mathbf{U}_o\mathbf{h}_{t-1} + \mathbf{b}_o)$$

5. Hidden State Update:

$$\mathbf{h}_t = \mathbf{o}_t \odot \tanh(\mathbf{C}_t)$$

These equations show how LSTMs can maintain and update information over long sequences, mitigating the vanishing gradient problem and allowing for the learning of long-term dependencies.

Advantages of LSTMs over Standard RNNs
LSTMs offer several advantages over standard RNNs, primarily due to their ability to handle long-term dependencies more effectively:

1. Mitigation of Vanishing Gradients: The gating mechanisms in LSTMs help maintain a constant error flow, reducing the vanishing gradient problem. This allows LSTMs to learn long-term dependencies that standard RNNs struggle with.
2. Selective Memory: LSTMs can selectively forget or update information in the cell state, providing a more robust and flexible memory mechanism compared to standard RNNs. This selective memory is crucial for tasks requiring context over extended sequences.
3. Enhanced Learning Capability: By maintaining gradients over long sequences, LSTMs can learn more complex temporal patterns and dependencies. This makes them suitable for tasks such as language modeling, machine translation, and time series prediction.

Example: Consider a sequence-to-sequence task, such as machine translation, where the input is a sequence of words in one language, and the output is a sequence of words in another language. An LSTM-based model can effectively capture the dependencies between words in the input sequence and generate the corresponding output sequence.

Implementation in TensorFlow:

```
import tensorflow as tf
from tensorflow.keras.models import Sequential
from tensorflow.keras.layers import LSTM, Dense

# Define the LSTM model
model = Sequential()
model.add(LSTM(50, activation='tanh', input_shape=(10, 1)))  # 10 time steps, 1 feature
model.add(Dense(1))

# Compile the model
model.compile(optimizer='adam', loss='mean_squared_error')

# Generate synthetic data for demonstration
import numpy as np
x_train = np.random.rand(1000, 10, 1)  # 1000 samples, 10 time steps, 1 feature
y_train = np.random.rand(1000, 1)      # 1000 target values

# Train the model
history = model.fit(x_train, y_train, epochs=20, batch_size=32)

# Evaluate the model
x_test = np.random.rand(100, 10, 1)
y_test = np.random.rand(100, 1)
loss = model.evaluate(x_test, y_test)
print(f'Test Loss: {loss}')
```

Implementation in PyTorch:

```
import torch
import torch.nn as nn
import torch.optim as optim

# Define the LSTM model
class LSTMModel(nn.Module):
    def __init__(self):
        super(LSTMModel, self).__init__()
        self.lstm = nn.LSTM(input_size=1, hidden_size=50, num_layers=1, batch_first=True)
        self.fc = nn.Linear(50, 1)

    def forward(self, x):
        h0 = torch.zeros(1, x.size(0), 50)  # Initial hidden state
        c0 = torch.zeros(1, x.size(0), 50)  # Initial cell state
        out, _ = self.lstm(x, (h0, c0))
        out = self.fc(out[:, -1, :])
        return out

# Instantiate the model
model = LSTMModel()

# Define the loss function and optimizer
criterion = nn.MSELoss()
optimizer = optim.Adam(model.parameters(), lr=0.001)

# Generate synthetic data for demonstration
x_train = torch.rand(1000, 10, 1)
y_train = torch.rand(1000, 1)

# Train the model
num_epochs = 20
for epoch in range(num_epochs):
    model.train()
    outputs = model(x_train)
    loss = criterion(outputs, y_train)

    optimizer.zero_grad()
```

```
    loss.backward()
    optimizer.step()

    if (epoch+1) % 5 == 0:
        print(f'Epoch [{epoch+1}/{num_epochs}], Loss: {loss.item():.4f}')

# Evaluate the model
model.eval()
x_test =

 torch.rand(100, 10, 1)
y_test = torch.rand(100, 1)
with torch.no_grad():
    predictions = model(x_test)
    test_loss = criterion(predictions, y_test)
print(f'Test Loss: {test_loss.item():.4f}')
```

5.3.2 Gated Recurrent Unit (GRU) Networks

Gated Recurrent Units (GRUs) are a more recent innovation in recurrent neural network architectures, introduced by Cho et al. in 2014. GRUs aim to address the vanishing gradient problem like LSTMs but with a simpler architecture. This section provides a detailed examination of GRU architecture, its mathematical formulation, and a comparison with LSTMs.

GRU Architecture and Components

The GRU architecture is designed to retain the benefits of LSTM networks while simplifying the internal structure. A GRU cell consists of two primary gates: the update gate and the reset gate, which regulate the flow of information through the cell.

Key components of a GRU cell include:

1. Update Gate ($\mathbf{z}_t$): The update gate determines how much of the past information needs to be passed along to the future. It combines the input at the current time step $\mathbf{x}_t$ and the hidden state from the previous time step $\mathbf{h}_{t-1}$.

$$\mathbf{z}_t = \sigma(\mathbf{W}_z\mathbf{x}_t + \mathbf{U}_z\mathbf{h}_{t-1} + \mathbf{b}_z)$$

2. Reset Gate ($\mathbf{r}_t$): The reset gate determines how much of the past information to forget. It also combines the current input and the previous hidden state.

$$\mathbf{r}_t = \sigma(\mathbf{W}_r\mathbf{x}_t + \mathbf{U}_r\mathbf{h}_{t-1} + \mathbf{b}_r)$$

3. Candidate Activation ($\tilde{\mathbf{h}}_t$): The candidate activation computes the new hidden state, incorporating the reset gate's influence to forget some parts of the past information.

$$\tilde{\mathbf{h}}_t = \tanh(\mathbf{W}_h\mathbf{x}_t + \mathbf{U}_h(\mathbf{r}_t \odot \mathbf{h}_{t-1}) + \mathbf{b}_h)$$

4. Hidden State ($\mathbf{h}_t$): The final hidden state is a combination of the previous hidden state and the candidate activation, modulated by the update gate.

$$\mathbf{h}_t = (1 - \mathbf{z}_t) \odot \mathbf{h}_{t-1} + \mathbf{z}_t \odot \tilde{\mathbf{h}}_t$$

Mathematical Formulation of GRUs

The GRU cell can be mathematically formulated through the following steps:

1. Update Gate Activation:

$$\mathbf{z}_t = \sigma(\mathbf{W}_z \mathbf{x}_t + \mathbf{U}_z \mathbf{h}_{t-1} + \mathbf{b}_z)$$

2. Reset Gate Activation:

$$\mathbf{r}_t = \sigma(\mathbf{W}_r \mathbf{x}_t + \mathbf{U}_r \mathbf{h}_{t-1} + \mathbf{b}_r)$$

3. Candidate Activation Calculation:

$$\tilde{\mathbf{h}}_t = \tanh(\mathbf{W}_h \mathbf{x}_t + \mathbf{U}_h (\mathbf{r}_t \odot \mathbf{h}_{t-1}) + \mathbf{b}_h)$$

4. Hidden State Update:

$$\mathbf{h}_t = (1 - \mathbf{z}_t) \odot \mathbf{h}_{t-1} + \mathbf{z}_t \odot \tilde{\mathbf{h}}_t$$

These equations describe how GRUs manage the flow of information through the cell and adjust their internal state to retain relevant information over long sequences.

Comparison of GRUs and LSTMs

While both GRUs and LSTMs are designed to solve the vanishing gradient problem and capture long-term dependencies, they have some key differences:

1. Architecture Complexity: LSTMs have three gates (input, forget, and output gates) and a more complex internal structure involving a separate cell state and hidden state. GRUs simplify this architecture by using only two gates (update and reset gates) and merging the cell state and hidden state.
2. Computational Efficiency: Due to their complexity, LSTMs generally require more computational resources and longer training times. GRUs, being simpler, are computationally more efficient and faster to train, making them preferable in scenarios where computational resources are limited.
3. Performance: LSTMs are often preferred for tasks that require learning long-term dependencies very precisely, such as language modeling and machine translation. GRUs perform comparably to LSTMs on many tasks and sometimes outperform them, particularly in cases with smaller datasets or when the task does not require capturing very long-term dependencies.

Example: Consider a sequence prediction task where the goal is to predict the next value in a time series. Both LSTMs and GRUs can be used to model such a sequence, but the choice between them can depend on the specific requirements of the task.

Implementation in TensorFlow:

```
import tensorflow as tf
from tensorflow.keras.models import Sequential
from tensorflow.keras.layers import GRU, Dense

# Define the GRU model
model = Sequential()
model.add(GRU(50, activation='tanh', input_shape=(10, 1)))  # 10 time steps, 1 feature
model.add(Dense(1))

# Compile the model
model.compile(optimizer='adam', loss='mean_squared_error')

# Generate synthetic data for demonstration
import numpy as np
x_train = np.random.rand(1000, 10, 1)  # 1000 samples, 10 time steps, 1 feature
y_train = np.random.rand(1000, 1)      # 1000 target values

# Train the model
history = model.fit(x_train, y_train, epochs=20, batch_size=32)

# Evaluate the model
x_test = np.random.rand(100, 10, 1)
y_test = np.random.rand(100, 1)
loss = model.evaluate(x_test, y_test)
print(f'Test Loss: {loss}')
```

Implementation in PyTorch:

```
import torch
import torch.nn as nn
import torch.optim as optim

# Define the GRU model
class GRUModel(nn.Module):
    def _ _init_ _(self):
        super(GRUModel, self)._ _init_ _()
        self.gru = nn.GRU(input_size=1, hidden_size=50, num_layers=1, batch_first=True)
        self.fc = nn.Linear(50, 1)

    def forward(self, x):
        h0 = torch.zeros(1, x.size(0), 50)  # Initial hidden state
        out, _ = self.gru(x, h0)
        out = self.fc(out[:, -1, :])
        return out

# Instantiate the model
model = GRUModel()

# Define the loss function and optimizer
criterion = nn.MSELoss()
optimizer = optim.Adam(model.parameters(), lr=0.001)

# Generate synthetic data for demonstration
x_train = torch.rand(1000, 10, 1)
y_train = torch.rand(1000, 1)

# Train the model
num_epochs = 20
for epoch in range(num_epochs):
    model.train()
```

```
    outputs = model(x_train)
    loss = criterion(outputs, y_train)

    optimizer.zero_grad()
    loss.backward()
    optimizer.step()

    if (epoch+1) % 5 == 0:
        print(f'Epoch [{epoch+1}/{num_epochs}], Loss: {loss.item():.4f}')

# Evaluate the model
model.eval()
x_test = torch.rand(100, 10, 1)
y_test = torch.rand(100, 1)
with torch.no_grad():
    predictions = model(x_test)
    test_loss = criterion(predictions, y_test)
print(f'Test Loss: {test_loss.item():.4f}')
```

5.4 Attention Mechanisms and Sequence-to-Sequence Models

Attention mechanisms have revolutionized sequence modeling tasks, significantly improving the performance of models in various applications such as machine translation, text summarization, and image captioning. This section introduces the concept of attention mechanisms, their necessity in sequence modeling, and the different types of attention mechanisms.

5.4.1 Introduction to Attention Mechanisms

Attention mechanisms allow models to focus on specific parts of the input sequence when generating each part of the output sequence. By dynamically weighting the importance of different input elements, attention mechanisms enable the model to capture relevant information more effectively, especially for long sequences where traditional models struggle with maintaining context.

The Need for Attention in Sequence Modeling

Traditional sequence-to-sequence models, like those based on RNNs or LSTM networks, often face challenges when dealing with long sequences. These challenges include:

1. Fixed-Length Context Vector: In standard sequence-to-sequence models, the encoder compresses the entire input sequence into a fixed-length context vector. This vector serves as the sole source of information for the decoder to generate the output sequence. As the input sequence length increases, this fixed-length representation may lose critical information, leading to poorer performance.

2. Vanishing Gradient Problem: Even with advanced architectures like LSTMs and GRUs, maintaining long-term dependencies can be difficult due to the vanishing gradient problem. This issue limits the model's ability to propagate information over long distances effectively.
3. Alignment Challenges: In tasks like machine translation, the alignment between input and output sequences is often non-monotonic. Standard sequence-to-sequence models may struggle to capture such complex alignments without additional mechanisms to guide the process.

Attention mechanisms address these challenges by allowing the model to dynamically focus on different parts of the input sequence at each decoding step. This selective focus helps the model to retain relevant information and improve its overall performance.

Types of Attention Mechanisms

Several types of attention mechanisms have been developed, each with its own mathematical formulation and application. The most commonly used attention mechanisms include:

1. Additive (Bahdanau) Attention: Proposed by Bahdanau et al. in 2015, additive attention calculates the attention weights using a feed-forward neural network. The attention score e_{ij} between the i-th decoder hidden state $\mathbf{s}_i$ and the j-th encoder hidden state $\mathbf{h}_j$ is computed as:

$$e_{ij} = \mathbf{v}_a^\top \tanh(\mathbf{W}_a \mathbf{s}_i + \mathbf{U}_a \mathbf{h}_j)$$

Here, $\mathbf{W}_a$ and $\mathbf{U}_a$ are learnable weight matrices, and $\mathbf{v}_a$ is a learnable vector. The attention weights α_{ij} are then obtained using a softmax function:

$$\alpha_{ij} = \frac{\exp(e_{ij})}{\sum_{k=1}^{T} \exp(e_{ik})}$$

The context vector $\mathbf{c}_i$ for the i-th decoder step is a weighted sum of the encoder hidden states:

$$\mathbf{c}_i = \sum_{j=1}^{T} \alpha_{ij} \mathbf{h}_j$$

2. Multiplicative (Luong) Attention: Introduced by Luong et al. in 2015, multiplicative attention computes the attention scores using a dot product between the decoder hidden state and the encoder hidden states. The attention score e_{ij} is given by:

$$e_{ij} = \mathbf{s}_i^\top \mathbf{W}_a \mathbf{h}_j$$

Here, $\mathbf{W}_a$ is a learnable weight matrix. The attention weights α_{ij} and the context vector $\mathbf{c}_i$ are computed similarly to additive attention.

3. Scaled Dot-Product Attention: Scaled dot-product attention, used in the Transformer model proposed by Vaswani et al. in 2017, scales the dot product of the query and key vectors by the square root of the dimensionality of the key vectors. The attention score e_{ij} is computed as:

$$e_{ij} = \frac{\mathbf{q}_i^\top \mathbf{k}_j}{\sqrt{d_k}}$$

Here, $\mathbf{q}_i$ is the query vector, $\mathbf{k}_j$ is the key vector, and d_k is the dimensionality of the key vectors. The attention weights α_{ij} and context vector $\mathbf{c}_i$ are computed using a softmax function and weighted sum, respectively:

$$\alpha_{ij} = \frac{\exp(e_{ij})}{\sum_{k=1}^{T} \exp(e_{ik})}$$

$$\mathbf{c}_i = \sum_{j=1}^{T} \alpha_{ij} \mathbf{v}_j$$

4. Self-Attention: Self-attention, also known as intra-attention, is a mechanism where the attention is applied within the same sequence. It allows each position in the sequence to attend to all positions, enabling the model to capture dependencies regardless of their distance. The scaled dot-product attention is commonly used in self-attention.

Example: Consider a machine translation task where the input sequence is a sentence in English, and the output sequence is the corresponding translation in French. Using attention mechanisms, the model can dynamically focus on different parts of the English sentence while generating each word in the French translation.

Implementation in TensorFlow:

```
import tensorflow as tf
from tensorflow.keras.layers import Layer, Dense

class AdditiveAttention(Layer):
    def _ _init_ _(self, units):
        super(AdditiveAttention, self)._ _init_ _()
        self.W_a = Dense(units)
        self.U_a = Dense(units)
        self.V_a = Dense(1)

    def call(self, query, values):
        query_with_time_axis = tf.expand_dims(query, 1)
        score = self.V_a(tf.nn.tanh(self.W_a(query_with_time_axis) + self.U_a(values)))
        attention_weights = tf.nn.softmax(score, axis=1)
        context_vector = attention_weights * values
        context_vector = tf.reduce_sum(context_vector, axis=1)
        return context_vector, attention_weights

# Example in a sequence-to-sequence model
units = 512
attention_layer = AdditiveAttention(units)

# Encoder hidden states (values) and decoder hidden state (query)
```

```
values = tf.random.normal([batch_size, sequence_length, hidden_dim])
query = tf.random.normal([batch_size, hidden_dim])

context_vector, attention_weights = attention_layer(query, values)
```

Implementation in PyTorch:

```
import torch
import torch.nn as nn

class AdditiveAttention(nn.Module):
    def __init__(self, hidden_dim):
        super(AdditiveAttention, self).__init__()
        self.W_a = nn.Linear(hidden_dim, hidden_dim)
        self.U_a = nn.Linear(hidden_dim, hidden_dim)
        self.V_a = nn.Linear(hidden_dim, 1)

    def forward(self, query, values):
        query_with_time_axis = query.unsqueeze(1)
        score = self.V_a(torch.tanh(self.W_a(query_with_time_axis) + self.U_a(values)))
        attention_weights = torch.softmax(score, dim=1)
        context_vector = attention_weights * values
        context_vector = torch.sum(context_vector, dim=1)
        return context_vector, attention_weights

# Example in a sequence-to-sequence model
hidden_dim = 512
attention_layer = AdditiveAttention(hidden_dim)

# Encoder hidden states (values) and decoder hidden state (query)
values = torch.randn(batch_size, sequence_length, hidden_dim)
query = torch.randn(batch_size, hidden_dim)

context_vector, attention_weights = attention_layer(query, values)
```

5.4.2 Sequence-to-Sequence Models

Sequence-to-sequence (Seq2Seq) models are a class of models designed to handle tasks where input and output sequences can have different lengths. These models have been successfully applied to various tasks, including machine translation, text summarization, and speech recognition. This section discusses the encoder–decoder architecture of Seq2Seq models and the incorporation of attention mechanisms to enhance their performance.

Encoder–Decoder Architecture
The encoder–decoder architecture is the foundation of Seq2Seq models. It consists of two main components: the encoder, which processes the input sequence, and the decoder, which generates the output sequence.

Encoder: The encoder reads the input sequence and compresses it into a fixed-length context vector, which represents the entire input sequence. The context vector is then used by the decoder to generate the output sequence. The encoder typically consists of a stack of RNN, LSTM, or GRU layers.

Let $\mathbf{x} = \{\mathbf{x}_1, \mathbf{x}_2, \ldots, \mathbf{x}_T\}$ be the input sequence. The encoder processes each input element sequentially, producing a hidden state $\mathbf{h}_t$ at each time step t:

$$\mathbf{h}_t = \text{EncoderRNN}(\mathbf{x}_t, \mathbf{h}_{t-1})$$

The final hidden state $\mathbf{h}_T$ is used as the context vector $\mathbf{c}$:

$$\mathbf{c} = \mathbf{h}_T$$

Decoder: The decoder generates the output sequence by taking the context vector and producing one element of the output sequence at a time. At each time step t, the decoder takes the previous hidden state and the previous output (or a start token for the first step) as inputs to generate the current output and update the hidden state:

$$\mathbf{h}_t^{\text{dec}} = \text{DecoderRNN}(\mathbf{y}_{t-1}, \mathbf{h}_{t-1}^{\text{dec}})$$

$$\mathbf{y}_t = \text{OutputLayer}(\mathbf{h}_t^{\text{dec}})$$

The process continues until the entire output sequence is generated. The final output sequence $\mathbf{y} = \{\mathbf{y}_1, \mathbf{y}_2, \ldots, \mathbf{y}_{T'}\}$ can be of a different length T' than the input sequence.

Example: Consider a simple Seq2Seq model for translating a sentence from English to French. The encoder reads the English sentence and produces a context vector. The decoder uses this context vector to generate the corresponding French sentence.

Implementation in TensorFlow:

```
import tensorflow as tf
from tensorflow.keras.layers import LSTM, Dense, Embedding

# Define the encoder
encoder_inputs = tf.keras.Input(shape=(None,))
encoder_embedding = Embedding(input_dim=input_vocab_size, output_dim=embedding_dim)
(encoder_inputs)
encoder_lstm = LSTM(units)(encoder_embedding)
encoder_outputs, state_h, state_c = LSTM(units, return_state=True)(encoder_lstm)
encoder_states = [state_h, state_c]

# Define the decoder
decoder_inputs = tf.keras.Input(shape=(None,))
decoder_embedding = Embedding(input_dim=output_vocab_size, output_dim=embedding_dim)
(decoder_inputs)
decoder_lstm = LSTM(units, return_sequences=True, return_state=True)
decoder_outputs, _, _ = decoder_lstm(decoder_embedding, initial_state=encoder_states)
decoder_dense = Dense(output_vocab_size, activation='softmax')
decoder_outputs = decoder_dense(decoder_outputs)

# Define the Seq2Seq model
model = tf.keras.Model([encoder_inputs, decoder_inputs], decoder_outputs)
model.compile(optimizer='adam', loss='sparse_categorical_crossentropy')
```

Implementation in PyTorch:

```
import torch
import torch.nn as nn

# Define the encoder
class Encoder(nn.Module):
    def _ _init_ _(self, input_dim, emb_dim, hid_dim, n_layers, dropout):
        super()._ _init_ _()
        self.embedding = nn.Embedding(input_dim, emb_dim)
        self.rnn = nn.LSTM(emb_dim, hid_dim, n_layers, dropout=dropout)
        self.dropout = nn.Dropout(dropout)

    def forward(self, src):
        embedded = self.dropout(self.embedding(src))
        outputs, (hidden, cell) = self.rnn(embedded)
        return hidden, cell

# Define the decoder
class Decoder(nn.Module):
    def _ _init_ _(self, output_dim, emb_dim, hid_dim, n_layers, dropout):
        super()._ _init_ _()
        self.embedding = nn.Embedding(output_dim, emb_dim)
        self.rnn = nn.LSTM(emb_dim, hid_dim, n_layers, dropout=dropout)
        self.fc_out = nn.Linear(hid_dim, output_dim)
        self.dropout = nn.Dropout(dropout)

    def forward(self, trg, hidden, cell):
        trg = trg.unsqueeze(0)
        embedded = self.dropout(self.embedding(trg))
        output, (hidden, cell) = self.rnn(embedded, (hidden, cell))
        prediction = self.fc_out(output.squeeze(0))
        return prediction, hidden, cell

# Define the Seq2Seq model
class Seq2Seq(nn.Module):
    def _ _init_ _(self, encoder, decoder, device):
        super()._ _init_ _()
        self.encoder = encoder
        self.decoder = decoder
        self.device = device

    def forward(self, src, trg, teacher_forcing_ratio=0.5):
        trg_len = trg.shape[0]
        trg_vocab_size = self.decoder.output_dim
        outputs = torch.zeros(trg_len, batch_size, trg_vocab_size).to(self.device)
        hidden, cell = self.encoder(src)
        input = trg[0, :]

        for t in range(1, trg_len):
            output, hidden, cell = self.decoder(input, hidden, cell)
            outputs[t] = output
            teacher_force = random.random() < teacher_forcing_ratio
            top1 = output.argmax(1)
            input = trg[t] if teacher_force else top1

        return outputs
```

Incorporating Attention into Sequence-to-Sequence Models

Incorporating attention mechanisms into Seq2Seq models significantly enhances their performance by allowing the decoder to focus on different parts of the input sequence at each decoding step. This selective focus helps in better alignment between the input and output sequences, especially for long sequences. The attention mechanism calculates a set of attention weights that indicate the importance of each input element for generating the current output element. These weights are then used to compute a context vector, which is a weighted sum of the encoder hidden states.

$$\mathbf{c}_i = \sum_{j=1}^{T} \alpha_{ij}\mathbf{h}_j$$

where α_{ij} are the attention weights calculated as:

$$\alpha_{ij} = \frac{\exp(e_{ij})}{\sum_{k=1}^{T} \exp(e_{ik})}$$

and e_{ij} is the alignment score between the i-th decoder state and the j-th encoder state.

Example: Consider an attention-based Seq2Seq model for machine translation. The encoder processes the input sentence and generates hidden states. The decoder, at each time step, uses the attention mechanism to focus on different parts of the input sentence to generate the translated output.

Implementation in TensorFlow:

```
class AttentionLayer(tf.keras.layers.Layer):
    def _ _init_ _(self, units):
        super(AttentionLayer, self)._ _init_ _()
        self.W1 = tf.keras.layers.Dense(units)
        self.W2 = tf.keras.layers.Dense(units)
        self.V = tf.keras.layers.Dense(1)

    def call(self, query, values):
        query_with_time_axis = tf.expand_dims(query, 1)
        score = self.V(tf.nn.tanh(self.W1(query_with_time_axis) + self.W2(values)))
        attention_weights = tf.nn.softmax(score, axis=1)
        context_vector = attention_weights * values
        context_vector = tf.reduce_sum(context_vector, axis=1)
        return context_vector, attention_weights

# Define the encoder with attention
encoder_inputs = tf.keras.Input(shape=(None,))
encoder_embedding = Embedding(input_dim=input_vocab_size, output_dim=embedding_dim)
(encoder_inputs)
encoder_lstm = LSTM(units, return_sequences=True, return_state=True)
encoder_outputs, state_h, state_c = encoder_lstm(encoder_embedding)

# Define the decoder with attention
decoder_inputs = tf.keras.Input(shape=(None,))
decoder_embedding = Embedding(input_dim=output_vocab_size, output_dim=embedding_dim)
(decoder_inputs)
decoder_lstm = LSTM(units, return_sequences=True, return_state=True)
decoder_outputs, _, _ = decoder_lstm(decoder_embedding, initial_state=[state_h, state_c])

# Attention layer
attention_layer = AttentionLayer(units)
context_vector, attention_weights = attention_layer(state_h, encoder_outputs)
decoder_combined_context = tf.concat([tf.expand_dims(context_vector, 1), decoder_outputs],
axis=-1)

# Final output layer
decoder_dense = Dense(output_vocab_size, activation='softmax')
decoder_outputs = decoder_dense(decoder_combined_context)
```

```
# Define the Seq2Seq model with attention
model = tf.keras.Model([encoder_inputs, decoder_inputs], decoder_outputs)
model.compile(optimizer='adam', loss='sparse_categorical_crossentropy')
```

Implementation in PyTorch:

```
class Attention(nn.Module):
    def _ _init_ _(self, hid_dim):
        super()._ _init_ _()
        self.attn = nn.Linear((hid_dim * 2) + hid_dim, hid_dim)
        self.v = nn.Linear(hid_dim, 1, bias=False)

    def forward(self, hidden, encoder_outputs):
        src_len = encoder_outputs.shape[0]
        hidden = hidden.unsqueeze(1).repeat(1, src_len, 1)
        energy = torch.tanh(self.attn(torch.cat((hidden, encoder_outputs), dim=2)))
        attention = self.v(energy).squeeze(2)
        return torch.softmax(attention, dim=1)

class DecoderWithAttention(nn.Module):
    def _ _init_ _(self, output_dim, emb_dim, hid_dim, n_layers, dropout, attention):
        super()._ _init_ _()
        self.output_dim = output_dim
        self.attention = attention
        self.embedding = nn.Embedding(output_dim, emb_dim)
        self.rnn = nn.LSTM((hid_dim * 2) + emb_dim, hid_dim, n_layers, dropout=dropout)
        self.fc_out = nn.Linear((hid_dim * 2) + hid_dim + emb_dim, output_dim)
        self.dropout = nn.Dropout(dropout)

    def forward(self, trg, hidden, cell, encoder_outputs):
        trg = trg.unsqueeze(0)
        embedded = self.dropout(self.embedding(trg))
        a = self.attention(hidden[-1], encoder_outputs)
        a = a.unsqueeze(1)
        weighted = torch.bmm(a, encoder_outputs.permute(1, 0, 2))
        rnn_input = torch.cat((embedded, weighted.permute(1, 0, 2)), dim=2)
        output, (hidden, cell) = self.rnn(rnn_input, (hidden, cell))
        embedded = embedded.squeeze(0)
        output = output.squeeze(0)
        weighted = weighted.squeeze(1)
        prediction = self.fc_out(torch.cat((output, weighted, embedded), dim=1))
        return prediction, hidden, cell, a.squeeze(1)
```

5.4.3 Applications of Attention Mechanisms

Attention mechanisms have transformed various fields within NLP and beyond. By allowing models to focus on relevant parts of the input data selectively, attention mechanisms improve performance on numerous tasks. This section delves into two primary applications of attention mechanisms: machine translation and text summarization.

Machine Translation
Machine translation is the task of automatically converting text from one language to another. Traditional Seq2Seq models faced challenges with long sentences and complex linguistic structures. The introduction of attention mechanisms significantly improved the performance of these models by addressing alignment issues and

enhancing the model's ability to handle long-term dependencies. In a machine translation task, the encoder processes the input sentence $\mathbf{x} = \{x_1, x_2, \ldots, x_T\}$ and generates a sequence of hidden states $\mathbf{h} = \{h_1, h_2, \ldots, h_T\}$. The decoder generates the translated sentence $\mathbf{y} = \{y_1, y_2, \ldots, y_{T'}\}$. At each decoding step t, the attention mechanism computes a context vector $\mathbf{c}_t$ as a weighted sum of the encoder hidden states:

$$\mathbf{c}_t = \sum_{j=1}^{T} \alpha_{tj} \mathbf{h}_j$$

The attention weights α_{tj} are computed using the alignment scores e_{tj}:

$$e_{tj} = \mathbf{v}_a^\top \tanh(\mathbf{W}_a \mathbf{s}_{t-1} + \mathbf{U}_a \mathbf{h}_j)$$
$$\alpha_{tj} = \frac{\exp(e_{tj})}{\sum_{k=1}^{T} \exp(e_{tk})}$$

where $\mathbf{s}_{t-1}$ is the decoder hidden state at the previous time step, $\mathbf{W}_a$ and $\mathbf{U}_a$ are learnable weight matrices, and $\mathbf{v}_a$ is a learnable vector. The context vector $\mathbf{c}_t$ is then used along with the previous decoder hidden state and the current input to generate the next hidden state and output:

$$\mathbf{s}_t = \text{DecoderRNN}(\mathbf{y}_{t-1}, \mathbf{s}_{t-1}, \mathbf{c}_t)$$
$$\mathbf{y}_t = g(\mathbf{W}_y \mathbf{s}_t + \mathbf{b}_y)$$

Example: Consider translating the English sentence "The cat is on the mat" to French. The encoder processes the English sentence and generates hidden states. The decoder, with the help of the attention mechanism, focuses on different parts of the English sentence to generate the French translation "Le chat est sur le tapis."

Implementation in TensorFlow:

```
import tensorflow as tf
from tensorflow.keras.layers import Layer, LSTM, Dense, Embedding

class BahdanauAttention(Layer):
    def _ _init_ _(self, units):
        super(BahdanauAttention, self)._ _init_ _()
        self.W1 = Dense(units)
        self.W2 = Dense(units)
        self.V = Dense(1)

    def call(self, query, values):
        query_with_time_axis = tf.expand_dims(query, 1)
        score = self.V(tf.nn.tanh(self.W1(query_with_time_axis) + self.W2(values)))
        attention_weights = tf.nn.softmax(score, axis=1)
        context_vector = attention_weights * values
        context_vector = tf.reduce_sum(context_vector, axis=1)
        return context_vector, attention_weights

# Define the encoder with attention
encoder_inputs = tf.keras.Input(shape=(None,))
encoder_embedding = Embedding(input_dim=input_vocab_size, output_dim=embedding_dim)
(encoder_inputs)
```

```
encoder_lstm = LSTM(units, return_sequences=True, return_state=True)
encoder_outputs, state_h, state_c = encoder_lstm(encoder_embedding)

# Define the decoder with attention
decoder_inputs = tf.keras.Input(shape=(None,))
decoder_embedding = Embedding(input_dim=output_vocab_size, output_dim=embedding_dim)
(decoder_inputs)
decoder_lstm = LSTM(units, return_sequences=True, return_state=True)
decoder_outputs, _, _ = decoder_lstm(decoder_embedding, initial_state=[state_h, state_c])

# Attention layer
attention_layer = BahdanauAttention(units)
context_vector, attention_weights = attention_layer(state_h, encoder_outputs)
decoder_combined_context = tf.concat([tf.expand_dims(context_vector, 1), decoder_outputs],
axis=-1)

# Final output layer
decoder_dense = Dense(output_vocab_size, activation='softmax')
decoder_outputs = decoder_dense(decoder_combined_context)

# Define the Seq2Seq model with attention
model = tf.keras.Model([encoder_inputs, decoder_inputs], decoder_outputs)
model.compile(optimizer='adam', loss='sparse_categorical_crossentropy')
```

Implementation in PyTorch:

```
import torch
import torch.nn as nn

class BahdanauAttention(nn.Module):
    def _ _init_ _(self, hidden_dim):
        super(BahdanauAttention, self)._ _init_ _()
        self.W1 = nn.Linear(hidden_dim, hidden_dim)
        self.W2 = nn.Linear(hidden_dim, hidden_dim)
        self.V = nn.Linear(hidden_dim, 1)

    def forward(self, query, values):
        query_with_time_axis = query.unsqueeze(1)
        score = self.V(torch.tanh(self.W1(query_with_time_axis) + self.W2(values)))
        attention_weights = torch.softmax(score, dim=1)
        context_vector = attention_weights * values
        context_vector = torch.sum(context_vector, dim=1)
        return context_vector, attention_weights

# Define the encoder with attention
class Encoder(nn.Module):
    def _ _init_ _(self, input_dim, emb_dim, hid_dim, n_layers, dropout):
        super()._ _init_ _()
        self.embedding = nn.Embedding(input_dim, emb_dim)
        self.rnn = nn.LSTM(emb_dim, hid_dim, n_layers, dropout=dropout)
        self.dropout = nn.Dropout(dropout)

    def forward(self, src):
        embedded = self.dropout(self.embedding(src))
        outputs, (hidden, cell) = self.rnn(embedded)
        return outputs, hidden, cell

# Define the decoder with attention
class Decoder(nn.Module):
    def _ _init_ _(self, output_dim, emb_dim, hid_dim, n_layers, dropout, attention):
        super()._ _init_ _()
        self.output_dim = output_dim
        self.attention = attention
        self.embedding = nn.Embedding(output_dim, emb_dim)
        self.rnn = nn.LSTM((hid_dim * 2) + emb_dim, hid_dim, n_layers, dropout=dropout)
        self.fc_out = nn.Linear((hid_dim * 2) + hid_dim + emb_dim, output_dim)
        self.dropout = nn.Dropout(dropout)
```

```
    def forward(self, trg, hidden, cell, encoder_outputs):
        trg = trg.unsqueeze(0)
        embedded = self.dropout(self.embedding(trg))
        a = self.attention(hidden[-1], encoder_outputs)
        a = a.unsqueeze(1)
        weighted = torch.bmm(a, encoder_outputs.permute(1, 0, 2))
        rnn_input = torch.cat((embedded, weighted.permute(1, 0, 2)), dim=2)
        output, (hidden, cell) = self.rnn(rnn_input, (hidden, cell))
        embedded = embedded.squeeze(0)
        output = output.squeeze(0)
        weighted = weighted.squeeze(1)
        prediction = self.fc_out(torch.cat((output, weighted, embedded), dim=1))
        return prediction, hidden, cell, a.squeeze(1)

# Define the Seq2Seq model with attention
class Seq2Seq(nn.Module):
    def _ _init_ _(self, encoder, decoder, device):
        super()._ _init_ _()
        self.encoder = encoder
        self.decoder = decoder
        self.device = device

    def forward(self, src, trg, teacher_forcing_ratio=0.5):
        trg_len = trg.shape[0]
        trg_vocab_size = self.decoder.output_dim
        outputs = torch.zeros(trg_len, batch_size, trg_vocab_size).to(self.device)
        encoder_outputs, hidden, cell = self.encoder(src)
        input = trg[0, :]

        for t in range(1, trg_len):
            output, hidden, cell, _ = self.decoder(input, hidden, cell, encoder_outputs)
            outputs[t] = output
            teacher_force = random.random() < teacher_forcing_ratio
            top1 = output.argmax(1)
            input = trg[t] if teacher_force else top1

        return outputs
```

Text Summarization

Text summarization aims to generate a concise and coherent summary of a larger body of text. This task can be divided into extractive summarization, where key sentences are selected from the source text, and abstractive summarization, where new sentences are generated to capture the main ideas of the source text. Attention mechanisms play a crucial role in abstractive summarization by helping the model focus on relevant parts of the source text while generating the summary. In text summarization, the input text $\mathbf{x} = \{x_1, x_2, \ldots, x_T\}$ is processed by the encoder to produce hidden states $\mathbf{h} = \{h_1, h_2, \ldots, h_T\}$. The decoder generates the summary $\mathbf{y} = \{y_1, y_2, \ldots, y_{T'}\}$ by leveraging the attention mechanism to compute context vectors at each decoding step. The context vector $\mathbf{c}_t$ is computed similarly to the machine translation task:

$$\mathbf{c}_t = \sum_{j=1}^{T} \alpha_{tj} \mathbf{h}_j$$

$$\alpha_{tj} = \frac{\exp(e_{tj})}{\sum_{k=1}^{T} \exp(e_{tk})}$$

$$e_{tj} = \mathbf{v}_a^\top \tanh(\mathbf{W}_a \mathbf{s}_{t-1} + \mathbf{U}_a \mathbf{h}_j)$$

The context vector $\mathbf{c}_t$ is then used along with the previous decoder hidden state and the current input to generate the next hidden state and output:

$$\mathbf{s}_t = \text{DecoderRNN}(\mathbf{y}_{t-1}, \mathbf{s}_{t-1}, \mathbf{c}_t)$$

$$\mathbf{y}_t = g(\mathbf{W}_y \mathbf{s}_t + \mathbf{b}_y)$$

Example: Consider summarizing a news article. The encoder processes the article and generates hidden states. The decoder, with the help of the attention mechanism, focuses on different parts of the article to generate a coherent summary that captures the main points.

Implementation in TensorFlow:

```
# Define the encoder with attention
encoder_inputs = tf.keras.Input(shape=(None,))
encoder_embedding = Embedding(input_dim=input_vocab_size, output_dim=embedding_dim)
(encoder_inputs)
encoder_lstm = LSTM(units, return_sequences=True, return_state=True)
encoder_outputs, state_h, state_c = encoder_lstm(encoder_embedding)

# Define the decoder with attention
decoder_inputs = tf.keras.Input(shape=(None,))
decoder_embedding = Embedding(input_dim=output_vocab_size, output_dim=embedding_dim)
(decoder_inputs)
decoder_lstm = LSTM(units, return_sequences=True, return_state=True)
decoder_outputs, _, _ = decoder_lstm(decoder_embedding, initial_state=[state_h, state_c])

# Attention layer
attention_layer = BahdanauAttention(units)
context_vector, attention_weights = attention_layer(state_h, encoder_outputs)
decoder_combined_context = tf.concat([tf.expand_dims(context_vector, 1), decoder_outputs],
axis=-1)

# Final output layer
decoder_dense = Dense(output_vocab_size, activation='softmax')
decoder_outputs = decoder_dense(decoder_combined_context)

# Define the Seq2Seq model with attention
model = tf.keras.Model([encoder_inputs, decoder_inputs], decoder_outputs)
model.compile(optimizer='adam', loss='sparse_categorical_crossentropy')
```

Implementation in PyTorch:

```
# Define the encoder with attention
class Encoder(nn.Module):
    def _ _init_ _(self, input_dim, emb_dim, hid_dim, n_layers, dropout):
        super()._ _init_ _()
        self.embedding = nn.Embedding(input_dim, emb_dim)
        self.rnn = nn.LSTM(emb_dim, hid_dim, n_layers, dropout=dropout)
        self.dropout = nn.Dropout(dropout)
```

```
    def forward(self, src):
        embedded = self.dropout(self.embedding(src))
        outputs, (hidden, cell) = self.rnn(embedded)
        return outputs, hidden, cell

# Define the decoder with attention
class Decoder(nn.Module):
    def _ _init_ _(self, output_dim, emb_dim, hid_dim, n_layers, dropout, attention):
        super()._ _init_ _()
        self.output_dim = output_dim
        self.attention = attention
        self.embedding = nn.Embedding(output_dim, emb_dim)
        self.rnn = nn.LSTM((hid_dim * 2) + emb_dim, hid_dim, n_layers, dropout=dropout)
        self.fc_out = nn.Linear((hid_dim * 2) + hid_dim + emb_dim, output_dim)
        self.dropout = nn.Dropout(dropout)

    def forward(self, trg, hidden, cell, encoder_outputs):
        trg = trg.unsqueeze(0)
        embedded = self.dropout(self.embedding(trg))
        a = self.attention(hidden[-1], encoder_outputs)
        a = a.unsqueeze(1)
        weighted = torch.bmm(a, encoder_outputs.permute(1, 0, 2))
        rnn_input = torch.cat((embedded, weighted.permute(1, 0, 2)), dim=2)
        output, (hidden, cell) = self.rnn(rnn_input, (hidden, cell))
        embedded = embedded.squeeze(0)
        output = output.squeeze(0)
        weighted = weighted.squeeze(1)
        prediction = self.fc_out(torch.cat((output, weighted, embedded), dim=1))
        return prediction, hidden, cell, a.squeeze(1)

# Define the Seq2Seq model with attention
class Seq2Seq(nn.Module):
    def _ _init_ _(self, encoder, decoder, device):
        super()._ _init_ _()
        self.encoder = encoder
        self.decoder = decoder
        self.device = device

    def forward(self, src, trg, teacher_forcing_ratio=0.5):
        trg_len = trg.shape[0]
        trg_vocab_size = self.decoder.output_dim
        outputs = torch.zeros(trg_len, batch_size, trg_vocab_size).to(self.device)
        encoder_outputs, hidden, cell = self.encoder(src)
        input = trg[0, :]

        for t in range(1, trg_len):
            output, hidden, cell, _ = self.decoder(input, hidden, cell, encoder_outputs)
            outputs[t] = output
            teacher_force = random.random() < teacher_forcing_ratio
            top1 = output.argmax(1)
            input = trg[t] if teacher_force else top1

        return outputs
```

5.5 Training and Optimization of RNNs

Training RNNs involves unique challenges due to their sequential nature and the dependence of each time step on previous time steps. Two critical techniques to address these challenges are Truncated Backpropagation Through Time (TBPTT) and gradient clipping. This section provides a detailed exploration of these techniques, their mathematical foundations, and practical implementations.

5.5.1 Training Techniques for RNNs

Truncated Backpropagation Through Time (TBPTT)
BPTT is the standard method for training RNNs. It involves unrolling the RNN over time and applying the backpropagation algorithm to compute gradients for each time step. However, for long sequences, BPTT can be computationally expensive and prone to the vanishing gradient problem. Truncated Backpropagation Through Time (TBPTT) is an efficient alternative that mitigates these issues by splitting the sequence into shorter segments. Let $\mathbf{x} = \{\mathbf{x}_1, \mathbf{x}_2, \ldots, \mathbf{x}_T\}$ be the input sequence and $\mathbf{y} = \{\mathbf{y}_1, \mathbf{y}_2, \ldots, \mathbf{y}_T\}$ be the target sequence. The RNN computes hidden states $\mathbf{h}_t$ and outputs $\hat{\mathbf{y}}_t$ at each time step t:

$$\mathbf{h}_t = f(\mathbf{W}_h\mathbf{x}_t + \mathbf{U}_h\mathbf{h}_{t-1} + \mathbf{b}_h)$$
$$\hat{\mathbf{y}}_t = g(\mathbf{W}_y\mathbf{h}_t + \mathbf{b}_y)$$

In BPTT, the loss $\mathcal{L}$ is computed over the entire sequence, and gradients are propagated back through all time steps:

$$\mathcal{L} = \sum_{t=1}^{T} \ell(\hat{\mathbf{y}}_t, \mathbf{y}_t)$$

$$\frac{\partial \mathcal{L}}{\partial \mathbf{W}_h} = \sum_{t=1}^{T} \frac{\partial \mathcal{L}}{\partial \mathbf{h}_t}\frac{\partial \mathbf{h}_t}{\partial \mathbf{W}_h}$$

In TBPTT, the sequence is divided into segments of length k. Gradients are computed and updated after each segment, reducing the computational load and mitigating the vanishing gradient problem. Let k be the segment index for each segment $[t_k, t_{k+1}]$:

$$\mathcal{L}_k = \sum_{t=t_k}^{t_{k+1}} \ell(\hat{\mathbf{y}}_t, \mathbf{y}_t)$$

$$\frac{\partial \mathcal{L}_k}{\partial \mathbf{W}_h} = \sum_{t=t_k}^{t_{k+1}} \frac{\partial \mathcal{L}_k}{\partial \mathbf{h}_t} \frac{\partial \mathbf{h}_t}{\partial \mathbf{W}_h}$$

Example: Consider an RNN trained on a sequence of length 1000, with TBPTT segment length $k = 100$.

Implementation in TensorFlow:

```
import tensorflow as tf
from tensorflow.keras.models import Sequential
from tensorflow.keras.layers import SimpleRNN, Dense

# Define the RNN model
model = Sequential()
model.add(SimpleRNN(50, activation='tanh', input_shape=(None, 1)))
model.add(Dense(1))

# Compile the model
model.compile(optimizer='adam', loss='mean_squared_error')

# Generate synthetic data for demonstration
import numpy as np
x_train = np.random.rand(100, 1000, 1)  # 100 samples, 1000 time steps, 1 feature
y_train = np.random.rand(100, 1000, 1)

# Train the model with TBPTT
for epoch in range(20):
    for i in range(0, 1000, 100):
        x_segment = x_train[:, i:i+100, :]
        y_segment = y_train[:, i:i+100, :]
        model.train_on_batch(x_segment, y_segment)

# Evaluate the model
x_test = np.random.rand(10, 1000, 1)
y_test = np.random.rand(10, 1000, 1)
loss = model.evaluate(x_test, y_test)
print(f'Test Loss: {loss}')
```

Implementation in PyTorch:

```
import torch
import torch.nn as nn
import torch.optim as optim

# Define the RNN model
class SimpleRNN(nn.Module):
    def _ _init_ _(self):
        super(SimpleRNN, self)._ _init_ _()
        self.rnn = nn.RNN(input_size=1, hidden_size=50, batch_first=True)
        self.fc = nn.Linear(50, 1)

    def forward(self, x):
        out, _ = self.rnn(x)
        out = self.fc(out)
        return out

# Instantiate the model, loss function, and optimizer
model = SimpleRNN()
criterion = nn.MSELoss()
optimizer = optim.Adam(model.parameters(), lr=0.001)

# Generate synthetic data for demonstration
x_train = torch.rand(100, 1000, 1)  # 100 samples, 1000 time steps, 1 feature
y_train = torch.rand(100, 1000, 1)
```

```
# Train the model with TBPTT
num_epochs = 20
segment_length = 100
for epoch in range(num_epochs):
    for i in range(0, 1000, segment_length):
        x_segment = x_train[:, i:i+segment_length, :]
        y_segment = y_train[:, i:i+segment_length, :]
        outputs = model(x_segment)
        loss = criterion(outputs, y_segment)

        optimizer.zero_grad()
        loss.backward()
        optimizer.step()

    if (epoch + 1) % 5 == 0:
        print(f'Epoch [{epoch+1}/{num_epochs}], Loss: {loss.item():.4f}')

# Evaluate the model
x_test = torch.rand(10, 1000, 1)
y_test = torch.rand(10, 1000, 1)
with torch.no_grad():
    predictions = model(x_test)
    test_loss = criterion(predictions, y_test)
print(f'Test Loss: {test_loss.item():.4f}')
```

Gradient Clipping

Gradient clipping is a technique used to prevent the exploding gradient problem, which occurs when gradients become excessively large during training, leading to unstable updates and poor convergence. Gradient clipping ensures that the gradients remain within a reasonable range by scaling them when they exceed a predefined threshold. Given a gradient vector $\mathbf{g}$ and a threshold τ, gradient clipping modifies the gradients as follows:

$$\mathbf{g} \leftarrow \mathbf{g} \cdot \min\left(1, \frac{\tau}{\|\mathbf{g}\|}\right)$$

This ensures that the norm of the gradient vector does not exceed the threshold τ.

Example: Consider an RNN trained on a sequence with gradient clipping applied during training.

Implementation in TensorFlow:

```
# Define the RNN model
model = Sequential()
model.add(SimpleRNN(50, activation='tanh', input_shape=(None, 1)))
model.add(Dense(1))

# Compile the model with gradient clipping
optimizer = tf.keras.optimizers.Adam(clipvalue=1.0)
model.compile(optimizer=optimizer, loss='mean_squared_error')

# Generate synthetic data for demonstration
x_train = np.random.rand(100, 1000, 1)  # 100 samples, 1000 time steps, 1 feature
y_train = np.random.rand(100, 1000, 1)

# Train the model
model.fit(x_train, y_train, epochs=20, batch_size=32)

# Evaluate the model
x_test = np.random.rand(10, 1000, 1)
```

```
y_test = np.random.rand(10, 1000, 1)
loss = model.evaluate(x_test, y_test)
print(f'Test Loss: {loss}')
```

Implementation in PyTorch:

```
# Define the RNN model
model = SimpleRNN()

# Define the loss function and optimizer with gradient clipping
criterion = nn.MSELoss()
optimizer = optim.Adam(model.parameters(), lr=0.001)

# Generate synthetic data for demonstration
x_train = torch.rand(100, 1000, 1)  # 100 samples, 1000 time steps, 1 feature
y_train = torch.rand(100, 1000, 1)

# Train the model with gradient clipping
num_epochs = 20
clip_value = 1.0
for epoch in range(num_epochs):
    for i in range(0, 1000, segment_length):
        x_segment = x_train[:, i:i+segment_length, :]
        y_segment

 = y_train[:, i:i+segment_length, :]
        outputs = model(x_segment)
        loss = criterion(outputs, y_segment)

        optimizer.zero_grad()
        loss.backward()
        nn.utils.clip_grad_value_(model.parameters(), clip_value)
        optimizer.step()

    if (epoch + 1) % 5 == 0:
        print(f'Epoch [{epoch+1}/{num_epochs}], Loss: {loss.item():.4f}')

# Evaluate the model
x_test = torch.rand(10, 1000, 1)
y_test = torch.rand(10, 1000, 1)
with torch.no_grad():
    predictions = model(x_test)
    test_loss = criterion(predictions, y_test)
print(f'Test Loss: {test_loss.item():.4f}')
```

5.5.2 Regularization Techniques

Regularization techniques are essential in preventing overfitting during the training of neural networks, including RNNs. Here we cover two key regularization techniques: recurrent dropout and zoneout and weight regularization.

Recurrent Dropout and Zoneout
Recurrent dropout is a variant of dropout specifically designed for RNNs. Unlike standard dropout, which is applied independently at each time step, recurrent dropout is applied consistently across all time steps, ensuring that the same units are dropped at each time step within a sequence.

In recurrent dropout, the same dropout mask **d** is applied at every time step:

$$\tilde{\mathbf{h}}_t = \mathbf{h}_t \odot \mathbf{d}$$

where **d** is a binary mask vector sampled once for the entire sequence.

Zoneout is another regularization technique for RNNs that stochastically keeps some hidden units unchanged instead of dropping them out. This technique allows some parts of the hidden state to remain unperturbed, encouraging the network to rely on the same information over multiple time steps. For each unit, with probability p, the unit's value is kept from the previous time step:

$$\tilde{\mathbf{h}}_t = \mathbf{h}_t \odot \mathbf{d} + \mathbf{h}_{t-1} \odot (1 - \mathbf{d})$$

where **d** is a binary mask vector with each element sampled from a Bernoulli distribution with probability p.

Example: Consider training an RNN with recurrent dropout and zoneout.

Implementation in TensorFlow:

```
# Define the RNN model with recurrent dropout
model = Sequential()
model.add(SimpleRNN(50, activation='tanh', input_shape=(None, 1), dropout=0.5,
recurrent_dropout=0.2))
model.add(Dense(1))

# Compile the model
optimizer = tf.keras.optimizers.Adam(learning_rate=0.001)
model.compile(optimizer=optimizer, loss='mean_squared_error')

# Generate synthetic data for demonstration
x_train = np.random.rand(100, 100, 1)  # 100 samples, 100 time steps, 1 feature
y_train = np.random.rand(100, 100, 1)

# Train the model
model.fit(x_train, y_train, epochs=20, batch_size=32)

# Evaluate the model
x_test = np.random.rand(10, 100, 1)
y_test = np.random.rand(10, 100, 1)
loss = model.evaluate(x_test, y_test)
print(f'Test Loss: {loss}')
```

Implementation in PyTorch:

```
# Define the RNN model with zoneout
class ZoneoutWrapper(nn.Module):
    def _ _init_ _(self, module, zoneout_prob=0.1):
        super(ZoneoutWrapper, self)._ _init_ _()
        self.module = module
        self.zoneout_prob = zoneout_prob

    def forward(self, x, hidden):
        next_hidden = self.module(x, hidden)
        if self.training:
            mask = torch.ones_like(next_hidden[0]).bernoulli_(1 - self.zoneout_prob)
            next_hidden = (mask * next_hidden[0] + (1 - mask) * hidden[0], next_hidden[1])
        return next_hidden

class SimpleRNN(nn.Module):
    def _ _init_ _(self):
```

```
        super(SimpleRNN, self).__init__()
        self.rnn = nn.RNN(input_size=1, hidden_size=50, batch_first=True)
        self.zoneout = ZoneoutWrapper(self.rnn, zoneout_prob=0.2)
        self.fc = nn.Linear(50, 1)

    def forward(self, x):
        out, hidden = self.zoneout(x, None)
        out = self.fc(out)
        return out

# Instantiate the model, loss function, and optimizer
model = SimpleRNN()
criterion = nn.MSELoss()
optimizer = optim.Adam(model.parameters(), lr=0.001)

# Generate synthetic data for demonstration
x_train = torch.rand(100, 100, 1)  # 100 samples, 100 time steps, 1 feature
y_train = torch.rand(100, 100, 1)

# Train the model
num_epochs = 20
for epoch in range(num_epochs):
    model.train()
    outputs = model(x_train)
    loss = criterion(outputs, y_train)

    optimizer.zero_grad()
    loss.backward()
    optimizer.step()

    if (epoch + 1) % 5 == 0:
        print(f'Epoch [{epoch+1}/{num_epochs}], Loss: {loss.item():.4f}')

# Evaluate the model
x_test = torch.rand(10, 100, 1)
y_test = torch.rand(10, 100, 1)
model.eval()
with torch.no_grad():
    predictions = model(x_test)
    test_loss = criterion(predictions, y_test)
print(f'Test Loss: {test_loss.item():.4f}')
```

Weight Regularization

Weight regularization techniques such as L1 and L2 regularization add a penalty to the loss function to discourage large weights, which can help prevent overfitting. These techniques introduce an additional term to the loss function that penalizes large weights.

Example: Consider training an RNN with L2 regularization.

Implementation in TensorFlow:

```
# Define the RNN model with L2 regularization
from tensorflow.keras.regularizers import l2

model = Sequential()
model.add(SimpleRNN(50, activation='tanh', input_shape=(None, 1),
kernel_regularizer=l2(0.01)))
model.add(Dense(1, kernel_regularizer=l2(0.01)))

# Compile the model
optimizer = tf.keras.optimizers.Adam(learning_rate=0.001)
model.compile(optimizer=optimizer, loss='mean_squared_error')

# Generate synthetic data for demonstration
x_train = np.random.rand(100, 100, 1)  # 100 samples, 100 time steps, 1 feature
```

```
y_train = np.random.rand(100, 100, 1)

# Train the model
model.fit(x_train, y_train, epochs=20, batch_size=32)

# Evaluate the model
x_test = np.random.rand(10, 100, 1)
y_test = np.random.rand(10, 100, 1)
loss = model.evaluate(x_test, y_test)
print(f'Test Loss: {loss}')
```

Implementation in PyTorch:

```
# Define the RNN model with L2 regularization
class SimpleRNN(nn.Module):
    def _ _init_ _(self):
        super(SimpleRNN, self)._ _init_ _()
        self.rnn = nn.RNN(input_size=1, hidden_size=50, batch_first=True)
        self.fc = nn.Linear(50, 1)

    def forward(self, x):
        out, _ = self.rnn(x)
        out = self.fc(out)
        return out

# Instantiate the model, loss function, and optimizer
model = SimpleRNN()
criterion = nn.MSELoss()
optimizer = optim.Adam(model.parameters(), lr=0.001, weight_decay=0.01)

# Generate synthetic data for demonstration
x_train = torch.rand(100, 100, 1)  # 100 samples, 100 time steps, 1 feature
y_train = torch.rand(100, 100, 1)

# Train the model
num_epochs = 20
for epoch in range(num_epochs):
    model.train()
    outputs = model(x_train)
    loss = criterion(outputs, y_train)

    optimizer.zero_grad()
    loss.backward()
    optimizer.step()

    if (epoch + 1) % 5 == 0:
        print(f'Epoch [{epoch+1}/{num_epochs}], Loss: {loss.item():.4f}')

# Evaluate the model
x_test = torch.rand(10, 100, 1)
y_test = torch.rand(10, 100, 1)
model.eval()
with torch.no_grad():
    predictions = model(x_test)
    test_loss = criterion(predictions, y_test)
print(f'Test Loss: {test_loss.item():.4f}')
```

5.6 Practical Applications of RNNs

RNNs are powerful models for handling sequential data, making them well-suited for a variety of practical applications. One of the most prominent applications of RNNs is in language translation. This section explores the construction and evaluation of RNN-based translation models.

5.6.1 *Language Translation*

Language translation involves converting text from one language to another. RNNs, particularly in the form of Seq2Seq models with attention mechanisms, have proven effective in this task. We will discuss how to build a translator using RNNs and how to evaluate the performance of translation models.

Building a Translator with RNNs

Building a translator with RNNs involves creating a Seq2Seq model that consists of an encoder and a decoder. The encoder processes the input sentence and generates a context vector, which the decoder uses to generate the translated output sentence.

Model Architecture:

1. Encoder: The encoder reads the input sentence word by word and generates a sequence of hidden states. The final hidden state represents the context vector, summarizing the entire input sentence. Let $\mathbf{x} = \{x_1, x_2, \ldots, x_T\}$ be the input sentence, where T is the length of the sentence. The encoder's hidden states $\mathbf{h}_t$ are computed as follows:
$$\mathbf{h}_t = \text{RNN}(\mathbf{W}_e \mathbf{x}_t + \mathbf{U}_e \mathbf{h}_{t-1})$$
Here, $\mathbf{W}_e$ and $\mathbf{U}_e$ are the weight matrices for the input and hidden states, respectively.
2. Decoder: The decoder generates the translated sentence word by word, using the context vector and the previously generated words. At each time step t, the decoder's hidden state and output are computed as:
$$\mathbf{s}_t = \text{RNN}(\mathbf{W}_d \mathbf{y}_{t-1} + \mathbf{U}_d \mathbf{s}_{t-1} + \mathbf{V}_d \mathbf{c})$$
$$\hat{\mathbf{y}}_t = \text{softmax}(\mathbf{W}_o \mathbf{s}_t)$$
Here, $\mathbf{y}_{t-1}$ is the previous word, $\mathbf{c}$ is the context vector from the encoder, and $\mathbf{W}_d$, $\mathbf{U}_d$, and $\mathbf{V}_d$ are the weight matrices.

The attention mechanism improves the performance of the Seq2Seq model by allowing the decoder to focus on different parts of the input sentence at each time step. The attention weights α_{tj} are computed as:

$$e_{tj} = \mathbf{v}_a^\top \tanh(\mathbf{W}_a \mathbf{s}_{t-1} + \mathbf{U}_a \mathbf{h}_j)$$
$$\alpha_{tj} = \frac{\exp(e_{tj})}{\sum_{k=1}^{T} \exp(e_{tk})}$$

The context vector $\mathbf{c}_t$ is then computed as a weighted sum of the encoder's hidden states:

$$\mathbf{c}_t = \sum_{j=1}^{T} \alpha_{tj} \mathbf{h}_j$$

Example: Consider implementing a Seq2Seq model with attention for English to French translation.

Implementation in TensorFlow:

```
import tensorflow as tf
from tensorflow.keras.layers import Embedding, LSTM, Dense, Input
from tensorflow.keras.models import Model

# Encoder
encoder_inputs = Input(shape=(None,))
encoder_embedding = Embedding(input_dim=input_vocab_size, output_dim=embedding_dim)
(encoder_inputs)
encoder_lstm = LSTM(units, return_sequences=True, return_state=True)
encoder_outputs, state_h, state_c = encoder_lstm(encoder_embedding)

# Attention
attention_dense = Dense(1, activation='tanh')
attention_weights = attention_dense(encoder_outputs)
attention_weights = tf.nn.softmax(attention_weights, axis=1)
context_vector = attention_weights * encoder_outputs
context_vector = tf.reduce_sum(context_vector, axis=1)

# Decoder
decoder_inputs = Input(shape=(None,))
decoder_embedding = Embedding(input_dim=output_vocab_size, output_dim=embedding_dim)
(decoder_inputs)
decoder_lstm = LSTM(units, return_sequences=True, return_state=True)
decoder_outputs, _, _ = decoder_lstm(decoder_embedding, initial_state=[state_h, state_c])
decoder_dense = Dense(output_vocab_size, activation='softmax')
decoder_outputs = decoder_dense(decoder_outputs)

# Seq2Seq model with attention
model = Model([encoder_inputs, decoder_inputs], decoder_outputs)
model.compile(optimizer='adam', loss='sparse_categorical_crossentropy')

# Generate synthetic data for demonstration
import numpy as np
x_train = np.random.randint(0, input_vocab_size, (100, 10))
y_train = np.random.randint(0, output_vocab_size, (100, 10))
decoder_input_data = np.zeros_like(y_train)

# Train the model
model.fit([x_train, decoder_input_data], y_train, epochs=20, batch_size=32)
```

Implementation in PyTorch:

```
import torch
import torch.nn as nn
import torch.optim as optim

class Encoder(nn.Module):
    def _ _init_ _(self, input_dim, emb_dim, hid_dim, n_layers, dropout):
        super()._ _init_ _()
        self.embedding = nn.Embedding(input_dim, emb_dim)
        self.rnn = nn.LSTM(emb_dim, hid_dim, n_layers, dropout=dropout, batch_first=True)

    def forward(self, src):
        embedded = self.embedding(src)
        outputs, (hidden, cell) = self.rnn(embedded)
        return outputs, hidden, cell

class Attention(nn.Module):
    def _ _init_ _(self, hid_dim):
        super()._ _init_ _()
        self.attn = nn.Linear(hid_dim * 2, 1)

    def forward(self, hidden, encoder_outputs):
        src_len = encoder_outputs.shape[1]
        hidden = hidden[-1].unsqueeze(1).repeat(1, src_len, 1)
        energy = torch.tanh(self.attn(torch.cat((hidden, encoder_outputs), dim=2)))
        attention = torch.softmax(energy.squeeze(2), dim=1)
        return attention

class Decoder(nn.Module):
    def _ _init_ _(self, output_dim, emb_dim, hid_dim, n_layers, dropout, attention):
        super()._ _init_ _()
        self.embedding = nn.Embedding(output_dim, emb_dim)
        self.rnn = nn.LSTM(emb_dim + hid_dim, hid_dim, n_layers, dropout=dropout,
        batch_first=True)
        self.fc_out = nn.Linear(hid_dim * 2, output_dim)
        self.attention = attention

    def forward(self, trg, hidden, cell, encoder_outputs):
        trg = trg.unsqueeze(1)
        embedded = self.embedding(trg)
        attention = self.attention(hidden, encoder_outputs).unsqueeze(1)
        context = torch.bmm(attention, encoder_outputs)
        rnn_input = torch.cat((embedded, context), dim=2)
        output, (hidden, cell) = self.rnn(rnn_input, (hidden, cell))
        output = self.fc_out(torch.cat((output, context), dim=2).squeeze(1))
        return output, hidden, cell

class Seq2Seq(nn.Module):
    def _ _init_ _(self, encoder, decoder, device):
        super()._ _init_ _()
        self.encoder = encoder
        self.decoder = decoder
        self.device = device

    def forward(self, src, trg, teacher_forcing_ratio=0.5):
        batch_size = trg.shape[0]
        trg_len = trg.shape[1]
        trg_vocab_size = self.decoder.fc_out.out_features

        outputs = torch.zeros(batch_size, trg_len, trg_vocab_size).to(self.device)
        encoder_outputs, hidden, cell = self.encoder(src)

        input = trg[:, 0]
        for t in range(1, trg_len):
            output, hidden, cell = self.decoder(input, hidden, cell, encoder_outputs)
            outputs[:, t, :] = output
```

```
            top1 = output.argmax(1)
            input = trg[:, t] if random.random() < teacher_forcing_ratio else top1

        return outputs

# Hyperparameters
input_dim = 1000
output_dim = 1000
emb_dim = 256
hid_dim = 512
n_layers = 2
dropout = 0.5
device = torch.device('cuda' if torch.cuda.is_available() else 'cpu')

# Model instantiation
attn = Attention(hid_dim)
enc = Encoder(input_dim, emb_dim, hid_dim, n_layers, dropout)
dec = Decoder(output_dim, emb_dim, hid_dim, n_layers, dropout, attn)
model = Seq2Seq(enc, dec, device).to(device)

# Optimizer and Loss
optimizer = optim.Adam(model.parameters())
criterion = nn.CrossEntropyLoss()

# Synthetic Data
x_train = torch.randint(0, input_dim, (100, 10)).to(device)
y_train = torch.randint(0, output_dim, (100, 10)).to(device)

# Training Loop
num_epochs = 20
for epoch in range(num_epochs):
    model.train()
    optimizer.zero_grad()
    output = model(x_train, y_train)
    output_dim = output.shape[-1]
    output = output[1:].view(-1, output_dim)
    y_train = y_train[1:].view(-1)
    loss = criterion(output, y_train)
    loss.backward()
    optimizer.step()
    if (epoch + 1) % 5 == 0:
        print(f'Epoch [{epoch + 1}/{num_epochs}], Loss: {loss.item():.4f}')
```

Evaluating the performance of translation models is crucial to ensure their effectiveness. Several metrics and techniques are used to assess the quality of machine translation:

BLEU Score: The Bilingual Evaluation Understudy (BLEU) score is a popular metric for evaluating machine-translated text. It measures the precision of n-grams between the machine-generated translation and one or more reference translations. The BLEU score is computed as:

$$\text{BLEU} = \text{BP} \cdot \exp\left(\sum_{n=1}^{N} w_n \log p_n\right)$$

where p_n is the precision of n-grams, w_n are the weights (typically equal), and BP is the brevity penalty to penalize short translations.

Example: Consider a reference sentence: “The cat is on the mat” and a machine-generated sentence: “The cat is mat”. The BLEU score for unigrams and bigrams would be calculated by comparing the n-grams of the generated sentence with the reference sentence.

Human Evaluation: While automated metrics like BLEU are useful, human evaluation is often necessary for a more nuanced assessment of translation quality. Human evaluators judge the translations based on fluency, adequacy, and overall quality.

Example: BLEU score calculation in Python:

```
from nltk.translate.bleu_score import sentence_bleu

reference = [['the', 'cat', 'is', 'on', 'the', 'mat']]
candidate = ['the', 'cat', 'is', 'mat']
score = sentence_bleu(reference, candidate)
print(f'BLEU Score: {score:.4f}')
```

5.6.2 *Sentiment Analysis*

Sentiment analysis is the process of determining the emotional tone behind a body of text. It is commonly used to analyze customer reviews, social media posts, and other text data to gauge public sentiment. RNNs are particularly well-suited for this task due to their ability to capture sequential dependencies in the data. This section explores how to implement sentiment analysis using RNNs and addresses the challenges and solutions associated with this task.

Implementing Sentiment Analysis with RNNs

To implement sentiment analysis with RNNs, we follow these steps:

1. Data Preprocessing: The text data must be preprocessed to convert it into a format suitable for RNN input. This involves tokenization, removing stop words, stemming/lemmatization, and padding sequences to ensure uniform input lengths.
2. Model Architecture: The RNN model for sentiment analysis typically includes an embedding layer, one or more recurrent layers (such as LSTM or GRU), and a fully connected output layer.

Let $\mathbf{x} = \{x_1, x_2, \ldots, x_T\}$ be the input text sequence, where T is the length of the sequence. The RNN processes the sequence and generates hidden states $\mathbf{h}_t$ at each time step t:

$$\mathbf{h}_t = f(\mathbf{W}_h\mathbf{x}_t + \mathbf{U}_h\mathbf{h}_{t-1} + \mathbf{b}_h)$$

Here, $\mathbf{W}_h$ and $\mathbf{U}_h$ are weight matrices, $\mathbf{b}_h$ is the bias vector, and f is the activation function (e.g., tanh or ReLU).

The final hidden state $\mathbf{h}_T$ is passed through a fully connected layer with a sigmoid activation to produce the sentiment score $\hat{y}$:

$$\hat{y} = \sigma(\mathbf{W}_o \mathbf{h}_T + \mathbf{b}_o)$$

where $\mathbf{W}_o$ and $\mathbf{b}_o$ are the weight matrix and bias vector for the output layer, and σ is the sigmoid activation function.

Example: Consider implementing a sentiment analysis model using LSTM in TensorFlow and PyTorch.

Implementation in TensorFlow:

```
import tensorflow as tf
from tensorflow.keras.models import Sequential
from tensorflow.keras.layers import Embedding, LSTM, Dense
from tensorflow.keras.preprocessing.text import Tokenizer
from tensorflow.keras.preprocessing.sequence import pad_sequences

# Sample data
texts = ["I love this product!", "This is the worst thing ever."]
labels = [1, 0]  # 1 for positive, 0 for negative

# Tokenize and pad sequences
tokenizer = Tokenizer(num_words=10000)
tokenizer.fit_on_texts(texts)
sequences = tokenizer.texts_to_sequences(texts)
x_data = pad_sequences(sequences, maxlen=10)
y_data = tf.convert_to_tensor(labels)

# Define the RNN model for sentiment analysis
model = Sequential()
model.add(Embedding(input_dim=10000, output_dim=64, input_length=10))
model.add(LSTM(128))
model.add(Dense(1, activation='sigmoid'))

# Compile the model
model.compile(optimizer='adam', loss='binary_crossentropy', metrics=['accuracy'])

# Train the model
model.fit(x_data, y_data, epochs=10, batch_size=2)

# Evaluate the model
loss, accuracy = model.evaluate(x_data, y_data)
print(f'Loss: {loss}, Accuracy: {accuracy}')
```

Implementation in PyTorch:

```
import torch
import torch.nn as nn
import torch.optim as optim
from torchtext.legacy.data import Field, LabelField, TabularDataset, BucketIterator

# Sample data
texts = ["I love this product!", "This is the worst thing ever."]
labels = [1, 0]  # 1 for positive, 0 for negative

# Define fields
TEXT = Field(tokenize='spacy', include_lengths=True)
LABEL = LabelField(dtype=torch.float)

# Create dataset
fields = [('text', TEXT), ('label', LABEL)]
```

```
examples = [torchtext.legacy.data.Example.fromlist([texts[i], labels[i]], fields) for i
in range(len(texts))]
dataset = torchtext.legacy.data.Dataset(examples, fields)

# Build vocabulary and create iterators
TEXT.build_vocab(dataset, max_size=10000)
LABEL.build_vocab(dataset)
train_iterator, valid_iterator = BucketIterator.splits((dataset, dataset), batch_size=2,
sort_within_batch=True, device=device)

# Define the RNN model for sentiment analysis
class SentimentRNN(nn.Module):
    def __init__(self, vocab_size, embedding_dim, hidden_dim, output_dim):
        super().__init__()
        self.embedding = nn.Embedding(vocab_size, embedding_dim)
        self.rnn = nn.LSTM(embedding_dim, hidden_dim, batch_first=True)
        self.fc = nn.Linear(hidden_dim, output_dim)
        self.sigmoid = nn.Sigmoid()

    def forward(self, text, text_lengths):
        embedded = self.embedding(text)
        packed_embedded = nn.utils.rnn.pack_padded_sequence(embedded, text_lengths,
        batch_first=True)
        packed_output, (hidden, cell) = self.rnn(packed_embedded)
        output, output_lengths = nn.utils.rnn.pad_packed_sequence(packed_output,
        batch_first=True)
        hidden = hidden[-1,:,:]
        return self.sigmoid(self.fc(hidden))

# Instantiate the model, loss function, and optimizer
vocab_size = len(TEXT.vocab)
embedding_dim = 64
hidden_dim = 128
output_dim = 1
model = SentimentRNN(vocab_size, embedding_dim, hidden_dim, output_dim)
criterion = nn.BCELoss()
optimizer = optim.Adam(model.parameters())

# Train the model
num_epochs = 10
for epoch in range(num_epochs):
    model.train()
    epoch_loss = 0
    epoch_acc = 0
    for batch in train_iterator:
        optimizer.zero_grad()
        text, text_lengths = batch.text
        predictions = model(text, text_lengths.cpu()).squeeze(1)
        loss = criterion(predictions, batch.label)
        loss.backward()
        optimizer.step()
        epoch_loss += loss.item()
    print(f'Epoch {epoch+1}/{num_epochs}, Loss: {epoch_loss/len(train_iterator):.4f}')

# Evaluate the model
model.eval()
with torch.no_grad():
    for batch in valid_iterator:
        text, text_lengths = batch.text
        predictions = model(text, text_lengths.cpu()).squeeze(1)
        loss = criterion(predictions, batch.label)
        print(f'Evaluation Loss: {loss.item():.4f}')
```

Challenges and Solutions

While RNNs are effective for sentiment analysis, several challenges need to be addressed:

1. Vanishing Gradients: RNNs suffer from vanishing gradients when processing long sequences. This can be mitigated using advanced RNN variants like LSTM or GRU, which are designed to retain information over longer time spans.
2. Handling Imbalanced Data: Sentiment analysis datasets are often imbalanced, with a higher proportion of neutral or positive reviews. Techniques such as oversampling, undersampling, and using weighted loss functions can help address this issue.
3. Feature Representation: The quality of the input representation significantly impacts the model's performance. Using pre-trained word embeddings (e.g., GloVe or Word2Vec) can enhance the representation quality. Additionally, transformer-based embeddings like BERT can be fine-tuned for sentiment analysis tasks.
4. Overfitting: Overfitting is a common issue in neural network training. Regularization techniques such as dropout, recurrent dropout, and L2 regularization can help prevent overfitting. Early stopping and model checkpoints can also be used to monitor and control overfitting during training.
5. Context and Ambiguity: Understanding context and resolving ambiguity in text can be challenging. Incorporating context-aware models like transformers and using attention mechanisms can improve the model's ability to capture contextual information and resolve ambiguities.

Example of Handling Imbalanced Data: In TensorFlow, the 'class_weight' parameter can be used to assign different weights to different classes during training:

```
# Calculate class weights
import numpy as np
from sklearn.utils import class_weight

class_weights = class_weight.compute_class_weight('balanced', np.unique(y_data),
y_data.numpy())
class_weight_dict = dict(enumerate(class_weights))

# Train the model with class weights
model.fit(x_data, y_data, epochs=10, batch_size=2, class_weight=class_weight_dict)
```

5.6.3 *Text Generation*

Text generation is a fascinating application of RNNs, where the model learns to generate coherent and contextually relevant text based on the patterns it has learned from the training data. This section delves into the intricacies of training RNNs for text generation and explores its applications in creative writing and content creation.

Training RNNs for Text Generation

Training RNNs for text generation involves several steps, including data preprocessing, model design, training, and generation of text. The goal is to create a model that can generate new sequences of text that are similar to the training data. Given a sequence of characters or words, the RNN predicts the next character or word in the sequence. Let $\mathbf{x} = \{x_1, x_2, \ldots, x_T\}$ be the input sequence, and $\mathbf{y} = \{y_1, y_2, \ldots, y_T\}$ be the target sequence (shifted by one time step). The RNN computes hidden states $\mathbf{h}_t$ and outputs $\hat{y}_t$ at each time step t:

$$\mathbf{h}_t = f(\mathbf{W}_h \mathbf{x}_t + \mathbf{U}_h \mathbf{h}_{t-1} + \mathbf{b}_h)$$

$$\hat{y}_t = \text{softmax}(\mathbf{W}_o \mathbf{h}_t + \mathbf{b}_o)$$

The model is trained to minimize the cross-entropy loss between the predicted and actual next characters or words:

$$\mathcal{L} = -\sum_{t=1}^{T} y_t \log(\hat{y}_t)$$

Example: Consider implementing a character-level RNN for text generation in TensorFlow and PyTorch.

Implementation in TensorFlow:

```
import tensorflow as tf
from tensorflow.keras.models import Sequential
from tensorflow.keras.layers import LSTM, Dense, Embedding
from tensorflow.keras.preprocessing.text import Tokenizer
from tensorflow.keras.preprocessing.sequence import pad_sequences
import numpy as np

# Sample data
text = "This is an example of text generation using RNNs."
tokenizer = Tokenizer(char_level=True)
tokenizer.fit_on_texts([text])
sequences = tokenizer.texts_to_sequences([text])[0]

# Prepare input-output pairs
x_data = []
y_data = []
seq_length = 5
for i in range(len(sequences) - seq_length):
    x_data.append(sequences[i:i+seq_length])
    y_data.append(sequences[i+seq_length])
x_data = np.array(x_data)
y_data = np.array(y_data)

# Define the RNN model for text generation
model = Sequential()
model.add(Embedding(input_dim=len(tokenizer.word_index) + 1, output_dim=50,
input_length=seq_length))
model.add(LSTM(100))
model.add(Dense(len(tokenizer.word_index) + 1, activation='softmax'))

# Compile the model
model.compile(optimizer='adam', loss='sparse_categorical_crossentropy')
```

```
# Train the model
model.fit(x_data, y_data, epochs=50, batch_size=32)

# Generate text
def generate_text(model, tokenizer, seq_length, seed_text, num_chars):
    result = seed_text
    for _ in range(num_chars):
        encoded = tokenizer.texts_to_sequences([result[-seq_length:]])[0]
        encoded = pad_sequences([encoded], maxlen=seq_length, truncating='pre')
        y_pred = model.predict(encoded, verbose=0)
        predicted_char = tokenizer.index_word[np.argmax(y_pred)]
        result += predicted_char
    return result

seed_text = "This "
generated_text = generate_text(model, tokenizer, seq_length, seed_text, 50)
print(generated_text)
```

Implementation in PyTorch:

```
import torch
import torch.nn as nn
import torch.optim as optim
import numpy as np
from torchtext.legacy.data import Field, TabularDataset, BucketIterator

# Sample data
text = "This is an example of text generation using RNNs."
chars = sorted(list(set(text)))
char_to_idx = {char: idx for idx, char in enumerate(chars)}
idx_to_char = {idx: char for idx, char in enumerate(chars)}

# Prepare input-output pairs
sequences = [char_to_idx[char] for char in text]
x_data = []
y_data = []
seq_length = 5
for i in range(len(sequences) - seq_length):
    x_data.append(sequences[i:i+seq_length])
    y_data.append(sequences[i+seq_length])
x_data = torch.tensor(x_data, dtype=torch.long)
y_data = torch.tensor(y_data, dtype=torch.long)

# Define the RNN model for text generation
class CharRNN(nn.Module):
    def _ _init_ _(self, vocab_size, embedding_dim, hidden_dim, output_dim):
        super(CharRNN, self)._ _init_ _()
        self.embedding = nn.Embedding(vocab_size, embedding_dim)
        self.rnn = nn.LSTM(embedding_dim, hidden_dim, batch_first=True)
        self.fc = nn.Linear(hidden_dim, output_dim)

    def forward(self, x):
        embedded = self.embedding(x)
        output, _ = self.rnn(embedded)
        output = self.fc(output[:, -1, :])
        return output

# Instantiate the model, loss function, and optimizer
vocab_size = len(chars)
embedding_dim = 50
hidden_dim = 100
output_dim = vocab_size
model = CharRNN(vocab_size, embedding_dim, hidden_dim, output_dim)
criterion = nn.CrossEntropyLoss()
optimizer = optim.Adam(model.parameters(), lr=0.001)

# Train the model
```

```
num_epochs = 50
for epoch in range(num_epochs):
    model.train()
    optimizer.zero_grad()
    output = model(x_data)
    loss = criterion(output, y_data)
    loss.backward()
    optimizer.step()
    if (epoch + 1) % 10 == 0:
        print(f'Epoch [{epoch + 1}/{num_epochs}], Loss: {loss.item():.4f}')

# Generate text
def generate_text(model, char_to_idx, idx_to_char, seq_length, seed_text, num_chars):
    result = seed_text
    model.eval()
    for _ in range(num_chars):
        input_seq = torch.tensor([char_to_idx[char] for char in result[-seq_length:]],
        dtype=torch.long).unsqueeze(0)
        with torch.no_grad():
            output = model(input_seq)
        predicted_char = idx_to_char[output.argmax(1).item()]
        result += predicted_char
    return result

seed_text = "This "
generated_text = generate_text(model, char_to_idx, idx_to_char, seq_length, seed_text, 50)
print(generated_text)
```

Applications in Creative Writing and Content Creation

RNNs can be employed in various applications in creative writing and content creation. By training on a diverse range of texts, RNNs can generate content that mimics the style and structure of the training data. Here are some notable applications:

1. Poetry and Story Generation: RNNs can be trained on collections of poems or stories to generate new, original works. By capturing the linguistic patterns and themes of the training texts, the model can create coherent and stylistically consistent poems or stories.
2. Script Writing: In the entertainment industry, RNNs can assist scriptwriters by generating dialogue and plot ideas. By training on existing scripts, the model can produce text that adheres to the conventions and tone of the genre.
3. Content Recommendation: RNNs can be used to generate personalized content recommendations based on user preferences. By analyzing user interactions and preferences, the model can suggest articles, books, or other forms of content that align with the user's interests.
4. Interactive Storytelling: RNNs can be integrated into interactive storytelling platforms, where the model generates real-time responses to user inputs. This creates dynamic and engaging narratives that adapt to user choices.

Example (Generating Poetry) Consider training an RNN to generate poetry based on a dataset of poems.

Implementation in TensorFlow:

```
# Sample data: A collection of poems
poems = [
    "The sun sets in the west,\nLeaving a trail of golden hues.
    \nThe night embraces the earth,\nWith stars like diamond dues.",
```

```
    "The moonlit path beckons,\nWhispering secrets of the night.
    \nThe trees sway in rhythm,\nDancing to the music of light."
]

# Tokenize and pad sequences
tokenizer = Tokenizer(char_level=True)
tokenizer.fit_on_texts(poems)
sequences = tokenizer.texts_to_sequences(poems)
x_data = []
y_data = []
seq_length = 10
for seq in sequences:
    for i in range(len(seq) - seq_length):
        x_data.append(seq[i:i+seq_length])
        y_data.append(seq[i+seq_length])
x_data = np.array(x_data)
y_data = np.array(y_data)

# Define the RNN model for poetry generation
model = Sequential()
model.add(Embedding(input_dim=len(tokenizer.word_index) + 1, output_dim=50,
input_length=seq_length))
model.add(LSTM(100))
model.add(Dense(len(tokenizer.word_index) + 1, activation='softmax'))

# Compile the model
model.compile(optimizer='adam', loss='sparse_categorical_crossentropy')

# Train the model
model.fit(x_data, y_data, epochs=50, batch_size=32)

# Generate poetry
seed_text = "The night"
generated_text = generate_text(model, tokenizer, seq_length, seed_text, 50)
print(generated_text)
```

Implementation in PyTorch:

```
# Sample data: A collection of poems
poems = [
    "The sun sets in the west,\nLeaving a trail of golden hues.
    \nThe night embraces the earth,\nWith stars like diamond dues.",
    "The moonlit path beckons,\nWhispering secrets of the night.
    \nThe trees sway in rhythm,\nDancing to the music of light."
]

# Prepare input-output pairs
sequences = [char_to_idx[char] for poem in poems for char in poem]
x_data = []
y_data = []
seq_length = 10
for i in range(len(sequences) - seq_length):
    x_data.append(sequences[i:i+seq_length])
    y_data.append(sequences[i+seq_length])
x_data = torch.tensor(x_data, dtype=torch.long)
y_data = torch.tensor(y_data, dtype=torch.long)

# Define the RNN model for poetry generation
class PoetryRNN(nn.Module):
    def __init__(self, vocab_size, embedding_dim, hidden_dim, output_dim):
        super(PoetryRNN, self).__init__()
        self.embedding = nn.Embedding(vocab_size, embedding_dim)
        self.rnn = nn.LSTM(embedding_dim, hidden_dim, batch_first=True)
        self.fc = nn.Linear(hidden_dim, output_dim)

    def forward(self, x):
        embedded = self.embedding(x)
```

```
        output, _ = self.rnn(embedded)
        output = self.fc(output[:, -1, :])
        return output

# Instantiate the model, loss function, and optimizer
model = PoetryRNN(vocab_size, embedding_dim, hidden_dim, output_dim)
criterion = nn.CrossEntropyLoss()
optimizer = optim.Adam(model.parameters(), lr=0.001)

# Train the model
num_epochs = 50
for epoch in range(num_epochs):
    model.train()
    optimizer.zero_grad()
    output = model(x_data)
    loss = criterion(output, y_data)
    loss.backward()
    optimizer.step()
    if (epoch + 1) % 10 == 0:
        print(f'Epoch [{epoch + 1}/{num_epochs}], Loss: {loss.item():.4f}')

# Generate poetry
seed_text = "The night"
generated_text = generate_text(model, char_to_idx, idx_to_char, seq_length, seed_text, 50)
print(generated_text)
```

5.6.4 Other Applications

RNNs have a wide range of applications beyond text processing and sentiment analysis. This section explores some of these applications, including speech recognition, time series forecasting, and video analysis. Each application leverages the ability of RNNs to capture sequential dependencies and temporal patterns in data.

Speech Recognition
Speech recognition involves converting spoken language into written text. RNNs, particularly LSTM and GRU networks, are effective in modeling the temporal dependencies in speech signals. Given an input sequence of acoustic features $\mathbf{x} = \{x_1, x_2, \ldots, x_T\}$ and the corresponding output sequence of phonemes or words $\mathbf{y} = \{y_1, y_2, \ldots, y_{T'}\}$, the goal is to maximize the likelihood of the output sequence given the input sequence:

$$P(\mathbf{y} \mid \mathbf{x}) = \prod_{t=1}^{T'} P(y_t \mid \mathbf{h}_t)$$

where $\mathbf{h}_t$ is the hidden state at time step t, computed as:

$$\mathbf{h}_t = f(\mathbf{W}_h \mathbf{x}_t + \mathbf{U}_h \mathbf{h}_{t-1} + \mathbf{b}_h)$$

Example: Consider implementing a simple speech recognition model using LSTM in TensorFlow.

Implementation in TensorFlow:

```
import tensorflow as tf
from tensorflow.keras.models import Sequential
from tensorflow.keras.layers import LSTM, Dense
from tensorflow.keras.preprocessing.sequence import pad_sequences
import numpy as np

# Sample data: sequences of acoustic features
x_train = np.random.rand(100, 100, 13)  # 100 samples, 100 time steps, 13 features
y_train = np.random.randint(0, 10, (100, 10))  # 100 samples, 10 phonemes

# Define the RNN model for speech recognition
model = Sequential()
model.add(LSTM(128, input_shape=(100, 13), return_sequences=True))
model.add(Dense(10, activation='softmax'))

# Compile the model
model.compile(optimizer='adam', loss='sparse_categorical_crossentropy', metrics=['accuracy'])

# Train the model
model.fit(x_train, y_train, epochs=20, batch_size=32)

# Evaluate the model
x_test = np.random.rand(10, 100, 13)
y_test = np.random.randint(0, 10, (10, 10))
loss, accuracy = model.evaluate(x_test, y_test)
print(f'Test Loss: {loss}, Test Accuracy: {accuracy}')
```

Implementation in PyTorch:

```
import torch
import torch.nn as nn
import torch.optim as optim

# Define the RNN model for speech recognition
class SpeechRNN(nn.Module):
    def _ _init_ _(self, input_dim, hidden_dim, output_dim):
        super(SpeechRNN, self)._ _init_ _()
        self.rnn = nn.LSTM(input_dim, hidden_dim, batch_first=True)
        self.fc = nn.Linear(hidden_dim, output_dim)

    def forward(self, x):
        out, _ = self.rnn(x)
        out = self.fc(out)
        return out

# Instantiate the model, loss function, and optimizer
input_dim = 13
hidden_dim = 128
output_dim = 10
model = SpeechRNN(input_dim, hidden_dim, output_dim)
criterion = nn.CrossEntropyLoss()
optimizer = optim.Adam(model.parameters(), lr=0.001)

# Generate synthetic data for demonstration
x_train = torch.rand(100, 100, input_dim)  # 100 samples, 100 time steps, 13 features
y_train = torch.randint(0, output_dim, (100, 100))

# Train the model
num_epochs = 20
for epoch in range(num_epochs):
    model.train()
    optimizer.zero_grad()
```

```
    outputs = model(x_train)
    loss = criterion(outputs.view(-1, output_dim), y_train.view(-1))
    loss.backward()
    optimizer.step()
    if (epoch + 1) % 5 == 0:
        print(f'Epoch [{epoch+1}/{num_epochs}], Loss: {loss.item():.4f}')

# Evaluate the model
x_test = torch.rand(10, 100, input_dim)
y_test = torch.randint(0, output_dim, (10, 100))
model.eval()
with torch.no_grad():
    outputs = model(x_test)
    loss = criterion(outputs.view(-1, output_dim), y_test.view(-1))
print(f'Test Loss: {loss.item():.4f}')
```

Time Series Forecasting

Time series forecasting involves predicting future values based on historical data. RNNs are well-suited for this task due to their ability to capture temporal dependencies and patterns in time series data. Given a time series $\mathbf{x} = \{x_1, x_2, \ldots, x_T\}$, the goal is to predict the future values $\hat{\mathbf{y}} = \{\hat{y}_{T+1}, \hat{y}_{T+2}, \ldots, \hat{y}_{T+H}\}$, where H is the forecast horizon. The RNN learns to model the conditional distribution of the future values given the past observations:

$$P(\hat{\mathbf{y}} \mid \mathbf{x}) = \prod_{t=T+1}^{T+H} P(\hat{y}_t \mid \mathbf{h}_t)$$

where $\mathbf{h}_t$ is the hidden state at time step t, computed as:

$$\mathbf{h}_t = f(\mathbf{W}_h \mathbf{x}_t + \mathbf{U}_h \mathbf{h}_{t-1} + \mathbf{b}_h)$$

Example: Consider implementing a time series forecasting model using LSTM in TensorFlow.

Implementation in TensorFlow:

```
import tensorflow as tf
from tensorflow.keras.models import Sequential
from tensorflow.keras.layers import LSTM, Dense
import numpy as np

# Sample data: sequences of time series values
x_train = np.random.rand(100, 50, 1)  # 100 samples, 50 time steps, 1 feature
y_train = np.random.rand(100, 10)  # 100 samples, 10 future values

# Define the RNN model for time series forecasting
model = Sequential()
model.add(LSTM(128, input_shape=(50, 1)))
model.add(Dense(10))

# Compile the model
model.compile(optimizer='adam', loss='mean_squared_error')

# Train the model
model.fit(x_train, y_train, epochs=20, batch_size=32)

# Evaluate the model
x_test = np.random.rand(10, 50, 1)
```

```
y_test = np.random.rand(10, 10)
loss = model.evaluate(x_test, y_test)
print(f'Test Loss: {loss}')
```

Implementation in PyTorch:

```
import torch
import torch.nn as nn
import torch.optim as optim

# Define the RNN model for time series forecasting
class TimeSeriesRNN(nn.Module):
    def __init__(self, input_dim, hidden_dim, output_dim):
        super(TimeSeriesRNN, self).__init__()
        self.rnn = nn.LSTM(input_dim, hidden_dim, batch_first=True)
        self.fc = nn.Linear(hidden_dim, output_dim)

    def forward(self, x):
        out, _ = self.rnn(x)
        out = self.fc(out[:, -1, :])
        return out

# Instantiate the model, loss function, and optimizer
input_dim = 1
hidden_dim = 128
output_dim = 10
model = TimeSeriesRNN(input_dim, hidden_dim, output_dim)
criterion = nn.MSELoss()
optimizer = optim.Adam(model.parameters(), lr=0.001)

# Generate synthetic data for demonstration
x_train = torch.rand(100, 50, input_dim)  # 100 samples, 50 time steps, 1 feature
y_train = torch.rand(100, output_dim)

# Train the model
num_epochs = 20
for epoch in range(num_epochs):
    model.train()
    optimizer.zero_grad()
    outputs = model(x_train)
    loss = criterion(outputs, y_train)
    loss.backward()
    optimizer.step()
    if (epoch + 1) % 5 == 0:
        print(f'Epoch [{epoch+1}/{num_epochs}], Loss: {loss.item():.4f}')

# Evaluate the model
x_test = torch.rand(10, 50, input_dim)
y_test = torch.rand(10, output_dim)
model.eval()
with torch.no_grad():
    outputs = model(x_test)
    loss = criterion(outputs, y_test)
print(f'Test Loss: {loss.item():.4f}')
```

Video Analysis

Video analysis involves processing and understanding video data, which includes both spatial and temporal dimensions. RNNs can be used to model the temporal dependencies in video sequences, enabling applications such as action recognition, video summarization, and anomaly detection. Given a sequence of video frames $\mathbf{X} = \{\mathbf{x}_1, \mathbf{x}_2, \ldots, \mathbf{x}_T\}$, where each frame $\mathbf{x}_t$ is represented by a feature vector, the goal is to model the temporal dynamics in the video. The RNN computes hidden states $\mathbf{h}_t$ at each time step t:

$$\mathbf{h}_t = f(\mathbf{W}_h\mathbf{x}_t + \mathbf{U}_h\mathbf{h}_{t-1} + \mathbf{b}_h)$$

The hidden states can be used for various downstream tasks, such as predicting actions or summarizing the video content.

Example: Consider implementing a simple action recognition model using LSTM in TensorFlow.

Implementation in TensorFlow:

```
import tensorflow as tf
from tensorflow.keras.models import Sequential
from tensorflow.keras.layers import LSTM, Dense
import numpy as np

# Sample data: sequences of video frame features
x_train = np.random.rand(100, 50, 2048)  # 100 samples, 50 frames, 2048 features
y_train = np.random.randint(0, 10, 100)  # 100 samples, 10 action classes

# Define the RNN model for video analysis
model = Sequential()
model.add(LSTM(128, input_shape=(50, 2048)))
model.add(Dense(10, activation='softmax'))

# Compile the model
model.compile(optimizer='adam', loss='sparse_categorical_crossentropy', metrics=['accuracy'])

# Train the model
model.fit(x_train, y_train, epochs=20, batch_size=32)

# Evaluate the model
x_test = np.random.rand(10, 50, 2048)
y_test = np.random.randint(0, 10, 10)
loss, accuracy = model.evaluate(x_test, y_test)
print(f'Test Loss: {loss}, Test Accuracy: {accuracy}')
```

Implementation in PyTorch:

```
import torch
import torch.nn as nn
import torch.optim as optim

# Define the RNN model for video analysis
class VideoRNN(nn.Module):
    def _ _init_ _(self, input_dim, hidden_dim, output_dim):
        super(VideoRNN, self)._ _init_ _()
        self.rnn = nn.LSTM(input_dim, hidden_dim, batch_first=True)
        self.fc = nn.Linear(hidden_dim, output_dim)

    def forward(self, x):
        out, _ = self.rnn(x)
        out = self.fc(out[:, -1, :])
        return out

# Instantiate the model, loss function, and optimizer
input_dim = 2048
hidden_dim = 128
output_dim = 10
model = VideoRNN(input_dim, hidden_dim, output_dim)
criterion = nn.CrossEntropyLoss()
optimizer = optim.Adam(model.parameters(), lr=0.001)

# Generate synthetic data for demonstration
x_train = torch.rand(100, 50, input_dim)  # 100 samples, 50 frames, 2048 features
```

```
y_train = torch.randint(0, output_dim, (100,))

# Train the model
num_epochs = 20
for epoch in range(num_epochs):
    model.train()
    optimizer.zero_grad()
    outputs = model(x_train)
    loss = criterion(outputs, y_train)
    loss.backward()
    optimizer.step()
    if (epoch + 1) % 5 == 0:
        print(f'Epoch [{epoch+1}/{num_epochs}], Loss: {loss.item():.4f}')

# Evaluate the model
x_test = torch.rand(10, 50, input_dim)
y_test = torch.randint(0, output_dim, (10,))
model.eval()
with torch.no_grad():
    outputs = model(x_test)
    loss = criterion(outputs, y_test)
print(f'Test Loss: {loss.item():.4f}')
```

5.7 Implementing RNNs with TensorFlow and PyTorch

5.7.1 Building RNNs with TensorFlow

In this section, we will go through a step-by-step implementation of RNNs using TensorFlow.

Step-by-Step Implementation
To implement an RNN in TensorFlow, we follow these key steps: data preparation, model building, model compilation, training, and evaluation. We will build a simple RNN to predict the next value in a sequence.

Step 1: Data Preparation: First, we need to prepare the data. For demonstration purposes, we will use a synthetic dataset of sequences.

```
import numpy as np
import tensorflow as tf
from tensorflow.keras.preprocessing.sequence import pad_sequences

# Generate synthetic data
def generate_sequence(n_samples, n_timesteps, n_features):
    x = np.random.rand(n_samples, n_timesteps, n_features)
    y = np.random.rand(n_samples, n_timesteps, n_features)
    return x, y

n_samples = 1000
n_timesteps = 50
n_features = 1
x_train, y_train = generate_sequence(n_samples, n_timesteps, n_features)
x_test, y_test = generate_sequence(100, n_timesteps, n_features)
```

Step 2: Model Building: Next, we build the RNN model using TensorFlow's Keras API. We will use an LSTM layer, which is a type of RNN that is particularly good at capturing long-term dependencies in sequences.

```
from tensorflow.keras.models import Sequential
from tensorflow.keras.layers import LSTM, Dense

# Build the RNN model
model = Sequential()
model.add(LSTM(50, activation='relu', input_shape=(n_timesteps, n_features)))
model.add(Dense(n_features))

# Display the model summary
model.summary()
```

Step 3: Model Compilation: Compile the model by specifying the loss function and the optimizer.

```
# Compile the model
model.compile(optimizer='adam', loss='mse')
```

Step 4: Model Training: Train the model using the training data.

```
# Train the model
model.fit(x_train, y_train, epochs=200, verbose=1, validation_data=(x_test, y_test))
```

Step 5: Model Evaluation: Evaluate the model's performance on the test data.

```
# Evaluate the model
loss = model.evaluate(x_test, y_test, verbose=0)
print(f'Test Loss: {loss}')
```

Training and Evaluation

Training Process: During training, the model learns to minimize the loss function using BPTT. The weights are updated using an optimizer, such as Adam, which adjusts the learning rate dynamically to improve convergence.

Evaluation Metrics: MSE is commonly used as the loss function for regression tasks, as it measures the average squared difference between the predicted and actual values. For classification tasks, categorical cross-entropy or binary cross-entropy is used depending on the output format.

Example: Consider the following synthetic sequence prediction problem where the goal is to predict the next value in the sequence. The model's performance can be evaluated using MSE on the test data.

```
# Generate synthetic sequence data for demonstration
x_train = np.random.rand(1000, 50, 1)
y_train = np.random.rand(1000, 50, 1)
x_test = np.random.rand(100, 50, 1)
y_test = np.random.rand(100, 50, 1)

# Define the RNN model
model = Sequential()
model.add(LSTM(50, activation='relu', input_shape=(50, 1)))
model.add(Dense(1))

# Compile the model
model.compile(optimizer='adam', loss='mse')

# Train the model
```

```
model.fit(x_train, y_train, epochs=20, verbose=1, validation_data=(x_test, y_test))

# Evaluate the model
loss = model.evaluate(x_test, y_test, verbose=0)
print(f'Test Loss: {loss}')
```

5.7.2 *Building RNNs with PyTorch*

In this section, we will go through a step-by-step implementation of RNNs using PyTorch.

Step-by-Step Implementation

To implement an RNN in PyTorch, we follow similar steps as in TensorFlow: data preparation, model building, model compilation, training, and evaluation. We will build a simple RNN to predict the next value in a sequence.

Step 1: Data Preparation

```
import torch
import torch.nn as nn
import numpy as np

# Generate synthetic data
def generate_sequence(n_samples, n_timesteps, n_features):
    x = np.random.rand(n_samples, n_timesteps, n_features)
    y = np.random.rand(n_samples, n_timesteps, n_features)
    return torch.tensor(x, dtype=torch.float32), torch.tensor(y, dtype=torch.float32)

n_samples = 1000
n_timesteps = 50
n_features = 1
x_train, y_train = generate_sequence(n_samples, n_timesteps, n_features)
x_test, y_test = generate_sequence(100, n_timesteps, n_features)
```

Step 2: Model Building

```
class SimpleRNN(nn.Module):
    def _ _init_ _(self, input_dim, hidden_dim, output_dim):
        super(SimpleRNN, self)._ _init_ _()
        self.lstm = nn.LSTM(input_dim, hidden_dim, batch_first=True)
        self.fc = nn.Linear(hidden_dim, output_dim)

    def forward(self, x):
        lstm_out, _ = self.lstm(x)
        out = self.fc(lstm_out)
        return out

# Initialize the model
input_dim = n_features
hidden_dim = 50
output_dim = n_features
model = SimpleRNN(input_dim, hidden_dim, output_dim)

# Print the model summary
print(model)
```

Step 3: Model Compilation

```
# Define loss function and optimizer
criterion = nn.MSELoss()
optimizer = torch.optim.Adam(model.parameters(), lr=0.001)
```

Step 4: Model Training

```
# Training the model
num_epochs = 200
for epoch in range(num_epochs):
    model.train()
    outputs = model(x_train)
    optimizer.zero_grad()
    loss = criterion(outputs, y_train)
    loss.backward()
    optimizer.step()
    if (epoch + 1) % 10 == 0:
        print(f'Epoch [{epoch + 1}/{num_epochs}], Loss: {loss.item():.4f}')
```

Step 5: Model Evaluation

```
# Evaluate the model
model.eval()
with torch.no_grad():
    outputs = model(x_test)
    test_loss = criterion(outputs, y_test)
print(f'Test Loss: {test_loss.item():.4f}')
```

Example: Consider the following synthetic sequence prediction problem where the goal is to predict the next value in the sequence. The model's performance can be evaluated using MSE on the test data.

```
import torch
import torch.nn as nn
import torch.optim as optim
import numpy as np

# Generate synthetic sequence data for demonstration
x_train = torch.rand(1000, 50, 1)  # 1000 samples, 50 time steps, 1 feature
y_train = torch.rand(1000, 50, 1)
x_test = torch.rand(100, 50, 1)
y_test = torch.rand(100, 50, 1)

# Define the RNN model
class SimpleRNN(nn.Module):
    def _ _init_ _(self, input_dim, hidden_dim, output_dim):
        super(SimpleRNN, self)._ _init_ _()
        self.lstm = nn.LSTM(input_dim, hidden_dim, batch_first=True)
        self.fc = nn.Linear(hidden_dim, output_dim)

    def forward(self, x):
        lstm_out, _ = self.lstm(x)
        out = self.fc(lstm_out)
        return out

# Initialize the model, loss function, and optimizer
input_dim = 1
hidden_dim = 50
output_dim = 1
model = SimpleRNN(input_dim, hidden_dim, output_dim)
criterion = nn.MSELoss()
optimizer = optim.Adam(model.parameters(), lr=0.001)

# Train the model
```

```
num_epochs = 200
for epoch in range(num_epochs):
    model.train()
    optimizer.zero_grad()
    outputs = model(x_train)
    loss = criterion(outputs, y_train)
    loss.backward()
    optimizer.step()
    if (epoch + 1) % 10 == 0:
        print(f'Epoch [{epoch+1}/{num_epochs}], Loss: {loss.item():.4f}')

# Evaluate the model
model.eval()
with torch.no_grad():
    outputs = model(x_test)
    test_loss = criterion(outputs, y_test)
print(f'Test Loss: {test_loss.item():.4f}')
```

5.8 Case Studies

RNNs have proven to be highly effective in various sequential data tasks. This section presents case studies and exercises that demonstrate the application of RNNs in different scenarios, including language translation, sentiment analysis, text generation, and a custom RNN design exercise.

Language Translation Case Study

Language translation is a classic problem in natural language processing, where the goal is to translate text from one language to another. RNNs, particularly sequence-to-sequence (Seq2Seq) models with attention mechanisms, are well-suited for this task.

Model Architecture:

1. Encoder–Decoder Architecture: The Seq2Seq model consists of an encoder and a decoder. The encoder processes the input sequence and compresses it into a context vector, which the decoder uses to generate the output sequence.

$$\begin{aligned}
\mathbf{h}_t^{\text{enc}} &= \text{LSTM}(\mathbf{x}_t, \mathbf{h}_{t-1}^{\text{enc}}) \\
\mathbf{c} &= \mathbf{h}_T^{\text{enc}} \\
\mathbf{h}_t^{\text{dec}} &= \text{LSTM}(\mathbf{y}_{t-1}, \mathbf{h}_{t-1}^{\text{dec}}, \mathbf{c})
\end{aligned}$$

2. Attention Mechanism: The attention mechanism allows the decoder to focus on different parts of the input sequence at each step. The attention weights are computed as:

$$\begin{aligned}
e_{tj} &= \mathbf{v}_a^\top \tanh(\mathbf{W}_a \mathbf{h}_{t-1}^{\text{dec}} + \mathbf{U}_a \mathbf{h}_j^{\text{enc}}) \\
\alpha_{tj} &= \frac{\exp(e_{tj})}{\sum_{k=1}^{T} \exp(e_{tk})} \\
\mathbf{c}_t &= \sum_{j=1}^{T} \alpha_{tj} \mathbf{h}_j^{\text{enc}}
\end{aligned}$$

Example: Consider implementing a Seq2Seq model with attention in PyTorch for translating English to French.

```
import torch
import torch.nn as nn
import torch.optim as optim

class Encoder(nn.Module):
    def _ _init_ _(self, input_dim, emb_dim, hid_dim, n_layers, dropout):
        super()._ _init_ _()
        self.embedding = nn.Embedding(input_dim, emb_dim)
        self.rnn = nn.LSTM(emb_dim, hid_dim, n_layers, dropout=dropout, batch_first=True)

    def forward(self, src):
        embedded = self.embedding(src)
        outputs, (hidden, cell) = self.rnn(embedded)
        return outputs, hidden, cell

class Attention(nn.Module):
    def _ _init_ _(self, hid_dim):
        super()._ _init_ _()
        self.attn = nn.Linear(hid_dim * 2, 1)

    def forward(self, hidden, encoder_outputs):
        src_len = encoder_outputs.shape[1]
        hidden = hidden[-1].unsqueeze(1).repeat(1, src_len, 1)
        energy = torch.tanh(self.attn(torch.cat((hidden, encoder_outputs), dim=2)))
        attention = torch.softmax(energy.squeeze(2), dim=1)
        return attention

class Decoder(nn.Module):
    def _ _init_ _(self, output_dim, emb_dim, hid_dim, n_layers, dropout, attention):
        super()._ _init_ _()
        self.embedding = nn.Embedding(output_dim, emb_dim)
        self.rnn = nn.LSTM(emb_dim + hid_dim, hid_dim, n_layers, dropout=dropout,
        batch_first=True)
        self.fc_out = nn.Linear(hid_dim * 2, output_dim)
        self.attention = attention

    def forward(self, trg, hidden, cell, encoder_outputs):
        trg = trg.unsqueeze(1)
        embedded = self.embedding(trg)
        attention = self.attention(hidden, encoder_outputs).unsqueeze(1)
        context = torch.bmm(attention, encoder_outputs)
        rnn_input = torch.cat((embedded, context), dim=2)
        output, (hidden, cell) = self.rnn(rnn_input, (hidden, cell))
        output = self.fc_out(torch.cat((output, context), dim=2).squeeze(1))
        return output, hidden, cell

class Seq2Seq(nn.Module):
    def _ _init_ _(self, encoder, decoder, device):
        super()._ _init_ _()
        self.encoder = encoder
        self.decoder = decoder
        self.device = device

    def forward(self, src, trg, teacher_forcing_ratio=0.5):
        batch_size = trg.shape[0]
        trg_len = trg.shape[1]
        trg_vocab_size = self.decoder.fc_out.out_features

        outputs = torch.zeros(batch_size, trg_len, trg_vocab_size).to(self.device)
        encoder_outputs, hidden, cell = self.encoder(src)
        input = trg[:, 0]

        for t in range(1, trg_len):
            output, hidden, cell = self.decoder(input, hidden, cell, encoder_outputs)
```

```
            outputs[:, t, :] = output
            top1 = output.argmax(1)
            input = trg[:, t] if random.random() < teacher_forcing_ratio else top1

        return outputs

# Hyperparameters and model instantiation
input_dim = len(SRC.vocab)
output_dim = len(TRG.vocab)
emb_dim = 256
hid_dim = 512
n_layers = 2
dropout = 0.5
device = torch.device('cuda' if torch.cuda.is_available() else 'cpu')

attn = Attention(hid_dim)
enc = Encoder(input_dim, emb_dim, hid_dim, n_layers, dropout)
dec = Decoder(output_dim, emb_dim, hid_dim, n_layers, dropout, attn)
model = Seq2Seq(enc, dec, device).to(device)

# Optimizer and Loss
optimizer = optim.Adam(model.parameters())
criterion = nn.CrossEntropyLoss()

# Training loop
num_epochs = 10
for epoch in range(num_epochs):
    model.train()
    epoch_loss = 0
    for batch in train_iterator:
        src = batch.src
        trg = batch.trg
        optimizer.zero_grad()
        output = model(src, trg)
        output_dim = output.shape[-1]
        output = output[1:].view(-1, output_dim)
        trg = trg[1:].view(-1)
        loss = criterion(output, trg)
        loss.backward()
        optimizer.step()
        epoch_loss += loss.item()
    print(f'Epoch {epoch+1}, Loss: {epoch_loss/len(train_iterator)}')

# Evaluate the model
model.eval()
test_loss = 0
with torch.no_grad():
    for batch in test_iterator:
        src = batch.src
        trg = batch.trg
        output = model(src, trg, 0)  # Turn off teacher forcing
        output_dim = output.shape[-1]
        output = output[1:].view(-1, output_dim)
        trg = trg[1:].view(-1)
        loss = criterion(output, trg)
        test_loss += loss.item()
print(f'Test Loss: {test_loss/len(test_iterator)}')
```

This case study demonstrates how to implement a Seq2Seq model with attention for language translation in PyTorch. The model is trained to translate sentences from English to French, leveraging the encoder–decoder architecture and attention mechanism to improve translation quality.

Sentiment Analysis Case Study

Sentiment analysis involves determining the sentiment expressed in a piece of text. RNNs, particularly LSTM and GRU networks, are effective in capturing the sequential dependencies in text data to predict sentiment.

Model Architecture:

1. Embedding Layer: Converts words into dense vectors of fixed size.

$$\mathbf{e}_t = \text{Embedding}(x_t)$$

2. Recurrent Layer: Processes the sequence of embeddings and captures temporal dependencies.

$$\mathbf{h}_t = \text{LSTM}(\mathbf{e}_t, \mathbf{h}_{t-1})$$

3. Fully Connected Layer: Outputs the sentiment prediction.

$$\hat{y} = \sigma(\mathbf{W}_o \mathbf{h}_T + \mathbf{b}_o)$$

Example: Consider implementing a sentiment analysis model using LSTM in TensorFlow.

```
import tensorflow as tf
from tensorflow.keras.models import Sequential
from tensorflow.keras.layers import Embedding, LSTM, Dense

# Sample data
texts = ["I love this product!", "This is the worst thing ever."]
labels = [1, 0]  # 1 for positive, 0 for negative

# Tokenize and pad sequences
tokenizer = tf.keras.preprocessing.text.Tokenizer(num_words=10000)
tokenizer.fit_on_texts(texts)
sequences = tokenizer.texts_to_sequences(texts)
x_data = tf.keras.preprocessing.sequence.pad_sequences(sequences, maxlen=10)
y_data = tf.convert_to_tensor(labels)

# Define the RNN model for sentiment analysis
model = Sequential()
model.add(Embedding(input_dim=10000, output_dim=64, input_length=10))
model.add(LSTM(128))
model.add(Dense(1, activation='sigmoid'))

# Compile the model
model.compile(optimizer='adam', loss='binary_crossentropy', metrics=['accuracy'])

# Train the model
model.fit(x_data, y_data, epochs=10, batch_size=2)

# Evaluate the model
loss, accuracy = model.evaluate(x_data, y_data)
print(f'Loss: {loss}, Accuracy: {accuracy}')
```

This case study illustrates how to build, train, and evaluate an LSTM-based model for sentiment analysis in TensorFlow. The model predicts the sentiment of text based on the sequential patterns captured in the input data.

Text Generation Case Study

Text generation involves generating new text based on a given input. RNNs can learn the sequential dependencies in text data and generate coherent and contextually relevant text.

Model Architecture:

1. Embedding Layer: Converts characters or words into dense vectors of fixed size.

$$\mathbf{e}_t = \text{Embedding}(x_t)$$

2. Recurrent Layer: Processes the sequence of embeddings and captures temporal dependencies.

$$\mathbf{h}_t = \text{LSTM}(\mathbf{e}_t, \mathbf{h}_{t-1})$$

3. Fully Connected Layer: Outputs the probability distribution of the next character or word.

$$\hat{y} = \text{softmax}(\mathbf{W}_o \mathbf{h}_T + \mathbf{b}_o)$$

Example: Consider implementing a character-level text generation model using LSTM in PyTorch.

```
import torch
import torch.nn as nn
import torch.optim as optim

# Sample data: a sequence of characters
text = "hello world"
chars = sorted(list(set(text)))
char_to_idx = {char: idx for idx, char in enumerate(chars)}
idx_to_char = {idx: char for char, idx in char_to_idx.items()}
vocab_size = len(chars)

# Prepare input-output pairs
sequences = [char_to_idx[char] for char in text]
x_data = []
y_data = []
seq_length = 5
for i in range(len(sequences) - seq_length):
    x_data.append(sequences[i:i+seq_length])
    y_data.append(sequences[i+seq_length])
x_data = torch.tensor(x_data, dtype=torch.long)
y_data = torch.tensor(y_data, dtype=torch.long)

# Define the RNN model for text generation
class CharRNN(nn.Module):
    def _ _init_ _(self, vocab_size, embedding_dim, hidden_dim, output_dim):
        super(CharRNN, self)._ _init_ _()
        self.embedding = nn.Embedding(vocab_size, embedding_dim)
        self.rnn = nn.LSTM(embedding_dim, hidden_dim, batch_first=True)
        self.fc = nn.Linear(hidden_dim, output_dim)
```

```
    def forward(self, x):
        embedded = self.embedding(x)
        output, _ = self.rnn(embedded)
        output = self.fc(output[:, -1, :])
        return output

# Instantiate the model, loss function, and optimizer
embedding_dim = 64
hidden_dim = 128
model = CharRNN(vocab_size, embedding_dim, hidden_dim, vocab_size)
criterion = nn.CrossEntropyLoss()
optimizer = optim.Adam(model.parameters(), lr=0.001)

# Train the model
num_epochs = 100
for epoch in range(num_epochs):
    model.train()
    optimizer.zero_grad()
    output = model(x_data)
    loss = criterion(output, y_data)
    loss.backward()
    optimizer.step()
    if (epoch + 1) % 10 == 0:
        print(f'Epoch [{epoch+1}/{num_epochs}], Loss: {loss.item():.4f}')

# Generate text
def generate_text(model, start_text, char_to_idx, idx_to_char, seq_length, num_chars):
    model.eval()
    input_seq = torch.tensor([char_to_idx[char] for char in start_text], dtype=torch.long).
    unsqueeze(0)
    generated_text = start_text
    for _ in range(num_chars):
        with torch.no_grad():
            output = model(input_seq)
            predicted_char_idx = output.argmax(1).item()
            predicted_char = idx_to_char[predicted_char_idx]
            generated_text += predicted_char
            input_seq = torch.cat((input_seq[:, 1:], torch.tensor([[predicted_char_idx]])),
            dim=1)
    return generated_text

start_text = "hello"
generated_text = generate_text(model, start_text, char_to_idx, idx_to_char, seq_length, 20)
print(generated_text)
```

This case study illustrates how to build, train, and evaluate a character-level LSTM-based model for text generation in PyTorch. The model learns to generate text by predicting the next character in a sequence based on the patterns it has learned from the training data.

5.9 Exercises

1. Given the following RNN cell equations:

$$\mathbf{h}_t = \tanh(\mathbf{W}_{xh}\mathbf{x}_t + \mathbf{W}_{hh}\mathbf{h}_{t-1} + \mathbf{b}_h)$$

$$\mathbf{y}_t = \mathbf{W}_{hy}\mathbf{h}_t + \mathbf{b}_y$$

where $\mathbf{W}_{xh} = \begin{pmatrix} 0.5 & 0.2 \\ 0.1 & 0.4 \end{pmatrix}$, $\mathbf{W}_{hh} = \begin{pmatrix} 0.3 & 0.6 \\ 0.7 & 0.9 \end{pmatrix}$, $\mathbf{W}_{hy} = \begin{pmatrix} 1.0 & 0.8 \end{pmatrix}$, $\mathbf{b}_h = \begin{pmatrix} 0.1 \\ 0.2 \end{pmatrix}$, and $\mathbf{b}_y = 0.3$. Assume $\mathbf{x}_1 = \begin{pmatrix} 0.6 \\ 0.9 \end{pmatrix}$ and the initial hidden state $\mathbf{h}_0 = \begin{pmatrix} 0 \\ 0 \end{pmatrix}$.

   (a) Compute the hidden state $\mathbf{h}_1$ and the output $\mathbf{y}_1$.
   (b) Calculate the gradient of the loss with respect to $\mathbf{W}_{hy}$, $\mathbf{W}_{xh}$, and $\mathbf{W}_{hh}$ using Backpropagation Through Time (BPTT). Assume that the loss function is mean squared error and the target output is $\hat{\mathbf{y}}_1 = 1.0$.

2. Consider an RNN with the following parameters:

$$\mathbf{W}_{xh} = \begin{pmatrix} 0.01 & 0.02 \\ 0.03 & 0.04 \end{pmatrix}, \quad \mathbf{W}_{hh} = \begin{pmatrix} 1.2 & 0.8 \\ 0.7 & 1.5 \end{pmatrix}$$

Given the input sequence $\mathbf{x}_t = \begin{pmatrix} 0.1 \\ 0.1 \end{pmatrix}$ for $t = 1, 2, \ldots, 50$, and the initial hidden state $\mathbf{h}_0 = \begin{pmatrix} 0 \\ 0 \end{pmatrix}$:

   (a) Compute the hidden state $\mathbf{h}_t$ for $t = 1, 2, \ldots, 50$.
   (b) Analyze the behavior of the hidden states. Explain and compute how the gradient vanishes or explodes over time.

3. Given the following LSTM cell parameters:

$$\mathbf{W}_f = \begin{pmatrix} 0.5 & 0.1 \\ 0.3 & 0.7 \end{pmatrix}, \quad \mathbf{W}_i = \begin{pmatrix} 0.6 & 0.2 \\ 0.4 & 0.8 \end{pmatrix}, \quad \mathbf{W}_o = \begin{pmatrix} 0.7 & 0.3 \\ 0.5 & 0.9 \end{pmatrix}, \quad \mathbf{W}_g = \begin{pmatrix} 0.8 & 0.4 \\ 0.6 & 1.0 \end{pmatrix}$$

$$\mathbf{b}_f = \begin{pmatrix} 0.1 \\ 0.2 \end{pmatrix}, \quad \mathbf{b}_i = \begin{pmatrix} 0.2 \\ 0.3 \end{pmatrix}, \quad \mathbf{b}_o = \begin{pmatrix} 0.3 \\ 0.4 \end{pmatrix}, \quad \mathbf{b}_g = \begin{pmatrix} 0.4 \\ 0.5 \end{pmatrix}$$

Given the input $\mathbf{x}_1 = \begin{pmatrix} 0.5 \\ 0.8 \end{pmatrix}$ and the initial states $\mathbf{h}_0 = \begin{pmatrix} 0 \\ 0 \end{pmatrix}$ and $\mathbf{c}_0 = \begin{pmatrix} 0 \\ 0 \end{pmatrix}$:

(a) Compute the gate activations $\mathbf{f}_1, \mathbf{i}_1, \mathbf{o}_1$, and $\mathbf{g}_1$.
(b) Calculate the new cell state $\mathbf{c}_1$ and the hidden state $\mathbf{h}_1$.

4. Given the following attention mechanism parameters:

$$\mathbf{Q} = \begin{pmatrix} 1 & 0 \\ 0 & 1 \end{pmatrix}, \quad \mathbf{K} = \begin{pmatrix} 1 & 0 \\ 0 & 1 \end{pmatrix}, \quad \mathbf{V} = \begin{pmatrix} 2 & 3 \\ 4 & 5 \end{pmatrix}$$

and the input sequence $\mathbf{x}_t = \begin{pmatrix} 0.6 \\ 0.8 \end{pmatrix}$:

(a) Compute the attention scores using the dot-product attention mechanism.
(b) Calculate the attention weights.
(c) Compute the context vector.

5. Given a sequence-to-sequence model with the following parameters:

$$\mathbf{W}_{\text{enc}} = \begin{pmatrix} 0.2 & 0.4 \\ 0.6 & 0.8 \end{pmatrix}, \quad \mathbf{W}_{\text{dec}} = \begin{pmatrix} 0.5 & 0.7 \\ 0.3 & 0.9 \end{pmatrix}$$

Input sequence: $\mathbf{x}_1 = \begin{pmatrix} 0.5 \\ 0.9 \end{pmatrix}$, $\mathbf{x}_2 = \begin{pmatrix} 0.6 \\ 0.8 \end{pmatrix}$, initial hidden states: $\mathbf{h}_{\text{enc}0} = \begin{pmatrix} 0 \\ 0 \end{pmatrix}$, $\mathbf{h}_{\text{dec}0} = \begin{pmatrix} 0 \\ 0 \end{pmatrix}$.

(a) Encode the input sequence and obtain the encoder hidden states.
(b) Compute the attention scores and weights.
(c) Decode the output sequence using the attention context vectors.

6. Consider a simple RNN with input sequence $\mathbf{x} = [1, 2, 3]$, hidden state $\mathbf{h} \in \mathbb{R}^2$, input-to-hidden weights $\mathbf{W}_x \in \mathbb{R}^{2\times1}$, hidden-to-hidden weights $\mathbf{W}_h \in \mathbb{R}^{2\times2}$, and bias $\mathbf{b} \in \mathbb{R}^2$:

$$\mathbf{W}_x = \begin{pmatrix} 0.5 \\ 1.0 \end{pmatrix}, \quad \mathbf{W}_h = \begin{pmatrix} 0.1 & 0.2 \\ 0.3 & 0.4 \end{pmatrix}, \quad \mathbf{b} = \begin{pmatrix} 0.1 \\ 0.2 \end{pmatrix}$$

(a) Compute the hidden states $\mathbf{h}_t$ for $t = 1, 2, 3$ using the recurrence relation $\mathbf{h}_t = \tanh(\mathbf{W}_x\mathbf{x}_t + \mathbf{W}_h\mathbf{h}_{t-1} + \mathbf{b})$. Assume $\mathbf{h}_0 = \begin{pmatrix} 0 \\ 0 \end{pmatrix}$.

7. Consider a Gated Recurrent Unit (GRU) with the following parameters: update gate $\mathbf{z}_t$, reset gate $\mathbf{r}_t$, candidate hidden state $\tilde{\mathbf{h}}_t$, and final hidden state $\mathbf{h}_t$:

$$\mathbf{z}_t = \sigma(\mathbf{W}_z\mathbf{x}_t + \mathbf{U}_z\mathbf{h}_{t-1} + \mathbf{b}_z), \quad \mathbf{r}_t = \sigma(\mathbf{W}_r\mathbf{x}_t + \mathbf{U}_r\mathbf{h}_{t-1} + \mathbf{b}_r)$$

$$\tilde{\mathbf{h}}_t = \tanh(\mathbf{W}_h\mathbf{x}_t + \mathbf{r}_t \odot (\mathbf{U}_h\mathbf{h}_{t-1}) + \mathbf{b}_h), \quad \mathbf{h}_t = (1 - \mathbf{z}_t) \odot \mathbf{h}_{t-1} + \mathbf{z}_t \odot \tilde{\mathbf{h}}_t$$

Let $\mathbf{x}_t = \begin{pmatrix} 1 \\ 0 \end{pmatrix}$, $\mathbf{h}_{t-1} = \begin{pmatrix} 0.5 \\ 0.5 \end{pmatrix}$, and all weight matrices and biases be initialized to ones.

(a) Compute $\mathbf{z}_t$, $\mathbf{r}_t$, $\tilde{\mathbf{h}}_t$, and $\mathbf{h}_t$.

8. Given an Long Short-Term Memory (LSTM) network with input sequence $\mathbf{x} = [1, 2]$, and the following parameters: input gate $\mathbf{i}_t$, forget gate $\mathbf{f}_t$, output gate $\mathbf{o}_t$, and candidate cell state $\tilde{\mathbf{c}}_t$:

$$\mathbf{i}_t = \sigma(\mathbf{W}_i \mathbf{x}_t + \mathbf{U}_i \mathbf{h}_{t-1} + \mathbf{b}_i), \quad \mathbf{f}_t = \sigma(\mathbf{W}_f \mathbf{x}_t + \mathbf{U}_f \mathbf{h}_{t-1} + \mathbf{b}_f)$$

$$\mathbf{o}_t = \sigma(\mathbf{W}_o \mathbf{x}_t + \mathbf{U}_o \mathbf{h}_{t-1} + \mathbf{b}_o), \quad \tilde{\mathbf{c}}_t = \tanh(\mathbf{W}_c \mathbf{x}_t + \mathbf{U}_c \mathbf{h}_{t-1} + \mathbf{b}_c)$$

The cell state $\mathbf{c}_t$ and hidden state $\mathbf{h}_t$ are updated as:

$$\mathbf{c}_t = \mathbf{f}_t \odot \mathbf{c}_{t-1} + \mathbf{i}_t \odot \tilde{\mathbf{c}}_t, \quad \mathbf{h}_t = \mathbf{o}_t \odot \tanh(\mathbf{c}_t)$$

Let $\mathbf{x}_t = 1$, $\mathbf{h}_0 = \mathbf{c}_0 = 0$, and all weight matrices and biases be initialized to ones.

(a) Compute the cell states $\mathbf{c}_t$ and hidden states $\mathbf{h}_t$ for $t = 1, 2$.

9. Consider an RNN with a single input, hidden, and output neuron. The input sequence is $\mathbf{x} = [1, 0]$, the initial hidden state is $\mathbf{h}_0 = 0$, and the weights and bias are initialized to $W_{xh} = 0.5$, $W_{hh} = 0.8$, and $b_h = 0.1$.

(a) Perform a forward pass to compute the hidden states and outputs.
(b) Compute the gradients of the loss with respect to the weights and bias using BPTT. Assume a mean squared error loss and that the target sequence is $\mathbf{y} = [0.5, 0]$.

10. Consider a simple attention mechanism with input sequence $\mathbf{x} = [x_1, x_2, x_3]$ and hidden states $\mathbf{h}_t$ computed using an RNN. The attention weights α_t are calculated as:

$$\alpha_t = \frac{\exp(e_t)}{\sum_{k=1}^{3} \exp(e_k)}, \quad e_t = \mathbf{v}^T \tanh(\mathbf{W}\mathbf{h}_t + \mathbf{b})$$

where $\mathbf{v} \in \mathbb{R}^2$, $\mathbf{W} \in \mathbb{R}^{2\times 2}$, and $\mathbf{b} \in \mathbb{R}^2$ are learnable parameters, initialized to ones.
Using the attention weights, express the context vector $\mathbf{c}$ as a weighted sum of the hidden states.

11. Consider a sequence-to-sequence model with encoder hidden states $\mathbf{h}_t^e$ and decoder hidden states $\mathbf{h}_t^d$. Let the encoder states be $\mathbf{h}_t^e = [0.1t, 0.2t]$ for $t = 1, 2, 3$ and the initial decoder state be $\mathbf{h}_0^d = [0, 0]$.

    (a) Using a simple dot-product attention mechanism, compute the alignment scores between the decoder hidden state at $t = 1$ and all encoder hidden states.
    (b) Compute the context vector for the decoder at $t = 1$ and update the decoder hidden state using a GRU. Assume the decoder input at $t = 1$ is $\mathbf{x}_1^d = [1]$, and all GRU parameters are initialized to ones.

12. Consider an RNN designed to smooth a noisy time series. The input sequence $\mathbf{x} = [0.5, 1.0, 1.5]$, hidden state $\mathbf{h} \in \mathbb{R}^2$, input-to-hidden weights $\mathbf{W}_x \in \mathbb{R}^{2\times1}$, hidden-to-hidden weights $\mathbf{W}_h \in \mathbb{R}^{2\times2}$, and bias $\mathbf{b} \in \mathbb{R}^2$:

$$\mathbf{W}_x = \begin{pmatrix} 0.1 \\ 0.2 \end{pmatrix}, \quad \mathbf{W}_h = \begin{pmatrix} 0.5 & 0.1 \\ 0.2 & 0.5 \end{pmatrix}, \quad \mathbf{b} = \begin{pmatrix} 0 \\ 0 \end{pmatrix}$$

    Compute the hidden states $\mathbf{h}_t$ for $t = 1, 2, 3$ using the recurrence relation $\mathbf{h}_t = \tanh(\mathbf{W}_x\mathbf{x}_t + \mathbf{W}_h\mathbf{h}_{t-1} + \mathbf{b})$. Assume $\mathbf{h}_0 = \begin{pmatrix} 0 \\ 0 \end{pmatrix}$.

13. Given an LSTM network, with the input sequence $\mathbf{x}_t = [1, 2]$ and initial hidden state $\mathbf{h}_0 = [0.5, 0.5]$ and cell state $\mathbf{c}_0 = [0.5, 0.5]$. The LSTM parameters are as follows:

$$\mathbf{W}_i = \mathbf{W}_f = \mathbf{W}_o = \mathbf{W}_c = \begin{pmatrix} 0.1 & 0.2 \\ 0.3 & 0.4 \end{pmatrix}, \quad \mathbf{U}_i = \mathbf{U}_f = \mathbf{U}_o = \mathbf{U}_c = \begin{pmatrix} 0.5 & 0.6 \\ 0.7 & 0.8 \end{pmatrix}, \quad \mathbf{b}_i = \mathbf{b}_f = \mathbf{b}_o = \mathbf{b}_c = \begin{pmatrix} 0.1 \\ 0.2 \end{pmatrix}$$

    (a) Compute the input gate $\mathbf{i}_t$, forget gate $\mathbf{f}_t$, output gate $\mathbf{o}_t$, and candidate cell state $\tilde{\mathbf{c}}_t$.
    (b) Compute the cell state $\mathbf{c}_t$ and hidden state $\mathbf{h}_t$ at $t = 1$.

14. Consider a GRU with input sequence $\mathbf{x}_t = [0.5, 1.0]$ and initial hidden state $\mathbf{h}_0 = [0.0, 0.0]$. The GRU parameters are:

$$\mathbf{W}_z = \mathbf{W}_r = \mathbf{W}_h = \begin{pmatrix} 0.1 & 0.2 \\ 0.3 & 0.4 \end{pmatrix}, \quad \mathbf{U}_z = \mathbf{U}_r = \mathbf{U}_h = \begin{pmatrix} 0.5 & 0.6 \\ 0.7 & 0.8 \end{pmatrix}, \quad \mathbf{b}_z = \mathbf{b}_r = \mathbf{b}_h = \begin{pmatrix} 0.1 \\ 0.2 \end{pmatrix}$$

    (a) Compute the update gate $\mathbf{z}_t$ and reset gate $\mathbf{r}_t$.
    (b) Compute the candidate hidden state $\tilde{\mathbf{h}}_t$ and final hidden state $\mathbf{h}_t$ at $t = 1$.

15. Consider an LSTM with input sequence $\mathbf{x} = [1, 0]$, initial hidden state $\mathbf{h}_0 = [0, 0]$, and initial cell state $\mathbf{c}_0 = [0, 0]$. The weights and biases are:

$$\mathbf{W}_i = \mathbf{W}_f = \mathbf{W}_o = \mathbf{W}_c = \begin{pmatrix} 0.5 & 0.1 \\ 0.2 & 0.5 \end{pmatrix}, \quad \mathbf{U}_i = \mathbf{U}_f = \mathbf{U}_o = \mathbf{U}_c = \begin{pmatrix} 0.8 & 0.3 \\ 0.4 & 0.7 \end{pmatrix}, \quad \mathbf{b}_i = \mathbf{b}_f = \mathbf{b}_o = \mathbf{b}_c = \begin{pmatrix} 0.1 \\ 0.2 \end{pmatrix}$$

(a) Perform the forward pass to compute the cell states and hidden states for each time step.
(b) Compute the gradients of the loss with respect to the weights and biases using BPTT, assuming a mean squared error loss with target sequence $\mathbf{y} = [0.5, 0]$.

Chapter 6
Transformer Architectures

He who refuses to do arithmetic is doomed to talk nonsense.

—John McCarthy, Programs with Common Sense

6.1 Introduction to Transformer Architectures

Transformer architectures have revolutionized the field of natural language processing (NLP) and have become the backbone of many state-of-the-art models. Unlike traditional RNNs and CNNs, transformers rely entirely on a mechanism known as self-attention to draw global dependencies between input and output. This allows transformers to process data in parallel, making them significantly faster and more efficient. The transformer model was introduced in the paper "Attention is All You Need" by Vaswani et al. (2017). The architecture consists of an encoder and a decoder, each composed of a stack of identical layers. The encoder maps an input sequence into a continuous representation, and the decoder uses this representation along with the output sequence (shifted right) to generate the output sequence. The key component of the transformer is the self-attention mechanism, which computes a weighted sum of the input values, with the weights being determined by the similarity between the inputs. This allows the model to focus on different parts of the sequence when producing each part of the output.

6.1.1 Historical Context and Development

The development of transformers marked a significant departure from previous architectures. Before transformers, RNNs and their variants like LSTMs and GRUs were the dominant models for sequence-to-sequence tasks. These models processed input

P. Singh and B. Raman, *Deep Learning Through the Prism of Tensors*, Studies in Big Data 162, https://doi.org/10.1007/978-981-97-8019-8_6

sequences sequentially, which limited their ability to parallelize computations and often led to difficulties in capturing long-range dependencies due to the vanishing gradient problem. CNNs, though effective in capturing local patterns, also struggled with long-range dependencies and required a fixed-size input, making them less flexible for tasks like language modeling. Transformers overcame these limitations by leveraging self-attention mechanisms, which allowed them to process sequences in parallel and capture global dependencies effectively. This breakthrough was enabled by several key innovations:

1. **Positional Encoding**: Since transformers do not have a built-in notion of sequence order, positional encodings are added to the input embeddings to provide information about the position of each token in the sequence.
2. **Multi-head Attention**: Instead of a single attention mechanism, multi-head attention allows the model to jointly attend to information from different representation subspaces at different positions.
3. **Feed-Forward Neural Networks**: Each encoder and decoder layer has a fully connected feed-forward network applied to each position separately and identically.
4. **Layer Normalization and Residual Connections**: These are used to stabilize and improve the training of deep networks.

6.1.2 *Impact of Transformers on NLP and Beyond*

Transformers have had a profound impact on NLP and have become the de facto standard for many tasks. Models based on transformers, such as Bidirectional Encoder Representations from Transformers (BERT), Generative Pre-trained Transformer (GPT), and Text-to-Text Transfer Transformer (T5), have achieved state-of-the-art results in a wide range of NLP benchmarks.

1. **BERT**: BERT is designed to pre-train deep bidirectional representations by jointly conditioning on both left and right contexts in all layers. It uses masked language modeling and next sentence prediction as pre-training tasks.

$$\text{Loss}_{\text{BERT}} = \text{MLM}_{\text{loss}} + \text{NSP}_{\text{loss}}$$

2. **GPT**: GPT uses a unidirectional transformer for language modeling, generating text by predicting the next token in a sequence. The model is fine-tuned for specific tasks using supervised learning.

$$\text{Loss}_{\text{GPT}} = \sum_{t=1}^{T} \log P(x_t \mid x_{<t})$$

3. **T5**: T5 frames all NLP tasks as a text-to-text problem, where both input and output are text strings. It leverages the flexibility of transformers to perform various tasks such as translation, summarization, and question answering.

$$\text{Loss}_{\text{T5}} = -\sum_{t=1}^{T} \log P(y_t \mid y_{<t}, x)$$

Transformers are also making significant strides in other domains, including computer vision, speech recognition, and even reinforcement learning. Vision Transformers (ViTs) have demonstrated that transformers can match or surpass the performance of CNNs on image classification tasks by treating images as sequences of patches. In speech recognition, transformers have been used to model long-range dependencies in audio signals, leading to improved performance over traditional RNN-based models. In reinforcement learning, transformers are being explored for their ability to handle long-term dependencies and multi-modal inputs, potentially leading to more robust and generalizable agents.

6.2 Self-attention Mechanisms

6.2.1 Introduction to Self-attention

Self-attention, also known as intra-attention, is a mechanism that allows a model to weigh the importance of different tokens in a sequence when generating a representation for a specific token. Unlike traditional attention mechanisms that focus on different parts of the input sequence relative to an output sequence, self-attention computes the attention scores within a single sequence. The self-attention mechanism is fundamental to the transformer architecture. It enables the model to capture dependencies between tokens, regardless of their distance in the sequence. This capability is crucial for understanding the context and relationships in tasks such as language modeling, machine translation, and text summarization.

Self-attention works by creating a weighted sum of the values (V) in the sequence, where the weights are determined by the similarity between the query (Q) and the keys (K) (see Fig. 6.1 for an overview of the encoder architecture). This allows the model to focus on relevant parts of the sequence when computing representations, thus capturing long-range dependencies more effectively than traditional RNNs or CNNs. The self-attention mechanism can be mathematically formulated as follows:

1. **Input Representations**: Let the input sequence be represented by $X \in \mathbb{R}^{n \times d}$, where n is the sequence length and d is the dimension of the embeddings.
2. **Projection to Queries, Keys, and Values**: The input X is linearly projected into three different spaces to obtain queries Q, keys K, and values V:

$$Q = XW^Q, \quad K = XW^K, \quad V = XW^V$$

Here, $W^Q, W^K, W^V \in \mathbb{R}^{d \times d_k}$ are learned projection matrices, and d_k is the dimension of the keys.

3. **Scaled Dot-Product Attention**: The attention scores are computed using the scaled dot-product attention mechanism:

$$\text{Attention}(Q, K, V) = \text{softmax}\left(\frac{QK^T}{\sqrt{d_k}}\right)V$$

The dot products between the queries and keys are scaled by the square root of the dimension of the keys $\sqrt{d_k}$. This scaling prevents the dot products from growing too large and helps stabilize the gradients during training.

4. **Attention Weights**: The attention weights are calculated using the softmax function, which ensures that the weights sum to one:

$$\alpha_{ij} = \frac{\exp(e_{ij})}{\sum_{k=1}^{n} \exp(e_{ik})}$$

where $e_{ij} = \frac{Q_i \cdot K_j}{\sqrt{d_k}}$ is the unnormalized attention score between query Q_i and key K_j.

5. **Weighted Sum of Values**: The final output of the self-attention mechanism is a weighted sum of the values V:

$$\text{Attention}(Q, K, V) = \sum_{j=1}^{n} \alpha_{ij} V_j$$

Example Consider a sequence of four tokens with embeddings of dimension $d = 64$. We can illustrate the self-attention mechanism with a simple example.

1. Input Sequence:

$$X = \begin{bmatrix} \mathbf{x}_1 \\ \mathbf{x}_2 \\ \mathbf{x}_3 \\ \mathbf{x}_4 \end{bmatrix} \in \mathbb{R}^{4 \times 64}$$

2. Projection Matrices:

$$W^Q, W^K, W^V \in \mathbb{R}^{64 \times 64}$$

3. Projected Queries, Keys, and Values:

$$Q = XW^Q, \quad K = XW^K, \quad V = XW^V$$

4. Compute Attention Scores:

$$\text{Scores} = QK^T/\sqrt{64}$$

5. Apply Softmax:

$$\alpha = \text{softmax}(\text{Scores})$$

6. Weighted Sum of Values:

$$\text{Output} = \alpha V$$

In practice, the self-attention mechanism is implemented efficiently using matrix operations, which take advantage of parallel computing capabilities of modern hardware. The transformer architecture further extends self-attention with multi-head attention, which allows the model to jointly attend to information from different representation subspaces at different positions.

$$\text{MultiHead}(Q, K, V) = \bigoplus(\text{head}_1, \text{head}_2, \ldots, \text{head}_h)W^O$$

Each head is an independent self-attention mechanism with its own projection matrices:

$$\text{head}_i = \text{Attention}(QW_i^Q, KW_i^K, VW_i^V)$$

Multi-head attention enhances the model's ability to focus on various aspects of the input sequence and capture a richer set of dependencies.

6.2.2 *Tensor Operations in Self-attention*

The self-attention mechanism is at the heart of transformer architectures, and its implementation involves several tensor operations. Understanding these operations is crucial for grasping how transformers process sequences efficiently and effectively.

Query, Key, and Value Tensors

In the self-attention mechanism, the input sequence is transformed into three distinct sets of tensors: queries (Q), keys (K), and values (V). These tensors are derived from the same input embeddings but through different linear projections. Given an input sequence represented as a matrix $X \in \mathbb{R}^{n \times d}$, where n is the sequence length and d is the dimension of the embeddings, we perform the following steps:

1. **Linear Projections**: The input matrix X is projected into queries, keys, and values using learned weight matrices W^Q, W^K, and W^V, respectively. Each projection matrix has dimensions $d \times d_k$, where d_k is the dimension of the queries and keys.

$$Q = XW^Q \quad \text{where} \quad W^Q \in \mathbb{R}^{d \times d_k}$$

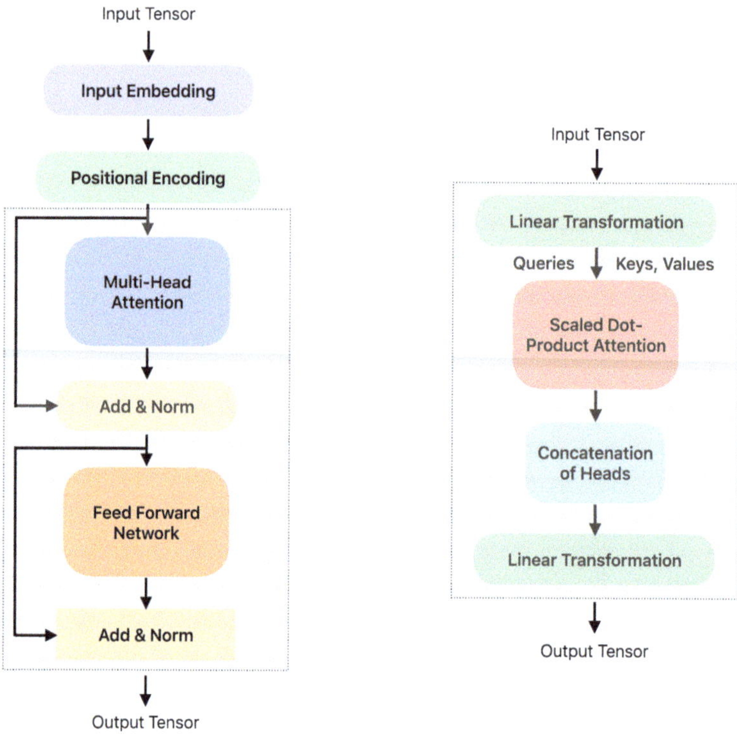

Fig. 6.1 Illustration of a transformer encoder (left) and multi-head attention (right)

$$K = XW^K \quad \text{where} \quad W^K \in \mathbb{R}^{d \times d_k}$$

$$V = XW^V \quad \text{where} \quad W^V \in \mathbb{R}^{d \times d_v}$$

The resulting tensors Q, K, and V each have dimensions $n \times d_k$ for queries and keys, and $n \times d_v$ for values, where typically $d_k = d_v = d/h$ for h attention heads.

2. **Tensor Dimensions**: The dimensions of these tensors are essential for the subsequent operations in the self-attention mechanism. For a single attention head, the tensors are:

$$Q, K \in \mathbb{R}^{n \times d_k}, \quad V \in \mathbb{R}^{n \times d_v}$$

For multi-head attention, each head performs its own set of linear projections and self-attention computations, followed by concatenation and another linear projection.

Scaled Dot-Product Attention

The scaled dot-product attention mechanism computes the attention scores between queries and keys, scales them, applies a softmax function to obtain the attention weights, and then uses these weights to create a weighted sum of the values.

1. **Dot Product of Queries and Keys**: The attention scores are computed as the dot product between queries and keys. Given $Q \in \mathbb{R}^{n \times d_k}$ and $K \in \mathbb{R}^{n \times d_k}$, the dot-product results in a matrix of scores:

$$\text{Scores} = QK^T \quad \text{where} \quad \text{Scores} \in \mathbb{R}^{n \times n}$$

2. **Scaling**: To prevent the dot-product values from becoming too large, which can adversely affect the gradients, the scores are scaled by the square root of the key dimension d_k:

$$\text{Scaled Scores} = \frac{QK^T}{\sqrt{d_k}}$$

3. **Softmax Function**: The scaled scores are passed through a softmax function to obtain the attention weights. The softmax function normalizes the scores to ensure that the weights sum to one:

$$\alpha_{ij} = \frac{\exp(\text{Scaled Scores}_{ij})}{\sum_{k=1}^{n} \exp(\text{Scaled Scores}_{ik})}$$

   The resulting matrix of attention weights $\alpha \in \mathbb{R}^{n \times n}$ indicates the importance of each key relative to each query.
4. **Weighted Sum of Values**: The final step involves computing the weighted sum of the value vectors. Given $V \in \mathbb{R}^{n \times d_v}$, the output of the attention mechanism is:

$$\text{Attention}(Q, K, V) = \alpha V \quad \text{where} \quad \text{Attention}(Q, K, V) \in \mathbb{R}^{n \times d_v}$$

   Each row of the output matrix is a weighted sum of the rows in V, where the weights are determined by the corresponding row in α.

Example Consider an input sequence of three tokens, each with an embedding dimension of 4. The projections and attention mechanism can be illustrated with the following steps:

1. Input Sequence:

$$X = \begin{bmatrix} x_{11} & x_{12} & x_{13} & x_{14} \\ x_{21} & x_{22} & x_{23} & x_{24} \\ x_{31} & x_{32} & x_{33} & x_{34} \end{bmatrix}$$

2. Linear Projections:

$$Q = XW^Q, \quad K = XW^K, \quad V = XW^V$$

3. Compute Scores:

$$\text{Scores} = QK^T = \begin{bmatrix} q_{11} & q_{12} & q_{13} \\ q_{21} & q_{22} & q_{23} \\ q_{31} & q_{32} & q_{33} \end{bmatrix} \begin{bmatrix} k_{11} & k_{21} & k_{31} \\ k_{12} & k_{22} & k_{32} \\ k_{13} & k_{23} & k_{33} \end{bmatrix}$$

4. Scale Scores:

$$\text{Scaled Scores} = \frac{\text{Scores}}{\sqrt{d_k}}$$

5. Apply Softmax:

$$\alpha = \text{softmax}(\text{Scaled Scores})$$

6. Weighted Sum of Values:

$$\text{Attention}(Q, K, V) = \alpha V$$

6.2.3 *Multi-head Attention*

The multi-head attention mechanism is a pivotal extension of the self-attention mechanism used in transformer architectures. It allows the model to attend to different parts of the sequence information from multiple perspectives or representation subspaces. The primary motivation behind multi-head attention is to enhance the model's ability to capture a wider range of dependencies and interactions within the input data. By employing multiple attention heads, the model can simultaneously attend to different parts of the sequence with varying degrees of focus and abstraction. This parallelism is crucial for understanding complex relationships and long-range dependencies in the data.

Benefits of Multi-head Attention

1. **Diverse Representations**: Multi-head attention allows the model to jointly consider information from different representation subspaces. Each attention head learns a distinct way to attend to the input sequence, providing a richer and more nuanced understanding of the data.
2. **Improved Learning**: By attending to different parts of the sequence concurrently, the model can capture various patterns and dependencies that a single-head attention mechanism might miss. This leads to better generalization and performance on tasks such as language modeling, machine translation, and more.
3. **Parallelization**: Multi-head attention can be efficiently implemented using matrix operations, enabling parallel computation across multiple heads. This results in faster training and inference times compared to sequential models like RNNs.

The multi-head attention mechanism can be broken down into the following steps:

1. **Linear Projections**: For each attention head, the input sequence $X \in \mathbb{R}^{n \times d}$ is linearly projected into queries Q, keys K, and values V using separate learned weight matrices W_i^Q, W_i^K, W_i^V for each head i:

$$Q_i = XW_i^Q, \quad K_i = XW_i^K, \quad V_i = XW_i^V$$

where $Q_i, K_i, V_i \in \mathbb{R}^{n \times d_k}$.

2. **Scaled Dot-Product Attention**: Each attention head independently computes the scaled dot-product attention. For head i:

$$\text{Attention}_i(Q_i, K_i, V_i) = \text{softmax}\left(\frac{Q_i K_i^T}{\sqrt{d_k}}\right) V_i$$

3. **Concatenation**: The outputs from all the attention heads are concatenated:

$$\bigoplus(\text{head}_1, \text{head}_2, \ldots, \text{head}_h) \in \mathbb{R}^{n \times (h \cdot d_k)}$$

4. **Final Linear Projection**: The concatenated output is then linearly projected using a weight matrix W^O:

$$\text{MultiHead}(Q, K, V) = \bigoplus(\text{head}_1, \text{head}_2, \ldots, \text{head}_h) W^O$$

where $W^O \in \mathbb{R}^{(h \cdot d_k) \times d}$.

Implementation in PyTorch

```
import torch
import torch.nn as nn

class MultiHeadAttention(nn.Module):
    def _ _init_ _(self, d_model, num_heads):
        super(MultiHeadAttention, self)._ _init_ _()
        assert d_model % num_heads == 0

        self.d_model = d_model
        self.num_heads = num_heads
        self.d_k = d_model // num_heads

        self.linear_q = nn.Linear(d_model, d_model)
        self.linear_k = nn.Linear(d_model, d_model)
        self.linear_v = nn.Linear(d_model, d_model)
        self.linear_out = nn.Linear(d_model, d_model)

    def forward(self, x):
        batch_size = x.size(0)

        # Linear projections
        q = self.linear_q(x).view(batch_size, -1, self.num_heads, self.d_k).transpose(1, 2)
        k = self.linear_k(x).view(batch_size, -1, self.num_heads, self.d_k).transpose(1, 2)
        v = self.linear_v(x).view(batch_size, -1, self.num_heads, self.d_k).transpose(1, 2)

        # Scaled dot-product attention
        attn_scores = torch.matmul(q, k.transpose(-2, -1)) / torch.sqrt(torch.tensor
        (self.d_k,
        dtype=torch.float32))
```

```
        attn_weights = nn.functional.softmax(attn_scores, dim=-1)
        attn_output = torch.matmul(attn_weights, v)

        # Concatenate and project
        attn_output = attn_output.transpose(1, 2).contiguous().view(batch_size, -1,
        self.d_model)
        output = self.linear_out(attn_output)

        return output

# Example
batch_size = 2
seq_length = 4
d_model = 8
num_heads = 2

x = torch.rand(batch_size, seq_length, d_model)
multi_head_attn = MultiHeadAttention(d_model, num_heads)
output = multi_head_attn(x)
print(output.shape)  # Output shape: (batch_size, seq_length, d_model)
```

In this example, the model projects the input sequence into queries, keys, and values, computes attention scores for each head, concatenates the outputs, and applies a final linear projection. This setup allows the transformer to process the sequence with multiple heads, each capturing different aspects of the data.

6.3 Architecture of Transformers

6.3.1 Positional Encoding

In transformer architectures, the notion of sequence order is crucial for capturing the temporal relationships between tokens. Since transformers do not inherently include any notion of sequence order in their structure, they require a mechanism to inject positional information into the model. This is achieved through positional encoding. Positional encoding serves to provide the model with information about the relative or absolute position of tokens in the sequence. This is essential because the self-attention mechanism within transformers processes tokens in parallel, unlike RNNs which process tokens sequentially. Without positional encoding, the transformer would lack awareness of the token order, significantly impairing its ability to understand and model the sequential nature of the data. The role of positional encoding can be summarized as follows:

1. **Injecting Positional Information**: Positional encodings enable the model to incorporate the order of tokens, which is critical for understanding the context and meaning in sequential data.
2. **Enabling Sequential Processing**: By adding positional encodings to the input embeddings, the transformer can differentiate between tokens based on their positions, allowing it to capture temporal dependencies and relationships.
3. **Facilitating Learning of Patterns**: Positional encodings help the model learn patterns and structures in the data that depend on the token order, such as grammatical rules in language or patterns in time series data.

The positional encoding used in transformers is designed to inject position information in a way that the model can easily learn to recognize and use. The approach described in the original transformer paper, "Attention is All You Need" by Vaswani et al. (2017), uses sine and cosine functions of different frequencies. The mathematical formulation for positional encoding is as follows: For each position (pos) and each dimension i, the positional encoding is defined as:

$$\text{PE}_{(\text{pos},2i)} = \sin\left(\frac{\text{pos}}{10{,}000^{2i/d_{\text{model}}}}\right)$$

$$\text{PE}_{(\text{pos},2i+1)} = \cos\left(\frac{\text{pos}}{10{,}000^{2i/d_{\text{model}}}}\right)$$

Here, (pos) is the position of the token in the sequence, i is the dimension index, and d_{model} is the dimension of the input embeddings. The use of sine and cosine functions of different frequencies allows the model to learn positional relationships easily. The positional encodings are computed once and added to the input embeddings. For a sequence of length n and embedding dimension d, the positional encoding matrix $\text{PE} \in \mathbb{R}^{n \times d}$ is constructed such that each row corresponds to the positional encoding of a token at a specific position.

Implementation

```
import torch
import math

class PositionalEncoding(nn.Module):
    def _ _init_ _(self, d_model, max_len=5000):
        super(PositionalEncoding, self)._ _init_ _()
        self.d_model = d_model

        # Create a matrix of shape (max_len, d_model) with positional encodings
        pe = torch.zeros(max_len, d_model)
        position = torch.arange(0, max_len, dtype=torch.float).unsqueeze(1)
        div_term = torch.exp(torch.arange(0, d_model, 2).float() * (-math.log(10000.0)
        / d_model))

        pe[:, 0::2] = torch.sin(position * div_term)
        pe[:, 1::2] = torch.cos(position * div_term)

        pe = pe.unsqueeze(0).transpose(0, 1)
        self.register_buffer('pe', pe)

    def forward(self, x):
        x = x + self.pe[:x.size(0), :]
        return x

# Example
d_model = 512
seq_length = 100
pos_encoding = PositionalEncoding(d_model)
input_embeddings = torch.randn(seq_length, 1, d_model)
output = pos_encoding(input_embeddings)
print(output.shape)  # Output shape: (seq_length, 1, d_model)
```

In this example, the 'PositionalEncoding' class computes the positional encodings using sine and cosine functions and adds them to the input embeddings. The 'forward' method ensures that the positional encodings are added to the input embeddings at each forward pass. The position matrix position is created with values ranging from 0 to max_len − 1, where max_len is the maximum length of the sequence. This matrix is then reshaped to facilitate element-wise operations with the division term. The division term div_term is calculated as $\exp(\text{torch.arange}(0, d\_\text{model}, 2) \times (-\log(10{,}000.0)/d\_\text{model}))$, which generates the scaling factors for the sine and cosine functions. The sine function is applied to the even indices, and the cosine function is applied to the odd indices of the positional encoding matrix. This alternating pattern ensures that each position in the sequence has a unique encoding.

6.3.2 Encoder–Decoder Structure

The transformer architecture is based on an encoder–decoder structure, which is particularly effective for sequence-to-sequence tasks such as machine translation. This structure consists of an encoder that processes the input sequence and a decoder that generates the output sequence. Each of these components is composed of several identical layers, and both employ self-attention mechanisms and feed-forward neural networks.

Encoder Architecture

The encoder in a transformer model is responsible for converting the input sequence into a continuous representation that captures the context and relationships within the data. Each encoder layer consists of two main components: a multi-head self-attention mechanism and a position-wise fully connected feed-forward network. Both of these components are followed by layer normalization and residual connections.

1. **Multi-head Self-attention**: The multi-head self-attention mechanism allows the encoder to focus on different parts of the input sequence simultaneously, capturing various aspects of the input data. For each head, the input sequence is projected into queries, keys, and values, and the scaled dot-product attention is computed.

$$\text{MultiHead}(Q, K, V) = \bigoplus(\text{head}_1, \text{head}_2, \ldots, \text{head}_h)W^O$$

   Each head head_i is computed as:

$$\text{head}_i = \text{Attention}(QW_i^Q, KW_i^K, VW_i^V)$$

2. **Feed-Forward Neural Network**: The output of the multi-head attention mechanism is passed through a position-wise feed-forward network, which consists of two linear transformations with a ReLU activation in between:

$$\text{FFN}(x) = \max(0, xW_1 + b_1)W_2 + b_2$$

3. **Layer Normalization and Residual Connections**: Layer normalization and residual connections are applied to stabilize the training process and allow for deeper networks.

$$\text{LayerNorm}(x + \text{Sublayer}(x))$$

Here, $\text{Sublayer}(x)$ refers to either the multi-head attention mechanism or the feed-forward network. The output of each encoder layer can be represented as:

$$\begin{aligned}\text{EncLayer}(x) = \text{LayerNorm}(&\text{LayerNorm}(x + \text{MultiHead}(x, x, x)) \\ &+ \text{FFN}(\text{LayerNorm}(x + \text{MultiHead}(x, x, x))))\end{aligned}$$

Given an input sequence X, the encoder produces an encoded representation H:

$$H = \text{Encoder}(X) = \text{EncLayer}_N(\ldots \text{EncLayer}_1(X)\ldots)$$

where N is the number of encoder layers.

Decoder Architecture

The decoder in a transformer model generates the output sequence based on the encoded representation from the encoder and the previously generated tokens in the output sequence. Each decoder layer consists of three main components: a masked multi-head self-attention mechanism, a multi-head attention mechanism over the encoder's output, and a position-wise feed-forward network (see Fig. 6.2 for an overview of the decoder architecture). These components are followed by layer normalization and residual connections.

1. **Masked Multi-head Self-attention**: The masked multi-head self-attention mechanism ensures that the model only attends to the tokens generated so far and not future tokens, preserving the autoregressive property.

$$\text{MaskedAttention}(Q, K, V) = \text{softmax}\left(\frac{QK^T}{\sqrt{d_k}} + \text{mask}\right)V$$

where the mask is a triangular matrix that prevents attention to future positions.

2. **Multi-head Attention over Encoder Output**: The decoder uses the encoded representation from the encoder as keys and values, allowing it to attend to relevant parts of the input sequence when generating the output.

$$\text{MultiHead}(Q, K, V) = \bigoplus(\text{head}_1, \text{head}_2, \ldots, \text{head}_h)W^O$$

Here, K and V come from the encoder's output, and Q comes from the previous decoder layer.

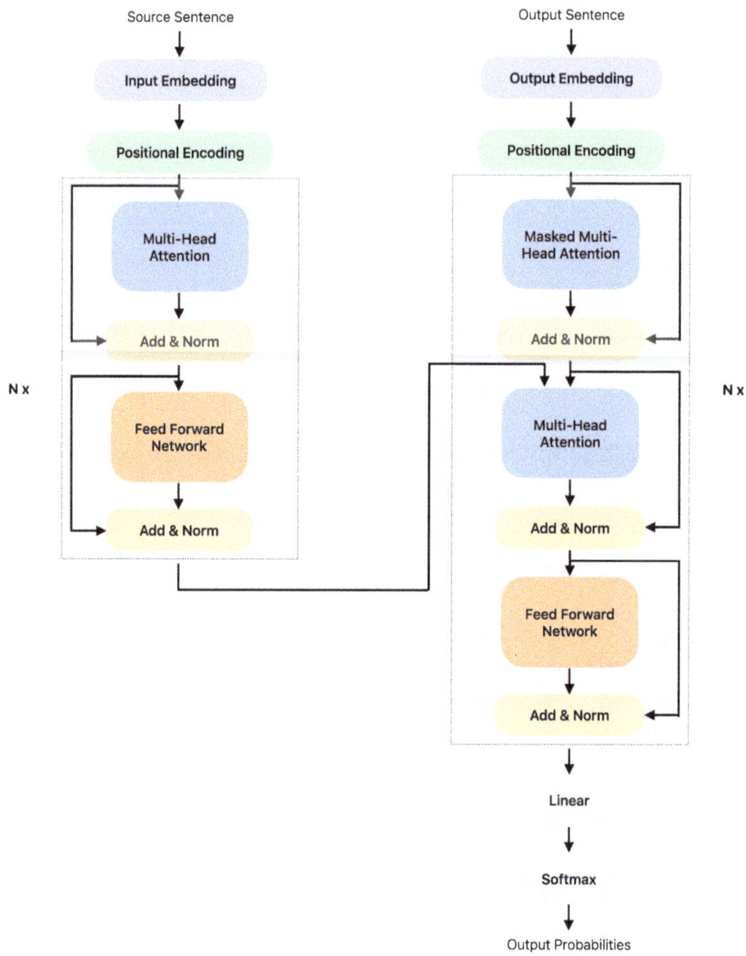

Fig. 6.2 Illustration of a transformer encoder–decoder architecture

3. **Feed-Forward Neural Network**: Similar to the encoder, the decoder layer includes a position-wise feed-forward network:

$$\text{FFN}(x) = \max(0, xW_1 + b_1)W_2 + b_2$$

4. **Layer Normalization and Residual Connections**: Layer normalization and residual connections are applied in the same manner as in the encoder.

   The output of each decoder layer can be represented as:

$$\begin{aligned}\text{DecLayer}(y, H) = \text{LayerNorm}\,(&\text{LayerNorm}\,(\text{LayerNorm}(y + \text{MaskedMultiHead}(y, y, y))\\&+ \text{MultiHead}(\text{LayerNorm}(y + \text{MaskedMultiHead}(y, y, y)), H, H))\\&+ \text{FFN}\,(\text{LayerNorm}\,(\text{LayerNorm}(y + \text{MaskedMultiHead}(y, y, y))\\&+ \text{MultiHead}(\text{LayerNorm}(y + \text{MaskedMultiHead}(y, y, y)), H, H)))\end{aligned}$$

Given the encoded representation H and the output sequence Y, the decoder produces the final output:

$$O = \text{Decoder}(Y, H) = \text{DecLayer}_N(\ldots \text{DecLayer}_1(Y, H)\ldots)$$

where N is the number of decoder layers.

Combined Encoder–Decoder Model

The transformer model combines the encoder and decoder components to form a powerful sequence-to-sequence model. The encoder processes the input sequence and produces a continuous representation, which the decoder uses along with the previously generated tokens to produce the output sequence.

1. **Input to Encoder**: The input sequence X is passed through the encoder to obtain the encoded representation H:

$$H = \text{Encoder}(X)$$

2. **Input to Decoder**: The output sequence Y (shifted right) and the encoded representation H are passed through the decoder to generate the final output:

$$O = \text{Decoder}(Y, H)$$

3. **Final Output**: The final output O is typically passed through a linear layer and a softmax function to produce the probability distribution over the target vocabulary for each position in the output sequence.

Implementation Consider a simple implementation of the transformer encoder and decoder in PyTorch as follows:

```
import torch
import torch.nn as nn

class TransformerEncoderLayer(nn.Module):
    def __init__(self, d_model, num_heads, d_ff, dropout=0.1):
        super(TransformerEncoderLayer, self).__init__()
        self.self_attn = nn.MultiheadAttention(d_model, num_heads, dropout=dropout)
        self.ffn = nn.Sequential(
            nn.Linear(d_model, d_ff),
            nn.ReLU(),
            nn.Linear(d_ff, d_model)
        )
        self.norm1 = nn.LayerNorm(d_model)
        self.norm2 = nn.LayerNorm(d_model)
        self.dropout = nn.Dropout(dropout)
```

```
    def forward(self, x):
        attn_output, _ = self.self_attn(x, x, x)
        x = self.norm1(x + self.dropout(attn_output))
        ffn_output = self.ffn(x)
        x = self.norm2(x + self.dropout(ffn_output))
        return x

class TransformerDecoderLayer(nn.Module):
    def _ _init_ _(self, d_model, num_heads, d_ff, dropout=0.1):
        super(TransformerDecoderLayer, self)._ _init_ _()
        self.self_attn = nn.MultiheadAttention(d_model, num_heads, dropout=dropout)
        self.cross_attn = nn.MultiheadAttention(d_model, num_heads, dropout=dropout)
        self.ffn = nn.Sequential(
            nn.Linear(d_model, d_ff),
            nn.ReLU(),
            nn.Linear(d_ff, d_model)
        )
        self.norm1 = nn.LayerNorm(d_model)
        self.norm2 = nn.LayerNorm(d_model)
        self.norm3 = nn.LayerNorm(d_model)
        self.dropout = nn.Dropout(dropout)

    def forward(self, x, memory):
        self_attn_output, _ = self.self_attn(x, x, x)
        x = self.norm1(x + self.dropout(self_attn_output))
        cross_attn_output, _ = self.cross_attn(x, memory, memory)
        x = self.norm2(x + self.dropout(cross_attn_output))
        ffn_output = self.ffn(x)
        x = self.norm3(x + self.dropout(ffn_output))
        return x

class Transformer(nn.Module):
    def _ _init_ _(self, d_model, num_heads, num_encoder_layers, num_decoder_layers,
    d_ff, input_vocab_size, output_vocab_size, max_seq_length, dropout=0.1):
        super(Transformer, self)._ _init_ _()
        self.encoder_embedding = nn.Embedding(input_vocab_size, d_model)
        self.decoder_embedding = nn.Embedding(output_vocab_size, d_model)
        self.positional_encoding = PositionalEncoding(d_model, max_len=max_seq_length)
        self.encoder_layers = nn.ModuleList([TransformerEncoderLayer(d_model, num_heads,
        d_ff, dropout) for _

 in range(num_encoder_layers)])
        self.decoder_layers = nn.ModuleList([TransformerDecoderLayer(d_model, num_heads,
        d_ff, dropout) for _ in range(num_decoder_layers)])
        self.linear = nn.Linear(d_model, output_vocab_size)

    def forward(self, src, tgt):
        src = self.encoder_embedding(src) * math.sqrt(self.d_model)
        src = self.positional_encoding(src)
        tgt = self.decoder_embedding(tgt) * math.sqrt(self.d_model)
        tgt = self.positional_encoding(tgt)

        for layer in self.encoder_layers:
            src = layer(src)
        memory = src

        for layer in self.decoder_layers:
            tgt = layer(tgt, memory)

        output = self.linear(tgt)
        return output
```

In this implementation, 'TransformerEncoderLayer' and 'TransformerDecoderLayer' represent individual encoder and decoder layers, respectively. The 'Transformer' class combines these layers to form the complete transformer model, incorporating positional encodings and embedding layers for both the input and output sequences.

6.4 Popular Transformer Models

6.4.1 Bidirectional Encoder Representations from Transformers (BERT)

Bidirectional Encoder Representations from Transformers (BERT) is a groundbreaking model in the field of NLP that has set new benchmarks for various NLP tasks. Developed by researchers at Google (Devlin et al. 2019), BERT utilizes a bidirectional approach to pre-train deep transformer models on large corpora of text, capturing both left and right contexts in all layers. This section delves into the architecture, training objectives, and applications of BERT.

Architecture
BERT's architecture is based on the transformer encoder, leveraging multiple layers of bidirectional self-attention. The bidirectional nature allows BERT to consider the context from both the left and the right sides of a word in all layers, unlike traditional left-to-right or right-to-left models. The architecture of BERT consists of:

Input Embeddings: BERT uses three types of embeddings.

Token Embeddings: Each token in the input sequence is mapped to a fixed-size vector.

Segment Embeddings: These embeddings help distinguish between sentences in tasks involving sentence pairs (e.g., question answering).

Position Embeddings: These embeddings encode the position of each token in the sequence, enabling the model to capture positional information.

The final input representation is the sum of these three embeddings.

Transformer Encoder Layers BERT consists of multiple transformer encoder layers. The base model, $BERT_{BASE}$, has 12 layers, while the larger model, $BERT_{LARGE}$, has 24 layers. Each encoder layer comprises:

Multi-head Self-attention: This mechanism allows each token to attend to every other token in the sequence.

Feed-Forward Neural Network: A position-wise feed-forward network processes the outputs from the self-attention mechanism.

Layer Normalization and Residual Connections: These stabilize and improve the training of deep networks.

Training Objectives
BERT's training involves two primary objectives designed to improve its ability to understand and generate natural language:

Masked Language Modeling (MLM) MLM is a form of cloze test where random tokens in the input sequence are masked, and the model is trained to predict these masked tokens. This allows BERT to learn bidirectional representations. Given an input sequence X, a subset of tokens X_m is randomly selected and replaced with a special token ([MASK]). The objective is to predict the original tokens X_m from the masked sequence.

$$\text{Loss}_{\text{MLM}} = -\sum_{i \in X_m} \log P(x_i \mid X_{\setminus m})$$

Next Sentence Prediction (NSP) NSP helps BERT understand the relationship between two sentences. The model is given pairs of sentences and trained to predict whether the second sentence is the actual next sentence or a random one. The input consists of sentence pairs (A, B), where 50. The objective is to predict a binary label indicating whether B is the true next sentence.

$$\text{Loss}_{\text{NSP}} = -\log P(\text{isNext} \mid A, B) - \log P(\text{isRandom} \mid A, B)$$

Implementation Consider a simplified implementation of BERT's masked language modeling objective in PyTorch.

```
import torch
import torch.nn as nn
import torch.optim as optim

class BERT(nn.Module):
    def __init__(self, vocab_size, d_model, num_heads, num_layers, max_seq_length):
        super(BERT, self).__init__()
        self.embedding = nn.Embedding(vocab_size, d_model)
        self.positional_encoding = PositionalEncoding(d_model, max_len=max_seq_length)
        self.encoder_layers = nn.ModuleList([TransformerEncoderLayer(d_model, num_heads,
        d_model * 4) for _ in range(num_layers)])
        self.linear = nn.Linear(d_model, vocab_size)

    def forward(self, x, mask):
        x = self.embedding(x) * torch.sqrt(torch.tensor(d_model, dtype=torch.float32))
        x = self.positional_encoding(x)
        for layer in self.encoder_layers:
```

```
            x = layer(x, mask)
        return self.linear(x)

# Example
vocab_size = 30522  # Vocabulary size for BERT
d_model = 768  # Embedding size for BERT_BASE
num_heads = 12  # Number of attention heads
num_layers = 12  # Number of encoder layers
max_seq_length = 512  # Maximum sequence length

model = BERT(vocab_size, d_model, num_heads, num_layers, max_seq_length)

# Training setup
optimizer = optim.Adam(model.parameters(), lr=3e-5)
criterion = nn.CrossEntropyLoss()

# Example input (batch of sequences with masked tokens)
input_ids = torch.randint(0, vocab_size, (2, max_seq_length))
attention_mask = torch.ones((2, max_seq_length))
masked_labels = torch.randint(0, vocab_size, (2, max_seq_length))

# Forward pass
outputs = model(input_ids, attention_mask)
loss = criterion(outputs.view(-1, vocab_size), masked_labels.view(-1))

# Backward pass and optimization
loss.backward()
optimizer.step()
```

In this example, the 'BERT' class implements a simplified version of BERT for masked language modeling. The model includes embedding layers, positional encodings, and transformer encoder layers. The forward pass computes the predictions for masked tokens, and the loss is computed using cross-entropy.

Applications and Use Cases

BERT has been successfully applied to a wide range of NLP tasks, achieving state-of-the-art performance in many benchmarks. Some key applications and use cases of BERT include:

1. **Question-Answering**: BERT has demonstrated exceptional performance in question-answering tasks. Given a context and a question, BERT can accurately extract the answer span from the context.

$$\text{Loss}_{\text{QA}} = -\log P(\text{start} \mid \text{context, question}) - \log P(\text{end} \mid \text{context, question})$$

2. **Named Entity Recognition (NER)**: BERT is used to identify and classify named entities (e.g., person names, organizations, locations) in text.

$$\text{Loss}_{\text{NER}} = -\sum_{i=1}^{n} \log P(y_i \mid x_i)$$

3. **Text Classification**: BERT can be fine-tuned for various text classification tasks, such as sentiment analysis, spam detection, and topic classification.

$$\text{Loss}_{\text{Classification}} = -\log P(y \mid X)$$

4. **Text Summarization**: BERT has been used to generate summaries of long texts, capturing the essential information while maintaining coherence and relevance.
5. **Machine Translation**: Although BERT is primarily an encoder, it can be integrated into transformer-based machine translation models to improve translation quality.

Example: Question Answering with BERT Consider using a pre-trained BERT model for a question-answering task. The model is fine-tuned on the SQuAD dataset, a benchmark for evaluating question-answering systems.

```
from transformers import BertForQuestionAnswering, BertTokenizer
import torch

# Load pre-trained BERT model and tokenizer
model = BertForQuestionAnswering.from_pretrained('bert-large-uncased
-whole-word-masking-finetuned-squad')
tokenizer = BertTokenizer.from_pretrained('bert-large-uncased-whole
-word-masking-finetuned-squad')

# Example context and question
context = "BERT stands for Bidirectional Encoder Representations from Transformers.
It is a popular NLP model."
question = "What does BERT stand for?"

# Tokenize input
inputs = tokenizer.encode_plus(question, context, return_tensors='pt')
input_ids = inputs['input_ids']
attention_mask = inputs['attention_mask']

# Get start and end scores
outputs = model(input_ids, attention_mask=attention_mask)
start_scores = outputs.start_logits
end_scores = outputs.end_logits

# Extract answer
start_index = torch.argmax(start_scores)
end_index = torch.argmax(end_scores) + 1
answer = tokenizer.convert_tokens_to_string(tokenizer.convert_ids_to_tokens
(input_ids[0][start_index:end_index]))
print(f'Answer: {answer}')
```

In this example, the 'BertForQuestionAnswering' class is used to load a pre-trained BERT model fine-tuned on the SQuAD dataset. The model is used to answer a question given a context by predicting the start and end positions of the answer span.

6.4.2 *Generative Pre-trained Transformer (GPT)*

Generative Pre-trained Transformer (GPT) is another influential model in the field of NLP, developed by OpenAI (Radford and Narasimhan 2018). It builds on the transformer architecture but focuses on generating coherent and contextually relevant text. Unlike BERT, which is designed for understanding and encoding text, GPT is optimized for generating text, making it suitable for a variety of language generation tasks.

Architecture

GPT is based on the transformer decoder architecture, which is designed to generate text in an autoregressive manner. This means that each token is generated one after another, with each token conditioned on the previously generated tokens. Key components of the GPT architecture include:

Input Embeddings: GPT uses token embeddings, positional embeddings, and segment embeddings (in some versions) to represent the input sequence. The token embeddings convert each token into a dense vector, while positional embeddings provide information about the position of each token in the sequence. Segment embeddings help distinguish between different parts of the input (e.g., different sentences in a dialogue).

Transformer Decoder Layers: GPT consists of multiple transformer decoder layers. The base model, GPT, has 12 layers, while larger versions like GPT-2 and GPT-3 have 48 layers. Each decoder layer includes:

Masked Multi-head Self-attention: This mechanism ensures that each token can only attend to previous tokens and itself, preserving the autoregressive property.

Feed-Forward Neural Network: A position-wise feed-forward network processes the output from the self-attention mechanism.

Layer Normalization and Residual Connections: These components help stabilize and improve the training process.

Training Objectives

GPT is trained using the language modeling objective, which involves predicting the next token in a sequence given the previous tokens. This training process helps the model learn the probability distribution of sequences in the training data.

1. **Language Modeling**: The primary objective is to maximize the likelihood of the target token x_t given the previous tokens $x_1, x_2, \ldots, x_{t-1}$:

$$\text{Loss}_{\text{LM}} = -\sum_{t=1}^{T} \log P(x_t \mid x_{<t})$$

   The model generates text by sampling from the probability distribution of the next token, conditioned on the previously generated tokens.
2. **Autoregressive Generation**: GPT generates text in an autoregressive manner, meaning that it generates one token at a time and uses that token as input for generating the next token. This approach ensures that the generated text is coherent and contextually relevant.

Implementation Consider a simplified implementation of GPT's language modeling objective in PyTorch.

```
import torch
import torch.nn as nn
import torch.optim as optim

class GPT(nn.Module):
    def __init__(self, vocab_size, d_model, num_heads, num_layers, max_seq_length):
        super(GPT, self).__init__()
        self.embedding = nn.Embedding(vocab_size, d_model)
        self.positional_encoding = PositionalEncoding(d_model, max_len=max_seq_length)
        self.decoder_layers = nn.ModuleList([TransformerDecoderLayer(d_model, num_heads,
        d_model * 4) for _ in range(num_layers)])
        self.linear = nn.Linear(d_model, vocab_size)

    def forward(self, x):
        x = self.embedding(x) * torch.sqrt(torch.tensor(d_model, dtype=torch.float32))
        x = self.positional_encoding(x)
        for layer in self.decoder_layers:
            x = layer(x, x)
        return self.linear(x)

# Example
vocab_size = 50257  # Vocabulary size for GPT-2
d_model = 768  # Embedding size for GPT
num_heads = 12  # Number of attention heads
num_layers = 12  # Number of decoder layers
max_seq_length = 1024  # Maximum sequence length

model = GPT(vocab_size, d_model, num_heads, num_layers, max_seq_length)

# Training setup
optimizer = optim.Adam(model.parameters(), lr=3e-5)
criterion = nn.CrossEntropyLoss()

# Example input (batch of sequences)
input_ids = torch.randint(0, vocab_size, (2, max_seq_length))

# Forward pass
outputs = model(input_ids)
loss = criterion(outputs.view(-1, vocab_size), input_ids.view(-1))

# Backward pass and optimization
loss.backward()
optimizer.step()
```

In this example, the 'GPT' class implements a simplified version of GPT for language modeling. The model includes embedding layers, positional encodings, and transformer decoder layers. The forward pass computes the predictions for the next token, and the loss is computed using cross-entropy.

Applications and Use Cases

GPT has been successfully applied to a wide range of NLP tasks, particularly those involving text generation. Some key applications and use cases of GPT include:

1. **Text Generation**: GPT can generate coherent and contextually relevant text given a prompt. This capability is useful for applications such as story generation, dialogue systems, and content creation.

$$\text{Loss}_{\text{TextGen}} = -\sum_{t=1}^{T} \log P(x_t \mid x_{<t})$$

2. **Language Translation**: GPT can be fine-tuned for language translation tasks, generating translations that are fluent and accurate. The autoregressive nature of GPT allows it to generate translations one token at a time, considering the context of the entire sentence.
3. **Text Summarization**: GPT can generate concise and relevant summaries of long texts. This application is particularly useful for summarizing articles, reports, and other lengthy documents.
4. **Question Answering**: GPT can generate answers to questions based on the context provided. This capability is useful for building conversational agents and virtual assistants.
5. **Code Generation**: GPT can generate code snippets given a natural language description of the desired functionality. This application is useful for automating code generation and assisting developers.

Example: Text Generation with GPT Consider using a pre-trained GPT model for text generation. The model generates text given an initial prompt by sampling from the probability distribution of the next token.

```
from transformers import GPT2LMHeadModel, GPT2Tokenizer
import torch

# Load pre-trained GPT-2 model and tokenizer
model = GPT2LMHeadModel.from_pretrained('gpt2')
tokenizer = GPT2Tokenizer.from_pretrained('gpt2')

# Example prompt
prompt = "Once upon a time"

# Encode input prompt
input_ids = tokenizer.encode(prompt, return_tensors='pt')

# Generate text
output = model.generate(input_ids, max_length=100, num_return_sequences=1)
generated_text = tokenizer.decode(output[0], skip_special_tokens=True)
print(f'Generated Text: {generated_text}')
```

In this example, the 'GPT2LMHeadModel' class is used to load a pre-trained GPT-2 model. The model generates text given an initial prompt by sampling from the probability distribution of the next token.

6.4.3 Transformer Variants and Improvements

As the field of NLP has evolved, several variants and improvements of the original transformer models like BERT and GPT have been developed. These models address specific limitations, optimize performance, and introduce new capabilities. Notable transformer variants include ALBERT, RoBERTa, DistilBERT, T5, and BART. This section explores the architecture, improvements, and applications of these models.

ALBERT (A Lite BERT for Self-supervised Learning of Language Representations) ALBERT is designed to improve the efficiency of BERT while maintaining its performance (Lan et al. 2019). It introduces two main changes: parameter sharing and factorized embedding parameterization.

1. **Parameter Sharing**: ALBERT reduces the number of parameters by sharing weights across layers. This technique significantly decreases the model size without affecting the model's capacity to learn representations.

$$W_{\text{shared}} = W_{\text{layer}_1} = W_{\text{layer}_2} = \ldots = W_{\text{layer}_N}$$

2. **Factorized Embedding Parameterization**: Instead of using a large vocabulary embedding matrix, ALBERT factorizes the embeddings into smaller matrices, separating the size of hidden layers from the size of embeddings.

$$E = E_1 \cdot E_2 \quad \text{where} \quad E_1 \in \mathbb{R}^{V \times d}, \ E_2 \in \mathbb{R}^{d \times H}$$

   Here, V is the vocabulary size, d is the embedding size, and H is the hidden size.

Robustly Optimized BERT Approach (RoBERTa) RoBERTa builds on BERT by optimizing the training process and using more data (Liu et al. 2019). It removes the next sentence prediction objective and introduces dynamic masking.

1. **Training Data**: RoBERTa is trained on a larger corpus compared to BERT, using datasets like Common Crawl and OpenWebText.
2. **Dynamic Masking**: Instead of fixing the masked tokens during preprocessing, RoBERTa applies masking dynamically during training, providing more robust learning.

$$\text{Mask}(X) \approx \text{DynamicMask}(X, t)$$

3. **Longer Training**: RoBERTa trains for more steps with larger batches, improving the model's performance.

DistilBERT DistilBERT aims to reduce the size of BERT while retaining 97% of its language understanding capabilities (Sanh et al. 2019). It uses knowledge distillation, where a smaller student model learns from a larger teacher model.

1. **Knowledge Distillation**: The student model is trained to mimic the teacher model's outputs, capturing essential knowledge with fewer parameters.

$$\text{Loss}_{\text{Distil}} = \alpha \cdot \text{Loss}_{\text{soft}} + (1 - \alpha) \cdot \text{Loss}_{\text{hard}}$$

   Here, $\text{Loss}_{\text{soft}}$ is the soft cross-entropy loss between the student and teacher logits, and $\text{Loss}_{\text{hard}}$ is the standard cross-entropy loss with true labels.
2. **Model Compression**: DistilBERT reduces the number of layers by half, speeding up inference and reducing memory usage.

Applications: These models are used in various NLP tasks such as text classification, sentiment analysis, question answering, and named entity recognition. The improvements in efficiency and robustness make them suitable for real-world applications where computational resources and latency are critical.

Text-to-Text Transfer Transformer (T5) T5 treats every NLP problem as a text-to-text problem, allowing a unified approach to different tasks (Raffel et al. 2019). It uses the same model, objective, training procedure, and decoding process for all tasks.

1. **Unified Text-to-Text Framework**: In T5, both inputs and outputs are text strings. For example, for a translation task, the input could be "translate English to French: How are you?" and the output would be "Comment ça va?".
2. **Pre-training Objective**: T5 uses a span corruption objective for pre-training, where contiguous spans of tokens are replaced with a single sentinel token, and the model is trained to reconstruct the original text.

$$\text{Loss}_{\text{T5}} = -\sum_{t=1}^{T} \log P(y_t \mid y_{<t}, x)$$

3. **Model Architecture**: T5 follows the standard transformer architecture with encoder–decoder layers, allowing it to handle diverse tasks with the same model structure.

Bidirectional and Autoregressive Transformers (BARTs) BART combines the strengths of bidirectional and autoregressive models (Lewis et al. 2019). It is particularly effective for text generation tasks.

1. **Denoising Autoencoder**: BART is pre-trained as a denoising autoencoder, where the input text is corrupted with noise (e.g., token masking, deletion, shuffling), and the model is trained to reconstruct the original text.

$$\text{Loss}_{\text{BART}} = -\log P(x \mid \text{Corrupt}(x))$$

2. **Bidirectional Encoder and Autoregressive Decoder**: The encoder in BART is bidirectional, allowing it to capture context from both directions. The decoder is autoregressive, generating text one token at a time based on the context.
3. **Flexible Pre-training**: BART's pre-training strategy makes it versatile for tasks like text generation, summarization, and machine translation.

Applications: T5 excels in tasks like translation, summarization, question answering, and classification due to its unified framework. It simplifies the process of adapting the model to new tasks. BART is particularly strong in text generation tasks, including summarization, dialogue generation, and text completion. Its denoising autoencoder approach enhances its ability to generate coherent and contextually relevant text.

Example: Fine-tuning T5 for Summarization Consider fine-tuning T5 for a text summarization task using the CNN/Daily Mail dataset.

```
from transformers import T5Tokenizer, T5ForConditionalGeneration, Trainer, TrainingArguments
import torch

# Load pre-trained T5 model and tokenizer
model = T5ForConditionalGeneration.from_pretrained('t5-base')
tokenizer = T5Tokenizer.from_pretrained('t5-base')

# Example input text
text = "The quick brown fox jumps over the lazy dog. The quick brown fox jumps over the
lazy dog."

# Encode input text
input_ids = tokenizer.encode("summarize: " + text, return_tensors='pt', max_length=512,
truncation=True)

# Fine-tune model
training_args = TrainingArguments(
    output_dir='./results',
    num_train_epochs=3,
    per_device_train_batch_size=4,
    save_steps=10_000,
    save_total_limit=2,
)
trainer = Trainer(
    model=model,
    args=training_args,
    train_dataset=...  # training dataset
    eval_dataset=...  # evaluation dataset
)
trainer.train()

# Generate summary
outputs = model.generate(input_ids, max_length=50, num_return_sequences=1)
summary = tokenizer.decode(outputs[0], skip_special_tokens=True)
print(f'Summary: {summary}')
```

In this example, the 'T5ForConditionalGeneration' class is used to load a pre-trained T5 model. The model is fine-tuned for summarization using the CNN/Daily Mail dataset. The fine-tuned model generates a summary given an input text.

6.5 Vision Transformers (ViT and SWIN)

Traditionally, CNNs have dominated the field of computer vision due to their ability to effectively capture spatial hierarchies in images. However, the advent of Vision Transformers (ViTs) marks a significant shift in this paradigm. Vision Transformers leverage the transformer architecture, originally designed for natural language processing, to handle image data. By treating image patches as tokens, ViTs apply self-attention mechanisms to model long-range dependencies and global context more effectively than CNNs. The Vision Transformer (ViT) introduced by Dosovitskiy et al. (2020) and subsequent improvements like the Swin Transformer (Liu et al. 2021) demonstrate the potential of transformers to excel in various vision tasks, including image classification, object detection, and segmentation.

6.5.1 Vision Transformer (ViT)

ViT represents a groundbreaking approach to image classification by directly applying transformer architectures to image patches. This section delves into the architecture, design choices, and performance of ViT.

Architecture and Design Choices

Patch Embedding The core idea of ViT is to divide an image into a sequence of fixed-size patches, each of which is treated as a token. For an input image $x \in \mathbb{R}^{H \times W \times C}$ (height H, width W, and channels C), the image is split into patches of size $P \times P$.

1. **Flattening Patches**: Each patch is flattened into a vector:

$$x_p = \text{Flatten}(x_{i,j}) \quad \text{for} \quad i, j \in \{1, \ldots, \frac{H}{P}\}$$

   Here, $x_{i,j}$ represents the patch at position (i, j).
2. **Linear Projection**: These flattened patches are then linearly projected into a fixed-dimensional embedding space using a trainable projection matrix E:

$$z_0^{i,j} = x_p^{i,j} \cdot E$$

   where $E \in \mathbb{R}^{(P \cdot P \cdot C) \times D}$, and D is the dimension of the embedded patches.

$$x_p = \text{Reshape}(x, [\frac{H}{P} \times \frac{W}{P}, P \times P \times C])$$

Position Embedding To retain positional information, which is crucial for vision tasks, positional embeddings are added to the patch embeddings. These embeddings are learnable parameters that encode the position of each patch in the sequence.

$$z_0 = [z_0^{1,1} + p_{1,1}, \ldots, z_0^{\frac{H}{P}, \frac{W}{P}} + p_{\frac{H}{P}, \frac{W}{P}}]$$

where $p_{i,j}$ represents the positional embedding for the patch at position (i, j).

$$z_0 = x_p \cdot E + p$$

Transformer Encoder The sequence of embedded patches, augmented with positional information, is then fed into a standard transformer encoder. The output of the transformer encoder is a sequence of contextualized patch embeddings. For classification tasks, a special classification token $[CLS]$ is prepended to the input sequence, and its final hidden state is used as the aggregate representation of the entire image.

Implementation

```
import torch
import torch.nn as nn

class VisionTransformer(nn.Module):
    def _ _init_ _(self, image_size, patch_size, num_layers, num_heads, hidden_dim,
    mlp_dim, num_classes):
        super(VisionTransformer, self)._ _init_ _()
        self.num_patches = (image_size // patch_size) ** 2
        self.patch_dim = patch_size * patch_size * 3
        self.hidden_dim = hidden_dim

        self.patch_embedding = nn.Linear(self.patch_dim, hidden_dim)
        self.position_embedding = nn.Parameter(torch.zeros(1, self.num_patches + 1,
        hidden_dim))
        self.cls_token = nn.Parameter(torch.zeros(1, 1, hidden_dim))
        self.transformer = nn.TransformerEncoder(
            nn.TransformerEncoderLayer(hidden_dim, num_heads, mlp_dim),
            num_layers
        )
        self.mlp_head = nn.Sequential(
            nn.LayerNorm(hidden_dim),
            nn.Linear(hidden_dim, num_classes)
        )

    def forward(self, x):
        B, C, H, W = x.shape
        x = x.reshape(B, self.num_patches, self.patch_dim)
        x = self.patch_embedding(x)

        cls_tokens = self.cls_token.expand(B, -1, -1)
        x = torch.cat((cls_tokens, x), dim=1)
        x = x + self.position_embedding

        x = self.transformer(x)
        cls_output = x[:, 0]

        return self.mlp_head(cls_output)

# Example
model = VisionTransformer(image_size=224, patch_size=16, num_layers=12, num_heads=12,
hidden_dim=768, mlp_dim=3072, num_classes=1000)
input_image = torch.randn(1, 3, 224, 224)
output = model(input_image)
print(output.shape)  # Output shape: (1, 1000)
```

In this example, the ‘VisionTransformer’ class implements the ViT model. The input image is divided into patches, embedded, and passed through the transformer encoder. The final hidden state of the classification token is used for classification.

Performance and Applications

ViT has demonstrated impressive performance on image classification benchmarks, often matching or exceeding the performance of state-of-the-art CNNs when trained on large datasets. Some key points about ViT’s performance and applications include:

1. **Large-Scale Training**: ViT performs exceptionally well when pre-trained on large-scale datasets like ImageNet-21k or JFT-300M and fine-tuned on smaller datasets. The large-scale training helps the model learn rich representations.
2. **Transfer Learning**: ViT can be fine-tuned on various downstream tasks, such as object detection and segmentation, by leveraging pre-trained weights from large datasets.

3. **Data Efficiency**: While ViT requires large datasets for pre-training to achieve state-of-the-art performance, it shows remarkable transfer learning capabilities, making it efficient for smaller datasets during fine-tuning.

Example: Image Classification with ViT Consider fine-tuning a pre-trained ViT model for image classification on a smaller dataset.

```
from transformers import ViTForImageClassification, ViTFeatureExtractor
from datasets import load_dataset
import torch
from torch.utils.data import DataLoader

# Load pre-trained ViT model and feature extractor
model = ViTForImageClassification.from_pretrained('google/vit-base-patch16-224')
feature_extractor = ViTFeatureExtractor.from_pretrained('google/vit-base-patch16-224')

# Load and preprocess dataset
dataset = load_dataset('cifar10')
def preprocess(example):
    example['pixel_values'] = feature_extractor(example['image'], return_tensors='pt')
    ['pixel_values']
    return example
dataset = dataset.map(preprocess, batched=True)

# Create data loaders
train_loader = DataLoader(dataset['train'], batch_size=32, shuffle=True)
eval_loader = DataLoader(dataset['test'], batch_size=32)

# Fine-tuning setup
optimizer = torch.optim.AdamW(model.parameters(), lr=5e-5)
criterion = torch.nn.CrossEntropyLoss()

# Training loop
for epoch in range(3

):
    model.train()
    for batch in train_loader:
        optimizer.zero_grad()
        outputs = model(pixel_values=batch['pixel_values'].squeeze())
        loss = criterion(outputs.logits, batch['label'])
        loss.backward()
        optimizer.step()
    print(f'Epoch {epoch} completed.')

# Evaluation
model.eval()
correct = 0
total = 0
with torch.no_grad():
    for batch in eval_loader:
        outputs = model(pixel_values=batch['pixel_values'].squeeze())
        _, predicted = torch.max(outputs.logits, 1)
        total += batch['label'].size(0)
        correct += (predicted == batch['label']).sum().item()
print(f'Accuracy: {correct / total * 100}%')
```

In this example, the 'ViTForImageClassification' class is used to load a pre-trained ViT model. The model is fine-tuned on the CIFAR-10 dataset, demonstrating the transfer learning capabilities of ViT.

6.5.2 *Shifted Window (SWIN) Transformer*

The Shifted Window (SWIN) Transformer introduces a novel approach to applying transformers in vision tasks. Unlike the ViT, which treats the entire image as a sequence of patches, SWIN Transformer processes images using a hierarchy of shifted windows (Liu et al. 2021). This approach enables the model to efficiently capture both local and global contexts, making it highly effective for various vision tasks, including image classification, object detection, and segmentation.

Architecture and Design Choices

Hierarchical Representation SWIN Transformer constructs hierarchical feature representations by merging image patches into larger ones progressively. This hierarchical structure mimics the way CNNs gradually reduce spatial resolution while increasing the number of channels.

1. **Patch Partitioning**: The input image $x \in \mathbb{R}^{H\times W\times C}$ is initially divided into non-overlapping patches of size 4×4. Each patch is treated as a token, resulting in a sequence of patches:

$$x_p = \text{Reshape}\left(x, \left[\frac{H}{4}, \frac{W}{4}, 4 \times 4 \times C\right]\right)$$

2. **Linear Embedding**: These patches are then projected into a lower-dimensional embedding space:

$$z_0 = x_p \cdot E$$

where $E \in \mathbb{R}^{(4\times 4\times C)\times d}$ and d is the dimension of the embeddings.

Shifted Window Multi-head Self-attention The core innovation in SWIN Transformer is the shifted window approach, which performs self-attention within local windows that are shifted between consecutive layers. This mechanism ensures that connections across windows are established, enabling the model to capture long-range dependencies without the computational overhead of global self-attention.

1. **Window-based Multi-head Self-attention (W-MSA)**: Self-attention is computed within non-overlapping windows, each of size $M \times M$. For a given window, the self-attention mechanism is applied as:

$$\text{Attention}(Q, K, V) = \text{softmax}\left(\frac{QK^T}{\sqrt{d_k}}\right)V$$

2. **Shifted Window Multi-head Self-attention (SW-MSA)**: To enable interactions across windows, the windows are shifted by a fixed number of pixels ($\frac{M}{2}$) in the next layer. This shifted approach ensures that each patch attends to patches in adjacent windows in alternating layers.

$$\text{Shift}\left(x, \frac{M}{2}\right) \rightarrow \text{W-MSA}(x) \rightarrow \text{Shift}\left(x, -\frac{M}{2}\right)$$

Layer Normalization and Feed-Forward Network Each SWIN Transformer block includes layer normalization and a position-wise feed-forward network, similar to standard transformer architectures.

Shifted Window Self-attention:

$$z_l = \text{LayerNorm}\left(z_{l-1} + \text{SW-MSA}\left(\text{Shift}\left(z_{l-1}, \frac{M}{2}\right)\right)\right) + \text{LayerNorm}(z_{l-1} + \text{FFN}(z_{l-1}))$$

for $l = 1, \ldots, L$, where L is the number of layers.

Implementation

```
import torch
import torch.nn as nn
import torch.nn.functional as F

class SWINTransformerBlock(nn.Module):
    def __init__(self, dim, num_heads, window_size, shift_size=0):
        super(SWINTransformerBlock, self).__init__()
        self.dim = dim
        self.num_heads = num_heads
        self.window_size = window_size
        self.shift_size = shift_size

        self.attn = nn.MultiheadAttention(dim, num_heads)
        self.norm1 = nn.LayerNorm(dim)
        self.norm2 = nn.LayerNorm(dim)

        self.ffn = nn.Sequential(
            nn.Linear(dim, 4 * dim),
            nn.GELU(),
            nn.Linear(4 * dim, dim)
        )

    def forward(self, x):
        H, W = x.size(2), x.size(3)
        if self.shift_size > 0:
            shifted_x = torch.roll(x, shifts=(-self.shift_size, -self.shift_size),
            dims=(2, 3))
        else:
            shifted_x = x

        x_windows = self.window_partition(shifted_x)
        x_windows = x_windows.view(-1, self.window_size * self.window_size, self.dim)

        attn_windows = self.attn(x_windows, x_windows, x_windows)[0]
        attn_windows = attn_windows.view(-1, self.window_size, self.window_size, self.dim)

        shifted_x = self.window_reverse(attn_windows, H, W)

        if self.shift_size > 0:
            x = torch.roll(shifted_x, shifts=(self.shift_size, self.shift_size), dims=(2, 3))
        else:
            x = shifted_x

        x = x + self.norm1(x)
        x = x + self.norm2(self.ffn(x))
        return x

    def window_partition(self, x):
```

```
        B, C, H, W = x.shape
        x = x.view(B, C, H // self.window_size, self.window_size, W // self.window_size,
        self.window_size)
        windows = x.permute(0, 2, 4, 1, 3, 5).contiguous().view(-1, C, self.window_size,
        self.window_size)
        return windows

    def window_reverse(self, windows, H, W):
        B = int(windows.shape[0] / (H * W / (self.window_size * self.window_size)))
        x = windows.view(B, H // self.window_size, W // self.window_size, self.dim,
        self.window_size, self.window_size)
        x = x.permute(0, 3, 1, 4, 2, 5).contiguous().view(B, self.dim, H, W)
        return x

# Example
model = SWINTransformerBlock(dim=96, num_heads=3, window_size=7, shift_size=3)
input_image = torch.randn(1, 96, 224, 224)
output = model(input_image)
print(output.shape)  # Output shape: (1, 96, 224, 224)
```

In this example, the ‘SWINTransformerBlock’ class implements a basic block of the SWIN Transformer, including window partitioning, shifted window attention, and feed-forward networks. The input image is processed through this block, demonstrating the model’s ability to handle image data with hierarchical window-based attention.

Performance and Applications

SWIN Transformer has shown state-of-the-art performance on various computer vision benchmarks, often outperforming traditional CNNs and other transformer-based models. Some key points about SWIN’s performance and applications include:

1. **Image Classification**: SWIN Transformer achieves high accuracy on image classification benchmarks like ImageNet, benefiting from its ability to capture both local and global contexts efficiently.
2. **Object Detection**: SWIN Transformer is highly effective for object detection tasks, such as those in the COCO dataset, due to its hierarchical feature representation and shifted window attention mechanism.
3. **Semantic Segmentation**: SWIN Transformer performs well in semantic segmentation tasks, providing fine-grained predictions while capturing the broader context of the image.
4. **Scalability**: SWIN Transformer’s hierarchical approach makes it scalable to larger images and higher resolutions, maintaining computational efficiency and performance.

Example: Object Detection with SWIN Transformer Consider using a pre-trained SWIN Transformer model for object detection.

```
from transformers import SwinForObjectDetection, SwinFeatureExtractor
from PIL import Image
import requests

# Load pre-trained SWIN model and feature extractor
model = SwinForObjectDetection.from_pretrained('microsoft/swin-base-patch4-window7-224')
feature_extractor = SwinFeatureExtractor.from_pretrained
('microsoft/swin-base-patch4-window7-224')
```

```
# Load and preprocess image
url = 'http://images.cocodataset.org/val2017/000000039769.jpg'
image = Image.open(requests.get(url, stream=True).raw)
inputs = feature_extractor(images=image, return_tensors='pt')

# Perform object detection
outputs = model(**inputs)
logits = outputs.logits

# Get predicted boxes
boxes = outputs.pred_boxes
print(f'Predicted boxes: {boxes}')
```

In this example, the 'SwinForObjectDetection' class is used to load a pre-trained SWIN Transformer model for object detection. The model processes an input image and outputs the predicted bounding boxes for objects within the image.

6.6 Advanced Topics in Transformers

6.6.1 Transfer Learning with Transformers

Transfer learning is a powerful technique in machine learning where a model pre-trained on a large dataset is fine-tuned on a smaller, task-specific dataset. Transformers, with their extensive pre-training on large corpora and flexible architecture, are particularly well-suited for transfer learning. This section explores the principles of transfer learning with transformers, focusing on pre-training and fine-tuning, as well as domain adaptation.

Pre-training and Fine-tuning

Pre-training involves training a transformer model on a large, diverse corpus to learn general language representations. This process typically employs self-supervised learning objectives that do not require labeled data. Common pre-training objectives include masked language modeling (MLM) and next sentence prediction (NSP) for models like BERT, and autoregressive language modeling for models like GPT.

1. **Masked Language Modeling (MLM)**: In MLM, a percentage of the input tokens are masked, and the model is trained to predict the original tokens. This objective helps the model learn bidirectional context representations.

$$\mathcal{L}_{\text{MLM}} = -\sum_{i \in \mathcal{M}} \log P(x_i \mid x_{\setminus i})$$

Here, $\mathcal{M}$ denotes the set of masked positions, and $x_{\setminus i}$ represents the input sequence with the i-th token masked.

2. **Next Sentence Prediction (NSP)**: In NSP, the model is trained to predict whether a given sentence follows another sentence. This objective helps the model understand sentence relationships and coherence.

$$\mathcal{L}_{\text{NSP}} = -\log P(\text{isNext} \mid A, B) - \log P(\text{isNotNext} \mid A, B)$$

   Here, A and B are sentence pairs, and the model predicts whether B follows A.
3. **Autoregressive Language Modeling**: In this approach, the model generates text one token at a time, predicting the next token based on the previous tokens.

$$\mathcal{L}_{\text{AR}} = -\sum_{t=1}^{T} \log P(x_t \mid x_{1:t-1})$$

Fine-tuning involves adapting the pre-trained model to a specific downstream task by training it on a smaller, task-specific dataset. This process typically requires adding task-specific layers to the pre-trained model and fine-tuning all or part of the model parameters.

1. **Task-specific Layers**: Depending on the task, additional layers such as classification heads, sequence labeling heads, or regression heads are added to the model.

$$\begin{aligned} h &= \text{Transformer}(x) \\ y &= \text{TaskHead}(h) \end{aligned}$$

2. **Fine-tuning Process**: The model is trained end-to-end on the task-specific dataset, usually with a lower learning rate to preserve the pre-trained knowledge while adapting to the new task.

$$\mathcal{L}_{\text{fine-tune}} = \sum_{(x,y)\in\mathcal{D}} \mathcal{L}_{\text{task}}(y, \text{TaskHead}(\text{Transformer}(x)))$$

Implementation Consider fine-tuning BERT for a text classification task using the IMDb movie reviews dataset.

```
from transformers import BertTokenizer, BertForSequenceClassification, Trainer,
TrainingArguments
from datasets import load_dataset

# Load pre-trained BERT model and tokenizer
model = BertForSequenceClassification.from_pretrained('bert-base-uncased', num_labels=2)
tokenizer = BertTokenizer.from_pretrained('bert-base-uncased')

# Load and preprocess dataset
dataset = load_dataset('imdb')
def preprocess(example):
    return tokenizer(example['text'], truncation=True, padding='max_length', max_length=128)
dataset = dataset.map(preprocess, batched=True)
dataset = dataset.rename_column('label', 'labels')
dataset.set_format(type='torch', columns=['input_ids', 'attention_mask', 'labels'])
```

```
# Fine-tuning setup
training_args = TrainingArguments(
    output_dir='./results',
    num_train_epochs=3,
    per_device_train_batch_size=16,
    evaluation_strategy="epoch",
    save_steps=10_000,
    save_total_limit=2,
)
trainer = Trainer(
    model=model,
    args=training_args,
    train_dataset=dataset['train'],
    eval_dataset=dataset['test']
)

# Train and evaluate
trainer.train()
trainer.evaluate()
```

In this example, the 'BertForSequenceClassification' class is used to load a pre-trained BERT model with a classification head. The model is fine-tuned on the IMDb dataset, demonstrating the transfer learning process from pre-training to fine-tuning.

Domain Adaptation

Domain adaptation involves fine-tuning a pre-trained model on a new domain or dataset that differs from the pre-training data. This process helps the model adapt to specific language use, vocabulary, and styles present in the target domain. Domain adaptation can be particularly beneficial when the target domain has limited labeled data.

1. **Adapting to a New Domain**: The model is fine-tuned on a small amount of labeled data from the target domain. This process helps the model adjust its representations to better fit the new domain's characteristics.
2. **Unsupervised Domain Adaptation**: In cases where labeled data is scarce, unsupervised domain adaptation techniques can be employed. These techniques involve aligning the distributions of the source and target domains using unlabeled data from the target domain.

$$\mathcal{L}_{\text{domain}} = \mathcal{L}_{\text{task}} + \lambda \cdot \mathcal{L}_{\text{unsupervised}}$$

Here, λ is a hyperparameter that balances the supervised task loss and the unsupervised domain adaptation loss.

Example: Fine-tuning BERT for Legal Texts: Consider fine-tuning BERT for a legal text classification task. The model is first pre-trained on a large corpus of general text, then fine-tuned on a smaller dataset of legal documents.

```
from transformers import BertTokenizer, BertForSequenceClassification,
Trainer, TrainingArguments
from datasets import load_dataset

# Load pre-trained BERT model and tokenizer
model = BertForSequenceClassification.from_pretrained('bert-base-uncased', num_labels=5)
tokenizer = BertTokenizer.from_pretrained('bert-base-uncased')
```

```
# Load and preprocess legal text dataset
dataset = load_dataset('lex_glue', 'ecthr_a')
def preprocess(example):
    return tokenizer(example['text'], truncation=True, padding='max_length', max_length=128)
dataset = dataset.map(preprocess, batched=True)
dataset = dataset.rename_column('label', 'labels')
dataset.set_format(type='torch', columns=['input_ids', 'attention_mask', 'labels'])

# Fine-tuning setup
training_args = TrainingArguments(
    output_dir='./results',
    num_train_epochs=3,
    per_device_train_batch_size=16,
    evaluation_strategy="epoch",
    save_steps=10_000,
    save_total_limit=2,
)
trainer = Trainer(
    model=model,
    args=training_args,
    train_dataset=dataset['train'],
    eval_dataset=dataset['test']
)

# Train and evaluate
trainer.train()
trainer.evaluate()
```

In this example, the 'BertForSequenceClassification' class is used to load a pre-trained BERT model. The model is then fine-tuned on a dataset of legal texts, demonstrating domain adaptation for a specific domain.

6.6.2 *Efficient Transformers*

As the scale and complexity of transformer models increase, so do the computational requirements. Efficient transformers aim to reduce these computational demands while maintaining or improving model performance. This section explores various methods for achieving efficiency, focusing on reducing computational complexity and employing sparse attention mechanisms.

Reducing Computational Complexity
Transformers are computationally intensive, primarily due to the quadratic complexity of the self-attention mechanism with respect to the input sequence length. Reducing this complexity is crucial for scaling transformers to longer sequences and larger datasets.

Linearized Attention Linearized attention mechanisms aim to reduce the quadratic complexity of the standard self-attention mechanism to linear complexity. One approach involves approximating the softmax operation to make the attention mechanism more efficient.

1. **Kernelized Attention**: By expressing the softmax function as a kernel, linearized attention methods approximate the dot-product attention using kernel functions. The attention computation can be formulated as:

$$\text{Attention}(Q, K, V) = \text{softmax}\left(\frac{QK^T}{\sqrt{d_k}}\right)V \approx \phi(Q)\left(\phi(K)^T V\right)$$

   Here, $\phi(\cdot)$ is a kernel function that approximates the softmax operation, allowing for efficient computation.
2. **Efficient Attention with Linear Complexity**: Linear Transformers, such as the Linear Transformer model, achieve linear complexity by decomposing the attention mechanism and reordering computations to avoid the quadratic bottleneck. The key idea is to use kernel functions to approximate the attention weights.

$$A(Q, K, V) = \phi(Q)\left(\phi(K)^T V\right) \approx QK^T V$$

3. **Local Attention**: Each token attends only to its local neighborhood, defined by a fixed window size. This method is effective for tasks where local context is crucial, such as language modeling and image processing.

$$\text{Attention}(Q, K, V) = \text{softmax}\left(\frac{QK^T \odot M_{\text{local}}}{\sqrt{d_k}}\right)V$$

   Here, M_{local} is a binary mask that defines the local attention window.
4. **Strided Attention**: Tokens attend to other tokens at fixed intervals, creating a strided pattern. This method captures both local and distant dependencies efficiently.

$$\text{Attention}(Q, K, V) = \text{softmax}\left(\frac{QK^T \odot M_{\text{strided}}}{\sqrt{d_k}}\right)V$$

   M_{strided} is a binary mask that defines the strided attention pattern.

Sparse Attention Mechanisms Sparse attention mechanisms reduce the number of tokens that each token attends to, thereby decreasing the computational load. These methods introduce sparsity patterns in the attention matrix to focus only on the most relevant tokens.

1. **Fixed Sparse Patterns**: Fixed sparse patterns predefine the attention structure, ensuring that each token attends to a fixed subset of tokens. This approach significantly reduces the number of attention calculations.

$$\text{Attention}(Q, K, V) = \text{softmax}\left(\frac{QK^T \odot M}{\sqrt{d_k}}\right)V$$

   Here, M is a binary mask matrix that defines the sparse attention pattern, and $\odot$ denotes element-wise multiplication.

2. **Learnable Sparse Patterns**: Learnable sparse patterns allow the model to learn which tokens to attend to during training. This approach provides more flexibility and can adapt to different tasks and data distributions.

$$\text{Attention}(Q, K, V) = \text{softmax}\left(\frac{QK^T \odot \sigma(W)}{\sqrt{d_k}}\right) V$$

   In this formulation, $\sigma(W)$ is a learnable mask matrix, with σ being an activation function such as sigmoid.

3. **Dynamic Convolution**: Dynamic convolution dynamically adjusts the receptive field for each token based on the input sequence, providing a flexible and efficient attention mechanism.

$$\text{DynamicConv}(x) = \sum_k w_k \cdot \text{Conv}(x, k)$$

   Here, w_k is learnable weights for different convolution kernels k.

Implementation Example: Linear Transformer Consider an implementation of the Linear Transformer with kernelized attention in PyTorch.

```
import torch
import torch.nn as nn
import torch.nn.functional as F

class LinearAttention(nn.Module):
    def __init__(self, d_model):
        super(LinearAttention, self).__init__()
        self.d_model = d_model

    def forward(self, Q, K, V):
        Q = torch.nn.functional.elu(Q) + 1
        K = torch.nn.functional.elu(K) + 1
        K_transpose_V = torch.einsum('bnd,bne->bde', K, V)
        attention = torch.einsum('bnd,bde->bne', Q, K_transpose_V)
        return attention / (Q.size(-1) ** 0.5)

class LinearTransformerLayer(nn.Module):
    def __init__(self, d_model, num_heads):
        super(LinearTransformerLayer, self).__init__()
        self.num_heads = num_heads
        self.d_model = d_model
        self.attn = LinearAttention(d_model)
        self.norm1 = nn.LayerNorm(d_model)
        self.norm2 = nn.LayerNorm(d_model)
        self.ffn = nn.Sequential(
            nn.Linear(d_model, 4 * d_model),
            nn.GELU(),
            nn.Linear(4 * d_model, d_model)
        )

    def forward(self, x):
        attn_output = self.attn(x, x, x)
        x = x + attn_output
        x = self.norm1(x)
        ffn_output = self.ffn(x)
        x = x + ffn_output
        x = self.norm2(x)
```

```
        return x

# Example
model = LinearTransformerLayer(d_model=512, num_heads=8)
input_tensor = torch.randn(32, 10, 512)  # Batch size 32, sequence length 10,
embedding dimension 512
output_tensor = model(input_tensor)
print(output_tensor.shape)  # Output shape: (32, 10, 512)
```

In this example, the 'LinearAttention' class implements an efficient attention mechanism using kernelized attention. The 'LinearTransformerLayer' class uses this attention mechanism along with layer normalization and feed-forward networks to build a Linear Transformer layer.

Implementation Example: Sparse Transformer Consider an implementation of a Sparse Transformer with learnable sparse attention in PyTorch.

```
import torch
import torch.nn as nn
import torch.nn.functional as F

class SparseAttention(nn.Module):
    def __init__(self, d_model, sparsity):
        super(SparseAttention, self).__init__()
        self.d_model = d_model
        self.sparsity = sparsity
        self.query_proj = nn.Linear(d_model, d_model)
        self.key_proj = nn.Linear(d_model, d_model)
        self.value_proj = nn.Linear(d_model, d_model)
        self.sparse_weights = nn.Parameter(torch.rand(d_model, d_model))

    def forward(self, Q, K, V):
        Q = self.query_proj(Q)
        K = self.key_proj(K)
        V = self.value_proj(V)

        sparse_mask = torch.sigmoid(self.sparse_weights)
        sparse_mask = sparse_mask > self.sparsity
        attention_scores = torch.matmul(Q, K.transpose(-1, -2)) * sparse_mask
        attention_weights = F.softmax(attention_scores, dim=-1)
        output = torch.matmul(attention_weights, V)
        return output

class SparseTransformerLayer(nn.Module):
    def __init__(self, d_model, num_heads, sparsity):
        super(SparseTransformerLayer, self).__init__()
        self.num_heads = num_heads
        self.d_model = d_model
        self.attn = SparseAttention(d_model, sparsity)
        self.norm1 = nn.LayerNorm(d_model)
        self.norm2 = nn.LayerNorm(d_model)
        self.ffn = nn.Sequential(
            nn.Linear(d_model, 4 * d_model),
            nn.GELU(),
            nn.Linear(4 * d_model, d_model)
        )

    def forward(self, x):
        attn_output = self.attn(x, x, x)
        x = x + attn_output
        x = self.norm1(x)
        ffn_output = self.ffn(x)
        x = x + ffn_output
        x = self.norm2(x)
```

```
        return x

# Example
model = SparseTransformerLayer(d_model=512, num_heads=8, sparsity=0.1)
input_tensor = torch.randn(32, 10, 512)  # Batch size 32, sequence length 10,
embedding dimension 512
output_tensor = model(input_tensor)
print(output_tensor.shape)  # Output shape: (32, 10, 512)
```

In this example, the 'SparseAttention' class implements a sparse attention mechanism with learnable sparsity. The 'SparseTransformerLayer' class uses this attention mechanism along with layer normalization and feed-forward networks to build a sparse transformer layer.

6.6.3 Interpretability of Transformers

Interpretability in transformer models is crucial for understanding how these complex architectures make decisions. Visualizing attention maps and understanding model decisions are two key approaches that help demystify the workings of transformers.

Visualizing Attention Maps

Attention maps provide a way to visualize the internal workings of transformer models by showing which parts of the input sequence the model focuses on when making predictions. This visualization helps interpret how the model distributes its attention across different tokens and can highlight important patterns and dependencies.

In transformer models, the attention mechanism computes a weight matrix that represents the importance of each token in the input sequence relative to every other token. To visualize the attention maps, we can extract the attention weights from a pre-trained transformer model and plot them. Each element A_{ij} in the attention matrix represents the attention weight between the i-th query and the j-th key.

Example: Visualizing Attention Maps in BERT Consider an example where we visualize the attention maps of a BERT model using the Hugging Face Transformers library and Matplotlib.

```
import torch
import matplotlib.pyplot as plt
from transformers import BertTokenizer, BertModel

# Load pre-trained BERT model and tokenizer
model = BertModel.from_pretrained('bert-base-uncased', output_attentions=True)
tokenizer = BertTokenizer.from_pretrained('bert-base-uncased')

# Example sentence
sentence = "The quick brown fox jumps over the lazy dog."

# Tokenize input and get attention weights
inputs = tokenizer(sentence, return_tensors='pt')
outputs = model(**inputs)
attention = outputs.attentions  # List of attention matrices for each layer

# Select attention map from the first layer and the first head
attention_map = attention[0][0, 0].detach().numpy()
```

```
# Plot attention map
plt.matshow(attention_map, cmap='viridis')
plt.colorbar()
plt.title('Attention Map')
plt.xlabel('Key Positions')
plt.ylabel('Query Positions')
plt.show()
```

In this example, the 'BertModel' is loaded with 'output_attentions = True' to obtain the attention weights. The attention map for the first layer and the first head is extracted and visualized using Matplotlib. By examining the attention maps, we can gain insights into which parts of the input sequence the model attends to when making predictions. For instance, in a machine translation task, we might observe that attention heads focus on corresponding words in the source and target sentences, indicating that the model has learned alignment between languages.

Understanding Model Decisions

Understanding model decisions involves interpreting how the output is derived from the input and identifying the contribution of different components and tokens to the final prediction. This understanding is crucial for debugging models, improving their performance, and ensuring their fairness and reliability. One way to understand model decisions is by analyzing the attention weights, which reveal how much each token contributes to the model's prediction. Higher attention weights indicate greater importance.

Saliency Maps: Saliency maps highlight the most important parts of the input that influence the model's decision. These maps are created by computing the gradient of the output with respect to the input tokens, indicating how changes in the input affect the prediction.

1. **Gradient-Based Saliency**: The gradient of the model's output y with respect to the input x can be computed as:

$$\text{Saliency}(x) = \left| \frac{\partial y}{\partial x} \right|$$

   This gradient highlights the tokens that have the most significant impact on the output.
2. **Integrated Gradients**: Integrated gradients are an advanced method that considers the accumulated gradients along the path from a baseline input to the actual input. It provides a more comprehensive view of the input's contribution to the output.

$$\text{IntegratedGrad}(x) = (x - x') \times \int_0^1 \frac{\partial f(x' + \alpha(x - x'))}{\partial x} \mathrm{d}\alpha$$

Here, x' is the baseline input, α is a scaling factor, and f is the model's prediction function.

Example: Computing Saliency Maps in BERT Consider an example where we compute and visualize saliency maps for a text classification task using a pre-trained BERT model.

```
import torch
from transformers import BertTokenizer, BertForSequenceClassification
from torch.autograd import grad

# Load pre-trained BERT model and tokenizer
model = BertForSequenceClassification.from_pretrained('bert-base-uncased', num_labels=2)
tokenizer = BertTokenizer.from_pretrained('bert-base-uncased')

# Example sentence
sentence = "The quick brown fox jumps over the lazy dog."
inputs = tokenizer(sentence, return_tensors='pt', truncation=True, padding=True)

# Forward pass and compute gradients
model.eval()
inputs.requires_grad_(True)
outputs = model(**inputs)
loss = outputs.logits[:, 1].sum()
model.zero_grad()
loss.backward()

# Compute saliency map
saliency = inputs.grad.abs().detach().numpy().sum(axis=2).squeeze()
tokens = tokenizer.convert_ids_to_tokens(inputs['input_ids'].squeeze().tolist())

# Plot saliency map
plt.bar(tokens, saliency)
plt.title('Saliency Map')
plt.xlabel('Tokens')
plt.ylabel('Saliency')
plt.show()
```

In this example, the 'BertForSequenceClassification' model is used to classify the input sentence. The gradient of the output with respect to the input tokens is computed to generate the saliency map, which is then visualized using a bar plot.

6.7 Practical Applications of Transformers

6.7.1 *Text Generation*

Text generation is a critical application of transformer models, leveraging their ability to capture contextual dependencies and generate coherent and contextually relevant text. This section delves into building and training generative models, as well as their applications in creative writing and content generation.

Building and Training Generative Models
Generative models aim to generate new data samples from the learned distribution of a given dataset. In the context of text generation, transformer-based generative

models such as GPT have shown remarkable performance in producing human-like text. Transformer-based generative models utilize autoregressive language modeling, where the probability of generating a sequence of tokens $x = (x_1, x_2, \ldots, x_T)$ is decomposed into a product of conditional probabilities:

$$P(x) = \prod_{t=1}^{T} P(x_t \mid x_{1:t-1})$$

Here, $P(x_t \mid x_{1:t-1})$ represents the probability of generating the token x_t given the preceding tokens $x_{1:t-1}$.

Generative models like GPT are based on the transformer decoder architecture. The key components include: Self-attention allows each token to attend to all previous tokens in the sequence, capturing dependencies and contextual information. To ensure that each token can only attend to previous tokens (and not future ones), a masking mechanism is applied to the attention scores.

$$A_{ij} = \begin{cases} \frac{Q_i K_j^T}{\sqrt{d_k}} & \text{if } j \leq i \\ -\infty & \text{if } j > i \end{cases}$$

The training process involves minimizing the negative log-likelihood of the predicted token probabilities given the ground truth sequence:

$$\mathcal{L} = -\sum_{t=1}^{T} \log P(x_t \mid x_{1:t-1})$$

Implementation Example: Training GPT for Text Generation Consider an example where we train a GPT model for text generation using the Hugging Face Transformers library.

```
from transformers import GPT2Tokenizer, GPT2LMHeadModel, Trainer, TrainingArguments
from datasets import load_dataset

# Load pre-trained GPT model and tokenizer
model = GPT2LMHeadModel.from_pretrained('gpt2')
tokenizer = GPT2Tokenizer.from_pretrained('gpt2')

# Load and preprocess dataset
dataset = load_dataset('wikitext', 'wikitext-2-raw-v1')
def preprocess(example):
    return tokenizer(example['text'], truncation=True, padding=True, max_length=512)
dataset = dataset.map(preprocess, batched=True)
dataset.set_format(type='torch', columns=['input_ids'])

# Training setup
training_args = TrainingArguments(
    output_dir='./results',
    num_train_epochs=3,
    per_device_train_batch_size=2,
    save_steps=10_000,
    save_total_limit=2,
)
```

```
trainer = Trainer(
    model=model,
    args=training_args,
    train_dataset=dataset['train'],
    eval_dataset=dataset['validation']
)

# Train model
trainer.train()
```

In this example, the 'GPT2LMHeadModel' class is used to load a pre-trained GPT-2 model. The model is fine-tuned on the Wikitext-2 dataset, demonstrating the training process for text generation.

Applications in Creative Writing and Content Generation
Generative models, particularly transformer-based models, have found significant applications in creative writing and content generation. These models can generate coherent and contextually relevant text, aiding writers, marketers, and content creators in various domains.

1. **Creative Writing**: Generative models can assist authors by providing writing prompts, continuing stories, or generating entire passages of text based on a given theme or style. This capability enhances creativity and helps overcome writer's block.
   Example:

   ```
   prompt = "Once upon a time in a distant land,"
   inputs = tokenizer.encode(prompt, return_tensors='pt')
   outputs = model.generate(inputs, max_length=100, num_return_sequences=1)
   generated_text = tokenizer.decode(outputs[0], skip_special_tokens=True)
   print(generated_text)
   ```

2. **Content Marketing**: Businesses can leverage generative models to create marketing content, including blog posts, social media updates, and product descriptions. This automation saves time and ensures a consistent brand voice.
   Example:

   ```
   prompt = "Introducing our latest product, a state-of-the-art smartphone with
   cutting-edge features."
   inputs = tokenizer.encode(prompt, return_tensors='pt')
   outputs = model.generate(inputs, max_length=100, num_return_sequences=1)
   generated_text = tokenizer.decode(outputs[0], skip_special_tokens=True)
   print(generated_text)
   ```

3. **Personal Assistants**: Generative models can be integrated into virtual personal assistants to generate personalized messages, emails, or summaries. This application improves productivity and user experience.

Example:

```
prompt = "Please summarize the following meeting notes:"
notes = "The meeting discussed the quarterly financial report. Sales increased by 15%,
and the marketing team proposed a new advertising strategy."
inputs = tokenizer.encode(prompt + notes, return_tensors='pt')
outputs = model.generate(inputs, max_length=100, num_return_sequences=1)
summary = tokenizer.decode(outputs[0], skip_special_tokens=True)
print(summary)
```

While generative models offer numerous benefits, there are challenges and ethical considerations to address:

1. **Bias and Fairness**: Generative models can inadvertently generate biased or harmful content if trained on biased datasets. Ensuring fairness and mitigating bias are critical for ethical AI deployment.
2. **Content Quality**: The quality of generated content can vary, and manual review may be necessary to ensure accuracy and relevance.
3. **Intellectual Property**: Using generative models in creative writing raises questions about authorship and intellectual property. Clear guidelines and policies are needed to address these issues.

6.7.2 *Text Summarization*

Text summarization is a crucial task in NLP that involves condensing a large body of text into a shorter version while retaining the key information and meaning. Transformer models have significantly advanced the state of text summarization, providing powerful tools for both extractive and abstractive summarizations. This section explores these two approaches and details the process of building summarization models.

Extractive Summarization
Extractive summarization involves selecting important sentences, phrases, or sections from the original text and combining them to create a summary. This method relies on identifying the most relevant parts of the text based on certain criteria, such as sentence importance scores. Extractive summarization can be framed as an optimization problem, where the goal is to select a subset of sentences that maximizes the relevance and coverage of the original document. Let S be the set of sentences in the document, and $\mathcal{S} \subseteq S$ be the subset of selected sentences. The objective function $f(\mathcal{S})$ can be defined to maximize relevance:

$$\mathcal{S}^* = \arg\max_{\mathcal{S} \subseteq S} f(\mathcal{S})$$

where $f(\mathcal{S})$ is a scoring function that measures the importance of the selected sentences.

Implementation Example Extractive summarization can be implemented using pre-trained transformer models like BERT. The BERTSUM model fine-tunes BERT for sentence classification, where each sentence is classified as being part of the summary or not.

```
from transformers import BertTokenizer, BertForSequenceClassification
import torch

# Load pre-trained BERT model and tokenizer
tokenizer = BertTokenizer.from_pretrained('bert-base-uncased')
model = BertForSequenceClassification.from_pretrained('bert-base-uncased', num_labels=2)

# Example document
document = [
    "The quick brown fox jumps over the lazy dog.",
    "The dog lies down lazily in the sun.",
    "Foxes are known for their agility and speed."
]

# Tokenize and classify each sentence
inputs = [tokenizer(sentence, return_tensors='pt') for sentence in document]
outputs = [model(**input) for input in inputs]
scores = [output.logits.argmax().item() for output in outputs]

# Select sentences with high scores as summary
summary = [sentence for sentence, score in zip(document, scores) if score == 1]
print("Summary:", summary)
```

Abstractive Summarization

Abstractive summarization involves generating new sentences that capture the essence of the original text. This approach requires the model to understand and paraphrase the content, making it more challenging but also more flexible than extractive summarization. It can be framed as a sequence-to-sequence learning problem, where the model generates a summary sequence $y = (y_1, y_2, \ldots, y_T)$ given an input sequence $x = (x_1, x_2, \ldots, x_S)$. The objective is to maximize the conditional probability $P(y \mid x)$:

$$P(y \mid x) = \prod_{t=1}^{T} P(y_t \mid y_{1:t-1}, x)$$

Here, $P(y_t \mid y_{1:t-1}, x)$ is the probability of generating the next token y_t given the previous tokens $y_{1:t-1}$ and the input sequence x.

Implementation Abstractive summarization can be implemented using pre-trained transformer models like BART or T5.

```
from transformers import BartTokenizer, BartForConditionalGeneration

# Load pre-trained BART model and tokenizer
model = BartForConditionalGeneration.from_pretrained('facebook/bart-large-cnn')
tokenizer = BartTokenizer.from_pretrained('facebook/bart-large-cnn')

# Example document
document = "The quick brown fox jumps over the lazy dog. The dog lies down lazily
in the sun.
Foxes are known for their agility and speed."
```

```
# Tokenize and summarize
inputs = tokenizer(document, return_tensors='pt', max_length=1024, truncation=True)
summary_ids = model.generate(inputs['input_ids'], num_beams=4, max_length=50,
early_stopping=True)
summary = tokenizer.decode(summary_ids[0], skip_special_tokens=True)
print("Summary:", summary)
```

Comparison and Use Cases

Extractive Summarization: Best suited for applications where the goal is to identify and highlight the most important parts of the text without altering the original wording. Common use cases include news article summarization and document highlighting.

Abstractive Summarization: More flexible and capable of generating more coherent and fluent summaries by rephrasing and condensing the information. Suitable for creating summaries that are easier to read and understand, such as executive summaries and abstracts.

Challenges and Considerations

1. Faithfulness and Coherence: Ensuring that the generated summaries are faithful to the original content and coherent is a major challenge. Abstractive models, in particular, need to avoid introducing factual inaccuracies.
2. Evaluation Metrics: Evaluating the quality of summaries is challenging. Common metrics include ROUGE (Recall-Oriented Understudy for Gisting Evaluation) for comparing the overlap of n-grams between the generated summary and reference summaries.

$$\text{ROUGE-N} = \frac{\sum_{S \in \text{ReferenceSummaries}} \sum_{gram_n \in S} \text{Count}_{\text{match}}(gram_n)}{\sum_{S \in \text{ReferenceSummaries}} \sum_{gram_n \in S} \text{Count}(gram_n)}$$

3. Scalability: Handling long documents efficiently is a significant challenge, especially for transformer models with quadratic complexity in self-attention.

6.7.3 *Question Answering*

Question-answering (QA) systems are designed to automatically answer questions posed by humans in natural language. Transformer-based models have revolutionized QA by enabling the understanding and processing of vast amounts of text to find accurate answers. This section discusses how to build QA systems with transformers and explores evaluation metrics and techniques.

Building QA Systems with Transformers

QA systems can be classified into different types, including extractive, abstractive, and multiple-choice. Extractive QA involves selecting a span of text from a document that directly answers the question. Abstractive QA generates a new answer, paraphrasing or summarizing information from the document. Multiple-choice QA involves selecting the correct answer from a list of options. Transformers, such as BERT, are particularly effective for extractive QA. The task can be formulated as identifying the start and end positions of the answer span within a given context. Let C be the context and Q be the question. The model outputs probabilities for each token being the start and end of the answer span:

$$P(\text{start}_i \mid C, Q) = \text{softmax}(W_s \cdot h_i)$$
$$P(\text{end}_j \mid C, Q) = \text{softmax}(W_e \cdot h_j)$$

Here, h_i and h_j are the hidden states corresponding to the i-th and j-th tokens, and W_s and W_e are learned weights.

The training objective for extractive QA is to minimize the negative log-likelihood of the correct start and end positions:

$$\mathcal{L} = -\left(\log P(\text{start}_{\text{true}} \mid C, Q) + \log P(\text{end}_{\text{true}} \mid C, Q)\right)$$

Implementation Example: Building a QA System with BERT Consider an example where we build an extractive QA system using a pre-trained BERT model.

```
from transformers import BertTokenizer, BertForQuestionAnswering
import torch

# Load pre-trained BERT model and tokenizer
tokenizer = BertTokenizer.from_pretrained
('bert-large-uncased-whole-word-masking-finetuned-squad')
model = BertForQuestionAnswering.from_pretrained
('bert-large-uncased-whole-word-masking-finetuned-squad')

# Example context and question
context = "The quick brown fox jumps over the lazy dog."
question = "What does the fox jump over?"

# Tokenize input
inputs = tokenizer(question, context, return_tensors='pt')
input_ids = inputs['input_ids'].tolist()[0]

# Get model outputs
outputs = model(**inputs)
start_scores = outputs.start_logits
end_scores = outputs.end_logits

# Get the most likely beginning and end of the answer span
start_idx = torch.argmax(start_scores)
end_idx = torch.argmax(end_scores)

# Convert tokens to answer
answer = tokenizer.convert_tokens_to_string(tokenizer.convert_ids_to_tokens
(input_ids[start_idx:end_idx+1]))
print("Answer:", answer)
```

In this example, the 'BertForQuestionAnswering' class is used to load a pre-trained BERT model fine-tuned on the SQuAD dataset. The model predicts the start and end positions of the answer span, which are then converted back to text using the tokenizer.

For multiple-choice QA, models like RoBERTa or ALBERT can be used. The task involves scoring each choice and selecting the one with the highest score. Let C be the context, Q be the question, and A_i be the i-th answer choice. The model computes a score for each choice:

$$\text{score}(A_i \mid C, Q) = W \cdot \bigoplus(h_C, h_Q, h_{A_i})$$

The training objective is to maximize the likelihood of the correct answer:

$$\mathcal{L} = -\log P(A_{\text{true}} \mid C, Q)$$

Example: Multiple-Choice QA with RoBERTa

```
from transformers import RobertaTokenizer, RobertaForMultipleChoice
import torch

# Load pre-trained RoBERTa model and tokenizer
tokenizer = RobertaTokenizer.from_pretrained('roberta-large')
model = RobertaForMultipleChoice.from_pretrained('roberta-large')

# Example context, question, and choices
context = "The quick brown fox jumps over the lazy dog."
question = "What does the fox jump over?"
choices = ["a fence", "the lazy dog", "a river", "a log"]

# Tokenize input
encoding = tokenizer([context] * len(choices), choices, truncation=True, padding=True,
return_tensors='pt')
input_ids = encoding['input_ids']
attention_mask = encoding['attention_mask']

# Get model outputs
outputs = model(input_ids=input_ids, attention_mask=attention_mask)
logits = outputs.logits

# Get the most likely answer
answer_idx = torch.argmax(logits)
answer = choices[answer_idx]
print("Answer:", answer)
```

In this example, the 'RobertaForMultipleChoice' class is used to load a pre-trained RoBERTa model. The model scores each choice, and the choice with the highest score is selected as the answer.

Evaluation Metrics and Techniques

Evaluating QA systems requires metrics that accurately reflect the model's performance in understanding and answering questions. Common metrics include Exact Match (EM), F1-score, and BLEU.

Exact Match (EM): EM measures the percentage of predictions that match the ground truth exactly. It is a strict metric suitable for tasks where precision is critical.

$$\text{EM} = \frac{1}{N} \sum_{i=1}^{N} \mathbb{I}(\hat{y}_i = y_i)$$

Here, N is the total number of questions, $\hat{y}_i$ is the predicted answer, y_i is the ground truth answer, and $\mathbb{I}$ is the indicator function.

Bilingual Evaluation Understudy (BLEU):
BLEU measures the similarity between the generated text and the reference text, based on n-gram overlaps. It is commonly used for evaluating machine translation and can be adapted for QA.

$$\text{BLEU} = \text{BP} \times \exp\left(\sum_{n=1}^{N} w_n \log p_n\right)$$

Here, BP is the brevity penalty, w_n is the weights for different n-grams, and p_n is the n-gram precision scores.

Example: Evaluating a QA Model with Exact Match and F1-Score

```
from transformers import BertTokenizer, BertForQuestionAnswering
from datasets import load_dataset, load_metric

# Load pre-trained BERT model and tokenizer
tokenizer = BertTokenizer.from_pretrained
('bert-large-uncased-whole-word-masking-finetuned-squad')
model = BertForQuestionAnswering.from_pretrained
('bert-large-uncased-whole-word-masking-finetuned-squad')

# Load dataset and evaluation metric
dataset = load_dataset('squad', split='validation')
metric = load_metric('squad')

# Evaluate model
for example in dataset:
    inputs = tokenizer(example['question'], example['context'], return_tensors='pt')
    outputs = model(**inputs)
    start_idx = torch.argmax(outputs.start_logits)
    end_idx = torch.argmax(outputs.end_logits)
    predicted_answer = tokenizer.decode(inputs['input_ids'][0][start_idx:end_idx+1])

    metric.add(prediction=predicted_answer, reference=example['answers']['text'][0])

# Compute metric
results = metric.compute()
print(f"Exact Match: {results['exact_match']}")
print(f"F1 Score: {results['f1']}")
```

In this example, the ‘squad’ dataset and ‘squad’ evaluation metric from the Hugging Face datasets library are used to evaluate the BERT model. The Exact Match and F1-scores are computed to assess the model’s performance.

6.7.4 Other Applications

Transformer models have proven to be highly versatile, excelling in a wide range of natural language processing tasks beyond question answering. This section explores three other prominent applications of transformers: language translation, sentiment analysis, and named entity recognition (NER).

Language Translation

Language translation involves converting text from one language to another while preserving the original meaning and context. Transformer models, particularly the transformer architecture introduced by Vaswani et al., have set new benchmarks in machine translation due to their ability to capture long-range dependencies and contextual information effectively. Language translation can be modeled as a sequence-to-sequence learning problem, where the model learns to map a source sequence $x = (x_1, x_2, \ldots, x_S)$ in the source language to a target sequence $y = (y_1, y_2, \ldots, y_T)$ in the target language. The objective is to maximize the conditional probability $P(y \mid x)$:

$$P(y \mid x) = \prod_{t=1}^{T} P(y_t \mid y_{1:t-1}, x)$$

Here, $P(y_t \mid y_{1:t-1}, x)$ is the probability of generating the token y_t given the previous tokens $y_{1:t-1}$ and the source sequence x.

The transformer model consists of an encoder–decoder architecture:

1. **Encoder**: The encoder processes the source sequence and generates a sequence of hidden states $h = (h_1, h_2, \ldots, h_S)$:

$$h = \text{Encoder}(x)$$

2. **Decoder**: The decoder generates the target sequence, attending to the encoder's hidden states at each step:

$$y_t = \text{Decoder}(y_{1:t-1}, h)$$

Implementation Example: Language Translation with MarianMT Consider an example where we use the MarianMT model for translating text from English to French.

```
from transformers import MarianMTModel, MarianTokenizer

# Load pre-trained MarianMT model and tokenizer
model_name = 'Helsinki-NLP/opus-mt-en-fr'
model = MarianMTModel.from_pretrained(model_name)
tokenizer = MarianTokenizer.from_pretrained(model_name)

# Example text
text = "The quick brown fox jumps over the lazy dog."

# Tokenize and translate
inputs = tokenizer(text, return_tensors='pt')
translated_ids = model.generate(inputs['input_ids'], max_length=50)
translation = tokenizer.decode(translated_ids[0], skip_special_tokens=True)
print("Translation:", translation)
```

In this example, the 'MarianMTModel' and 'MarianTokenizer' are used to translate an English sentence into French.

Sentiment Analysis

Sentiment analysis involves determining the sentiment expressed in a piece of text, typically classifying it as positive, negative, or neutral. This task is essential for applications such as customer feedback analysis, social media monitoring, and market research. It can be framed as a text classification problem, where the goal is to assign a sentiment label y to a given text x. The model learns to predict the probability distribution over the sentiment labels:

$$P(y \mid x) = \text{softmax}(W \cdot h + b)$$

Here, h is the hidden representation of the text, W is the weight matrix, and b is the bias term.

Implementation Example: Sentiment Analysis with BERT Consider an example where we use the BERT model for sentiment analysis on movie reviews:

```
from transformers import BertTokenizer, BertForSequenceClassification
import torch

# Load pre-trained BERT model and tokenizer
model = BertForSequenceClassification.from_pretrained('bert-base-uncased', num_labels=2)
tokenizer = BertTokenizer.from_pretrained('bert-base-uncased')

# Example text
text = "The movie was fantastic! I really enjoyed it."

# Tokenize input
inputs = tokenizer(text, return_tensors='pt')
outputs = model(**inputs)
predictions = torch.softmax(outputs.logits, dim=-1)
label = torch.argmax(predictions).item()

# Map label to sentiment
sentiment = "positive" if label == 1 else "negative"
print("Sentiment:", sentiment)
```

In this example, the 'BertForSequenceClassification' model is used to classify the sentiment of a movie review.

Named Entity Recognition (NER)

NER involves identifying and classifying named entities (such as people, organizations, locations, dates, etc.) in a text. NER is crucial for tasks like information extraction, knowledge graph construction, and improving search relevance. NER can be formulated as a sequence labeling problem, where each token in the text is assigned a label representing the entity type. Let $x = (x_1, x_2, \ldots, x_T)$ be the input sequence and $y = (y_1, y_2, \ldots, y_T)$ be the corresponding labels. The model learns to predict the label sequence:

$$P(y \mid x) = \prod_{t=1}^{T} P(y_t \mid x)$$

Here, $P(y_t \mid x)$ is the probability of the t-th token being assigned the label y_t.

Implementation Example: NER with BERT Consider an example where we use the BERT model for NER:

```
from transformers import BertTokenizer, BertForTokenClassification
import torch

# Load pre-trained BERT model and tokenizer
model = BertForTokenClassification.from_pretrained
('dbmdz/bert-large-cased-finetuned-conll03-english')
tokenizer = BertTokenizer.from_pretrained('bert-base-cased')

# Example text
text = "John Doe works at Acme Corporation in New York."

# Tokenize input
inputs = tokenizer(text, return_tensors='pt')
outputs = model(**inputs)
predictions = torch.argmax(outputs.logits, dim=2)

# Map predictions to labels
labels = {0: 'O', 1: 'B-MISC', 2: 'I-MISC', 3: 'B-PER', 4: 'I-PER', 5: 'B-ORG',
6: 'I-ORG', 7: 'B-LOC', 8: 'I-LOC'}
tokens = tokenizer.convert_ids_to_tokens(inputs['input_ids'][0])
named_entities = [(token, labels[pred.item()]) for token, pred in
zip(tokens, predictions[0]) if labels[pred.item()] != 'O']
print("Named Entities:", named_entities)
```

In this example, the 'BertForTokenClassification' model is used to identify named entities in a sentence.

6.8 Implementing Transformers with TensorFlow and PyTorch

6.8.1 Building Transformers with TensorFlow

1. **Encoder Layer**:

```
import tensorflow as tf
from tensorflow.keras.layers import Dense, LayerNormalization, Dropout
from tensorflow.keras import Model

class MultiHeadAttention(tf.keras.layers.Layer):
    def _ _init_ _(self, d_model, num_heads):
        super(MultiHeadAttention, self)._ _init_ _()
        self.num_heads = num_heads
        self.d_model = d_model

        assert d_model % self.num_heads == 0

        self.depth = d_model // self.num_heads

        self.wq = Dense(d_model)
        self.wk = Dense(d_model)
        self.wv = Dense(d_model)
        self.dense = Dense(d_model)

    def split_heads(self, x, batch_size):
        x = tf.reshape(x, (batch_size, -1, self.num_heads, self.depth))
```

```
        return tf.transpose(x, perm=[0, 2, 1, 3])

    def call(self, v, k, q, mask):
        batch_size = tf.shape(q)[0]

        q = self.wq(q)
        k = self.wk(k)
        v = self.wv(v)

        q = self.split_heads(q, batch_size)
        k = self.split_heads(k, batch_size)
        v = self.split_heads(v, batch_size)

        scaled_attention, _ = self.scaled_dot_product_attention(q, k, v, mask)
        scaled_attention = tf.transpose(scaled_attention, perm=[0, 2, 1, 3])
        concat_attention = tf.reshape(scaled_attention, (batch_size, -1, self.d_model))

        output = self.dense(concat_attention)
        return output

    def scaled_dot_product_attention(self, q, k, v, mask):
        matmul_qk = tf.matmul(q, k, transpose_b=True)
        dk = tf.cast(tf.shape(k)[-1], tf.float32)
        scaled_attention_logits = matmul_qk / tf.math.sqrt(dk)

        if mask is not None:
            scaled_attention_logits += (mask * -1e9)

        attention_weights = tf.nn.softmax(scaled_attention_logits, axis=-1)
        output = tf.matmul(attention_weights, v)
        return output, attention_weights

class EncoderLayer(tf.keras.layers.Layer):
    def __init__(self, d_model, num_heads, dff, rate=0.1):
        super(EncoderLayer, self).__init__()

        self.mha = MultiHeadAttention(d_model, num_heads)
        self.ffn = tf.keras.Sequential([
            Dense(dff, activation='relu'),
            Dense(d_model)
        ])

        self.layernorm1 = LayerNormalization(epsilon=1e-6)
        self.layernorm2 = LayerNormalization(epsilon=1e-6)

        self.dropout1 = Dropout(rate)
        self.dropout2 = Dropout(rate)

    def call(self, x, training, mask):
        attn_output = self.mha(x, x, x, mask)
        attn_output = self.dropout1(attn_output, training=training)
        out1 = self.layernorm1(x + attn_output)

        ffn_output = self.ffn(out1)
        ffn_output = self.dropout2(ffn_output, training=training)
        out2 = self.layernorm2(out1 + ffn_output)

        return out2
```

2. **Decoder Layer**:

```
class DecoderLayer(tf.keras.layers.Layer):
    def _ _init_ _(self, d_model, num_heads, dff, rate=0.1):
        super(DecoderLayer, self)._ _init_ _()

        self.mha1 = MultiHeadAttention(d_model, num_heads)
```

```
        self.mha2 = MultiHeadAttention(d_model, num_heads)

        self.ffn = tf.keras.Sequential([
            Dense(dff, activation='relu'),
            Dense(d_model)
        ])

        self.layernorm1 = LayerNormalization(epsilon=1e-6)
        self.layernorm2 = LayerNormalization(epsilon=1e-6)
        self.layernorm3 = LayerNormalization(epsilon=1e-6)

        self.dropout1 = Dropout(rate)
        self.dropout2 = Dropout(rate)
        self.dropout3 = Dropout(rate)

    def call(self, x, enc_output, training, look_ahead_mask, padding_mask):
        attn1, attn_weights_block1 = self.mha1(x, x, x, look_ahead_mask)
        attn1 = self.dropout1(attn1, training=training)
        out1 = self.layernorm1(x + attn1)

        attn2, attn_weights_block2 = self.mha2(
            enc_output, enc_output, out1, padding_mask)
        attn2 = self.dropout2(attn2, training=training)
        out2 = self.layernorm2(out1 + attn2)

        ffn_output = self.ffn(out2)
        ffn_output = self.dropout3(ffn_output, training=training)
        out3 = self.layernorm3(out2 + ffn_output)

        return out3, attn_weights_block1, attn_weights_block2
```

3. **Transformer Model**:

```
class Transformer(tf.keras.Model):
    def _ _init_ _(self, num_layers, d_model, num_heads, dff, input_vocab_size,
                 target_vocab_size, pe_input, pe_target, rate=0.1):
        super(Transformer, self)._ _init_ _()

        self.encoder = [EncoderLayer(d_model, num_heads, dff, rate)
        for _ in range(num_layers)]
        self.decoder = [DecoderLayer(d_model, num_heads, dff, rate)
        for _ in range(num_layers)]

        self.final_layer = Dense(target_vocab_size)

    def call(self, inp, tar, training, enc_padding_mask, look_ahead_mask,
    dec_padding_mask):
        enc_output = inp

        for i in range(len(self.encoder)):
            enc_output = self.encoder[i](enc_output, training, enc_padding_mask)

        dec_output = tar
        for i in range(len(self.decoder)):
            dec_output, _, _ = self.decoder[i](dec_output, enc_output, training,
            look_ahead_mask, dec_padding_mask)

        final_output = self.final_layer(dec_output)

        return final_output
```

Note: This is a simplified implementation. For a full implementation, we need to include positional encoding, complete masking functions, and embedding layers.

Training and Evaluation

The training loop involves passing the input through the encoder and decoder, computing the loss, and updating the model weights using backpropagation. The loss function for sequence-to-sequence tasks typically involves computing the cross-entropy loss between the predicted and true sequences.

```
loss_object = tf.keras.losses.SparseCategoricalCrossentropy(from_logits=True,
reduction='none')

def loss_function(real, pred):
    mask = tf.math.logical_not(tf.math.equal(real, 0))
    loss_ = loss_object(real, pred)

    mask = tf.cast(mask, dtype=loss_.dtype)
    loss_ *= mask

    return tf.reduce_sum(loss_)/tf.reduce_sum(mask)
```

An optimizer like Adam with learning rate scheduling is commonly used for training transformers.

```
class CustomSchedule(tf.keras.optimizers.schedules.LearningRateSchedule):
    def _ _init_ _(self, d_model, warmup_steps=4000):
        super(CustomSchedule, self)._ _init_ _()

        self.d_model = d_model
        self.d_model = tf.cast(self.d_model, tf.float32)

        self.warmup_steps = warmup_steps

    def _ _call_ _(self, step):
        arg1 = tf.math.rsqrt(step)
        arg2 = step * (self.warmup_steps ** -1.5)

        return tf.math.rsqrt(self.d_model) * tf.math.minimum(arg1, arg2)

learning_rate = CustomSchedule(d_model=512)
optimizer = tf.keras.optimizers.Adam(learning_rate, beta_1=0.9, beta_2=0.98, epsilon=1e-9)
```

Define the training step function to update the model weights.

```
@tf.function
def train_step(inp, tar):
    tar_inp = tar[:, :-1]
    tar_real = tar[:, 1:]

    with tf.GradientTape() as tape:
        predictions = transformer

(inp, tar_inp, True, enc_padding_mask, look_ahead_mask, dec_padding_mask)
        loss = loss_function(tar_real, predictions)

    gradients = tape.gradient(loss, transformer.trainable_variables)
    optimizer.apply_gradients(zip(gradients, transformer.trainable_variables))

    train_loss(loss)
    train_accuracy(tar_real, predictions)
```

Implement the complete training loop with epoch management.

```
for epoch in range(EPOCHS):
    start = time.time()

    train_loss.reset_states()
    train_accuracy.reset_states()

    for (batch, (inp, tar)) in enumerate(dataset):
        train_step(inp, tar)

    print(f'Epoch {epoch + 1} Loss {train_loss.result()} Accuracy {train_accuracy.result()}')
    print(f'Time taken for 1 epoch: {time.time() - start} secs\n')
```

TensorBoard for Visualization
TensorBoard is a powerful tool for visualizing training metrics, such as loss and accuracy, as well as model graphs and distributions.

Setting up TensorBoard:

```
import tensorflow as tf
import datetime

# Define log directory
log_dir = "logs/fit/" + datetime.datetime.now().strftime("%Y%m%d-%H%M%S")
tensorboard_callback = tf.keras.callbacks.TensorBoard(log_dir=log_dir, histogram_freq=1)

# Training the model with TensorBoard callback
model.fit(x_train, y_train, epochs=10, callbacks=[tensorboard_callback])
```

Launching TensorBoard: To visualize the training process, launch TensorBoard from the command line:

```
tensorboard --logdir logs/fit
```

Once TensorBoard is running, we can view the training metrics and graphs in our web browser, to obtain insights into the model's performance and behavior during training.

6.8.2 Building Transformers with PyTorch

1. **Encoder Layer**:

```
import torch
import torch.nn as nn
import torch.nn.functional as F

class MultiHeadAttention(nn.Module):
    def __init__(self, d_model, num_heads):
        super(MultiHeadAttention, self).__init__()
        self.num_heads = num_heads
        self.d_model = d_model

        assert d_model % num_heads == 0

        self.depth = d_model // num_heads

        self.wq = nn.Linear(d_model, d_model)
```

```
        self.wk = nn.Linear(d_model, d_model)
        self.wv = nn.Linear(d_model, d_model)
        self.dense = nn.Linear(d_model, d_model)

    def split_heads(self, x, batch_size):
        x = x.view(batch_size, -1, self.num_heads, self.depth)
        return x.transpose(1, 2)

    def forward(self, q, k, v, mask=None):
        batch_size = q.size(0)

        q = self.wq(q)
        k = self.wk(k)
        v = self.wv(v)

        q = self.split_heads(q, batch_size)
        k = self.split_heads(k, batch_size)
        v = self.split_heads(v, batch_size)

        attn_output, _ = self.scaled_dot_product_attention(q, k, v, mask)
        attn_output = attn_output.transpose(1, 2).contiguous().view(batch_size, -1,
        self.d_model)
        output = self.dense(attn_output)
        return output

    def scaled_dot_product_attention(self, q, k, v, mask=None):
        dk = k.size(-1)
        scores = torch.matmul(q, k.transpose(-2, -1)) / torch.sqrt(dk)

        if mask is not None:
            scores = scores.masked_fill(mask == 0, -1e9)

        attn_weights = F.softmax(scores, dim=-1)
        output = torch.matmul(attn_weights, v)
        return output, attn_weights

class EncoderLayer(nn.Module):
    def __init__(self, d_model, num_heads, dff, dropout=0.1):
        super(EncoderLayer, self).__init__()
        self.mha = MultiHeadAttention(d_model, num_heads)
        self.ffn = nn.Sequential(
            nn.Linear(d_model, dff),
            nn.ReLU(),
            nn.Linear(dff, d_model)
        )
        self.layernorm1 = nn.LayerNorm(d_model)
        self.layernorm2 = nn.LayerNorm(d_model)
        self.dropout1 = nn.Dropout(dropout)
        self.dropout2 = nn.Dropout(dropout)

    def forward(self, x, mask=None):
        attn_output = self.mha(x, x, x, mask)
        out1 = self.layernorm1(x + self.dropout1(attn_output))
        ffn_output = self.ffn(out1)
        out2 = self.layernorm2(out1 + self.dropout2(ffn_output))
        return out2
```

2. **Decoder Layer**:

```
class DecoderLayer(nn.Module):
    def _ _init_ _(self, d_model, num_heads, dff, dropout=0.1):
        super(DecoderLayer, self)._ _init_ _()
        self.mha1 = MultiHeadAttention(d_model, num_heads)
        self.mha2 = MultiHeadAttention(d_model, num_heads)
        self.ffn = nn.Sequential(
            nn.Linear(d_model, dff),
```

```
            nn.ReLU(),
            nn.Linear(dff, d_model)
        )
        self.layernorm1 = nn.LayerNorm(d_model)
        self.layernorm2 = nn.LayerNorm(d_model)
        self.layernorm3 = nn.LayerNorm(d_model)
        self.dropout1 = nn.Dropout(dropout)
        self.dropout2 = nn.Dropout(dropout)
        self.dropout3 = nn.Dropout(dropout)

    def forward(self, x, enc_output, look_ahead_mask=None, padding_mask=None):
        attn1, _ = self.mha1(x, x, x, look_ahead_mask)
        out1 = self.layernorm1(x + self.dropout1(attn1))

        attn2, _ = self.mha2(enc_output, enc_output, out1, padding_mask)
        out2 = self.layernorm2(out1 + self.dropout2(attn2))

        ffn_output = self.ffn(out2)
        out3 = self.layernorm3(out2 + self.dropout3(ffn_output))
        return out3
```

3. Transformer Model:

```
class Transformer(nn.Module):
    def _ _init_ _(self, num_layers, d_model, num_heads, dff, input_vocab_size,
    target_vocab_size, dropout=0.1):
        super(Transformer, self)._ _init_ _()
        self.encoder_layers = nn.ModuleList([EncoderLayer
        (d_model, num_heads, dff, dropout)
        for _ in range(num_layers)])
        self.decoder_layers = nn.ModuleList([DecoderLayer
        (d_model, num_heads, dff, dropout)
        for _ in range(num_layers)])
        self.final_layer = nn.Linear(d_model, target_vocab_size)

    def forward(self, src, tgt, src_mask=None, tgt_mask=None, src_tgt_mask=None):
        enc_output = src
        for layer in self.encoder_layers:
            enc_output = layer(enc_output, src_mask)

        dec_output = tgt
        for layer in self.decoder_layers:
            dec_output = layer(dec_output, enc_output, tgt_mask, src_tgt_mask)

        final_output = self.final_layer(dec_output)
        return final_output
```

Training and Evaluation

```
criterion = nn.CrossEntropyLoss()

def loss_function(pred, real):
    mask = real != 0
    loss_ = criterion(pred.permute(0, 2, 1), real)
    mask = mask.float()
    loss_ = loss_ * mask
    return loss_.sum() / mask.sum()


optimizer = torch.optim.Adam(model.parameters(), lr=0.001, betas=(0.9, 0.98), eps=1e-9)
```

Define the training step function to update the model weights:

```
def train_step(src, tgt, model, optimizer, criterion):
    model.train()
    optimizer.zero_grad()

    output = model(src, tgt)
    tgt_out = tgt[:, 1:]
    loss = loss_function(output, tgt_out)
    loss.backward()
    optimizer.step()

    return loss.item()
```

Implement the complete training loop with epoch management:

```
EPOCHS = 10
for epoch in range(EPOCHS):
    total_loss = 0
    for batch, (src, tgt) in enumerate(dataloader):
        loss = train_step(src, tgt, model, optimizer, criterion)
        total_loss += loss

    print(f'Epoch {epoch+1}, Loss: {total_loss/len(dataloader)}')
```

Visualizing with TorchVision
While TorchVision is primarily used for image-related tasks, it can be used to visualize the attention maps and other components of the transformer model.

Example: Visualizing Attention Maps To visualize attention maps, extract the attention weights from the model and plot them using Matplotlib.

```
import matplotlib.pyplot as plt

def plot_attention(attention, sentence, predicted_sentence):
    fig = plt.figure(figsize=(10, 10))
    ax = fig.add_subplot(1, 1, 1)

    attention = attention.squeeze(0).cpu().detach().numpy()
    ax.matshow(attention, cmap='viridis')

    fontdict = {'fontsize': 10}

    ax.set_xticks(range(len(sentence)))
    ax.set_yticks(range(len(predicted_sentence)))

    ax.set_xticklabels(sentence, fontdict=fontdict, rotation=90)
    ax.set_yticklabels(predicted_sentence, fontdict=fontdict)

    plt.show()

# Example
# Assuming 'attention_weights' is a tensor with attention weights
sentence = ["This", "is", "a", "test", "."]
predicted_sentence = ["Ceci", "est", "un", "test", "."]
plot_attention(attention_weights, sentence, predicted_sentence)
```

6.9 Exercises

1. Given the following input sequences and self-attention parameters:

$$\mathbf{Q} = \begin{pmatrix} 1 & 0 \\ 0 & 1 \end{pmatrix}, \quad \mathbf{K} = \begin{pmatrix} 1 & 2 \\ 0 & 1 \end{pmatrix}, \quad \mathbf{V} = \begin{pmatrix} 2 & 0 \\ 1 & 3 \end{pmatrix}$$

   (a) Compute the attention scores using the scaled dot-product attention mechanism.
   (b) Calculate the attention weights.
   (c) Compute the context vector.

   Consider a multi-head attention mechanism with two heads, where the input sequence is $\mathbf{X} = \begin{pmatrix} 1 & 2 \\ 3 & 4 \end{pmatrix}$. The projection matrices for the heads are:

   Head 1:

$$\mathbf{W}_Q^{(1)} = \begin{pmatrix} 1 & 0 \\ 0 & 1 \end{pmatrix}, \quad \mathbf{W}_K^{(1)} = \begin{pmatrix} 0 & 1 \\ 1 & 0 \end{pmatrix}, \quad \mathbf{W}_V^{(1)} = \begin{pmatrix} 1 & 1 \\ 1 & -1 \end{pmatrix}$$

   Head 2:

$$\mathbf{W}_Q^{(2)} = \begin{pmatrix} 0 & 1 \\ 1 & 0 \end{pmatrix}, \quad \mathbf{W}_K^{(2)} = \begin{pmatrix} 1 & 0 \\ 0 & 1 \end{pmatrix}, \quad \mathbf{W}_V^{(2)} = \begin{pmatrix} 0 & 1 \\ -1 & 1 \end{pmatrix}$$

2. (a) Compute the query, key, and value matrices for both heads.
   (b) Calculate the attention scores and weights for both heads.
   (c) Compute the concatenated output of the multi-head attention.
3. Given the positional encoding function for dimension $d = 4$ and sequence length $n = 3$:

$$\text{PE}_{(\text{pos},2i)} = \sin\left(\frac{\text{pos}}{10{,}000^{2i/d}}\right), \quad \text{PE}_{(\text{pos},2i+1)} = \cos\left(\frac{\text{pos}}{10{,}000^{2i/d}}\right)$$

   (a) Compute the positional encodings for positions $pos = 0, 1, 2$.
   (b) Construct the positional encoding matrix for the given sequence length and dimension.

4. Consider a self-attention mechanism where the input sequence $\mathbf{X} \in \mathbb{R}^{4\times3}$ consists of four tokens, each represented by a three-dimensional embedding. The matrices for the query, key, and value are $\mathbf{W}_Q, \mathbf{W}_K, \mathbf{W}_V \in \mathbb{R}^{3\times3}$:

$$\mathbf{X} = \begin{pmatrix} 1 & 0 & 1 \\ 0 & 1 & 1 \\ 1 & 1 & 0 \\ 0 & 0 & 1 \end{pmatrix}, \quad \mathbf{W}_Q = \begin{pmatrix} 1 & 0 & 0 \\ 0 & 1 & 0 \\ 0 & 0 & 1 \end{pmatrix}, \quad \mathbf{W}_K = \begin{pmatrix} 0 & 1 & 0 \\ 1 & 0 & 0 \\ 0 & 0 & 1 \end{pmatrix}, \quad \mathbf{W}_V = \begin{pmatrix} 1 & 0 & 0 \\ 0 & 1 & 0 \\ 0 & 0 & 1 \end{pmatrix}$$

(a) Compute the query, key, and value matrices for **X**.
(b) Calculate the attention scores and the resulting output of the self-attention mechanism.

5. Given a sequence length of 5 and a model dimension of 4, compute the positional encodings using the formula:

$$\mathrm{PE}_{(\mathrm{pos},2i)} = \sin\left(\frac{\mathrm{pos}}{10{,}000^{\frac{2i}{d}}}\right), \quad \mathrm{PE}_{(\mathrm{pos},2i+1)} = \cos\left(\frac{\mathrm{pos}}{10{,}000^{\frac{2i}{d}}}\right)$$

where (pos) is the position and d is the model dimension. Compute the positional encoding matrix $\mathbf{PE} \in \mathbb{R}^{5\times 4}$.

6. Consider a word embedding matrix $\mathbf{E} \in \mathbb{R}^{10\times 4}$, where each row represents a four-dimensional embedding of a vocabulary of size 10. The input sequence of word indices is $\mathbf{w} = [2, 5, 3, 8]$.

(a) Extract the embeddings corresponding to the input sequence **w**.
(b) Compute the transformed embeddings using a transformation matrix $\mathbf{W}_T \in \mathbb{R}^{4\times 4}$:

$$\mathbf{W}_T = \begin{pmatrix} 1 & 0 & 0 & 0 \\ 0 & 1 & 0 & 0 \\ 0 & 0 & 1 & 0 \\ 0 & 0 & 0 & 1 \end{pmatrix}$$

7. Consider a multi-head self-attention mechanism with 2 heads for an input sequence $\mathbf{X} \in \mathbb{R}^{3\times 3}$:

$$\mathbf{X} = \begin{pmatrix} 1 & 2 & 3 \\ 4 & 5 & 6 \\ 7 & 8 & 9 \end{pmatrix}$$

The query, key, and value weight matrices for each head are:

$$\mathbf{W}_Q^{(1)} = \mathbf{W}_K^{(1)} = \mathbf{W}_V^{(1)} = \mathbf{I}_3, \quad \mathbf{W}_Q^{(2)} = \mathbf{W}_K^{(2)} = \mathbf{W}_V^{(2)} = 2 \cdot \mathbf{I}_3$$

(a) Compute the outputs of the self-attention mechanism for each head.
(b) Concatenate the outputs and apply a final linear transformation with weight matrix $\mathbf{W}_O \in \mathbb{R}^{6\times 3}$:

$$\mathbf{W}_O = \begin{pmatrix} 1 & 0 & 0 \\ 0 & 1 & 0 \\ 0 & 0 & 1 \\ 1 & 1 & 0 \\ 0 & 1 & 1 \\ 1 & 0 & 1 \end{pmatrix}$$

8. Let $\mathbf{X} \in \mathbb{R}^{5\times4}$ represent a sequence of five tokens, each with a four-dimensional embedding. The attention weights matrix $\mathbf{A} \in \mathbb{R}^{5\times5}$ is given by:

$$\mathbf{A} = \begin{pmatrix} 0.1 & 0.2 & 0.3 & 0.2 & 0.2 \\ 0.3 & 0.1 & 0.2 & 0.2 & 0.2 \\ 0.2 & 0.3 & 0.1 & 0.2 & 0.2 \\ 0.2 & 0.2 & 0.3 & 0.1 & 0.2 \\ 0.2 & 0.2 & 0.2 & 0.3 & 0.1 \end{pmatrix}$$

   (a) Compute the context vector for each token by performing the matrix multiplication $\mathbf{AX}$.
   (b) Apply a linear transformation using weight matrix $\mathbf{W}_C \in \mathbb{R}^{4\times4}$:

$$\mathbf{W}_C = \begin{pmatrix} 1 & 0 & 0 & 0 \\ 0 & 1 & 0 & 0 \\ 0 & 0 & 1 & 0 \\ 0 & 0 & 0 & 1 \end{pmatrix}$$

   and write down the final transformed context vectors.

9. Consider a cross-attention mechanism where the query tensor $\mathbf{Q} \in \mathbb{R}^{3\times4}$, key tensor $\mathbf{K} \in \mathbb{R}^{4\times4}$, and value tensor $\mathbf{V} \in \mathbb{R}^{4\times4}$ are given by:

$$\mathbf{Q} = \begin{pmatrix} 1 & 0 & 1 & 0 \\ 0 & 1 & 0 & 1 \\ 1 & 1 & 1 & 1 \end{pmatrix}, \quad \mathbf{K} = \begin{pmatrix} 1 & 2 & 3 & 4 \\ 4 & 3 & 2 & 1 \\ 1 & 3 & 2 & 4 \\ 4 & 2 & 1 & 3 \end{pmatrix}, \quad \mathbf{V} = \begin{pmatrix} 0 & 1 & 0 & 1 \\ 1 & 0 & 1 & 0 \\ 0 & 1 & 1 & 0 \\ 1 & 0 & 0 & 1 \end{pmatrix}$$

   (a) Compute the attention scores by performing the dot product of $\mathbf{Q}$ and $\mathbf{K}^T$. Apply a softmax function to the resulting scores.
   (b) Using the computed attention scores, calculate the output of the cross-attention mechanism by multiplying the scores with $\mathbf{V}$.

10. Given an input sequence $\mathbf{X} \in \mathbb{R}^{3\times4}$ and a mask tensor $\mathbf{M} \in \mathbb{R}^{3\times3}$ where:

$$\mathbf{X} = \begin{pmatrix} 1 & 2 & 3 & 4 \\ 5 & 6 & 7 & 8 \\ 9 & 10 & 11 & 12 \end{pmatrix}, \quad \mathbf{M} = \begin{pmatrix} 0 & -\infty & -\infty \\ 0 & 0 & -\infty \\ 0 & 0 & 0 \end{pmatrix}$$

   (a) Compute the query, key, and value matrices using identity matrices for weights $\mathbf{W}_Q, \mathbf{W}_K, \mathbf{W}_V \in \mathbb{R}^{4\times4}$.
   (b) Apply the mask $\mathbf{M}$ to the attention scores and compute the final masked self-attention output.

11. Consider a feed-forward neural network with one hidden layer. The input tensor $\mathbf{X} \in \mathbb{R}^{3\times 2}$ is:

$$\mathbf{X} = \begin{pmatrix} 1 & 2 \\ 3 & 4 \\ 5 & 6 \end{pmatrix}$$

The weight matrices for the hidden layer and output layer are:

$$\mathbf{W}_1 \in \mathbb{R}^{2\times 3} = \begin{pmatrix} 0.5 & 0.2 & -0.3 \\ -0.1 & 0.4 & 0.6 \end{pmatrix}, \quad \mathbf{W}_2 \in \mathbb{R}^{3\times 1} = \begin{pmatrix} 0.7 \\ -0.5 \\ 0.8 \end{pmatrix}$$

(a) Compute the hidden layer activations using ReLU activation function.
(b) Compute the output of the network.

12. Let the query tensor $\mathbf{Q} \in \mathbb{R}^{2\times 3}$, key tensor $\mathbf{K} \in \mathbb{R}^{3\times 3}$, and value tensor $\mathbf{V} \in \mathbb{R}^{3\times 3}$ be given by:

$$\mathbf{Q} = \begin{pmatrix} 1 & 0 & 2 \\ 0 & 1 & 3 \end{pmatrix}, \quad \mathbf{K} = \begin{pmatrix} 1 & 1 & 1 \\ 2 & 2 & 2 \\ 3 & 3 & 3 \end{pmatrix}, \quad \mathbf{V} = \begin{pmatrix} 1 & 0 & 1 \\ 0 & 1 & 0 \\ 1 & 1 & 1 \end{pmatrix}$$

The positional encodings for the keys are $\mathbf{P} \in \mathbb{R}^{3\times 3}$:

$$\mathbf{P} = \begin{pmatrix} 0 & 0.5 & 1 \\ 0.5 & 1 & 1.5 \\ 1 & 1.5 & 2 \end{pmatrix}$$

(a) Incorporate the positional encodings into the key tensor and compute the attention scores with the query tensor.
(b) Calculate the cross-attention output using the updated key tensor.

13. Consider a feed-forward neural network with one hidden layer and dropout. The input tensor $\mathbf{X} \in \mathbb{R}^{4\times 3}$ is:

$$\mathbf{X} = \begin{pmatrix} 1 & 2 & 3 \\ 4 & 5 & 6 \\ 7 & 8 & 9 \\ 10 & 11 & 12 \end{pmatrix}$$

The weight matrices for the hidden layer and output layer are:

$$\mathbf{W}_1 \in \mathbb{R}^{3\times 3} = \begin{pmatrix} 0.1 & 0.2 & 0.3 \\ 0.4 & 0.5 & 0.6 \\ 0.7 & 0.8 & 0.9 \end{pmatrix}, \quad \mathbf{W}_2 \in \mathbb{R}^{3\times 2} = \begin{pmatrix} 0.2 & 0.1 \\ 0.4 & 0.3 \\ 0.6 & 0.5 \end{pmatrix}$$

(a) Compute the hidden layer activations using a dropout rate of 0.5. Assume a specific dropout mask $\mathbf{M} \in \mathbb{R}^{4\times 3}$:

$$\mathbf{M} = \begin{pmatrix} 1 & 0 & 1 \\ 1 & 1 & 0 \\ 0 & 1 & 1 \\ 1 & 0 & 0 \end{pmatrix}$$

(b) Compute the output of the network after applying the dropout mask.

14. Consider a grayscale image of size 8×8 represented as a matrix $\mathbf{I} \in \mathbb{R}^{8\times 8}$:

$$\mathbf{I} = \begin{pmatrix} 1 & 2 & 3 & 4 & 5 & 6 & 7 & 8 \\ 9 & 10 & 11 & 12 & 13 & 14 & 15 & 16 \\ 17 & 18 & 19 & 20 & 21 & 22 & 23 & 24 \\ 25 & 26 & 27 & 28 & 29 & 30 & 31 & 32 \\ 33 & 34 & 35 & 36 & 37 & 38 & 39 & 40 \\ 41 & 42 & 43 & 44 & 45 & 46 & 47 & 48 \\ 49 & 50 & 51 & 52 & 53 & 54 & 55 & 56 \\ 57 & 58 & 59 & 60 & 61 & 62 & 63 & 64 \end{pmatrix}$$

(a) Divide the image into non-overlapping patches of size 2×2. Write down the resulting patches as vectors.
(b) Flatten each patch and concatenate them to form a matrix of patch embeddings $\mathbf{P} \in \mathbb{R}^{16\times 4}$.

15. Given an image $\mathbf{I} \in \mathbb{R}^{6\times 6\times 3}$ representing a 6×6 RGB image, split it into non-overlapping patches of size 2×2. Assume the linear transformation matrix $\mathbf{W} \in \mathbb{R}^{12\times 6}$ is given by:

$$\mathbf{W} = \begin{pmatrix} 1 & 0 & 0 & 0 & 0 & 0 \\ 0 & 1 & 0 & 0 & 0 & 0 \\ 0 & 0 & 1 & 0 & 0 & 0 \\ 0 & 0 & 0 & 1 & 0 & 0 \\ 0 & 0 & 0 & 0 & 1 & 0 \\ 0 & 0 & 0 & 0 & 0 & 1 \\ 1 & 0 & 0 & 0 & 0 & 0 \\ 0 & 1 & 0 & 0 & 0 & 0 \\ 0 & 0 & 1 & 0 & 0 & 0 \\ 0 & 0 & 0 & 1 & 0 & 0 \\ 0 & 0 & 0 & 0 & 1 & 0 \\ 0 & 0 & 0 & 0 & 0 & 1 \end{pmatrix}$$

(a) Flatten each 2×2 patch into a vector and compute the patch embeddings by applying the linear transformation $\mathbf{W}$.
(b) Form the matrix of patch embeddings $\mathbf{P} \in \mathbb{R}^{9\times 12}$.

Chapter 7
Attention Mechanisms Beyond Transformers

We are just an advanced breed of monkeys on a minor planet of a very average star. But we can understand the Universe. That makes us something very special.

—Stephen Hawking, A Brief History of Time

7.1 Introduction to Attention Mechanisms

Attention mechanisms have become a fundamental component in modern machine learning, particularly in the realm of natural language processing (NLP) and computer vision. The core idea behind attention is to dynamically focus on different parts of the input data, allowing models to handle long-range dependencies more effectively and enhance interpretability. Mathematically, the attention mechanism can be described as a function that maps a query and a set of key-value pairs to an output, where the query, keys, and values are all vectors. The output is computed as a weighted sum of the values, with the weights (attention scores) computed based on the compatibility between the query and corresponding key.

7.1.1 Historical Context and Development

The concept of attention mechanisms was first introduced in the context of machine translation by Bahdanau et al. (2014) in their seminal paper "Neural Machine Translation by Jointly Learning to Align and Translate". This paper proposed the use of an alignment model to improve the performance of encoder–decoder architectures in translating longer sentences. The alignment model allowed the decoder to

P. Singh and B. Raman, *Deep Learning Through the Prism of Tensors*, Studies in Big Data 162, https://doi.org/10.1007/978-981-97-8019-8_7

selectively focus on parts of the input sentence, effectively handling variable-length sequences and improving translation quality. Following this breakthrough, attention mechanisms were quickly adopted and extended to various other tasks. Notable advancements include:

1. Global and Local Attention: Luong et al. (2015) proposed global and local attention mechanisms in their work "Effective Approaches to Attention-based Neural Machine Translation." Global attention considers all hidden states of the encoder, while local attention focuses on a specific subset, enhancing efficiency.
2. Self-Attention: Vaswani et al. (2017) introduced the concept of self-attention in their transformative paper "Attention is All You Need," which forms the foundation of the Transformer architecture. Self-attention mechanisms compute attention scores within the same sequence, enabling parallelization and capturing dependencies across distant tokens.
3. Multi-Head Attention: An extension of self-attention, multi-head attention was introduced by Vaswani et al. (2017) to allow the model to jointly attend to information from different representation subspaces. Formally, given queries, keys, and values $Q, K, V \in \mathbb{R}^{n \times d_{model}}$, multi-head attention projects them into h different subspaces using learned linear projections $W_i^Q, W_i^K, W_i^V \in \mathbb{R}^{d_{model} \times d_k}$, computes scaled dot-product attention in each subspace, and concatenates the results:

$$\text{Multi-Head}(Q, K, V) = \bigoplus(\text{head}_1, \text{head}_2, \ldots, \text{head}_h)W^O$$

where $\text{head}_i = \text{Attention}(QW_i^Q, KW_i^K, VW_i^V)$.

7.1.2 Importance of Attention Mechanisms in Machine Learning

Attention mechanisms play a critical role in enhancing the performance and interpretability of machine learning models. Their importance can be highlighted through several key benefits:

1. Handling Long-Range Dependencies: Traditional models like RNNs struggle with long-range dependencies due to issues like vanishing gradients. Attention mechanisms address this by allowing the model to directly focus on relevant parts of the input, regardless of their distance from the current position.
2. Parallelization: Self-attention mechanisms, as used in Transformers, enable parallel processing of input sequences, significantly speeding up training and inference compared to sequential models like RNNs.
3. Interpretability: Attention scores provide insights into the model's decision-making process by highlighting which parts of the input the model is focusing on. This transparency is valuable for debugging and understanding model behavior.

4. Versatility: Attention mechanisms are not limited to NLP tasks. They have been successfully applied in computer vision (e.g., image captioning), speech recognition, and even reinforcement learning, demonstrating their adaptability across different domains.
5. Improved Performance: By dynamically weighting different parts of the input, attention mechanisms allow models to capture more nuanced patterns and relationships, leading to state-of-the-art performance in various tasks, such as machine translation, text summarization, and question answering.

Example: Consider a machine translation task where the goal is to translate a sentence from English to French. Using an attention mechanism, the model can align words in the source sentence with corresponding words in the target sentence, even if their positions differ significantly. For instance, translating "The cat is on the mat" to "Le chat est sur le tapis" requires aligning "cat" with "chat" and "mat" with "tapis," which may be far apart in their respective sentences.

7.2 Types of Attention Mechanisms

7.2.1 *Additive Attention*

Additive attention, also known as Bahdanau attention after its introduction by Dzmitry Bahdanau in the context of neural machine translation, is a mechanism that computes the alignment scores between the target hidden state and each source hidden state in a sequence. Unlike dot-product attention, which directly computes the dot product between query and key vectors, additive attention employs a feed-forward neural network to derive the attention scores. This method can better capture complex relationships between inputs. Additive attention works by calculating an alignment score between the hidden state of the decoder at time step t (denoted as $\mathbf{s}_t$) and the hidden states of the encoder $\mathbf{h}_i$. The alignment score $e_{t,i}$ is calculated using a feed-forward neural network with a single hidden layer. The network takes the concatenation of $\mathbf{s}_t$ and $\mathbf{h}_i$ as input:

$$e_{t,i} = \mathbf{v}^{\mathrm{T}} \tanh(\mathbf{W}_1 \mathbf{s}_t + \mathbf{W}_2 \mathbf{h}_i + \mathbf{b})$$

where $\mathbf{W}_1$ and $\mathbf{W}_2$ are weight matrices, $\mathbf{b}$ is the bias vector, $\mathbf{v}$ is a weight vector, tanh is the hyperbolic tangent activation function. The alignment scores $e_{t,i}$ are then normalized using the softmax function to obtain the attention weights $\alpha_{t,i}$:

$$\alpha_{t,i} = \frac{\exp(e_{t,i})}{\sum_{k=1}^{T} \exp(e_{t,k})}$$

The context vector $\mathbf{c}_t$ for the decoder at time step t is computed as a weighted sum of the encoder hidden states:

$$\mathbf{c}_t = \sum_{i=1}^{T} \alpha_{t,i} \mathbf{h}_i$$

The context vector $\mathbf{c}_t$ is then combined with the decoder hidden state $\mathbf{s}_t$ to produce the final output of the decoder.

Applications and Use Cases

Additive attention is particularly useful in tasks where the relationship between the query and key vectors is complex and non-linear. It has been successfully applied in various NLP tasks, including:

1. Neural Machine Translation (NMT): In the seminal paper by Bahdanau et al. (2014), additive attention was used to improve the alignment between source and target sentences in NMT. The model could focus on different parts of the source sentence when generating each word in the target sentence, significantly enhancing translation quality, especially for long sentences.
2. Text Summarization: Additive attention helps summarization models focus on the most relevant parts of the document when generating summaries. By aligning the target summary tokens with the important segments of the source text, the model produces more coherent and informative summaries.
3. Image Captioning: In image captioning, additive attention can be used to generate descriptions for images by focusing on different regions of the image when generating each word of the caption. This allows the model to produce detailed and contextually accurate descriptions.

Example (Neural Machine Translation with Additive Attention): Consider an example where we use a simple encoder–decoder architecture with additive attention for translating sentences from English to French.

1. Encoder:

```
import torch
import torch.nn as nn

class Encoder(nn.Module):
    def __init__(self, input_dim, emb_dim, hid_dim, n_layers, dropout):
        super().__init__()
        self.embedding = nn.Embedding(input_dim, emb_dim)
        self.rnn = nn.GRU(emb_dim, hid_dim, n_layers, dropout=dropout)
        self.dropout = nn.Dropout(dropout)

    def forward(self, src):
        embedded = self.dropout(self.embedding(src))
        outputs, hidden = self.rnn(embedded)
        return outputs, hidden
```

2. Additive Attention Layer:

```
class Attention(nn.Module):
    def __init__(self, hid_dim):
        super().__init__()
        self.attn = nn.Linear(hid_dim * 2,hid_dim)
        self.v = nn.Parameter(torch.rand(hid_dim))

    def forward(self, hidden, encoder_outputs):
        src_len = encoder_outputs.shape[0]
        hidden = hidden[-1].unsqueeze(1).repeat(1, src_len, 1)
        energy = torch.tanh(self.attn(torch.cat((hidden, encoder_outputs),
                 dim=2)))
        attention = torch.sum(self.v * energy, dim=2)
        return torch.softmax(attention, dim=1)
```

3. Decoder:

```
class Decoder(nn.Module):
    def __init__(self, output_dim, emb_dim, hid_dim, n_layers, dropout
    , attention):
        super().__init__()
        self.output_dim = output_dim
        self.attention = attention
        self.embedding = nn.Embedding(output_dim, emb_dim)
        self.rnn = nn.GRU((hid_dim * 2) + emb_dim, hid_dim, n_layers,
                   dropout=dropout)
        self.fc_out = nn.Linear((hid_dim * 2) + emb_dim + hid_dim, output_dim)
        self.dropout = nn.Dropout(dropout)

    def forward(self, input, hidden, encoder_outputs):
        input = input.unsqueeze(0)
        embedded = self.dropout(self.embedding(input))
        a = self.attention(hidden, encoder_outputs)
        a = a.unsqueeze(1)
        weighted = torch.bmm(a, encoder_outputs.permute(1, 0, 2))
        rnn_input = torch.cat((embedded, weighted.permute(1, 0, 2)), dim=2)
        output, hidden = self.rnn(rnn_input, hidden.unsqueeze(0))
        embedded = embedded.squeeze(0)
        output = output.squeeze(0)
        weighted = weighted.squeeze(1)
        prediction = self.fc_out(torch.cat((output, weighted, embedded),dim=1))
        return prediction, hidden.squeeze(0)
```

4. Training and Evaluation: To train the model, define the loss function and optimizer, then iterate through the dataset to update the model parameters. Evaluate the model using metrics such as BLEU score to measure the translation quality.

7.2.2 Multiplicative (Dot-Product) Attention

Multiplicative, or dot-product, attention is a more computationally efficient mechanism compared to additive attention. It involves calculating the attention scores by taking the dot product of the query and key vectors. This form of attention was popularized by the Transformer architecture introduced by Vaswani et al. (2017) in the paper "Attention is All You Need". Dot-product attention is particularly efficient on modern hardware, as it can leverage matrix multiplication operations optimized for parallel processing. The core idea of multiplicative attention is to compute the alignment scores between the query and key vectors using the dot product. Given a query matrix Q, a key matrix K, and a value matrix V, the attention mechanism can be formulated as follows:

1. Calculate the Dot-Product Attention Scores: The attention scores are computed by taking the dot product of the query Q with the transpose of the key matrix K:

$$\text{scores} = QK^{\mathrm{T}}$$

2. Scale the Scores: To maintain stability in the gradients, the scores are scaled by the square root of the dimensionality of the key vectors, d_k:

$$\text{scaled scores} = \frac{QK^{\mathrm{T}}}{\sqrt{d_k}}$$

3. Apply the Softmax Function: The scaled scores are then passed through the softmax function to obtain the attention weights. This ensures that the weights are positive and sum to one, making them interpretable as probabilities:

$$\text{attention weights} = \text{softmax}\left(\frac{QK^{\mathrm{T}}}{\sqrt{d_k}}\right)$$

4. Compute the Context Vector: The context vector is calculated as a weighted sum of the value vectors V, using the attention weights:

$$\text{context} = \text{attention weights} \cdot V$$

Putting it all together, the dot-product attention mechanism can be represented as:

$$\text{Attention}(Q, K, V) = \text{softmax}\left(\frac{QK^{\mathrm{T}}}{\sqrt{d_k}}\right)V$$

Example: Consider a simple implementation of dot-product attention using PyTorch:

```
import torch
import torch.nn as nn

class DotProductAttention(nn.Module):
    def __init__(self, d_k):
        super(DotProductAttention, self).__init__()
        self.sqrt_dk = torch.sqrt(torch.tensor(d_k, dtype=torch.float32))

    def forward(self, Q, K, V, mask=None):
        scores = torch.matmul(Q, K.transpose(-2, -1)) / self.sqrt_dk
        if mask is not None:
```

```
        scores = scores.masked_fill(mask == 0, float('-inf'))
    attention_weights = torch.softmax(scores, dim=-1)
    context = torch.matmul(attention_weights, V)
    return context, attention_weights
```

In this implementation, 'Q', 'K', and 'V' are the query, key, and value matrices, respectively. The 'mask' is optional and can be used to mask out certain positions, ensuring that the attention mechanism does not attend to padding tokens or future tokens (in the case of autoregressive models).

Applications and Use Cases

Multiplicative attention has been widely adopted in various applications due to its computational efficiency and effectiveness in capturing dependencies within sequences. Some of the prominent use cases include:

1. Machine Translation: In the Transformer model, multiplicative attention is used in both the encoder and decoder layers. The encoder uses self-attention mechanisms to process the input sentence, while the decoder uses both self-attention and encoder–decoder attention to generate the translation.

   Example: Transformer-based Machine Translation

```
from transformers import BertTokenizer, BertForSequenceClassification
import torch

# Load pre-trained model and tokenizer
tokenizer = BertTokenizer.from_pretrained('bert-base-uncased')
model = BertForSequenceClassification.from_pretrained('bert-base-uncased')

# Example text
text = "The quick brown fox jumps over the lazy dog."

# Tokenize input
inputs = tokenizer(text, return_tensors='pt')
outputs = model(**inputs)

# Get attention weights from the model
attention_weights = outputs.attentions
```

2. Text Summarization: Multiplicative attention is used in text summarization models to focus on the most relevant parts of the input document when generating summaries. Models like BERTSUM and T5 utilize this mechanism to produce coherent and concise summaries.

   Example: Text Summarization with BERTSUM

```
from transformers import BertTokenizer, BertForSequenceClassification
import torch

# Load pre-trained BERTSUM model and tokenizer
tokenizer = BertTokenizer.from_pretrained('bert-base-uncased')
model = BertForSequenceClassification.from_pretrained('bert-base-uncased')

# Example document
document = "The quick brown fox jumps over the lazy dog.
            The dog lies down lazily in the sun.
            Foxes are known for their agility and speed."
```

```
# Tokenize input
inputs = tokenizer(document, return_tensors='pt')
outputs = model(**inputs)

# Get summarization output
summary = tokenizer.decode(outputs.logits.argmax(dim=-1),
          skip_special_tokens=True)
print("Summary:", summary)
```

3. Question Answering: In question answering tasks, multiplicative attention helps the model to focus on the relevant parts of the context when answering questions. Models like BERT and RoBERTa fine-tuned on SQuAD datasets use this mechanism to achieve state-of-the art performance

   Example: Question Answering with BERT

```
from transformers import BertTokenizer, BertForQuestionAnswering
import torch

# Load pre-trained BERT model and tokenizer
model = BertForQuestionAnswering.from_pretrained
        ('bert-large-uncased-whole-word-masking-finetuned-squad')
tokenizer = BertTokenizer.from_pretrained
            ('bert-large-uncased-whole-word-masking-finetuned-squad')

# Example context and question
context = "The quick brown fox jumps over the lazy dog."
question = "What does the fox jump over?"

# Tokenize input
inputs = tokenizer(question, context, return_tensors='pt')
outputs = model(**inputs)

# Get start and end indices
start_idx = torch.argmax(outputs.start_logits)
end_idx = torch.argmax(outputs.end_logits)

# Extract answer
answer = tokenizer.decode(inputs['input_ids'][0][start_idx:end_idx+1])
print("Answer:", answer)
```

7.2.3 *Hierarchical Attention*

Hierarchical attention mechanisms extend the idea of attention by applying it at multiple levels of granularity. This approach is particularly useful for tasks involving long documents or complex data structures, where capturing hierarchical relationships is crucial. The hierarchical attention mechanism operates in a multi-stage process, where each stage focuses on different levels of the data hierarchy, such as words within sentences and sentences within documents. The hierarchical attention mechanism was introduced to address the limitations of flat attention models in handling long documents and multi-level data structures. The core idea is to apply attention mechanisms at different levels of the data hierarchy, enabling the model to capture important information at each level and aggregate it effectively.

1. Word-level Attention: At the word level, the model focuses on identifying important words within each sentence. Given a sentence with T words, let $\mathbf{h}_t$ represent the hidden state of the word t. The word-level attention score α_t is computed as:

$$e_t = \mathbf{v}_w^{\mathrm{T}} \tanh(\mathbf{W}_w \mathbf{h}_t + \mathbf{b}_w)$$

$$\alpha_t = \frac{\exp(e_t)}{\sum_{t'} \exp(e_{t'})}$$

The context vector $\mathbf{c}_s$ for the sentence is then computed as a weighted sum of the hidden states of the words:

$$\mathbf{c}_s = \sum_t \alpha_t \mathbf{h}_t$$

2. Sentence-level Attention: At the sentence level, the model identifies important sentences within the document. Given a document with N sentences, let $\mathbf{s}_n$ represent the sentence representation (context vector) obtained from the word-level attention. The sentence-level attention score β_n is computed as:

$$e_n = \mathbf{v}_s^{\mathrm{T}} \tanh(\mathbf{W}_s \mathbf{s}_n + \mathbf{b}_s)$$

$$\beta_n = \frac{\exp(e_n)}{\sum_{n'} \exp(e_{n'})}$$

The document representation $\mathbf{d}$ is then computed as a weighted sum of the sentence representations:

$$\mathbf{d} = \sum_n \beta_n \mathbf{s}_n$$

Example: Consider an implementation of a hierarchical attention mechanism using PyTorch:

```
import torch
import torch.nn as nn

class WordAttention(nn.Module):
    def __init__(self, hidden_dim):
        super(WordAttention, self).__init__()
        self.hidden_dim = hidden_dim
        self.attn = nn.Linear(hidden_dim, hidden_dim)
        self.context_vector = nn.Parameter(torch.rand(hidden_dim))

    def forward(self, h):
        u = torch.tanh(self.attn(h))
        attn_scores = torch.matmul(u, self.context_vector)
        attn_weights = torch.softmax(attn_scores, dim=1)
        context = torch.sum(attn_weights.unsqueeze(-1) * h, dim=1)
        return context, attn_weights
```

```
class SentenceAttention(nn.Module):
    def __init__(self, hidden_dim):
        super(SentenceAttention, self).__init__()
        self.hidden_dim = hidden_dim
        self.attn = nn.Linear(hidden_dim, hidden_dim)
        self.context_vector = nn.Parameter(torch.rand(hidden_dim))

    def forward(self, s):
        u = torch.tanh(self.attn(s))
        attn_scores = torch.matmul(u, self.context_vector)
        attn_weights = torch.softmax(attn_scores, dim=1)
        context = torch.sum(attn_weights.unsqueeze(-1) * s, dim=1)
        return context, attn_weights

class HierarchicalAttentionNetwork(nn.Module):
    def __init__(self, vocab_size, embed_size, hidden_dim, num_classes):
        super(HierarchicalAttentionNetwork, self).__init__()
        self.embedding = nn.Embedding(vocab_size, embed_size)
        self.word_gru = nn.GRU(embed_size, hidden_dim, bidirectional=True,
                               batch_first=True)
        self.word_attention = WordAttention(hidden_dim * 2)
        self.sent_gru = nn.GRU(hidden_dim * 2,hidden_dim, bidirectional=True,
                               batch_first=True)
        self.sent_attention = SentenceAttention(hidden_dim * 2)
        self.fc = nn.Linear(hidden_dim * 2, num_classes)

    def forward(self, x):
        batch_size, num_sentences, num_words = x.size()
        x = x.view(batch_size * num_sentences, num_words)
        x = self.embedding(x)
        h, _ = self.word_gru(x)
        word_contexts, _ = self.word_attention(h)
        word_contexts = word_contexts.view(batch_size, num_sentences, -1)
        s, _ = self.sent_gru(word_contexts)
        doc_context, _ = self.sent_attention(s)
        out = self.fc(doc_context)
        return out
```

In this implementation, 'WordAttention' and 'SentenceAttention' classes handle attention at the word and sentence levels, respectively. The 'HierarchicalAttention-Network' class combines these components to process documents hierarchically.

Applications and Use Cases

Hierarchical attention mechanisms are particularly useful in tasks involving long documents or hierarchical data structures. Some notable applications include:

1. Document Classification: Hierarchical attention networks (HAN) are effective for classifying long documents by capturing important words within sentences and important sentences within documents. This hierarchical approach allows the model to focus on the most relevant parts of the document, improving classification performance.

```
# Assuming we have a preprocessed dataset of documents
vocab_size = 5000
embed_size = 300
hidden_dim = 128
num_classes = 10

model = HierarchicalAttentionNetwork(vocab_size, embed_size, hidden_dim,
        num_classes)
# Define loss function and optimizer
```

```
criterion = nn.CrossEntropyLoss()
optimizer = torch.optim.Adam(model.parameters(), lr=0.001)

# Training loop
for epoch in range(epochs):
    for inputs, labels in dataloader:
        outputs = model(inputs)
        loss = criterion(outputs, labels)
        optimizer.zero_grad()
        loss.backward()
        optimizer.step()
```

2. Text Summarization: Hierarchical attention can be applied to generate summaries of long documents by focusing on key sentences and key words within those sentences. This approach allows the model to produce more coherent and informative summaries.

```
from transformers import BertTokenizer, BertForSequenceClassification
import torch

# Load pre-trained BERTSUM model and tokenizer
tokenizer = BertTokenizer.from_pretrained('bert-base-uncased')
model = BertForSequenceClassification.from_pretrained('bert-base-uncased')

# Example document
document = "The quick brown fox jumps over the lazy dog.
            The dog lies down lazily in the sun.
            Foxes are known for their agility and speed."

# Tokenize input
inputs = tokenizer(document, return_tensors='pt')
outputs = model(**inputs)

# Get summarization output
summary = tokenizer.decode(outputs.logits.argmax(dim=-1),
          skip_special_tokens=True)
print("Summary:", summary)
```

3. Sentiment Analysis: For tasks like sentiment analysis of long reviews or articles, hierarchical attention mechanisms can effectively capture the sentiment by aggregating word-level and sentence-level information. This approach ensures that important expressions contributing to the overall sentiment are given appropriate weight.

```
from transformers import BertTokenizer, BertForSequenceClassification
import torch

# Load pre-trained BERT model and tokenizer
model = BertForSequenceClassification.from_pretrained('bert-base-uncased')
tokenizer = BertTokenizer.from_pretrained('bert-base-uncased')

# Example review
review = "The movie was fantastic!
          The plot was engaging and the characters were well-developed."

# Tokenize input
inputs = tokenizer(review, return_tensors='pt')
outputs = model(**inputs)

# Get sentiment prediction
sentiment = torch.argmax(outputs.logits, dim=1).item()
sentiment_label = "Positive" if sentiment == 1 else "Negative"
print("Sentiment:", sentiment_label)
```

7.3 Attention in Different Contexts

7.3.1 *Visual Attention in Computer Vision*

Visual attention mechanisms have significantly advanced the field of computer vision by enabling models to focus on important regions or features in an image, thereby improving the performance of tasks like image classification, object detection, and image captioning. Visual attention can be broadly categorized into spatial attention and channel attention mechanisms.

Spatial Attention Mechanisms

Spatial attention mechanisms focus on identifying the most relevant spatial regions within an image. This is achieved by generating an attention map that highlights important areas while suppressing irrelevant ones. Mathematically, given a feature map $F \in \mathbb{R}^{C\times H\times W}$ where C is the number of channels, H is the height, and W is the width, the spatial attention mechanism can be described as follows:

Spatial Attention Map: Compute the attention map $M_s \in \mathbb{R}^{H\times W}$ by applying a 2D convolution followed by a softmax function:

$$M_s = \text{softmax}(f(F))$$

where f is a function representing a convolutional layer.

Weighted Feature Map: Apply the attention map to the feature map to obtain the weighted feature map F_s:

$$F_s = M_s \odot F$$

Here, $\odot$ denotes element-wise multiplication.

Channel Attention Mechanisms

Channel attention mechanisms focus on identifying the most relevant feature channels in the feature map. This is achieved by generating an attention vector that highlights important channels while suppressing irrelevant ones. Given a feature map $F \in \mathbb{R}^{C\times H\times W}$, the channel attention mechanism can be described as follows:

Channel Attention Vector: Compute the attention vector $M_c \in \mathbb{R}^C$ by applying global average pooling followed by fully connected layers and a sigmoid activation function:

$$M_c = \sigma(W_2\text{ReLU}(W_1\text{GAP}(F)))$$

where W_1 and W_2 are weight matrices, GAP denotes global average pooling, and σ is the sigmoid function.

Weighted Feature Map:
Apply the attention vector to the feature map to obtain the weighted feature map F_c:

$$F_c = M_c \odot F$$

Example: Consider an implementation of spatial and channel attention mechanisms using PyTorch:

```
import torch
import torch.nn as nn

class SpatialAttention(nn.Module):
    def __init__(self):
        super(SpatialAttention, self).__init__()
        self.conv = nn.Conv2d(2, 1, kernel_size=7, padding=3)
        self.sigmoid = nn.Sigmoid()

    def forward(self, x):
        avg_out = torch.mean(x, dim=1, keepdim=True)
        max_out, _ = torch.max(x, dim=1, keepdim=True)
        x = torch.cat([avg_out, max_out], dim=1)
        x = self.conv(x)
        return self.sigmoid(x)

class ChannelAttention(nn.Module):
    def __init__(self, in_channels, reduction_ratio=16):
        super(ChannelAttention, self).__init__()
        self.avg_pool = nn.AdaptiveAvgPool2d(1)
        self.fc = nn.Sequential(
            nn.Linear(in_channels, in_channels // reduction_ratio, bias=False),
            nn.ReLU(),
            nn.Linear(in_channels // reduction_ratio, in_channels, bias=False),
            nn.Sigmoid()
        )

    def forward(self, x):
        b, c, _, _ = x.size()
        y = self.avg_pool(x).view(b, c)
        y = self.fc(y).view(b, c, 1, 1)
        return x * y.expand_as(x)
```

Applications

1. Image Classification: Visual attention mechanisms enhance image classification models by focusing on the most relevant regions and channels in an image. For instance, in a model like ResNet, spatial and channel attention can be integrated into the residual blocks to improve feature representation and classification accuracy.
   Example:

```
class ResidualBlockWithAttention(nn.Module):
    def __init__(self, in_channels, out_channels, stride=1):
        super(ResidualBlockWithAttention, self).__init__()
        self.conv1 = nn.Conv2d(in_channels, out_channels, kernel_size=3,
                               stride=stride, padding=1, bias=False)
        self.bn1 = nn.BatchNorm2d(out_channels)
        self.relu = nn.ReLU(inplace=True)
```

```
        self.conv2 = nn.Conv2d(out_channels, out_channels, kernel_size=3,
                               padding=1, bias=False)
        self.bn2 = nn.BatchNorm2d(out_channels)
        self.spatial_att = SpatialAttention()
        self.channel_att = ChannelAttention(out_channels)
        self.downsample = nn.Sequential()
        if stride != 1 or in_channels != out_channels:
            self.downsample = nn.Sequential(
                nn.Conv2d(in_channels, out_channels, kernel_size=1,
                          stride=stride, bias=False),
                nn.BatchNorm2d(out_channels)
            )

    def forward(self, x):
        residual = self.downsample(x)
        out = self.relu(self.bn1(self.conv1(x)))
        out = self.bn2(self.conv2(out))
        out = self.spatial_att(out) * out
        out = self.channel_att(out) * out
        out += residual
        out = self.relu(out)
        return out
```

2. Object Detection: In object detection tasks, visual attention mechanisms help models to focus on relevant parts of the image, improving the accuracy of detecting objects. For example, spatial attention can highlight potential object regions, while channel attention can enhance important feature channels that contribute to detecting objects.
   Example: In models like You Only Look Once (YOLO) or Faster R-CNN, attention mechanisms can be integrated into the feature extraction backbone to improve detection performance.

```
class AttentionBackbone(nn.Module):
    def __init__(self, base_model):
        super(AttentionBackbone, self).__init__()
        self.base_model = base_model
        self.spatial_att = SpatialAttention()
        self.channel_att = ChannelAttention(base_model.output_channels)

    def forward(self, x):
        x = self.base_model(x)
        x = self.spatial_att(x) * x
        x = self.channel_att(x) * x
        return x
```

7.3.2 Cross-Modal Attention in Multimodal Learning

Cross-modal attention mechanisms are designed to integrate and align information from different modalities, such as vision and language. These mechanisms enable models to perform tasks that require understanding and processing information from multiple sources, leveraging the strengths of each modality to improve performance and achieve more comprehensive understanding.

Attention Mechanisms in Multimodal Models

In multimodal learning, cross-modal attention mechanisms facilitate the interaction between different types of data, such as images and text. The goal is to create a shared representation that captures the relationships and dependencies between modalities. This process involves computing attention scores that indicate the relevance of elements from one modality to another. Consider two input sequences from different modalities: an image represented by a set of visual features $V \in \mathbb{R}^{N \times d_v}$ and a text sequence represented by a set of word embeddings $T \in \mathbb{R}^{M \times d_t}$, where N and M are the lengths of the visual and text sequences, respectively, and d_v and d_t are their respective dimensions.

1. Compute Query, Key, and Value Matrices: For cross-modal attention, we compute query, key, and value matrices for each modality. Let W_Q^{T}, W_K^V, W_V^V be learned weight matrices for text queries, visual keys, and visual values, respectively:

$$Q = TW_Q^{\mathrm{T}}, \quad K = VW_K^V, \quad V = VW_V^V$$

2. Calculate Attention Scores: The attention scores are computed by taking the dot product of the text queries and visual keys, followed by scaling and applying the softmax function:

$$\text{scores} = \frac{QK^{\mathrm{T}}}{\sqrt{d_k}}$$

$$\text{attention weights} = \text{softmax}\left(\frac{QK^{\mathrm{T}}}{\sqrt{d_k}}\right)$$

3. Compute Context Vectors: The context vectors are obtained by multiplying the attention weights with the visual value matrix:

$$\text{context} = \text{attention weights} \cdot V$$

Example: Consider an implementation of cross-modal attention using PyTorch:

```
import torch
import torch.nn as nn

class CrossModalAttention(nn.Module):
    def __init__(self, d_model, num_heads):
        super(CrossModalAttention, self).__init__()
        self.num_heads = num_heads
        self.d_model = d_model

        assert d_model % num_heads == 0

        self.depth = d_model // num_heads

        self.wq = nn.Linear(d_model, d_model)
        self.wk = nn.Linear(d_model, d_model)
        self.wv = nn.Linear(d_model, d_model)
        self.dense = nn.Linear(d_model, d_model)
```

```
    def split_heads(self, x, batch_size):
        x = x.view(batch_size, -1, self.num_heads, self.depth)
        return x.transpose(1, 2)

    def forward(self, text, vision):
        batch_size = text.size(0)

        q = self.wq(text)
        k = self.wk(vision)
        v = self.wv(vision)

        q = self.split_heads(q, batch_size)
        k = self.split_heads(k, batch_size)
        v = self.split_heads(v, batch_size)

        scores = torch.matmul(q, k.transpose(-2, -1))/torch.sqrt(self.depth)
        attention_weights = torch.softmax(scores, dim=-1)
        context = torch.matmul(attention_weights, v)
        context = context.transpose(1, 2).contiguous().view(batch_size, -1,
                  self.d_model)

        output = self.dense(context)
        return output, attention_weights
```

Applications in Vision-Language Tasks

1. Visual Question Answering (VQA): In Visual Question Answering tasks, cross-modal attention mechanisms enable the model to focus on relevant parts of the image based on the question. This interaction between the visual and textual modalities improves the model's ability to generate accurate answers.
   Example:

```
class VQAModel(nn.Module):
    def __init__(self, vision_dim, text_dim, hidden_dim, num_classes):
        super(VQAModel, self).__init__()
        self.text_encoder = nn.LSTM(text_dim, hidden_dim, batch_first=True)
        self.vision_encoder = nn.Conv2d(3, hidden_dim, kernel_size=3,
                              stride=1, padding=1)
        self.cross_modal_attention = CrossModalAttention(hidden_dim,
                                     num_heads=8)
        self.classifier = nn.Linear(hidden_dim, num_classes)

    def forward(self, text, vision):
        text_out, _ = self.text_encoder(text)
        vision_out = self.vision_encoder(vision).view(vision.size(0),
                     vision.size(1), -1).permute(0, 2, 1)
        context, _ = self.cross_modal_attention(text_out, vision_out)
        output = self.classifier(context.mean(dim=1))
        return output
```

2. Image Captioning: In Image Captioning tasks, cross-modal attention mechanisms help models generate descriptive captions by focusing on relevant regions of the image while generating each word in the caption.
   Example:

```
class ImageCaptioningModel(nn.Module):
    def __init__(self, vision_dim, text_dim, hidden_dim, vocab_size):
        super(ImageCaptioningModel, self).__init__()
        self.vision_encoder = nn.Conv2d(3, hidden_dim, kernel_size=3,
                              stride=1, padding=1)
```

```
        self.text_decoder = nn.LSTM(text_dim, hidden_dim, batch_first=True)
        self.cross_modal_attention = CrossModalAttention(hidden_dim,
                                             num_heads=8)
        self.fc = nn.Linear(hidden_dim, vocab_size)

    def forward(self, vision, text):
        vision_out = self.vision_encoder(vision).view(vision.size(0),
                     vision.size(1), -1).permute(0, 2, 1)
        text_out, _ = self.text_decoder(text)
        context, _ = self.cross_modal_attention(text_out, vision_out)
        output = self.fc(context)
        return output
```

3. Multimodal Sentiment Analysis: In Multimodal Sentiment Analysis, cross-modal attention mechanisms allow the model to integrate information from both text and visual modalities, leading to more accurate sentiment predictions.
   Example:

```
class MultimodalSentimentAnalysisModel(nn.Module):
    def __init__(self, vision_dim, text_dim, hidden_dim, num_classes):
        super(MultimodalSentimentAnalysisModel, self).__init__()
        self.text_encoder = nn.LSTM(text_dim, hidden_dim, batch_first=True)
        self.vision_encoder = nn.Conv2d(3, hidden_dim, kernel_size=3,
                              stride=1, padding=1)
        self.cross_modal_attention = CrossModalAttention(hidden_dim,
                                             num_heads=8)
        self.classifier = nn.Linear(hidden_dim, num_classes)

    def forward(self, text, vision):
        text_out, _ = self.text_encoder(text)
        vision_out = self.vision_encoder(vision).view(vision.size(0),
                     vision.size(1), -1).permute(0, 2, 1)
        context, _ = self.cross_modal_attention(text_out, vision_out)
        output = self.classifier(context.mean(dim=1))
        return output
```

7.4 Advanced Attention Mechanisms

7.4.1 Dynamic and Adaptive Attention

Dynamic and adaptive attention mechanisms are designed to adjust the focus of attention based on the input data or the model's state. Unlike static attention mechanisms, which apply a fixed method to compute attention scores, dynamic and adaptive mechanisms can change their behavior in response to different contexts, leading to more flexible and efficient models.

Adaptive Attention Mechanisms

Adaptive attention mechanisms adjust their parameters or structure based on the input data or intermediate representations. This adaptability allows the model to focus more precisely on relevant information, improving performance on tasks with varying contextual needs. Consider an input sequence X and an attention mechanism that generates context vectors $\mathbf{c}_t$ for each time step t. An adaptive attention mechanism can be formalized as follows:

1. Dynamic Weighting: Let $\mathbf{h}_t$ represent the hidden state at time step t. The attention scores $e_{t,i}$ are computed dynamically based on the hidden state and input vectors $\mathbf{x}_i$:

$$e_{t,i} = f(\mathbf{h}_t, \mathbf{x}_i)$$

   where f is a function that computes the attention scores. This function can be parameterized to adapt based on the input or the model's state.
2. Adaptive Attention Scores: The attention scores are then normalized using a softmax function to obtain the attention weights $\alpha_{t,i}$:

$$\alpha_{t,i} = \frac{\exp(e_{t,i})}{\sum_j \exp(e_{t,j})}$$

3. Context Vector: The context vector $\mathbf{c}_t$ is computed as a weighted sum of the input vectors $\mathbf{x}_i$:

$$\mathbf{c}_t = \sum_i \alpha_{t,i}\mathbf{x}_i$$

   Example: Consider an implementation of adaptive attention using PyTorch:

```
import torch
import torch.nn as nn

class AdaptiveAttention(nn.Module):
    def __init__(self, input_dim, hidden_dim):
        super(AdaptiveAttention, self).__init__()
        self.input_dim = input_dim
        self.hidden_dim = hidden_dim
        self.attn = nn.Linear(input_dim + hidden_dim, hidden_dim)
        self.v = nn.Parameter(torch.rand(hidden_dim))

    def forward(self, hidden, inputs):
        batch_size = inputs.size(0)
        max_len = inputs.size(1)

        hidden = hidden.unsqueeze(1).repeat(1, max_len, 1)
        attn_energies = self.score(hidden, inputs)

        return torch.softmax(attn_energies, dim=1)

    def score(self, hidden, inputs):
        energy = torch.tanh(self.attn(torch.cat([hidden, inputs], dim=2)))
        energy = energy.transpose(1, 2)
        v = self.v.repeat(inputs.size(0), 1).unsqueeze(1)
        energy = torch.bmm(v, energy)
        return energy.squeeze(1)
```

Applications and Use Cases

1. Neural Machine Translation: Adaptive attention mechanisms can dynamically adjust the focus based on the context of the translation task, leading to more accurate translations, especially for complex sentences.

Example:

```
class TranslationModel(nn.Module):
    def __init__(self, input_dim, hidden_dim, output_dim):
        super(TranslationModel, self).__init__()
        self.encoder = nn.LSTM(input_dim, hidden_dim, batch_first=True)
        self.decoder = nn.LSTM(hidden_dim, hidden_dim, batch_first=True)
        self.attention = AdaptiveAttention(hidden_dim, hidden_dim)
        self.fc = nn.Linear(hidden_dim, output_dim)

    def forward(self, src, trg):
        encoder_outputs, (hidden, cell) = self.encoder(src)
        outputs, (hidden, cell) = self.decoder(trg, (hidden, cell))
        attn_weights = self.attention(hidden[-1], encoder_outputs)
        context = attn_weights.bmm(encoder_outputs)
        output = self.fc(context)
        return output
```

2. Text Summarization: In text summarization, adaptive attention mechanisms can help the model to dynamically adjust its focus on different parts of the document, leading to more coherent and relevant summaries.
   Example:

```
class SummarizationModel(nn.Module):
    def __init__(self, input_dim, hidden_dim, output_dim):
        super(SummarizationModel, self).__init__()
        self.encoder = nn.LSTM(input_dim, hidden_dim, batch_first=True)
        self.decoder = nn.LSTM(hidden_dim, hidden_dim, batch_first=True)
        self.attention = AdaptiveAttention(hidden_dim, hidden_dim)
        self.fc = nn.Linear(hidden_dim, output_dim)

    def forward(self, src, trg):
        encoder_outputs, (hidden, cell) = self.encoder(src)
        outputs, (hidden, cell) = self.decoder(trg, (hidden, cell))
        attn_weights = self.attention(hidden[-1], encoder_outputs)
        context = attn_weights.bmm(encoder_outputs)
        output = self.fc(context)
        return output
```

3. Speech Recognition: Adaptive attention mechanisms can improve speech recognition models by dynamically focusing on different parts of the audio signal, leading to more accurate transcriptions.
   Example:

```
class SpeechRecognitionModel(nn.Module):
    def __init__(self, input_dim, hidden_dim, output_dim):
        super(SpeechRecognitionModel, self).__init__()
        self.encoder = nn.LSTM(input_dim, hidden_dim, batch_first=True)
        self.decoder = nn.LSTM(hidden_dim, hidden_dim, batch_first=True)
        self.attention = AdaptiveAttention(hidden_dim, hidden_dim)
        self.fc = nn.Linear(hidden_dim, output_dim)

    def forward(self, src, trg):
        encoder_outputs, (hidden, cell) = self.encoder(src)
        outputs, (hidden, cell) = self.decoder(trg, (hidden, cell))
        attn_weights = self.attention(hidden[-1], encoder_outputs)
        context = attn_weights.bmm(encoder_outputs)
        output = self.fc(context)
        return output
```

7.4.2 *Sparse and Localized Attention*

Sparse and localized attention mechanisms have been developed to address the high computational and memory costs associated with traditional dense attention mechanisms, especially in large-scale models. These techniques focus on reducing the number of attention operations by limiting the scope of attention to relevant subsets of the input, thereby improving efficiency while maintaining or even enhancing model performance.

Reducing Computational Complexity

In traditional dense attention mechanisms, every element in the input sequence attends to every other element, resulting in a quadratic complexity in terms of both computation and memory. This can be prohibitively expensive for long sequences or large-scale models. Sparse and localized attention mechanisms aim to mitigate this by reducing the number of pairwise interactions that need to be computed. Given an input sequence $X \in \mathbb{R}^{N \times d}$, where N is the sequence length and d is the dimensionality of the embeddings, the dense attention mechanism computes the attention scores and context vectors as follows:

1. Dense Attention Scores:
$$\text{scores} = \frac{QK^{\mathrm{T}}}{\sqrt{d_k}}$$
Here, $Q, K, V \in \mathbb{R}^{N \times d}$ are the query, key, and value matrices derived from the input X.
2. Sparse Attention Scores: In sparse attention, we introduce a sparsity pattern $S \in \{0, 1\}^{N \times N}$ that defines which pairs of elements can attend to each other. The attention scores are then computed only for the allowed pairs:
$$\text{scores}_{\text{sparse}} = S \odot \frac{QK^{\mathrm{T}}}{\sqrt{d_k}}$$
where $\odot$ denotes element-wise multiplication.
3. Localized Attention: Localized attention restricts the attention computation to a fixed window size w, reducing the complexity to $O(Nw)$:
$$\text{scores}_{\text{localized}} = \frac{Q_{i:i+w} K_{i:i+w}^{\mathrm{T}}}{\sqrt{d_k}}$$
This formulation ensures that each element only attends to its local neighborhood.

Example: Consider an implementation of sparse and localized attention using PyTorch:

```
import torch
import torch.nn as nn

class SparseAttention(nn.Module):
    def __init__(self, d_model, num_heads, sparsity_pattern):
        super(SparseAttention, self).__init__()
        self.num_heads = num_heads
        self.d_model = d_model
        self.sparsity_pattern = sparsity_pattern

        assert d_model % num_heads == 0

        self.depth = d_model // num_heads

        self.wq = nn.Linear(d_model, d_model)
        self.wk = nn.Linear(d_model, d_model)
        self.wv = nn.Linear(d_model, d_model)
        self.dense = nn.Linear(d_model, d_model)

    def split_heads(self, x, batch_size):
        x = x.view(batch_size, -1, self.num_heads, self.depth)
        return x.transpose(1, 2)

    def forward(self, x):
        batch_size = x.size(0)

        q = self.wq(x)
        k = self.wk(x)
        v = self.wv(x)

        q = self.split_heads(q, batch_size)
        k = self.split_heads(k, batch_size)
        v = self.split_heads(v, batch_size)

        scores = torch.matmul(q, k.transpose(-2, -1))/torch.sqrt(self.depth)
        scores = scores * self.sparsity_pattern
        attention_weights = torch.softmax(scores, dim=-1)
        context = torch.matmul(attention_weights, v)
        context = context.transpose(1, 2).contiguous().view(batch_size,
                  -1, self.d_model)

        output = self.dense(context)
        return output, attention_weights

class LocalizedAttention(nn.Module):
    def __init__(self, d_model, num_heads, window_size):
        super(LocalizedAttention, self).__init__()
        self.num_heads = num_heads
        self.d_model = d_model
        self.window_size = window_size

        assert d_model % num_heads == 0

        self.depth = d_model // num_heads

        self.wq = nn.Linear(d_model, d_model)
        self.wk = nn.Linear(d_model, d_model)
        self.wv = nn.Linear(d_model, d_model)
        self.dense = nn.Linear(d_model, d_model)

    def split_heads(self, x, batch_size):
        x = x.view(batch_size, -1, self.num_heads, self.depth)
        return x.transpose(1, 2)
```

```
    def forward(self, x):
        batch_size = x.size(0)
        seq_length = x.size(1)

        q = self.wq(x)
        k = self.wk(x)
        v = self.wv(x)

        q = self.split_heads(q, batch_size)
        k = self.split_heads(k, batch_size)
        v = self.split_heads(v, batch_size)

        scores = torch.zeros(batch_size, self.num_heads, seq_length,
                 seq_length).to(x.device)
        for i in range(seq_length):
            start = max(0, i - self.window_size // 2)
            end = min(seq_length, i + self.window_size // 2 + 1)
            scores[:, :, i, start:end] = torch.matmul(q[:, :, i:i+1, :],
                                         k[:, :, start:end, :]
                                         .transpose(-2, -1))/torch.sqrt
                                          (self.depth)

        attention_weights = torch.softmax(scores, dim=-1)
        context = torch.matmul(attention_weights, v)
        context = context.transpose(1, 2).contiguous().view(batch_size, -1,
                  self.d_model)

        output = self.dense(context)
        return output, attention_weights
```

Applications in Large-Scale Models

1. NLP: Sparse and localized attention mechanisms are used in large-scale language models to handle long text sequences efficiently. For instance, the Longformer and BigBird models utilize sparse attention patterns to manage long documents effectively.
   Example:

```
from transformers import LongformerModel, LongformerTokenizer

tokenizer = LongformerTokenizer.from_pretrained
            ('allenai/longformer-base-4096')
model = LongformerModel.from_pretrained
        ('allenai/longformer-base-4096')

text = "This is a long document example. " * 100
inputs = tokenizer(text, return_tensors='pt')
outputs = model(**inputs)
```

2. Vision Transformers: In vision transformers, sparse attention can be used to focus on relevant patches of the image, reducing computational load while maintaining accuracy. For example, the Swin Transformer employs a hierarchical approach with localized self-attention to efficiently process high-resolution images.
   Example:

```
from timm.models import swin_transformer

model = swin_transformer.swin_base_patch4_window7_224(pretrained=True)
```

3. Graph Neural Networks: In graph neural networks, sparse attention can be used to handle large-scale graphs by focusing on the most relevant nodes and edges, improving scalability and performance.
   Example:

```
from torch_geometric.nn import GATConv

class GATModel(torch.nn.Module):
    def __init__(self, in_channels, out_channels):
        super(GATModel, self).__init__()
        self.conv1 = GATConv(in_channels, 8, heads=8, dropout=0.6)
        self.conv2 = GATConv(8 * 8, out_channels, heads=1, concat=False,
                             dropout=0.6)

    def forward(self, x, edge_index):
        x = self.conv1(x, edge_index)
        x = torch.relu(x)
        x = self.conv2(x, edge_index)
        return x
```

7.5 Practical Applications of Attention Mechanisms

7.5.1 Image and Video Analysis

Attention mechanisms have revolutionized the field of image and video analysis by enabling models to dynamically focus on the most relevant parts of the input data. This selective focus enhances the ability of models to perform complex tasks such as image classification and video understanding with higher accuracy and efficiency.

Attention in Image Classification

In image classification, attention mechanisms help models to identify and focus on important regions of an image that are crucial for making classification decisions. This approach mimics human visual attention, where the focus is drawn to salient parts of a scene, allowing for more efficient and accurate recognition. Consider an image represented as a feature map $F \in \mathbb{R}^{C \times H \times W}$, where C is the number of channels, and H and W are the height and width of the feature map. The attention mechanism computes an attention map $A \in \mathbb{R}^{H \times W}$ that highlights important regions.

1. Attention Score Computation: Compute the attention scores $e_{i,j}$ for each spatial location (i, j) using a convolutional layer followed by a non-linear activation function:

$$e_{i,j} = \text{ReLU}(W_a * F_{i,j} + b_a)$$

   where W_a and b_a are the weights and bias of the convolutional layer, and $*$ denotes convolution.

2. Attention Weights: Normalize the attention scores to obtain attention weights $\alpha_{i,j}$ using the softmax function:

$$\alpha_{i,j} = \frac{\exp(e_{i,j})}{\sum_{k,l} \exp(e_{k,l})}$$

3. Context Vector: Compute the context vector $\mathbf{c}$ as a weighted sum of the feature map F:

$$\mathbf{c} = \sum_{i,j} \alpha_{i,j} F_{i,j}$$

Example: Consider an implementation of attention in image classification using PyTorch:

```
import torch
import torch.nn as nn

class AttentionLayer(nn.Module):
    def __init__(self, in_channels):
        super(AttentionLayer, self).__init__()
        self.conv = nn.Conv2d(in_channels, 1, kernel_size=1)
        self.relu = nn.ReLU()
        self.softmax = nn.Softmax(dim=2)

    def forward(self, x):
        batch_size, _, h, w = x.size()
        attention_scores = self.relu(self.conv(x)).view(batch_size, -1)
        attention_weights = self.softmax(attention_scores)
                            .view(batch_size, 1, h, w)
        context = x * attention_weights
        return context

class ImageClassificationModel(nn.Module):
    def __init__(self, num_classes):
        super(ImageClassificationModel, self).__init__()
        self.features = nn.Sequential(
            nn.Conv2d(3, 64, kernel_size=3, padding=1),
            nn.ReLU(),
            nn.MaxPool2d(kernel_size=2, stride=2),
            AttentionLayer(64),
            nn.Conv2d(64, 128, kernel_size=3, padding=1),
            nn.ReLU(),
            nn.MaxPool2d(kernel_size=2, stride=2)
        )
        self.classifier = nn.Sequential(
            nn.Linear(128 * 8 * 8, 256),
            nn.ReLU(),
            nn.Linear(256, num_classes)
        )

    def forward(self, x):
        x = self.features(x)
        x = x.view(x.size(0), -1)
        x = self.classifier(x)
        return x
```

Applications

1. Attention mechanisms can improve object recognition by focusing on salient objects within an image, reducing the influence of background noise.
2. In complex scenes, attention mechanisms help models to identify and classify multiple objects and their relationships accurately.

Attention in Video Understanding
Video understanding involves analyzing a sequence of frames to extract meaningful information. Attention mechanisms enhance video understanding by focusing on important frames and regions within frames, allowing models to capture temporal dynamics and spatial features effectively. Given a video represented as a sequence of feature maps $\{F_t\}_{t=1}^{\mathrm{T}}$, where T is the number of frames, attention mechanisms compute spatial and temporal attention maps to highlight important regions and moments.

1. Temporal Attention: Compute the temporal attention scores e_t for each frame t:

$$e_t = \mathrm{ReLU}(W_t \cdot F_t + b_t)$$

   Normalize the scores to obtain temporal attention weights α_t:

$$\alpha_t = \frac{\exp(e_t)}{\sum_{k=1}^{\mathrm{T}} \exp(e_k)}$$

2. Spatial Attention: For each frame t, compute the spatial attention map A_t:

$$A_t = \mathrm{softmax}(\mathrm{ReLU}(W_s * F_t + b_s))$$

3. Context Vector: Compute the context vector $\mathbf{c}$ as a weighted sum of the spatially attended frames:

$$\mathbf{c} = \sum_{t=1}^{\mathrm{T}} \alpha_t \sum_{i,j} A_{t,i,j} F_{t,i,j}$$

   Example: Consider an implementation of attention in video understanding using PyTorch:

```
import torch
import torch.nn as nn

class TemporalAttentionLayer(nn.Module):
    def __init__(self, in_channels, seq_len):
        super(TemporalAttentionLayer, self).__init__()
        self.fc = nn.Linear(in_channels, seq_len)
        self.relu = nn.ReLU()
        self.softmax = nn.Softmax(dim=1)

    def forward(self, x):
        batch_size, seq_len, _ = x.size()
```

```
        attention_scores = self.relu(self.fc(x.mean(dim=2)))
        attention_weights = self.softmax(attention_scores)
        context = torch.bmm(attention_weights.unsqueeze(1), x).squeeze(1)
        return context

class VideoClassificationModel(nn.Module):
    def __init__(self, num_classes, seq_len):
        super(VideoClassificationModel, self).__init__()
        self.features = nn.Sequential(
            nn.Conv3d(3, 64, kernel_size=(3, 3, 3), padding=1),
            nn.ReLU(),
            nn.MaxPool3d(kernel_size=(1, 2, 2)),
            nn.Conv3d(64, 128, kernel_size=(3, 3, 3), padding=1),
            nn.ReLU(),
            nn.MaxPool3d(kernel_size=(1, 2, 2))
        )
        self.temporal_attention = TemporalAttentionLayer(128 * 16 * 16, seq_len)
        self.classifier = nn.Sequential(
            nn.Linear(128 * 16 * 16, 256),
            nn.ReLU(),
            nn.Linear(256, num_classes)
        )

    def forward(self, x):
        batch_size, seq_len, _, _, _ = x.size()
        x = x.view(batch_size * seq_len, 3, 32, 32, 32)
        x = self.features(x)
        x = x.view(batch_size, seq_len, -1)
        x = self.temporal_attention(x)


        x = x.view(batch_size, seq_len, -1)
        x = self.temporal_attention(x)
        x = x.view(x.size(0), -1)
        x = self.classifier(x)
        return x
```

Applications

1. Action Recognition: Attention mechanisms can identify key frames and regions within frames that are critical for recognizing specific actions, improving the accuracy of action recognition models.
2. Video Summarization: By focusing on the most informative parts of a video, attention mechanisms can help create concise and relevant summaries, highlighting the main events and reducing redundancy.

7.5.2 *Speech and Audio Processing*

Attention mechanisms have significantly advanced the field of speech and audio processing by enabling models to focus on relevant parts of the input signal. This selective focus is particularly beneficial for tasks such as speech recognition and audio event detection, where the ability to highlight important temporal regions can enhance the performance and accuracy of models.

Speech Recognition

In speech recognition, attention mechanisms help models to focus on important parts of the audio signal that correspond to spoken words, thereby improving the transcription accuracy. This approach is especially useful for handling long audio sequences where certain segments may carry more information than others. Consider an input audio signal represented as a sequence of acoustic feature vectors $X \in \mathbb{R}^{T \times d}$, where T is the number of time steps, and d is the dimensionality of the feature vectors. The attention mechanism computes attention scores and context vectors as follows:

1. Attention Score Computation: Compute the attention scores e_t for each time step t using a feed-forward neural network followed by a non-linear activation function:

$$e_t = \tanh(W_a h_t + b_a)$$

   where W_a and b_a are the weights and bias of the network, and h_t is the hidden state at time step t.
2. Attention Weights: Normalize the attention scores to obtain attention weights α_t:

$$\alpha_t = \frac{\exp(e_t)}{\sum_{k=1}^{\mathrm{T}} \exp(e_k)}$$

3. Context Vector: Compute the context vector $\mathbf{c}$ as a weighted sum of the hidden states:

$$\mathbf{c} = \sum_{t=1}^{\mathrm{T}} \alpha_t h_t$$

Example: Consider an implementation of attention in speech recognition using PyTorch:

```
import torch
import torch.nn as nn

class AttentionLayer(nn.Module):
    def __init__(self, hidden_dim):
        super(AttentionLayer, self).__init__()
        self.attn = nn.Linear(hidden_dim, hidden_dim)
        self.v = nn.Parameter(torch.rand(hidden_dim))
        self.softmax = nn.Softmax(dim=1)

    def forward(self, hidden_states):
        attn_scores = torch.tanh(self.attn(hidden_states))
        attn_scores = torch.matmul(attn_scores, self.v)
        attn_weights = self.softmax(attn_scores)
        context = torch.sum(attn_weights.unsqueeze(2) * hidden_states, dim=1)
        return context, attn_weights

class SpeechRecognitionModel(nn.Module):
    def __init__(self, input_dim, hidden_dim, output_dim):
        super(SpeechRecognitionModel, self).__init__()
        self.lstm = nn.LSTM(input_dim, hidden_dim, batch_first=True)
        self.attention = AttentionLayer(hidden_dim)
        self.fc = nn.Linear(hidden_dim, output_dim)
```

```
    def forward(self, x):
        lstm_out, _ = self.lstm(x)
        context, attn_weights = self.attention(lstm_out)
        output = self.fc(context)
        return output, attn_weights
```

Applications

1. Automatic Speech Recognition (ASR): Attention mechanisms enhance ASR systems by allowing them to focus on phonemically rich segments of the audio signal, improving the transcription of spoken words into text.
2. Voice Command Recognition: In voice controlled systems, attention mechanisms help in accurately recognizing voice commands by focusing on the critical parts of the spoken input.

Audio Event Detection

In audio event detection, attention mechanisms enable models to identify and focus on important segments of the audio signal that correspond to specific events, such as alarms, music, or speech. This capability is crucial for accurately detecting and classifying various audio events in diverse environments. Given an audio signal represented as a sequence of feature vectors $X \in \mathbb{R}^{T \times d}$, the attention mechanism computes attention scores and context vectors as follows:

1. Attention Score Computation: Compute the attention scores e_t for each time step t using a feed-forward neural network:

$$e_t = \tanh(W_a h_t + b_a)$$

   where W_a and b_a are the weights and bias, and h_t is the hidden state at time step t.
2. Attention Weights: Normalize the attention scores to obtain attention weights α_t:

$$\alpha_t = \frac{\exp(e_t)}{\sum_{k=1}^{\mathrm{T}} \exp(e_k)}$$

3. Context Vector: Compute the context vector $\mathbf{c}$ as a weighted sum of the hidden states:

$$\mathbf{c} = \sum_{t=1}^{\mathrm{T}} \alpha_t h_t$$

Example: Consider an implementation of attention in audio event detection using PyTorch:

```
import torch
import torch.nn as nn

class AttentionLayer(nn.Module):
    def __init__(self, hidden_dim):
        super(AttentionLayer, self).__init__()
        self.attn = nn.Linear(hidden_dim, hidden_dim)
        self.v = nn.Parameter(torch.rand(hidden_dim))
        self.softmax = nn.Softmax(dim=1)

    def forward(self, hidden_states):
        attn_scores = torch.tanh(self.attn(hidden_states))
        attn_scores = torch.matmul(attn_scores, self.v)
        attn_weights = self.softmax(attn_scores)
        context = torch.sum(attn_weights.unsqueeze(2) * hidden_states,dim=1)
        return context, attn_weights

class AudioEventDetectionModel(nn.Module):
    def __init__(self, input_dim, hidden_dim, output_dim):
        super(AudioEventDetectionModel, self).__init__()
        self.lstm = nn.LSTM(input_dim, hidden_dim, batch_first=True)
        self.attention = AttentionLayer(hidden_dim)
        self.fc = nn.Linear(hidden_dim, output_dim)

    def forward(self, x):
        lstm_out, _ = self.lstm(x)
        context, attn_weights = self.attention(lstm_out)
        output = self.fc(context)
        return output, attn_weights
```

Applications

1. Environmental Sound Recognition: Attention mechanisms can identify important segments in environmental sound recordings, improving the detection and classification of sounds like sirens, bird calls, and footsteps.
2. Anomaly Detection: In industrial settings, attention mechanisms help in detecting anomalous sounds, such as machinery malfunctions, by focusing on unusual audio patterns.

7.5.3 Multimodal Applications

Multimodal applications leverage attention mechanisms to integrate and align information from different modalities, such as images and text. These mechanisms enhance the model's ability to understand and process complex inputs by focusing on the most relevant parts of each modality. This section explores three key applications: image captioning, visual question answering (VQA), and multimodal sentiment analysis.

Image Captioning
Image captioning involves generating descriptive textual captions for images. Attention mechanisms play a crucial role by allowing the model to focus on specific

regions of an image while generating each word in the caption. This dynamic focus ensures that the generated caption accurately describes the visual content. Consider an image represented as a set of feature vectors $V \in \mathbb{R}^{N \times d_v}$, where N is the number of regions and d_v is the dimensionality of the visual features. The text is represented as a sequence of word embeddings $T \in \mathbb{R}^{M \times d_t}$, where M is the number of words and d_t is the dimensionality of the embeddings.

1. Attention Score Computation: Compute the attention scores $e_{t,i}$ for each word t and image region i:

$$e_{t,i} = f(T_t, V_i)$$

   where f is a scoring function, typically implemented as a feed-forward neural network.
2. Attention Weights: Normalize the attention scores to obtain attention weights $\alpha_{t,i}$:

$$\alpha_{t,i} = \frac{\exp(e_{t,i})}{\sum_{j=1}^{N} \exp(e_{t,j})}$$

3. Context Vector: Compute the context vector $\mathbf{c}_t$ for each word t:

$$\mathbf{c}_t = \sum_{i=1}^{N} \alpha_{t,i} V_i$$

   Example: Consider an implementation of attention in image captioning using PyTorch:

```
import torch
import torch.nn as nn

class AttentionLayer(nn.Module):
    def __init__(self, hidden_dim, visual_dim):
        super(AttentionLayer, self).__init__()
        self.attn = nn.Linear(hidden_dim + visual_dim, hidden_dim)
        self.v = nn.Parameter(torch.rand(hidden_dim))
        self.softmax = nn.Softmax(dim=1)

    def forward(self, hidden, visual_features):
        combined = torch.cat([hidden.unsqueeze(1).repeat(1,
                   visual_features.size(1), 1), visual_features], dim=2)
        attn_scores = torch.tanh(self.attn(combined))
        attn_scores = torch.matmul(attn_scores, self.v)
        attn_weights = self.softmax(attn_scores)
        context = torch.sum(attn_weights.unsqueeze(2) * visual_features,
                  dim=1)
        return context, attn_weights

class ImageCaptioningModel(nn.Module):
    def __init__(self, vocab_size, visual_dim, hidden_dim):
        super(ImageCaptioningModel, self).__init__()
        self.embedding = nn.Embedding(vocab_size, hidden_dim)
        self.lstm = nn.LSTM(hidden_dim, hidden_dim, batch_first=True)
        self.attention = AttentionLayer(hidden_dim, visual_dim)
        self.fc = nn.Linear(hidden_dim, vocab_size)
```

```
    def forward(self, captions, visual_features):
        embeddings = self.embedding(captions)
        lstm_out, _ = self.lstm(embeddings)
        context, attn_weights = self.attention(lstm_out, visual_features)
        output = self.fc(context)
        return output, attn_weights
```

Applications

1. Generating Descriptions for Images: Attention mechanisms enable models to generate accurate and detailed descriptions of images, improving the performance of image captioning systems.
2. Assisting Visually Impaired Users: Image captioning systems powered by attention mechanisms can help visually impaired users by providing textual descriptions of their surroundings.

Visual Question Answering (VQA)

Visual question answering involves answering questions about the content of an image. Attention mechanisms help by allowing the model to focus on relevant regions of the image based on the question, improving the accuracy and relevance of the answers. Given an image represented as feature vectors $V \in \mathbb{R}^{N \times d_v}$ and a question represented as word embeddings $Q \in \mathbb{R}^{M \times d_q}$:

1. Attention Score Computation: Compute the attention scores $e_{i,j}$ for each image region i and question word j:

$$e_{i,j} = f(V_i, Q_j)$$

   where f is a scoring function.
2. Attention Weights: Normalize the attention scores to obtain attention weights $\alpha_{i,j}$:

$$\alpha_{i,j} = \frac{\exp(e_{i,j})}{\sum_{k=1}^{N} \exp(e_{k,j})}$$

3. Context Vector: Compute the context vector $\mathbf{c}_i$ for each image region i:

$$\mathbf{c}_i = \sum_{j=1}^{M} \alpha_{i,j} Q_j$$

Example: Consider an implementation of attention in VQA using PyTorch:

```
import torch
import torch.nn as nn

class VQAAttentionLayer(nn.Module):
    def __init__(self, visual_dim, question_dim, hidden_dim):
        super(VQAAttentionLayer, self).__init__()
        self.attn = nn.Linear(visual_dim + question_dim, hidden_dim)
        self.v = nn.Parameter(torch.rand(hidden_dim))
```

```
            self.softmax = nn.Softmax(dim=1)

        def forward(self, visual_features, question_features):
            combined = torch.cat([visual_features, question_features
                        .unsqueeze(1).repeat(1, visual_features.size(1), 1)]
                        , dim=2)
            attn_scores = torch.tanh(self.attn(combined))
            attn_scores = torch.matmul(attn_scores, self.v)
            attn_weights = self.softmax(attn_scores)
            context = torch.sum(attn_weights.unsqueeze(2) * visual_features,
                      dim=1)
            return context, attn_weights

    class VQAModel(nn.Module):
        def __init__(self, vocab_size, visual_dim, question_dim, hidden_dim,
        num_answers):
            super(VQAModel, self).__init__()
            self.question_embedding = nn.Embedding(vocab_size, question_dim)
            self.question_lstm = nn.LSTM(question_dim, hidden_dim,
                                 batch_first=True)
            self.attention = VQAAttentionLayer(visual_dim, hidden_dim,
                             hidden_dim)
            self.fc = nn.Linear(hidden_dim, num_answers)

        def forward(self, questions, visual_features):
            question_embeddings = self.question_embedding(questions)
            question_features, _ = self.question_lstm(question_embeddings)
            context, attn_weights = self.attention(visual_features,
                                    question_features[:, -1])
            output = self.fc(context)
            return output, attn_weights
```

Applications

1. Interactive Question Answering: Attention mechanisms enable VQA systems to interactively answer questions about images, improving user engagement and satisfaction.
2. Educational Tools: VQA systems can be used as educational tools to help students learn about visual content through interactive questioning.

Multimodal Sentiment Analysis

Multimodal sentiment analysis involves analyzing and predicting sentiment by integrating information from different modalities, such as text and images. Attention mechanisms help by focusing on the most relevant parts of each modality, leading to more accurate sentiment predictions. Given an image represented as feature vectors $V \in \mathbb{R}^{N \times d_v}$ and text represented as word embeddings $T \in \mathbb{R}^{M \times d_t}$:

1. Attention Score Computation: Compute the attention scores $e_{i,j}$ for each image region i and text word j:

$$e_{i,j} = f(V_i, T_j)$$

where f is a scoring function.

2. Attention Weights: Normalize the attention scores to obtain attention weights $\alpha_{i,j}$:

$$\alpha_{i,j} = \frac{\exp(e_{i,j})}{\sum_{k=1}^{N} \exp(e_{k,j})}$$

3. Context Vector: Compute the context vector $\mathbf{c}_i$ for each image region i:

$$\mathbf{c}_i = \sum_{j=1}^{M} \alpha_{i,j} T_j$$

Example: Consider an implementation of attention in multimodal sentiment analysis using PyTorch:

```
import torch
import torch.nn as nn

class MultimodalAttentionLayer(nn.Module):
    def __init__(self, visual

_dim, text_dim, hidden_dim):
        super(MultimodalAttentionLayer, self).__init__()
        self.attn = nn.Linear(visual_dim + text_dim, hidden_dim)
        self.v = nn.Parameter(torch.rand(hidden_dim))
        self.softmax = nn.Softmax(dim=1)

    def forward(self, visual_features, text_features):
        combined = torch.cat([visual_features, text_features.unsqueeze(1)
                  .repeat(1, visual_features.size(1), 1)], dim=2)
        attn_scores = torch.tanh(self.attn(combined))
        attn_scores = torch.matmul(attn_scores, self.v)
        attn_weights = self.softmax(attn_scores)
        context = torch.sum(attn_weights.unsqueeze(2)
                  * visual_features, dim=1)
        return context, attn_weights

class MultimodalSentimentAnalysisModel(nn.Module):
    def __init__(self, vocab_size, visual_dim, text_dim, hidden_dim,
    num_classes):
        super(MultimodalSentimentAnalysisModel, self).__init__()
        self.text_embedding = nn.Embedding(vocab_size, text_dim)
        self.text_lstm = nn.LSTM(text_dim, hidden_dim, batch_first=True)
        self.attention = MultimodalAttentionLayer(visual_dim, hidden_dim,
                         hidden_dim)
        self.fc = nn.Linear(hidden_dim, num_classes)

    def forward(self, texts, visual_features):
        text_embeddings = self.text_embedding(texts)
        text_features, _ = self.text_lstm(text_embeddings)
        context, attn_weights = self.attention(visual_features,
                                text_features[:, -1])
        output = self.fc(context)
        return output, attn_weights
```

Applications

1. Social Media Analysis: Multimodal sentiment analysis can be used to analyze posts on social media platforms, combining text and image content to predict user sentiment accurately.

2. Customer Feedback Analysis: By integrating textual reviews and accompanying images, businesses can gain deeper insights into customer sentiment and preferences.

7.6 Case Studies

7.6.1 Image Classification Case Study

In this case study, we focus on the application of attention mechanisms to enhance image classification tasks. The dataset used is the CIFAR-10 dataset, which consists of 60,000 32 × 32 color images in 10 different classes.

Objective: To build an attention-based CNN that improves classification accuracy by focusing on important regions of the images.

Model Architecture: The model architecture includes convolutional layers followed by an attention mechanism and fully connected layers for classification.

1. Convolutional Layer: Let $X \in \mathbb{R}^{B \times C \times H \times W}$ be the input tensor, where B is the batch size, C is the number of channels, and H and W are the height and width of the images.
   The convolutional operation can be represented as:
   $$Y = W * X + b$$
   where W is the weight tensor and b is the bias.
2. Attention Mechanism: Compute the attention scores e for the feature maps:
   $$e = \tanh(W_a * Y + b_a)$$
   Normalize the attention scores using a softmax function:
   $$\alpha = \frac{\exp(e)}{\sum_k \exp(e_k)}$$
   Compute the context vector:
   $$C = \alpha \odot Y$$
3. Fully Connected Layer: Flatten the context vector and pass it through fully connected layers:
   $$z = \text{ReLU}(W_f \cdot \text{flatten}(C) + b_f)$$

Compute the final output:

$$\hat{y} = \text{softmax}(W_o \cdot z + b_o)$$

Implementation:

```
import torch
import torch.nn as nn
import torch.optim as optim
import torchvision.transforms as transforms
import torchvision.datasets as datasets
import torchvision.models as models

class AttentionLayer(nn.Module):
    def __init__(self, in_channels):
        super(AttentionLayer, self).__init__()
        self.conv = nn.Conv2d(in_channels, 1, kernel_size=1)
        self.relu = nn.ReLU()
        self.softmax = nn.Softmax(dim=2)

    def forward(self, x):
        batch_size, _, h, w = x.size()
        attention_scores = self.relu(self.conv(x)).view(batch_size, -1)
        attention_weights = self.softmax(attention_scores).view(batch_size,
                            1, h, w)
        context = x * attention_weights
        return context

class AttentionCNN(nn.Module):
    def __init__(self, num_classes):
        super(AttentionCNN, self).__init__()
        self.features = nn.Sequential(
            nn.Conv2d(3, 64, kernel_size=3, padding=1),
            nn.ReLU(),
            nn.MaxPool2d(kernel_size=2, stride=2),
            AttentionLayer(64),
            nn.Conv2d(64, 128, kernel_size=3, padding=1),
            nn.ReLU(),
            nn.MaxPool2d(kernel_size=2, stride=2)
        )
        self.classifier = nn.Sequential(
            nn.Linear(128 * 8 * 8, 256),
            nn.ReLU(),
            nn.Linear(256, num_classes)
        )

    def forward(self, x):
        x = self.features(x)
        x = x.view(x.size(0), -1)
        x = self.classifier(x)
        return x

transform = transforms.Compose([
    transforms.ToTensor(),
    transforms.Normalize((0.5, 0.5, 0.5), (0.5, 0.5, 0.5))
])

trainset = datasets.CIFAR10(root='./data', train=True, download=True,
           transform=transform)
trainloader = torch.utils.data.DataLoader(trainset, batch_size=100,
              shuffle=True, num_workers=2)

testset = datasets.CIFAR10(root='./data', train=False, download=True,
          transform=transform)
testloader = torch.utils.data.DataLoader(testset, batch_size=100,
```

```
                shuffle=False, num_workers=2)

model = AttentionCNN(num_classes=10)
criterion = nn.CrossEntropyLoss()
optimizer = optim.Adam(model.parameters(), lr=0.001)

for epoch in range(10):
    model.train()
    running_loss = 0.0
    for inputs, labels in trainloader:
        optimizer.zero_grad()
        outputs = model(inputs)
        loss = criterion(outputs, labels)
        loss.backward()
        optimizer.step()
        running_loss += loss.item()
    print(f'Epoch {epoch + 1}, Loss: {running_loss / len(trainloader)}')

model.eval()
correct = 0
total = 0
with torch.no_grad():
    for inputs, labels in testloader:
        outputs = model(inputs)
        _, predicted = torch.max(outputs, 1)
        total += labels.size(0)
        correct += (predicted == labels).sum().item()
print(f'Accuracy: {100 * correct / total}%')
```

This case study demonstrates how attention mechanisms can enhance image classification by focusing on important regions. The implementation highlights the integration of an attention layer within a CNN and the subsequent improvement in classification accuracy.

7.6.2 Multimodal Learning Case Study

This case study focuses on the application of attention mechanisms in multimodal learning, specifically integrating text and image data for a visual question answering (VQA) task.

Objective: To build a multimodal model that answers questions about images by leveraging attention mechanisms to align and integrate information from both modalities.

Model Architecture: The model consists of separate encoders for image and text data, followed by an attention mechanism that integrates the two modalities for answering questions.

1. Text Encoder: Encode the question Q using a bidirectional LSTM:

$$h_t = \text{LSTM}(Q_t, h_{t-1})$$

2. Image Encoder: Encode the image I using a CNN to extract feature maps V:

$$V = \text{CNN}(I)$$

3. Attention Mechanism: Compute the attention scores and context vectors for integrating image and text features:

$$e_{i,j} = f(V_i, h_j)$$

$$\alpha_{i,j} = \frac{\exp(e_{i,j})}{\sum_k \exp(e_{k,j})}$$

$$\mathbf{c}_i = \sum_j \alpha_{i,j} h_j$$

4. Output Layer: Use the integrated context vector to predict the answer:

$$\hat{y} = \text{softmax}(W \cdot \mathbf{c} + b)$$

Implementation: Consider an implementation of attention in VQA using PyTorch:

```
import torch
import torch.nn as nn
import torchvision.models as models

class VQAAttentionLayer(nn.Module):
    def __init__(self, visual_dim, question_dim, hidden_dim):
        super(VQAAttentionLayer, self).__init__()
        self.attn = nn.Linear(visual_dim + question_dim, hidden_dim)
        self.v = nn.Parameter(torch.rand(hidden_dim))
        self.softmax = nn.Softmax(dim=1)

    def forward(self, visual_features, question_features):
        combined = torch.cat([visual_features, question_features.unsqueeze(1)
                   .repeat(1, visual_features.size(1), 1)], dim=2)
        attn_scores = torch.tanh(self.attn(combined))
        attn_scores = torch.matmul(attn_scores, self.v)
        attn_weights = self.softmax(attn_scores)
        context = torch.sum(attn_weights.unsqueeze(2) * visual_features,
                  dim=1)
        return context, attn_weights

class VQAModel(nn.Module):
    def __init__(self, vocab_size, visual_dim, question_dim, hidden_dim,
    num_answers):
        super(VQAModel, self).__init__()
        self.question_embedding = nn.Embedding(vocab_size, question_dim)
        self.question_lstm = nn.LSTM(question_dim, hidden_dim,
                                     batch_first=True)
        self.attention = VQAAttentionLayer(visual_dim, hidden_dim,
                                           hidden_dim)
        self.fc = nn.Linear(hidden_dim, num_answers)
        self.cnn = models.resnet18(pretrained=True)
        self.cnn.fc = nn.Identity()  # Remove the final classification layer

    def forward(self, questions, images):
        # Extract image features
```

```
        visual_features = self.cnn(images)
        visual_features = visual_features.unsqueeze(1)

        # Extract question features
        question_embeddings = self.question_embedding(questions)
        question_features, _ = self.question_lstm(question_embeddings)

        # Apply attention
        context, attn_weights = self.att
ention(visual_features, question_features[:, -1])

        # Classification
        output = self.fc(context)
        return output, attn_weights
```

This case study highlights the use of attention mechanisms to integrate and align information from multiple modalities, improving the performance of tasks such as VQA. The model demonstrates the effectiveness of attention in handling complex multimodal data.

7.7 Exercises

1. Given the following parameters for an additive attention mechanism:

$$\mathbf{W}_1 = \begin{pmatrix} 0.1 & 0.2 \\ 0.3 & 0.4 \end{pmatrix}, \quad \mathbf{W}_2 = \begin{pmatrix} 0.5 & 0.6 \\ 0.7 & 0.8 \end{pmatrix}, \quad \mathbf{v} = \begin{pmatrix} 1 \\ 1 \end{pmatrix}$$

   Input sequences:

$$\mathbf{h}_t = \begin{pmatrix} 0.1 \\ 0.2 \end{pmatrix}, \quad \mathbf{s}_{t-1} = \begin{pmatrix} 0.3 \\ 0.4 \end{pmatrix}$$

   (a) Compute the score using the additive attention mechanism.
   (b) Calculate the attention weights.
   (c) Compute the context vector.

2. Consider the following input sequences and parameters for multiplicative attention:

$$\mathbf{Q} = \begin{pmatrix} 1 & 2 \\ 3 & 4 \end{pmatrix}, \quad \mathbf{K} = \begin{pmatrix} 2 & 0 \\ 1 & 3 \end{pmatrix}, \quad \mathbf{V} = \begin{pmatrix} 0.5 & 1 \\ 2 & 1.5 \end{pmatrix}$$

   (a) Compute the attention scores using the dot-product attention mechanism.
   (b) Calculate the attention weights using the softmax function.
   (c) Compute the context vector for the given query matrix $\mathbf{Q}$.

3. Given the following parameters for hierarchical attention:

$$\mathbf{W}_{h1} = \begin{pmatrix} 0.2 & 0.4 \\ 0.6 & 0.8 \end{pmatrix}, \quad \mathbf{W}_{h2} = \begin{pmatrix} 0.1 & 0.3 \\ 0.5 & 0.7 \end{pmatrix}$$

Input sequences:

$$\mathbf{h}_{t1} = \begin{pmatrix} 0.3 \\ 0.6 \end{pmatrix}, \quad \mathbf{h}_{t2} = \begin{pmatrix} 0.4 \\ 0.5 \end{pmatrix}, \quad \mathbf{h}_{t3} = \begin{pmatrix} 0.7 \\ 0.8 \end{pmatrix}$$

(a) Compute the hierarchical attention scores for each level.
(b) Calculate the combined attention weights.
(c) Compute the final context vector for hierarchical attention.

4. Given a visual attention mechanism applied to image data:
Image patches (flattened):

$$\mathbf{X}_1 = \begin{pmatrix} 0.1 \\ 0.2 \\ 0.3 \end{pmatrix}, \quad \mathbf{X}_2 = \begin{pmatrix} 0.4 \\ 0.5 \\ 0.6 \end{pmatrix}, \quad \mathbf{X}_3 = \begin{pmatrix} 0.7 \\ 0.8 \\ 0.9 \end{pmatrix}$$

Attention parameters:

$$\mathbf{W}_Q = \begin{pmatrix} 0.2 & 0.1 \\ 0.4 & 0.3 \\ 0.6 & 0.5 \end{pmatrix}, \quad \mathbf{W}_K = \begin{pmatrix} 0.1 & 0.3 \\ 0.2 & 0.4 \\ 0.5 & 0.7 \end{pmatrix}, \quad \mathbf{W}_V = \begin{pmatrix} 0.6 & 0.5 \\ 0.7 & 0.8 \\ 0.9 & 1.0 \end{pmatrix}$$

(a) Compute the query, key, and value matrices for the image patches.
(b) Calculate the attention scores and weights.
(c) Compute the attended output for the image patches.

5. Given the following parameters for cross-modal attention between text and image modalities:
Text query vector:

$$\mathbf{Q}_{\text{text}} = \begin{pmatrix} 0.5 \\ 0.6 \\ 0.7 \end{pmatrix}$$

Image key and value vectors:

$$\mathbf{K}_{\text{image}} = \begin{pmatrix} 0.1 & 0.2 \\ 0.3 & 0.4 \\ 0.5 & 0.6 \end{pmatrix}, \quad \mathbf{V}_{\text{image}} = \begin{pmatrix} 0.4 & 0.3 \\ 0.2 & 0.1 \\ 0.6 & 0.5 \end{pmatrix}$$

(a) Compute the cross-modal attention scores between the text query and image keys.

(b) Calculate the attention weights using the softmax function.
(c) Compute the context vector using the image value vectors.

6. Given a dynamic attention mechanism with the following parameters:

$$\mathbf{W}_{\text{dyn}} = \begin{pmatrix} 0.2 & 0.4 \\ 0.6 & 0.8 \end{pmatrix}, \quad \mathbf{b}_{\text{dyn}} = \begin{pmatrix} 0.1 \\ 0.2 \end{pmatrix}$$

Input sequences:

$$\mathbf{h}_t = \begin{pmatrix} 0.3 \\ 0.6 \end{pmatrix}, \quad \mathbf{s}_{t-1} = \begin{pmatrix} 0.4 \\ 0.5 \end{pmatrix}$$

(a) Compute the dynamic attention scores.
(b) Calculate the attention weights.
(c) Compute the context vector dynamically.

7. Consider an additive attention mechanism where the query tensor $\mathbf{Q} \in \mathbb{R}^{3\times 4}$, key tensor $\mathbf{K} \in \mathbb{R}^{3\times 4}$, and value tensor $\mathbf{V} \in \mathbb{R}^{3\times 4}$ are given by:

$$\mathbf{Q} = \begin{pmatrix} 1 & 2 & 3 & 4 \\ 2 & 3 & 4 & 5 \\ 3 & 4 & 5 & 6 \end{pmatrix}, \quad \mathbf{K} = \begin{pmatrix} 0.1 & 0.2 & 0.3 & 0.4 \\ 0.2 & 0.3 & 0.4 & 0.5 \\ 0.3 & 0.4 & 0.5 & 0.6 \end{pmatrix}, \quad \mathbf{V} = \begin{pmatrix} 1 & 0 & 1 & 0 \\ 0 & 1 & 0 & 1 \\ 1 & 1 & 1 & 1 \end{pmatrix}$$

The additive attention scoring function is defined as:

$$e_{ij} = \mathbf{v}^{\mathrm{T}} \tanh(\mathbf{W}_1 \mathbf{Q}_i + \mathbf{W}_2 \mathbf{K}_j)$$

where $\mathbf{v} \in \mathbb{R}^{4\times 1}$, $\mathbf{W}_1, \mathbf{W}_2 \in \mathbb{R}^{4\times 4}$ are weight matrices. Let:

$$\mathbf{v} = \begin{pmatrix} 0.5 \\ 0.5 \\ 0.5 \\ 0.5 \end{pmatrix}, \quad \mathbf{W}_1 = \mathbf{W}_2 = \mathbf{I}_4$$

(a) Compute the attention scores e_{ij} for each pair of $\mathbf{Q}_i$ and $\mathbf{K}_j$.
(b) Apply a softmax function to the attention scores and compute the final weighted sum of the value vectors $\mathbf{V}$.

8. Given the same query, key, and value tensors from the previous question, assume the weight matrices $\mathbf{W}_1$ and $\mathbf{W}_2$ are:

$$\mathbf{W}_1 = \begin{pmatrix} 0.1 & 0.2 & 0.3 & 0.4 \\ 0.2 & 0.3 & 0.4 & 0.5 \\ 0.3 & 0.4 & 0.5 & 0.6 \\ 0.4 & 0.5 & 0.6 & 0.7 \end{pmatrix}, \quad \mathbf{W}_2 = \begin{pmatrix} 0.7 & 0.6 & 0.5 & 0.4 \\ 0.6 & 0.5 & 0.4 & 0.3 \\ 0.5 & 0.4 & 0.3 & 0.2 \\ 0.4 & 0.3 & 0.2 & 0.1 \end{pmatrix}$$

(a) Compute the modified attention scores e_{ij} using the new weight matrices.
(b) Calculate the new attention weights and the resulting output after applying these weights to the value tensor $\mathbf{V}$.

9. In a sequence-to-sequence model, the query tensor $\mathbf{Q} \in \mathbb{R}^{1\times 4}$ represents the decoder state, while the key tensor $\mathbf{K} \in \mathbb{R}^{3\times 4}$ and value tensor $\mathbf{V} \in \mathbb{R}^{3\times 4}$ represent the encoder outputs:

$$\mathbf{Q} = \begin{pmatrix} 1 & 1 & 1 & 1 \end{pmatrix}, \quad \mathbf{K} = \begin{pmatrix} 2 & 2 & 2 & 2 \\ 3 & 3 & 3 & 3 \\ 4 & 4 & 4 & 4 \end{pmatrix}, \quad \mathbf{V} = \begin{pmatrix} 0 & 1 & 0 & 1 \\ 1 & 0 & 1 & 0 \\ 0 & 0 & 0 & 0 \end{pmatrix}$$

Let the weight matrices be identity matrices and the scoring vector $\mathbf{v} = \begin{pmatrix} 1 \\ 1 \\ 1 \\ 1 \end{pmatrix}$.

(a) Compute the attention scores for the query tensor against each key tensor.
(b) Determine the attention weights and compute the context vector for the decoder.

10. For the query tensor $\mathbf{Q} \in \mathbb{R}^{2\times 3}$, key tensor $\mathbf{K} \in \mathbb{R}^{2\times 3}$, and value tensor $\mathbf{V} \in \mathbb{R}^{2\times 3}$:

$$\mathbf{Q} = \begin{pmatrix} 1 & 0 & 2 \\ 0 & 1 & 3 \end{pmatrix}, \quad \mathbf{K} = \begin{pmatrix} 0.5 & 1 & 1.5 \\ 1 & 0.5 & 2 \end{pmatrix}, \quad \mathbf{V} = \begin{pmatrix} 1 & 1 & 1 \\ 2 & 2 & 2 \end{pmatrix}$$

Let the weight matrices be identity matrices and the scoring vector $\mathbf{v} = \begin{pmatrix} 0.3 \\ 0.3 \\ 0.4 \end{pmatrix}$.

(a) Compute the attention scores using the ReLU activation function in the scoring function instead of the tanh function.
(b) Compute the attention weights and the final output using these new scores.

11. Consider a document consisting of 3 sentences, each represented by a sequence of 4 word embeddings. The word embeddings for each sentence form a tensor $\mathbf{E} \in \mathbb{R}^{3\times 4\times 5}$ (3 sentences, 4 words per sentence, 5-dimensional embeddings):

$$\mathbf{E}_1 = \begin{pmatrix} 1 & 0 & 1 & 0 & 1 \\ 0 & 1 & 0 & 1 & 0 \\ 1 & 1 & 1 & 1 & 1 \\ 0 & 0 & 0 & 1 & 1 \end{pmatrix}, \quad \mathbf{E}_2 = \begin{pmatrix} 1 & 1 & 0 & 0 & 1 \\ 0 & 1 & 1 & 0 & 0 \\ 1 & 0 & 0 & 1 & 1 \\ 0 & 1 & 1 & 1 & 0 \end{pmatrix}, \quad \mathbf{E}_3 = \begin{pmatrix} 0 & 0 & 1 & 1 & 1 \\ 1 & 0 & 0 & 1 & 0 \\ 1 & 1 & 0 & 0 & 1 \\ 0 & 1 & 0 & 1 & 1 \end{pmatrix}$$

The word-level attention mechanism is defined by query, key, and value matrices $\mathbf{Q}_w, \mathbf{K}_w, \mathbf{V}_w \in \mathbb{R}^{5\times 5}$:

$$\mathbf{Q}_w = \mathbf{K}_w = \mathbf{V}_w = \mathbf{I}_5$$

(a) Compute the word-level attention scores and the resulting word context vectors for each sentence.
(b) Aggregate the word context vectors to form sentence representations. Then, apply sentence-level attention using the same weight matrices to compute the final document representation.

12. Given the same document tensor **E** as in the previous question, let the word-level and sentence-level weight matrices be:

$$\mathbf{Q}_w = \begin{pmatrix} 0.1 & 0.2 & 0.3 & 0.4 & 0.5 \\ 0.5 & 0.4 & 0.3 & 0.2 & 0.1 \\ 0.3 & 0.3 & 0.3 & 0.3 & 0.3 \\ 0.6 & 0.7 & 0.8 & 0.9 & 1.0 \\ 1.0 & 0.9 & 0.8 & 0.7 & 0.6 \end{pmatrix}, \quad \mathbf{K}_w = \begin{pmatrix} 0.5 & 0.4 & 0.3 & 0.2 & 0.1 \\ 0.1 & 0.2 & 0.3 & 0.4 & 0.5 \\ 0.6 & 0.7 & 0.8 & 0.9 & 1.0 \\ 0.3 & 0.3 & 0.3 & 0.3 & 0.3 \\ 1.0 & 0.9 & 0.8 & 0.7 & 0.6 \end{pmatrix},$$

$$\mathbf{V}_w = \begin{pmatrix} 0.3 & 0.3 & 0.3 & 0.3 & 0.3 \\ 0.6 & 0.7 & 0.8 & 0.9 & 1.0 \\ 1.0 & 0.9 & 0.8 & 0.7 & 0.6 \\ 0.1 & 0.2 & 0.3 & 0.4 & 0.5 \\ 0.5 & 0.4 & 0.3 & 0.2 & 0.1 \end{pmatrix}$$

(a) Compute the word-level attention scores using these new weight matrices and the resulting word context vectors for each sentence.
(b) Use the following sentence-level weight matrices to compute sentence-level attention and derive the final document representation:

$$\mathbf{Q}_s = \begin{pmatrix} 0.2 & 0.3 & 0.4 \\ 0.3 & 0.2 & 0.1 \\ 0.4 & 0.3 & 0.2 \end{pmatrix}, \quad \mathbf{K}_s = \begin{pmatrix} 0.1 & 0.2 & 0.3 \\ 0.2 & 0.1 & 0.4 \\ 0.3 & 0.4 & 0.1 \end{pmatrix}, \quad \mathbf{V}_s = \begin{pmatrix} 0.3 & 0.4 & 0.5 \\ 0.5 & 0.4 & 0.3 \\ 0.4 & 0.5 & 0.3 \end{pmatrix}$$

13. Given the document tensor **E** as before, introduce positional encodings to the word embeddings. Assume the positional encoding for each word in a sentence is:

$$\mathbf{P}_1 = \begin{pmatrix} 0 & 0.1 & 0.2 & 0.3 & 0.4 \\ 0.4 & 0.3 & 0.2 & 0.1 & 0 \\ 0.1 & 0.2 & 0.3 & 0.4 & 0.5 \\ 0.5 & 0.4 & 0.3 & 0.2 & 0.1 \end{pmatrix}, \quad \mathbf{P}_2 = \begin{pmatrix} 0.2 & 0.1 & 0.4 & 0.3 & 0 \\ 0 & 0.3 & 0.2 & 0.1 & 0.4 \\ 0.3 & 0.2 & 0.1 & 0 & 0.4 \\ 0.4 & 0.3 & 0.2 & 0.1 & 0.2 \end{pmatrix},$$

$$\mathbf{P}_3 = \begin{pmatrix} 0.4 & 0.3 & 0.2 & 0.1 & 0 \\ 0 & 0.1 & 0.2 & 0.3 & 0.4 \\ 0.1 & 0.2 & 0.3 & 0.4 & 0.5 \\ 0.5 & 0.4 & 0.3 & 0.2 & 0.1 \end{pmatrix}$$

(a) Add the positional encodings to the word embeddings in **E**.
(b) Compute the word-level attention scores and the resulting word context vectors for each sentence using the updated embeddings.

14. Consider a scenario where some words in the document tensor $\mathbf{E}$ are masked (e.g., due to padding or irrelevant tokens). Let the mask tensor $\mathbf{M} \in \mathbb{R}^{3\times4}$ be:

$$\mathbf{M} = \begin{pmatrix} 1 & 1 & 1 & 0 \\ 1 & 1 & 0 & 0 \\ 1 & 1 & 1 & 1 \end{pmatrix}$$

The masked positions are indicated by 0, and the valid positions by 1.

(a) Apply the mask to the word embeddings in $\mathbf{E}$ and compute the masked word-level attention scores.
(b) Calculate the word context vectors for each sentence, ignoring the masked positions, and then perform sentence-level attention to derive the final document representation.

15. Consider an image feature map represented as a tensor $\mathbf{F} \in \mathbb{R}^{4\times4\times3}$ (height 4, width 4, and 3 channels):

$$\mathbf{F}_{:,:,1} = \begin{pmatrix} 1 & 2 & 3 & 4 \\ 5 & 6 & 7 & 8 \\ 9 & 10 & 11 & 12 \\ 13 & 14 & 15 & 16 \end{pmatrix}, \quad \mathbf{F}_{:,:,2} = \begin{pmatrix} 2 & 4 & 6 & 8 \\ 10 & 12 & 14 & 16 \\ 18 & 20 & 22 & 24 \\ 26 & 28 & 30 & 32 \end{pmatrix},$$

$$\mathbf{F}_{:,:,3} = \begin{pmatrix} 3 & 6 & 9 & 12 \\ 15 & 18 & 21 & 24 \\ 27 & 30 & 33 & 36 \\ 39 & 42 & 45 & 48 \end{pmatrix}$$

The spatial attention mechanism involves computing an attention map $\mathbf{A} \in \mathbb{R}^{4\times4}$ using a convolution operation followed by a sigmoid activation:

$$\mathbf{A} = \sigma(\text{Conv2D}(\mathbf{F}))$$

Assume the convolution filter is $\mathbf{W} \in \mathbb{R}^{1\times1\times3\times1}$ with weights $\mathbf{W} = \begin{pmatrix} 0.1 \\ 0.2 \\ 0.3 \end{pmatrix}$.

(a) Perform the convolution operation to compute the intermediate feature map.
(b) Apply the sigmoid function to obtain the attention map $\mathbf{A}$.

16. Consider the same feature map tensor $\mathbf{F} \in \mathbb{R}^{4\times4\times3}$ as given above.
The channel attention mechanism involves computing a channel attention vector $\mathbf{c} \in \mathbb{R}^3$ using global average pooling followed by a fully connected layer:

$$\mathbf{c}_i = \sigma(\mathbf{W}_c \cdot \text{GAP}(\mathbf{F}_{:,:,i}) + \mathbf{b}_c)$$

where GAP is the global average pooling operation, $\mathbf{W}_c \in \mathbb{R}^{1\times 1}$ and $\mathbf{b}_c \in \mathbb{R}^1$ are the weights and bias for the fully connected layer, respectively, and σ is the sigmoid activation function. Assume $\mathbf{W}_c = 0.5$ and $\mathbf{b}_c = 0.1$.

(a) Compute the global average pooling for each channel to get a vector of mean values.
(b) Compute the channel attention vector $\mathbf{c}$ using the given weights and bias.

17. Consider the feature map tensor $\mathbf{F} \in \mathbb{R}^{4\times 4\times 3}$ from the previous questions. The combined attention mechanism first applies channel attention and then spatial attention. Let the channel attention vector $\mathbf{c} \in \mathbb{R}^3$ be given by:

$$\mathbf{c} = \begin{pmatrix} 0.7 \\ 0.8 \\ 0.6 \end{pmatrix}$$

(a) Apply the channel attention to the feature map $\mathbf{F}$ by element-wise multiplication of each channel with the corresponding value in $\mathbf{c}$.
(b) Compute the spatial attention map $\mathbf{A}$ using the combined feature map after channel attention.

18. Given the feature map tensor $\mathbf{F} \in \mathbb{R}^{4\times 4\times 3}$ and attention maps $\mathbf{A} \in \mathbb{R}^{4\times 4}$ and $\mathbf{c} \in \mathbb{R}^3$ from the previous questions, the hierarchical attention mechanism involves the following steps:

(a) Apply the channel attention vector $\mathbf{c}$ to the feature map $\mathbf{F}$ to get an intermediate feature map.
(b) Apply the spatial attention map $\mathbf{A}$ to the intermediate feature map by element-wise multiplication.
(c) Compute the final attention-weighted feature map and provide its dimensions and values.

19. Consider a text sequence represented by a tensor $\mathbf{T} \in \mathbb{R}^{4\times 5}$ (4 words, 5-dimensional embeddings) and an image feature map represented by a tensor $\mathbf{I} \in \mathbb{R}^{3\times 3\times 6}$ (3×3 spatial dimensions, 6 channels):

$$\mathbf{T} = \begin{pmatrix} 1 & 0 & 2 & 1 & 0 \\ 0 & 1 & 0 & 2 & 1 \\ 1 & 1 & 1 & 0 & 0 \\ 0 & 0 & 1 & 1 & 1 \end{pmatrix}, \quad \mathbf{I}_{:,:,1} = \begin{pmatrix} 1 & 2 & 3 \\ 4 & 5 & 6 \\ 7 & 8 & 9 \end{pmatrix},$$

$$\mathbf{I}_{:,:,2} = \begin{pmatrix} 9 & 8 & 7 \\ 6 & 5 & 4 \\ 3 & 2 & 1 \end{pmatrix}, \quad \mathbf{I}_{:,:,3} = \begin{pmatrix} 1 & 3 & 5 \\ 7 & 9 & 11 \\ 13 & 15 & 17 \end{pmatrix},$$

$$\mathbf{I}_{:,:,4} = \begin{pmatrix} 2 & 4 & 6 \\ 8 & 10 & 12 \\ 14 & 16 & 18 \end{pmatrix}, \quad \mathbf{I}_{:,:,5} = \begin{pmatrix} 1 & 0 & 1 \\ 0 & 1 & 0 \\ 1 & 0 & 1 \end{pmatrix}, \quad \mathbf{I}_{:,:,6} = \begin{pmatrix} 2 & 4 & 6 \\ 8 & 10 & 12 \\ 14 & 16 & 18 \end{pmatrix}$$

The weight matrices are given by:

$$\mathbf{W}_Q = \begin{pmatrix} 1 & 0 & 0 & 0 & 0 \\ 0 & 1 & 0 & 0 & 0 \\ 0 & 0 & 1 & 0 & 0 \\ 0 & 0 & 0 & 1 & 0 \\ 0.1 & 0 & 0 & 0 & 1 \end{pmatrix}, \quad \mathbf{W}_K = \begin{pmatrix} 1 & 0 & 0 & 0 & 0 \\ 0 & 1 & 0 & 0 & 0 \\ 0 & 0 & 1 & 0 & 0 \\ 0 & 0 & 0 & 1 & 0 \\ 0 & 0 & 0 & 0 & 1 \\ 0.1 & 0 & 0 & 0 & 0 \end{pmatrix},$$

$$\mathbf{W}_V = \begin{pmatrix} 1 & 0 & 0 & 0 & 0 \\ 0 & 1 & 0 & 0 & 0 \\ 0 & 0 & 1 & 0 & 0 \\ 0 & 0 & 0 & 1 & 0 \\ 0 & 0 & 0 & 0 & 1 \\ 0.1 & 0 & 0 & 0 & 0 \end{pmatrix}$$

(a) Compute the query matrix $\mathbf{Q} = \mathbf{T}\mathbf{W}_Q$, the key matrix $\mathbf{K} = \mathbf{I}\mathbf{W}_K$, and the value matrix $\mathbf{V} = \mathbf{I}\mathbf{W}_V$ where $\mathbf{W}_Q \in \mathbb{R}^{5\times5}$, $\mathbf{W}_K \in \mathbb{R}^{6\times5}$, and $\mathbf{W}_V \in \mathbb{R}^{6\times5}$ are given as identity matrices. Note: For simplicity, flatten $\mathbf{I}$ to match the dimensions for matrix multiplication.

(b) Calculate the attention scores, apply the softmax function, and compute the attended values.

20. Given an audio feature tensor $\mathbf{A} \in \mathbb{R}^{10\times8}$ (10 timesteps, 8-dimensional features) and a visual feature tensor $\mathbf{V} \in \mathbb{R}^{4\times4\times8}$ 4×4 spatial dimensions, 8 channels:

$$\mathbf{A} = \begin{pmatrix} 0.1 & 0.2 & 0.3 & 0.4 & 0.5 & 0.6 & 0.7 & 0.8 \\ 0.2 & 0.3 & 0.4 & 0.5 & 0.6 & 0.7 & 0.8 & 0.9 \\ 0.3 & 0.4 & 0.5 & 0.6 & 0.7 & 0.8 & 0.9 & 1.0 \\ 0.4 & 0.5 & 0.6 & 0.7 & 0.8 & 0.9 & 1.0 & 1.1 \\ 0.5 & 0.6 & 0.7 & 0.8 & 0.9 & 1.0 & 1.1 & 1.2 \\ 0.6 & 0.7 & 0.8 & 0.9 & 1.0 & 1.1 & 1.2 & 1.3 \\ 0.7 & 0.8 & 0.9 & 1.0 & 1.1 & 1.2 & 1.3 & 1.4 \\ 0.8 & 0.9 & 1.0 & 1.1 & 1.2 & 1.3 & 1.4 & 1.5 \\ 0.9 & 1.0 & 1.1 & 1.2 & 1.3 & 1.4 & 1.5 & 1.6 \\ 1.0 & 1.1 & 1.2 & 1.3 & 1.4 & 1.5 & 1.6 & 1.7 \end{pmatrix},$$

$$\mathbf{V}_{:,:,1} = \begin{pmatrix} 0.1 & 0.2 & 0.3 & 0.4 \\ 0.5 & 0.6 & 0.7 & 0.8 \\ 0.9 & 1.0 & 1.1 & 1.2 \\ 1.3 & 1.4 & 1.5 & 1.6 \end{pmatrix}, \mathbf{V}_{:,:,2} = \begin{pmatrix} 1.6 & 1.5 & 1.4 & 1.3 \\ 1.2 & 1.1 & 1.0 & 0.9 \\ 0.8 & 0.7 & 0.6 & 0.5 \\ 0.4 & 0.3 & 0.2 & 0.1 \end{pmatrix},$$

$$\mathbf{V}_{:,:,3} = \begin{pmatrix} 0.1 & 0.3 & 0.5 & 0.7 \\ 0.9 & 1.1 & 1.3 & 1.5 \\ 1.7 & 1.9 & 2.1 & 2.3 \\ 2.5 & 2.7 & 2.9 & 3.1 \end{pmatrix}, \mathbf{V}_{:,:,4} = \begin{pmatrix} 3.1 & 2.9 & 2.7 & 2.5 \\ 2.3 & 2.1 & 1.9 & 1.7 \\ 1.5 & 1.3 & 1.1 & 0.9 \\ 0.7 & 0.5 & 0.3 & 0.1 \end{pmatrix},$$

$$\mathbf{V}_{:,:,5} = \begin{pmatrix} 0.2 & 0.4 & 0.6 & 0.8 \\ 1.0 & 1.2 & 1.4 & 1.6 \\ 1.8 & 2.0 & 2.2 & 2.4 \\ 2.6 & 2.8 & 3.0 & 3.2 \end{pmatrix}, \mathbf{V}_{:,:,6} = \begin{pmatrix} 3.2 & 3.0 & 2.8 & 2.6 \\ 2.4 & 2.2 & 2.0 & 1.8 \\ 1.6 & 1.4 & 1.2 & 1.0 \\ 0.8 & 0.6 & 0.4 & 0.2 \end{pmatrix},$$

$$\mathbf{V}_{:,:,7} = \begin{pmatrix} 0.3 & 0.6 & 0.9 & 1.2 \\ 1.5 & 1.8 & 2.1 & 2.4 \\ 2.7 & 3.0 & 3.3 & 3.6 \\ 3.9 & 4.2 & 4.5 & 4.8 \end{pmatrix}, \mathbf{V}_{:,:,8} = \begin{pmatrix} 0.8 & 0.7 & 0.6 & 0.5 \\ 0.4 & 0.3 & 0.2 & 0.1 \\ 0.9 & 0.8 & 0.7 & 0.6 \\ 0.5 & 0.4 & 0.3 & 0.2 \end{pmatrix}$$

The weight matrices are given by:

$$\mathbf{W}_Q = \begin{pmatrix} 1 & 0 & 0 & 0 & 0 & 0 & 0 & 0 \\ 0 & 1 & 0 & 0 & 0 & 0 & 0 & 0 \\ 0 & 0 & 1 & 0 & 0 & 0 & 0 & 0 \\ 0 & 0 & 0 & 1 & 0 & 0 & 0 & 0 \\ 0 & 0 & 0 & 0 & 1 & 0 & 0 & 0 \\ 0 & 0 & 0 & 0 & 0 & 1 & 0 & 0 \\ 0 & 0 & 0 & 0 & 0 & 0 & 1 & 0 \\ 0.1 & 0 & 0 & 0 & 0 & 0 & 0 & 1 \end{pmatrix}, \quad \mathbf{W}_K = \begin{pmatrix} 1 & 0 & 0 & 0 & 0 \\ 0 & 1 & 0 & 0 & 0 \\ 0 & 0 & 1 & 0 & 0 \\ 0 & 0 & 0 & 1 & 0 \\ 0 & 0 & 0 & 0 & 1 \\ 0 & 0 & 0 & 0 & 0 \\ 0 & 0 & 0 & 0 & 0 \\ 0.1 & 0 & 0 & 0 & 0 \end{pmatrix},$$

$$\mathbf{W}_V = \begin{pmatrix} 1 & 0 & 0 & 0 & 0 \\ 0 & 1 & 0 & 0 & 0 \\ 0 & 0 & 1 & 0 & 0 \\ 0 & 0 & 0 & 1 & 0 \\ 0 & 0 & 0 & 0 & 1 \\ 0 & 0 & 0 & 0 & 0 \\ 0 & 0 & 0 & 0 & 0 \\ 0.1 & 0 & 0 & 0 & 0 \end{pmatrix}$$

(a) Compute the query matrix from **A** and the key and value matrices from **V** using appropriate transformations to match the dimensions for cross-modal attention.

(b) Calculate the attention scores and the final attended values.

21. Given text embeddings $\mathbf{T} \in \mathbb{R}^{3\times 6}$, audio embeddings $\mathbf{A} \in \mathbb{R}^{5\times 6}$, and image embeddings $\mathbf{I} \in \mathbb{R}^{4\times 4\times 6}$:

$$\mathbf{T} = \begin{pmatrix} 0.1 & 0.2 & 0.3 & 0.4 & 0.5 & 0.6 \\ 0.6 & 0.5 & 0.4 & 0.3 & 0.2 & 0.1 \\ 0.3 & 0.4 & 0.5 & 0.6 & 0.7 & 0.8 \end{pmatrix}, \quad \mathbf{A} = \begin{pmatrix} 0.2 & 0.3 & 0.4 & 0.5 & 0.6 & 0.7 \\ 0.7 & 0.6 & 0.5 & 0.4 & 0.3 & 0.2 \\ 0.4 & 0.5 & 0.6 & 0.7 & 0.8 & 0.9 \\ 0.5 & 0.6 & 0.7 & 0.8 & 0.9 & 1.0 \\ 0.4 & 0.5 & 0.6 & 0.7 & 0.8 & 0.9 \end{pmatrix}$$

$$\mathbf{I}_{:,:,1} = \begin{pmatrix} 0.3 & 0.2 & 0.1 & 0.4 \\ 0.5 & 0.6 & 0.7 & 0.8 \\ 0.9 & 1.0 & 1.1 & 1.2 \\ 1.3 & 1.4 & 1.5 & 1.6 \end{pmatrix}, \quad \mathbf{I}_{:,:,2} = \begin{pmatrix} 0.4 & 0.3 & 0.2 & 0.1 \\ 0.6 & 0.5 & 0.4 & 0.3 \\ 0.8 & 0.7 & 0.6 & 0.5 \\ 1.0 & 0.9 & 0.8 & 0.7 \end{pmatrix},$$

$$\mathbf{I}_{:,:,3} = \begin{pmatrix} 0.5 & 0.4 & 0.3 & 0.2 \\ 0.7 & 0.6 & 0.5 & 0.4 \\ 0.9 & 0.8 & 0.7 & 0.6 \\ 1.1 & 1.0 & 0.9 & 0.8 \end{pmatrix}, \quad \mathbf{I}_{:,:,4} = \begin{pmatrix} 0.6 & 0.5 & 0.4 & 0.3 \\ 0.8 & 0.7 & 0.6 & 0.5 \\ 1.0 & 0.9 & 0.8 & 0.7 \\ 1.2 & 1.1 & 1.0 & 0.9 \end{pmatrix},$$

$$\mathbf{I}_{:,:,5} = \begin{pmatrix} 0.7 & 0.6 & 0.5 & 0.4 \\ 0.9 & 0.8 & 0.7 & 0.6 \\ 1.1 & 1.0 & 0.9 & 0.8 \\ 1.3 & 1.2 & 1.1 & 1.0 \end{pmatrix}, \quad \mathbf{I}_{:,:,6} = \begin{pmatrix} 1.1 & 1.0 & 0.9 & 0.8 \\ 0.7 & 0.6 & 0.5 & 0.4 \\ 0.3 & 0.2 & 0.1 & 0.0 \\ 0.9 & 0.8 & 0.7 & 0.6 \end{pmatrix}$$

The weight matrices are given by:

$$\mathbf{W}_Q = \begin{pmatrix} 1 & 0 & 0 & 0 & 0 & 0 \\ 0 & 1 & 0 & 0 & 0 & 0 \\ 0 & 0 & 1 & 0 & 0 & 0 \\ 0 & 0 & 0 & 1 & 0 & 0 \\ 0 & 0 & 0 & 0 & 1 & 0 \\ 0.1 & 0 & 0 & 0 & 0 & 1 \end{pmatrix}, \quad \mathbf{W}_K = \begin{pmatrix} 1 & 0 & 0 & 0 & 0 & 0 \\ 0 & 1 & 0 & 0 & 0 & 0 \\ 0 & 0 & 1 & 0 & 0 & 0 \\ 0 & 0 & 0 & 1 & 0 & 0 \\ 0 & 0 & 0 & 0 & 1 & 0 \\ 0.1 & 0 & 0 & 0 & 0 & 1 \end{pmatrix},$$

$$\mathbf{W}_V = \begin{pmatrix} 1 & 0 & 0 & 0 & 0 & 0 \\ 0 & 1 & 0 & 0 & 0 & 0 \\ 0 & 0 & 1 & 0 & 0 & 0 \\ 0 & 0 & 0 & 1 & 0 & 0 \\ 0 & 0 & 0 & 0 & 1 & 0 \\ 0.1 & 0 & 0 & 0 & 0 & 1 \end{pmatrix}$$

(a) Compute the query matrix from **T** and the key and value matrices from **A** using transformations to match dimensions.
(b) Compute another set of key and value matrices from **I** (flattened to match dimensions).
(c) Calculate the attention scores for both text-audio and text-image pairs, then aggregate the attended values.

22. Consider an image feature map $\mathbf{F} \in \mathbb{R}^{4\times4\times6}$ (4×4 spatial dimensions, 6 channels) and a caption represented by a sequence of word embeddings $\mathbf{T} \in \mathbb{R}^{5\times6}$ (5 words, 6-dimensional embeddings):

$$\mathbf{F}_{:,:,1} = \begin{pmatrix} 1 & 2 & 3 & 4 \\ 5 & 6 & 7 & 8 \\ 9 & 10 & 11 & 12 \\ 13 & 14 & 15 & 16 \end{pmatrix}, \quad \mathbf{F}_{:,:,2} = \begin{pmatrix} 2 & 4 & 6 & 8 \\ 10 & 12 & 14 & 16 \\ 18 & 20 & 22 & 24 \\ 26 & 28 & 30 & 32 \end{pmatrix},$$

$$\mathbf{F}_{:,:,3} = \begin{pmatrix} 3 & 6 & 9 & 12 \\ 15 & 18 & 21 & 24 \\ 27 & 30 & 33 & 36 \\ 39 & 42 & 45 & 48 \end{pmatrix}, \mathbf{F}_{:,:,4} = \begin{pmatrix} 4 & 8 & 12 & 16 \\ 20 & 24 & 28 & 32 \\ 36 & 40 & 44 & 48 \\ 52 & 56 & 60 & 64 \end{pmatrix},$$

$$\mathbf{F}_{:,:,5} = \begin{pmatrix} 5 & 10 & 15 & 20 \\ 25 & 30 & 35 & 40 \\ 45 & 50 & 55 & 60 \\ 65 & 70 & 75 & 80 \end{pmatrix}, \quad \mathbf{F}_{:,:,6} = \begin{pmatrix} 6 & 12 & 18 & 24 \\ 30 & 36 & 42 & 48 \\ 54 & 60 & 66 & 72 \\ 78 & 84 & 90 & 96 \end{pmatrix}$$

$$\mathbf{T} = \begin{pmatrix} 0.1 & 0.2 & 0.3 & 0.4 & 0.5 & 0.6 \\ 0.6 & 0.5 & 0.4 & 0.3 & 0.2 & 0.1 \\ 0.3 & 0.4 & 0.5 & 0.6 & 0.7 & 0.8 \\ 0.2 & 0.3 & 0.4 & 0.5 & 0.6 & 0.7 \\ 0.8 & 0.7 & 0.6 & 0.5 & 0.4 & 0.3 \end{pmatrix}$$

The adaptive attention mechanism involves using both the visual features and the previous word embedding to generate attention scores:

$$e_{ij} = \mathbf{v}^{\mathrm{T}} \tanh(\mathbf{W}_v \mathbf{F}_{ij} + \mathbf{W}_h \mathbf{T}_{t-1} + \mathbf{b}_a)$$

where $\mathbf{v} \in \mathbb{R}^6$, $\mathbf{W}_v \in \mathbb{R}^{6\times6}$, $\mathbf{W}_h \in \mathbb{R}^{6\times6}$, and $\mathbf{b}_a \in \mathbb{R}^6$.

Let $\mathbf{v} = \begin{pmatrix} 0.1 \\ 0.2 \\ 0.3 \\ 0.4 \\ 0.5 \\ 0.6 \end{pmatrix}$, $\mathbf{W}_v = \mathbf{W}_h = \mathbf{I}_6$, and $\mathbf{b}_a = \begin{pmatrix} 0.1 \\ 0.1 \\ 0.1 \\ 0.1 \\ 0.1 \\ 0.1 \end{pmatrix}$.

(a) Compute the attention scores for each spatial location in $\mathbf{F}$ given $\mathbf{T}_{t-1} = \begin{pmatrix} 0.3 \\ 0.4 \\ 0.5 \\ 0.6 \\ 0.7 \\ 0.8 \end{pmatrix}$.

(b) Apply the softmax function to obtain the attention weights and compute the attended visual feature vector.

23. Given a question represented by a sequence of word embeddings $\mathbf{Q} \in \mathbb{R}^{4\times 6}$ (4 words, 6-dimensional embeddings) and an image feature map $\mathbf{F} \in \mathbb{R}^{3\times 3\times 6}$:

$$\mathbf{Q} = \begin{pmatrix} 0.1 & 0.2 & 0.3 & 0.4 & 0.5 & 0.6 \\ 0.6 & 0.5 & 0.4 & 0.3 & 0.2 & 0.1 \\ 0.3 & 0.4 & 0.5 & 0.6 & 0.7 & 0.8 \\ 0.2 & 0.3 & 0.4 & 0.5 & 0.6 & 0.7 \end{pmatrix},$$

$$\mathbf{F}_{:,:,1} = \begin{pmatrix} 1 & 2 & 3 \\ 4 & 5 & 6 \\ 7 & 8 & 9 \end{pmatrix}, \quad \mathbf{F}_{:,:,2} = \begin{pmatrix} 2 & 3 & 4 \\ 5 & 6 & 7 \\ 8 & 9 & 10 \end{pmatrix},$$

$$\mathbf{F}_{:,:,3} = \begin{pmatrix} 3 & 4 & 5 \\ 6 & 7 & 8 \\ 9 & 10 & 11 \end{pmatrix}, \mathbf{F}_{:,:,4} = \begin{pmatrix} 4 & 5 & 6 \\ 7 & 8 & 9 \\ 10 & 11 & 12 \end{pmatrix},$$

$$\mathbf{F}_{:,:,5} = \begin{pmatrix} 5 & 6 & 7 \\ 8 & 9 & 10 \\ 11 & 12 & 13 \end{pmatrix}, \mathbf{F}_{:,:,6} = \begin{pmatrix} 6 & 5 & 4 \\ 3 & 2 & 1 \\ 0 & -1 & -2 \end{pmatrix}$$

The adaptive attention mechanism involves computing a joint attention score using both the question and image features:

$$e_{ij} = \mathbf{v}^{\mathrm{T}} \tanh(\mathbf{W}_q \mathbf{Q}_k + \mathbf{W}_v \mathbf{F}_{ij} + \mathbf{b}_a)$$

where $\mathbf{v} \in \mathbb{R}^6$, $\mathbf{W}_q \in \mathbb{R}^{6\times 6}$, $\mathbf{W}_v \in \mathbb{R}^{6\times 6}$, and $\mathbf{b}_a \in \mathbb{R}^6$.

Let $\mathbf{v} = \begin{pmatrix} 0.1 \\ 0.2 \\ 0.3 \\ 0.4 \\ 0.5 \\ 0.6 \end{pmatrix}$, $\mathbf{W}_q = \mathbf{W}_v = \mathbf{I}_6$, and $\mathbf{b}_a = \begin{pmatrix} 0.1 \\ 0.1 \\ 0.1 \\ 0.1 \\ 0.1 \\ 0.1 \end{pmatrix}$.

(a) Compute the attention scores for each spatial location in $\mathbf{F}$ given $\mathbf{Q}_k = \begin{pmatrix} 0.3 \\ 0.4 \\ 0.5 \\ 0.6 \\ 0.7 \\ 0.8 \end{pmatrix}$ for each word k in the question.

(b) Apply the softmax function to obtain the attention weights and compute the attended visual feature vector for the entire question.

24. Consider a sequence of word embeddings $\mathbf{E} \in \mathbb{R}^{6\times 8}$ (6 words, 8-dimensional embeddings):

$$\mathbf{E} = \begin{pmatrix} 0.1 & 0.2 & 0.3 & 0.4 & 0.5 & 0.6 & 0.7 & 0.8 \\ 0.8 & 0.7 & 0.6 & 0.5 & 0.4 & 0.3 & 0.2 & 0.1 \\ 0.2 & 0.3 & 0.4 & 0.5 & 0.6 & 0.7 & 0.8 & 0.9 \\ 0.9 & 0.8 & 0.7 & 0.6 & 0.5 & 0.4 & 0.3 & 0.2 \\ 0.3 & 0.4 & 0.5 & 0.6 & 0.7 & 0.8 & 0.9 & 1.0 \\ 1.0 & 0.9 & 0.8 & 0.7 & 0.6 & 0.5 & 0.4 & 0.3 \end{pmatrix}$$

Sparse attention limits attention to a subset of tokens. For simplicity, assume each token attends to at most 3 tokens, determined by predefined sparsity pattern:

$$\text{Attention Mask} = \begin{pmatrix} 1 & 1 & 0 & 0 & 0 & 0 \\ 1 & 1 & 1 & 0 & 0 & 0 \\ 0 & 1 & 1 & 1 & 0 & 0 \\ 0 & 0 & 1 & 1 & 1 & 0 \\ 0 & 0 & 0 & 1 & 1 & 1 \\ 0 & 0 & 0 & 0 & 1 & 1 \end{pmatrix}$$

The query, key, and value matrices are given by $\mathbf{Q} = \mathbf{K} = \mathbf{V} = \mathbf{E}$.

(a) Compute the sparse attention scores by element-wise multiplying the attention mask with the attention scores derived from $\mathbf{Q}$ and $\mathbf{K}$.
(b) Apply the softmax function to the sparse attention scores and compute the attended values.

25. Given an input sequence $\mathbf{I} \in \mathbb{R}^{7\times 8}$ (7 timesteps, 8-dimensional embeddings) and an output sequence $\mathbf{O} \in \mathbb{R}^{4\times 8}$ (4 timesteps, 8-dimensional embeddings):

$$\mathbf{I} = \begin{pmatrix} 0.1 & 0.2 & 0.3 & 0.4 & 0.5 & 0.6 & 0.7 & 0.8 \\ 0.8 & 0.7 & 0.6 & 0.5 & 0.4 & 0.3 & 0.2 & 0.1 \\ 0.2 & 0.3 & 0.4 & 0.5 & 0.6 & 0.7 & 0.8 & 0.9 \\ 0.9 & 0.8 & 0.7 & 0.6 & 0.5 & 0.4 & 0.3 & 0.2 \\ 0.3 & 0.4 & 0.5 & 0.6 & 0.7 & 0.8 & 0.9 & 1.0 \\ 1.0 & 0.9 & 0.8 & 0.7 & 0.6 & 0.5 & 0.4 & 0.3 \\ 0.4 & 0.5 & 0.6 & 0.7 & 0.8 & 0.9 & 1.0 & 0.1 \end{pmatrix},$$

$$\mathbf{O} = \begin{pmatrix} 0.3 & 0.4 & 0.5 & 0.6 & 0.7 & 0.8 & 0.9 & 1.0 \\ 1.0 & 0.9 & 0.8 & 0.7 & 0.6 & 0.5 & 0.4 & 0.3 \\ 0.2 & 0.3 & 0.4 & 0.5 & 0.6 & 0.7 & 0.8 & 0.9 \\ 0.9 & 0.8 & 0.7 & 0.6 & 0.5 & 0.4 & 0.3 & 0.2 \end{pmatrix}$$

Localized attention restricts the attention to a window around each target token. Assume each token in the output sequence attends to a window of size 3 in the input sequence.

(a) Compute the localized attention scores for the first token in the output sequence attending to the first three tokens in the input sequence.
(b) Apply the softmax function to the localized attention scores and compute the attended value.

26. Given a hierarchical structure with a document consisting of 3 paragraphs, each represented by a sequence of 4 sentence embeddings. Each sentence is represented by a sequence of 5 word embeddings, $\mathbf{W} \in \mathbb{R}^{3\times4\times5\times8}$ (3 paragraphs, 4 sentences per paragraph, 5 words per sentence, 8-dimensional embeddings):

$$\mathbf{W}_{1,1:2,:,:} = \left(\begin{pmatrix} 0.1 & 0.2 & 0.3 & 0.4 & 0.5 & 0.6 & 0.7 & 0.8 \\ 0.8 & 0.7 & 0.6 & 0.5 & 0.4 & 0.3 & 0.2 & 0.1 \\ 0.2 & 0.3 & 0.4 & 0.5 & 0.6 & 0.7 & 0.8 & 0.9 \\ 0.9 & 0.8 & 0.7 & 0.6 & 0.5 & 0.4 & 0.3 & 0.2 \\ 0.3 & 0.4 & 0.5 & 0.6 & 0.7 & 0.8 & 0.9 & 1.0 \end{pmatrix}, \right.$$

$$\left. \begin{pmatrix} 0.4 & 0.3 & 0.2 & 0.1 & 0.2 & 0.3 & 0.4 & 0.5 \\ 0.5 & 0.4 & 0.3 & 0.2 & 0.1 & 0.2 & 0.3 & 0.4 \\ 0.2 & 0.1 & 0.2 & 0.3 & 0.4 & 0.5 & 0.6 & 0.7 \\ 0.7 & 0.6 & 0.5 & 0.4 & 0.3 & 0.2 & 0.1 & 0.2 \\ 0.3 & 0.2 & 0.3 & 0.4 & 0.5 & 0.6 & 0.7 & 0.8 \end{pmatrix} \right),$$

$$\mathbf{W}_{1,3:4,:,:} = \left(\begin{pmatrix} 0.5 & 0.4 & 0.3 & 0.2 & 0.1 & 0.2 & 0.3 & 0.4 \\ 0.4 & 0.3 & 0.2 & 0.1 & 0.2 & 0.3 & 0.4 & 0.5 \\ 0.3 & 0.2 & 0.3 & 0.4 & 0.5 & 0.6 & 0.7 & 0.8 \\ 0.8 & 0.7 & 0.6 & 0.5 & 0.4 & 0.3 & 0.2 & 0.1 \\ 0.2 & 0.1 & 0.2 & 0.3 & 0.4 & 0.5 & 0.6 & 0.7 \end{pmatrix}, \right.$$

$$\left. \begin{pmatrix} 0.3 & 0.2 & 0.1 & 0.2 & 0.3 & 0.4 & 0.5 & 0.6 \\ 0.6 & 0.5 & 0.4 & 0.3 & 0.2 & 0.1 & 0.2 & 0.3 \\ 0.4 & 0.3 & 0.2 & 0.1 & 0.2 & 0.3 & 0.4 & 0.5 \\ 0.7 & 0.6 & 0.5 & 0.4 & 0.3 & 0.2 & 0.1 & 0.2 \\ 0.3 & 0.2 & 0.1 & 0.2 & 0.3 & 0.4 & 0.5 & 0.6 \end{pmatrix} \right),$$

$$\mathbf{W}_{2,1:2,:,:} = \left(\begin{pmatrix} 0.2 & 0.3 & 0.4 & 0.5 & 0.6 & 0.7 & 0.8 & 0.9 \\ 0.9 & 0.8 & 0.7 & 0.6 & 0.5 & 0.4 & 0.3 & 0.2 \\ 0.3 & 0.4 & 0.5 & 0.6 & 0.7 & 0.8 & 0.9 & 1.0 \\ 1.0 & 0.9 & 0.8 & 0.7 & 0.6 & 0.5 & 0.4 & 0.3 \\ 0.4 & 0.5 & 0.6 & 0.7 & 0.8 & 0.9 & 1.0 & 1.1 \end{pmatrix}, \right.$$

$$\begin{pmatrix} 0.5 & 0.4 & 0.3 & 0.2 & 0.3 & 0.4 & 0.5 & 0.6 \\ 0.6 & 0.5 & 0.4 & 0.3 & 0.2 & 0.3 & 0.4 & 0.5 \\ 0.3 & 0.2 & 0.3 & 0.4 & 0.5 & 0.6 & 0.7 & 0.8 \\ 0.8 & 0.7 & 0.6 & 0.5 & 0.4 & 0.3 & 0.2 & 0.3 \\ 0.4 & 0.3 & 0.4 & 0.5 & 0.6 & 0.7 & 0.8 & 0.9 \end{pmatrix}\Bigg),$$

$$\mathbf{W}_{2,3:4,:,:} = \Bigg(\begin{pmatrix} 0.6 & 0.5 & 0.4 & 0.3 & 0.2 & 0.3 & 0.4 & 0.5 \\ 0.5 & 0.4 & 0.3 & 0.2 & 0.3 & 0.4 & 0.5 & 0.6 \\ 0.4 & 0.3 & 0.4 & 0.5 & 0.6 & 0.7 & 0.8 & 0.9 \\ 0.9 & 0.8 & 0.7 & 0.6 & 0.5 & 0.4 & 0.3 & 0.2 \\ 0.3 & 0.2 & 0.3 & 0.4 & 0.5 & 0.6 & 0.7 & 0.8 \end{pmatrix},$$

$$\begin{pmatrix} 0.4 & 0.3 & 0.2 & 0.3 & 0.4 & 0.5 & 0.6 & 0.7 \\ 0.7 & 0.6 & 0.5 & 0.4 & 0.3 & 0.2 & 0.3 & 0.4 \\ 0.5 & 0.4 & 0.3 & 0.2 & 0.3 & 0.4 & 0.5 & 0.6 \\ 0.8 & 0.7 & 0.6 & 0.5 & 0.4 & 0.3 & 0.2 & 0.3 \\ 0.4 & 0.3 & 0.2 & 0.3 & 0.4 & 0.5 & 0.6 & 0.7 \end{pmatrix}\Bigg),$$

$$\mathbf{W}_{3,1:2,:,:} = \Bigg(\begin{pmatrix} 0.3 & 0.4 & 0.5 & 0.6 & 0.7 & 0.8 & 0.9 & 1.0 \\ 1.0 & 0.9 & 0.8 & 0.7 & 0.6 & 0.5 & 0.4 & 0.3 \\ 0.4 & 0.5 & 0.6 & 0.7 & 0.8 & 0.9 & 1.0 & 1.1 \\ 1.1 & 1.0 & 0.9 & 0.8 & 0.7 & 0.6 & 0.5 & 0.4 \\ 0.5 & 0.6 & 0.7 & 0.8 & 0.9 & 1.0 & 1.1 & 1.2 \end{pmatrix},$$

$$\begin{pmatrix} 0.6 & 0.5 & 0.4 & 0.3 & 0.4 & 0.5 & 0.6 & 0.7 \\ 0.7 & 0.6 & 0.5 & 0.4 & 0.3 & 0.4 & 0.5 & 0.6 \\ 0.4 & 0.3 & 0.4 & 0.5 & 0.6 & 0.7 & 0.8 & 0.9 \\ 0.9 & 0.8 & 0.7 & 0.6 & 0.5 & 0.4 & 0.3 & 0.4 \\ 0.5 & 0.4 & 0.5 & 0.6 & 0.7 & 0.8 & 0.9 & 1.0 \end{pmatrix}\Bigg),$$

$$\mathbf{W}_{3,3:4,:,:} = \Bigg(\begin{pmatrix} 0.7 & 0.6 & 0.5 & 0.4 & 0.3 & 0.4 & 0.5 & 0.6 \\ 0.6 & 0.5 & 0.4 & 0.3 & 0.4 & 0.5 & 0.6 & 0.7 \\ 0.5 & 0.4 & 0.5 & 0.6 & 0.7 & 0.8 & 0.9 & 1.0 \\ 1.0 & 0.9 & 0.8 & 0.7 & 0.6 & 0.5 & 0.4 & 0.3 \\ 0.4 & 0.3 & 0.4 & 0.5 & 0.6 & 0.7 & 0.8 & 0.9 \end{pmatrix},$$

$$\begin{pmatrix} 0.5 & 0.4 & 0.3 & 0.4 & 0.5 & 0.6 & 0.7 & 0.8 \\ 0.8 & 0.7 & 0.6 & 0.5 & 0.4 & 0.3 & 0.4 & 0.5 \\ 0.6 & 0.5 & 0.4 & 0.3 & 0.4 & 0.5 & 0.6 & 0.7 \\ 0.9 & 0.8 & 0.7 & 0.6 & 0.5 & 0.4 & 0.3 & 0.4 \\ 0.5 & 0.4 & 0.3 & 0.4 & 0.5 & 0.6 & 0.7 & 0.8 \end{pmatrix}\Bigg)$$

The hierarchical sparse attention mechanism involves performing sparse attention at the word level, followed by sentence-level and paragraph-level sparse attention.

(a) Compute the word-level sparse attention scores for each sentence, restricting each word to attend to at most 3 neighboring words.
(b) Aggregate the attended word embeddings to form sentence embeddings, then compute the sentence-level sparse attention scores, restricting each sentence to attend to at most 2 neighboring sentences.
(c) Aggregate the attended sentence embeddings to form paragraph embeddings, then compute the paragraph-level sparse attention scores.

27. Given an image represented by a feature map $\mathbf{F} \in \mathbb{R}^{6\times6\times3}$ (6×6 spatial dimensions, 3 channels):

$$\mathbf{F}_{:,:,1} = \begin{pmatrix} 0.1 & 0.2 & 0.3 & 0.4 & 0.5 & 0.6 \\ 0.6 & 0.5 & 0.4 & 0.3 & 0.2 & 0.1 \\ 0.2 & 0.3 & 0.4 & 0.5 & 0.6 & 0.7 \\ 0.7 & 0.6 & 0.5 & 0.4 & 0.3 & 0.2 \\ 0.3 & 0.4 & 0.5 & 0.6 & 0.7 & 0.8 \\ 0.8 & 0.7 & 0.6 & 0.5 & 0.4 & 0.3 \end{pmatrix}, \quad \mathbf{F}_{:,:,2} = \begin{pmatrix} 0.2 & 0.4 & 0.6 & 0.8 & 1.0 & 1.2 \\ 1.2 & 1.0 & 0.8 & 0.6 & 0.4 & 0.2 \\ 0.4 & 0.6 & 0.8 & 1.0 & 1.2 & 1.4 \\ 1.4 & 1.2 & 1.0 & 0.8 & 0.6 & 0.4 \\ 0.6 & 0.8 & 1.0 & 1.2 & 1.4 & 1.6 \\ 1.6 & 1.4 & 1.2 & 1.0 & 0.8 & 0.6 \end{pmatrix},$$

$$\mathbf{F}_{:,:,3} = \begin{pmatrix} 0.3 & 0.6 & 0.9 & 1.2 & 1.5 & 1.8 \\ 1.8 & 1.5 & 1.2 & 0.9 & 0.6 & 0.3 \\ 0.6 & 0.9 & 1.2 & 1.5 & 1.8 & 2.1 \\ 2.1 & 1.8 & 1.5 & 1.2 & 0.9 & 0.6 \\ 0.9 & 1.2 & 1.5 & 1.8 & 2.1 & 2.4 \\ 2.4 & 2.1 & 1.8 & 1.5 & 1.2 & 0.9 \end{pmatrix}$$

Localized attention in image processing restricts the attention to a local window around each spatial location. Assume a 3×3 window centered around each pixel.

(a) Compute the localized attention scores for the central pixel (3, 3) attending to its 8 neighboring pixels. Use the attention mechanism formula:

$$e_{ij} = \mathbf{v}^{\mathrm{T}} \tanh(\mathbf{W}_v \mathbf{F}_{ij} + \mathbf{W}_h \mathbf{F}_{33} + \mathbf{b}_a)$$

where $\mathbf{v} \in \mathbb{R}^3$, $\mathbf{W}_v \in \mathbb{R}^{3\times3}$, $\mathbf{W}_h \in \mathbb{R}^{3\times3}$, and $\mathbf{b}_a \in \mathbb{R}^3$.

Let $\mathbf{v} = \begin{pmatrix} 0.1 \\ 0.2 \\ 0.3 \end{pmatrix}$, $\mathbf{W}_v = \mathbf{W}_h = \mathbf{I}_3$, and $\mathbf{b}_a = \begin{pmatrix} 0.1 \\ 0.1 \\ 0.1 \end{pmatrix}$.

(b) Apply the softmax function to the localized attention scores and compute the attended feature value for the central pixel.

Chapter 8
Graph Neural Networks: Extending Deep Learning to Graphs

I do not fear computers. I fear the lack of them.

—Isaac Asimov, I, Robot

8.1 Introduction to Graph Neural Networks

Graph Neural Networks (GNNs) are a class of neural networks designed to perform inference on data structured as graphs. Unlike traditional neural networks that operate on grid-like structures (e.g., images or sequences), GNNs are capable of directly handling graph-structured data, making them suitable for a wide range of applications such as social network analysis, molecular chemistry, and recommendation systems. A graph G is formally defined as $G = (V, E)$, where V is a set of nodes (or vertices) and E is a set of edges. Each edge $e \in E$ is a pair (v_i, v_j) indicating a connection between nodes v_i and v_j. Additionally, graphs can be attributed, meaning that each node and edge can have associated features. These features are typically represented as vectors. GNNs leverage the structure of graphs by iteratively updating node representations based on their neighbors' features and the graph topology. The primary idea is to capture the dependencies between nodes through message passing mechanisms.

8.1.1 Historical Context and Development

The concept of neural networks operating on graphs dates back to early works in the 1990s. One of the pioneering approaches was introduced by Gori et al. (2005) and Scarselli et al. (2009), who proposed Graph Neural Networks (GNNs) that use

P. Singh and B. Raman, *Deep Learning Through the Prism of Tensors*, Studies in Big Data 162, https://doi.org/10.1007/978-981-97-8019-8_8

recursive neural networks to process graph data. These early models laid the groundwork for more sophisticated GNN architectures that followed. In the mid-2010s, significant advancements were made with the introduction of Graph Convolutional Networks (GCNs) by Kipf and Welling (2017), which extended the concept of convolution from grid-like data to graphs. GCNs efficiently aggregate and transform node features using a localized convolutional operation. Further innovations, such as GraphSAGE (Hamilton et al. 2017), Graph Attention Networks (GAT) (Velickovic et al. 2017a), and Message Passing Neural Networks (MPNN) (Gilmer et al. 2017), expanded the capabilities and applications of GNNs. These models introduced various techniques to improve the expressiveness and scalability of GNNs, including attention mechanisms, inductive learning, and efficient message passing.

8.1.2 Importance of GNNs in Various Domains

GNNs have become increasingly important across multiple domains due to their ability to capture complex relationships and dependencies in graph-structured data. Some notable applications include:

1. **Social Network Analysis**: GNNs are used to analyze social networks by modeling users as nodes and their interactions as edges. Tasks such as community detection, link prediction, and user recommendation benefit from the relational information captured by GNNs.
2. **Molecular Chemistry**: In molecular chemistry, molecules are naturally represented as graphs with atoms as nodes and bonds as edges. GNNs help predict molecular properties, drug discovery, and protein interaction by learning the intricate relationships within molecular structures.
3. **Recommendation Systems**: GNNs enhance recommendation systems by modeling users and items as a bipartite graph. Collaborative filtering and personalized recommendations are improved by capturing the interactions between users and items.
4. **Knowledge Graphs**: Knowledge graphs represent entities and their relationships in a graph structure. GNNs are applied to tasks such as entity classification, link prediction, and knowledge graph completion by effectively modeling the relational data.
5. **Computer Vision**: GNNs are used in computer vision tasks like scene graph generation and object detection by representing objects and their spatial relationships as graphs. This approach provides a higher-level understanding of visual scenes.

8.1.3 Mathematical Details of GNNs

The operation of GNNs can be formalized through the concept of message passing. For a node v_i in a graph, the message passing framework involves two main steps: message aggregation and node update.

Message Aggregation Each node v_i aggregates messages from its neighbors $\mathcal{N}(v_i)$:

$$m_i^{(t)} = \sum_{v_j \in \mathcal{N}(v_i)} M(h_i^{(t-1)}, h_j^{(t-1)}, e_{ij})$$

where $h_i^{(t-1)}$ and $h_j^{(t-1)}$ are the features of nodes v_i and v_j at iteration $t-1$, e_{ij} is the edge feature, and M is a message function.

Node Update The node's feature is then updated using the aggregated message:

$$h_i^{(t)} = U(h_i^{(t-1)}, m_i^{(t)})$$

where U is an update function.

Implementation Consider a simple GNN layer implementation in PyTorch:

```
import torch
import torch.nn as nn
import torch.nn.functional as F

class GNNLayer(nn.Module):
    def __init__(self, in_features, out_features):
        super(GNNLayer, self).__init__()
        self.linear = nn.Linear(in_features, out_features)

    def forward(self, x, adj):
        # Message passing
        h = torch.matmul(adj, x)
        # Node update
        h = self.linear(h)
        return F.relu(h)

class GNN(nn.Module):
    def __init__(self, in_features, hidden_features, out_features):
        super(GNN, self).__init__()
        self.gnn1 = GNNLayer(in_features, hidden_features)
        self.gnn2 = GNNLayer(hidden_features, out_features)

    def forward(self, x, adj):
        h = self.gnn1(x, adj)
        h = self.gnn2(h, adj)
        return h
```

8.2 Fundamentals of Graph Theory

Graph theory is a branch of mathematics that studies the properties and applications of graphs. A graph G is a mathematical structure used to model pairwise relations between objects. It consists of two sets: a set V of nodes (or vertices) and a set E of edges. Formally, a graph is defined as $G = (V, E)$. Nodes, also known as vertices, represent the entities in a graph. For example, in a social network, nodes could represent individuals. Edges represent the connections or relationships between nodes. An edge $e \in E$ is a pair (v_i, v_j) indicating a connection between nodes v_i and v_j. Edges can be directed or undirected, and weighted or unweighted. In a directed graph (or digraph), edges have a direction, indicating a one way relationship between nodes. A directed edge is represented as an ordered pair (v_i, v_j), meaning there is a connection from node v_i to node v_j. In an undirected graph, edges do not have a direction, indicating a mutual relationship between nodes. An undirected edge is represented as an unordered pair $\{v_i, v_j\}$.

A graph G can be represented mathematically using an adjacency matrix A or an adjacency list. An adjacency matrix $A \in \mathbb{R}^{|V| \times |V|}$ is a square matrix where A_{ij} indicates the presence (and possibly the weight) of an edge between nodes v_i and v_j. For an undirected graph, the adjacency matrix is symmetric.

$$A_{ij} = \begin{cases} 1 & \text{if } (v_i, v_j) \in E \\ 0 & \text{otherwise} \end{cases}$$

An adjacency list represents a graph as an array of lists. Each list corresponds to a node and contains the nodes adjacent to it. This representation is more space-efficient for sparse graphs. Different representations capture various aspects of graph topology and are used in different contexts depending on the operations and algorithms being applied.

Types of Graphs: Directed, Undirected, Weighted, and Unweighted In directed graphs, also known as digraphs, each edge has a direction, indicating a one-way relationship. These graphs are useful for representing asymmetric relationships, such as web links (where a link from page A to page B does not imply a link from B to A) or citation networks (where a paper cites another paper). Formally, a directed graph is represented as $G = (V, E)$, where E is a set of ordered pairs.

$$E = \{(v_i, v_j) \mid v_i, v_j \in V\}$$

In undirected graphs, edges have no direction, representing mutual relationships. These graphs are suitable for modeling symmetric relationships, such as friendships in social networks (where if person A is friends with person B, then person B is also friends with person A). Formally, an undirected graph is represented as $G = (V, E)$, where E is a set of unordered pairs.

$$E = \{\{v_i, v_j\} \mid v_i, v_j \in V\}$$

Weighted graphs assign a weight to each edge, representing the strength or capacity of the relationship between nodes. This is useful in scenarios such as road networks (where weights represent distances or travel times) or communication networks (where weights represent bandwidth or latency). Formally, a weighted graph is represented as $G = (V, E, W)$, where W is a set of weights associated with each edge.

$$W = \{w_{ij} \mid (v_i, v_j) \in E\}$$

Unweighted graphs do not assign weights to edges, representing relationships of equal strength or importance. These graphs are a special case of weighted graphs where all weights are equal (typically 1). Formally, an unweighted graph is represented as $G = (V, E)$, with no additional weight information.

Examples

1. **Social Network**: In a social network graph, nodes represent individuals, and edges represent friendships. An undirected edge $\{v_i, v_j\}$ indicates that both individuals v_i and v_j are friends.
2. **Citation Network**: In a citation network, nodes represent academic papers, and directed edges represent citations. A directed edge (v_i, v_j) indicates that paper v_i cites paper v_j.
3. **Transportation Network**: In a transportation network, nodes represent locations, and weighted edges represent roads with distances as weights. An undirected weighted edge $\{v_i, v_j\}$ with weight w_{ij} indicates a bidirectional road between locations v_i and v_j with distance w_{ij}.

Example Consider a simple graph with 3 nodes and edges (1–2), (2–3), and (1–3). The adjacency matrix is:

$$A = \begin{bmatrix} 0 & 1 & 1 \\ 1 & 0 & 1 \\ 1 & 1 & 0 \end{bmatrix}$$

Example For the same graph with edges (1–2), (2–3), and (1–3), the adjacency list is:

Node	Adjacent Nodes
1	2, 3
2	1, 3
3	1, 2

Incidence Matrix The incidence matrix B of a graph G with n nodes and m edges is an $n \times m$ matrix where each row represents a node and each column represents an edge. The element B_{ij} indicates the incidence of node v_i on edge e_j. For undirected graphs:

$$B_{ij} = \begin{cases} 1 & \text{if node } v_i \text{ is incident on edge } e_j \\ 0 & \text{otherwise} \end{cases}$$

For directed graphs, the incidence matrix indicates the direction of the edge:

$$B_{ij} = \begin{cases} 1 & \text{if node } v_i \text{ is the head of edge } e_j \\ -1 & \text{if node } v_i \text{ is the tail of edge } e_j \\ 0 & \text{otherwise} \end{cases}$$

Example For a graph with 3 nodes and 3 directed edges $e_1 = (1 \to 2)$, $e_2 = (2 \to 3)$, and $e_3 = (1 \to 3)$, the incidence matrix is:

$$B = \begin{bmatrix} 1 & 0 & 1 \\ -1 & 1 & 0 \\ 0 & -1 & -1 \end{bmatrix}$$

Graph Laplacian The graph Laplacian L is a matrix representation of a graph that is used in various applications, such as spectral clustering and graph signal processing. The Laplacian matrix is defined as:

$$L = D - A$$

where D is the degree matrix and A is the adjacency matrix. The degree matrix D is a diagonal matrix where D_{ii} is the degree of node v_i, i.e., the number of edges connected to v_i.

$$L_{ij} = \begin{cases} d_i & \text{if } i = j \\ -1 & \text{if } i \neq j \text{ and } (v_i, v_j) \in E \\ 0 & \text{otherwise} \end{cases}$$

Properties:

1. The Laplacian matrix is symmetric and positive semi-definite.
2. The eigenvalues of L provide important information about the graph's structure.
3. The smallest eigenvalue of L is always 0, and the multiplicity of the eigenvalue 0 corresponds to the number of connected components in the graph.

Example: For a simple graph with 3 nodes and edges (1–2), (2–3), and (1–3), the degree matrix D and the Laplacian matrix L are:

$$D = \begin{bmatrix} 2 & 0 & 0 \\ 0 & 2 & 0 \\ 0 & 0 & 2 \end{bmatrix}$$

$$L = D - A = \begin{bmatrix} 2 & -1 & -1 \\ -1 & 2 & -1 \\ -1 & -1 & 2 \end{bmatrix}$$

Applications:

1. **Spectral Clustering**: The eigenvalues and eigenvectors of the Laplacian matrix are used in spectral clustering to partition the graph into clusters.
2. **Graph Signal Processing**: The Laplacian matrix is used to analyze and process signals on graphs, such as in denoising and smoothing.
3. **Random Walks**: The Laplacian matrix is related to the transition matrix of a random walk on the graph, which is useful for studying the graph's connectivity and flow properties.

8.3 Representing Graph Data with Tensors

When dealing with graph-structured data, we represent graphs and their attributes as tensors to leverage the computational power and flexibility of deep learning frameworks like TensorFlow and PyTorch.

8.3.1 *Tensor Representations of Graphs*

Graphs are typically represented by several types of tensors, each capturing different aspects of the graph structure and node/edge attributes. The main types of tensors used in graph representation include node feature tensors, edge feature tensors, and adjacency matrices.

Node features are attributes associated with each node in the graph. These features can be scalar values, vectors, or even higher-dimensional tensors depending on the complexity of the data. For instance, in a social network graph, node features might include user attributes such as age, location, and interests. Let n be the number of nodes in the graph and d be the dimensionality of the node features. The node feature tensor X is of shape (n, d), where X_i represents the feature vector of node v_i.

$$X = \begin{pmatrix} x_{11} & x_{12} & \cdots & x_{1d} \\ x_{21} & x_{22} & \cdots & x_{2d} \\ \vdots & \vdots & \ddots & \vdots \\ x_{n1} & x_{n2} & \cdots & x_{nd} \end{pmatrix}$$

Edge features are attributes associated with each edge in the graph. These features can represent properties like weight, type, or any other relevant information about the connection between nodes. In a transportation network, edge features might include distance, travel time, and capacity. Let m be the number of edges in the graph and d' be the dimensionality of the edge features. The edge feature tensor E is of shape (m, d'), where E_j represents the feature vector of edge e_j.

$$E = \begin{pmatrix} e_{11} & e_{12} & \cdots & e_{1d'} \\ e_{21} & e_{22} & \cdots & e_{2d'} \\ \vdots & \vdots & \ddots & \vdots \\ e_{m1} & e_{m2} & \cdots & e_{md'} \end{pmatrix}$$

Example: Consider a graph with 3 nodes and 3 edges. Suppose each node has 2 features and each edge has 1 feature.

Node features:

$$X = \begin{pmatrix} 1.0 & 0.5 \\ 0.8 & 0.7 \\ 0.9 & 0.3 \end{pmatrix}$$

Edge features:

$$E = \begin{pmatrix} 2.0 \\ 1.5 \\ 2.5 \end{pmatrix}$$

Graph Tensors in Deep Learning Frameworks Deep learning frameworks like TensorFlow and PyTorch provide powerful tools for handling and manipulating tensors. Representing graphs with tensors in these frameworks allows us to apply neural network operations efficiently. The adjacency matrix A of a graph can be represented as a tensor. For a graph with n nodes, A is an $n \times n$ tensor where A_{ij} indicates the presence or weight of an edge between nodes v_i and v_j.

Example in PyTorch:

```
import torch

# Adjacency matrix for a graph with 3 nodes
adjacency_matrix = torch.tensor([
    [0, 1, 1],
    [1, 0, 1],
    [1, 1, 0]
], dtype=torch.float32)

# Node feature tensor for a graph with 3 nodes, each with 2 features
node_features = torch.tensor([
    [1.0, 0.5],
    [0.8, 0.7],
    [0.9, 0.3]
], dtype=torch.float32)
```

```
# Edge feature tensor for a graph with 3 edges, each with 1 feature
edge_features = torch.tensor([
    [2.0],
    [1.5],
    [2.5]
], dtype=torch.float32)
```

In a Graph Neural Network (GNN), we use these tensors to perform node and edge updates. Typically, a GNN layer will take the node feature tensor and the adjacency matrix as inputs and produce updated node features.

Example GNN Layer in PyTorch:

```
import torch.nn as nn
import torch.nn.functional as F

class GNNLayer(nn.Module):
    def __init__(self, in_features, out_features):
        super(GNNLayer, self).__init__()
        self.linear = nn.Linear(in_features, out_features)

    def forward(self, node_features, adjacency_matrix):
        # Message passing
        h = torch.matmul(adjacency_matrix, node_features)
        # Node update
        h = self.linear(h)
        return F.relu(h)

# Define a GNN with one layer
class GNN(nn.Module):
    def __init__(self, in_features, hidden_features, out_features):
        super(GNN, self).__init__()
        self.gnn1 = GNNLayer(in_features, hidden_features)
        self.gnn2 = GNNLayer(hidden_features, out_features)

    def forward(self, node_features, adjacency_matrix):
        h = self.gnn1(node_features, adjacency_matrix)
        h = self.gnn2(h, adjacency_matrix)
        return h

# Instantiate the GNN model
gnn = GNN(in_features=2, hidden_features=4, out_features=2)
output = gnn(node_features, adjacency_matrix)
print(output)
```

In this example, the 'GNNLayer' performs a simple message passing operation by multiplying the adjacency matrix with the node features, followed by a linear transformation and a ReLU activation. The 'GNN' model stacks two such layers to learn more complex representations.

8.3.2 Tensor Operations on Graphs

In the context of GNNs, tensor operations play a crucial role in processing and transforming graph-structured data. The primary operations include aggregation and combination of node features and performing graph convolutions. These operations enable GNNs to learn meaningful representations from graph data by leveraging the graph topology and node/edge attributes.

Aggregation Operations Aggregation operations involve collecting and combining information from the neighbors of a node. This step is essential for propagating information across the graph and capturing the dependencies between nodes. Aggregation can be performed in various ways, such as summation, mean, or maximum.

Summation: The summation aggregation for a node v_i collects features from its neighbors $\mathcal{N}(v_i)$ and sums them up. If $h_j^{(t-1)}$ represents the feature vector of node v_j at the $(t-1)$-th layer, the aggregated feature for node v_i at the t-th layer is:

$$m_i^{(t)} = \sum_{v_j \in \mathcal{N}(v_i)} h_j^{(t-1)}$$

Mean: The mean aggregation computes the average of the neighbors' features.

$$m_i^{(t)} = \frac{1}{|\mathcal{N}(v_i)|} \sum_{v_j \in \mathcal{N}(v_i)} h_j^{(t-1)}$$

Maximum: The maximum aggregation takes the element-wise maximum of the neighbors' features:

$$m_i^{(t)} = \max_{v_j \in \mathcal{N}(v_i)} h_j^{(t-1)}$$

Combination Operations: After aggregation, the combined feature is obtained by combining the aggregated message $m_i^{(t)}$ with the node's own feature $h_i^{(t-1)}$. This combination can be performed using various functions, such as concatenation followed by a linear transformation and non-linear activation.

Concatenation: Concatenate the node's feature with the aggregated message and apply a linear transformation:

$$h_i^{(t)} = \sigma\left(W \cdot [h_i^{(t-1)} \| m_i^{(t)}] + b\right)$$

where W is a weight matrix, b is a bias vector, σ is an activation function (e.g., ReLU), and $\|$ denotes concatenation.

Addition: Add the node's feature to the aggregated message:

$$h_i^{(t)} = \sigma\left(W \cdot (h_i^{(t-1)} + m_i^{(t)}) + b\right)$$

Example Consider a simple graph with nodes v_1, v_2, v_3 and the following adjacency matrix A:

$$A = \begin{pmatrix} 0 & 1 & 1 \\ 1 & 0 & 1 \\ 1 & 1 & 0 \end{pmatrix}$$

Let the initial node features be:

$$H^{(0)} = \begin{pmatrix} 1 & 2 \\ 2 & 3 \\ 3 & 1 \end{pmatrix}$$

Performing summation aggregation and concatenation combination:

Summation Aggregation:

$$m_1^{(1)} = h_2^{(0)} + h_3^{(0)} = \begin{pmatrix} 2 & 3 \end{pmatrix} + \begin{pmatrix} 3 & 1 \end{pmatrix} = \begin{pmatrix} 5 & 4 \end{pmatrix}$$

$$m_2^{(1)} = h_1^{(0)} + h_3^{(0)} = \begin{pmatrix} 1 & 2 \end{pmatrix} + \begin{pmatrix} 3 & 1 \end{pmatrix} = \begin{pmatrix} 4 & 3 \end{pmatrix}$$

$$m_3^{(1)} = h_1^{(0)} + h_2^{(0)} = \begin{pmatrix} 1 & 2 \end{pmatrix} + \begin{pmatrix} 2 & 3 \end{pmatrix} = \begin{pmatrix} 3 & 5 \end{pmatrix}$$

Concatenation Combination: Assume weight matrix W and bias b are identity and zero matrices for simplicity:

$$h_1^{(1)} = \sigma\left(W \cdot [h_1^{(0)} \| m_1^{(1)}] + b\right) = \sigma\left(\begin{pmatrix} 1 & 2 & 5 & 4 \end{pmatrix}\right)$$

Applying a ReLU activation:

$$h_1^{(1)} = \begin{pmatrix} 1 & 2 & 5 & 4 \end{pmatrix}$$

Graph Convolutions Graph convolutions generalize the concept of convolution from grid-like data (e.g., images) to graph-structured data. The convolution operation in GNNs aggregates and transforms node features based on the graph topology, allowing the model to capture local patterns and dependencies. For a node v_i with neighbors $\mathcal{N}(v_i)$, the graph convolution operation can be expressed as:

$$h_i^{(t)} = \sigma\left(\sum_{v_j \in \mathcal{N}(v_i) \cup \{v_i\}} \frac{1}{c_{ij}} W^{(t)} h_j^{(t-1)} + b^{(t)}\right)$$

where $W^{(t)}$ is the weight matrix at layer t, $b^{(t)}$ is the bias, c_{ij} is a normalization constant (e.g., $\sqrt{d_i d_j}$ for symmetric normalization), and σ is an activation function.

Example

```
import torch
import torch.nn as nn
import torch.nn.functional as F

class GraphConvolution(nn.Module):
    def __init__(self, in_features, out_features):
        super(GraphConvolution, self).__init__()
        self.linear = nn.Linear(in_features, out_features)

    def forward(self, node_features, adjacency_matrix):
        # Degree matrix for normalization
        degree_matrix = adjacency_matrix.sum(dim=1).unsqueeze(1)
        normalized_adjacency_matrix = adjacency_matrix / degree_matrix

        # Message passing
        h = torch.matmul(normalized_adjacency_matrix, node_features)

        # Node update
        h = self.linear(h)
        return F.relu(h)

# Define a GNN with one graph convolution layer
class GCN(nn.Module):
    def __init__(self, in_features, hidden_features, out_features):
        super(GCN, self).__init__()
        self.gconv1 = GraphConvolution(in_features, hidden_features)
        self.gconv2 = GraphConvolution(hidden_features, out_features)

    def forward(self, node_features, adjacency_matrix):
        h = self.gconv1(node_features, adjacency_matrix)
        h = self.gconv2(h, adjacency_matrix)
        return h

# Instantiate the GCN model
gcn = GCN(in_features=2, hidden_features=4, out_features=2)
output = gcn(node_features, adjacency_matrix)
print(output)
```

In the above example, the ‘GraphConvolution’ class defines a graph convolution layer that performs message passing and node updates. The degree matrix is used for normalization to ensure that the aggregated messages are appropriately scaled. The ‘GCN’ class stacks two graph convolution layers to learn hierarchical representations of the graph data.

8.4 Graph Neural Network Architectures

8.4.1 Graph Convolutional Networks (GCNs)

Graph Convolutional Networks (GCNs) are a powerful class of neural networks designed specifically for graph-structured data. They extend the concept of convolution from traditional grid-like data (e.g., images) to irregular graph data, enabling the capture of local node features and their dependencies within the graph. GCNs operate by iteratively aggregating and transforming node features through a series

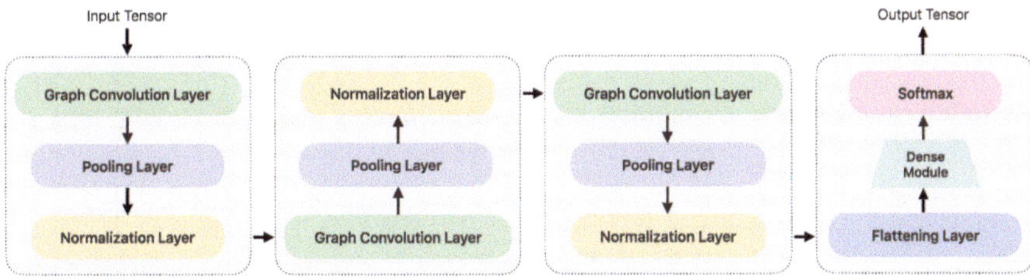

Fig. 8.1 Illustration of a GCN architecture

of graph convolutional layers. Each layer updates the node features by combining information from neighboring nodes, effectively performing a localized aggregation and transformation (see Fig. 8.1 for an overview of the architecture). A typical GCN layer consists of the following components:

1. **Message Passing**: Each node aggregates information from its neighbors to form an aggregated message. This step captures the local structure and feature information of the graph.
2. **Node Update**: The aggregated message is combined with the node's own features and transformed using a linear transformation followed by a non-linear activation function. This step updates the node's representation by incorporating the local neighborhood information.

Multiple GCN layers are stacked to enable the network to capture multi-hop neighborhood information. The deeper the network, the larger the receptive field for each node, allowing the GCN to capture more complex patterns and dependencies.

Example Consider a GCN with two layers. The first layer aggregates and transforms the node features based on their immediate neighbors, while the second layer aggregates and transforms the updated features based on the neighbors in the extended neighborhood.

Message Passing For a node v_i with neighbors $\mathcal{N}(v_i)$, the aggregated message at layer t is computed as:

$$m_i^{(t)} = \sum_{v_j \in \mathcal{N}(v_i) \cup \{v_i\}} \frac{1}{c_{ij}} W^{(t)} h_j^{(t-1)}$$

where $h_j^{(t-1)}$ is the feature vector of node v_j at layer $t-1$, $W^{(t)}$ is the weight matrix at layer t, and c_{ij} is a normalization constant (e.g., $\sqrt{d_i d_j}$ for symmetric normalization, where d_i and d_j are the degrees of nodes v_i and v_j).

Node Update The node's feature is updated by applying a non-linear activation function σ to the aggregated message:

$$h_i^{(t)} = \sigma\left(m_i^{(t)} + b^{(t)}\right)$$

where $b^{(t)}$ is a bias term. Combining the message passing and node update steps, the operation of a GCN layer can be compactly written as:

$$H^{(t)} = \sigma\left(\tilde{D}^{-1/2}\tilde{A}\tilde{D}^{-1/2}H^{(t-1)}W^{(t)} + B^{(t)}\right)$$

where $\tilde{A} = A + I$ is the adjacency matrix with added self-loops, $\tilde{D}$ is the degree matrix of $\tilde{A}$, $H^{(t-1)}$ is the matrix of node features at layer $t-1$, $W^{(t)}$ is the weight matrix at layer t, and $B^{(t)}$ is the bias matrix at layer t.

Example

```
import torch
import torch.nn as nn
import torch.nn.functional as F

class GCNLayer(nn.Module):
    def __init__(self, in_features, out_features):
        super(GCNLayer, self).__init__()
        self.linear = nn.Linear(in_features, out_features)

    def forward(self, node_features, adjacency_matrix):
        degree_matrix = adjacency_matrix.sum(dim=1).unsqueeze(1)
        normalized_adjacency_matrix = adjacency_matrix / degree_matrix

        # Message passing
        h = torch.matmul(normalized_adjacency_matrix, node_features)

        # Node update
        h = self.linear(h)
        return F.relu(h)

class GCN(nn.Module):
    def __init__(self, in_features, hidden_features, out_features):
        super(GCN, self).__init__()
        self.gconv1 = GCNLayer(in_features, hidden_features)
        self.gconv2 = GCNLayer(hidden_features, out_features)

    def forward(self, node_features, adjacency_matrix):
        h = self.gconv1(node_features, adjacency_matrix)
        h = self.gconv2(h, adjacency_matrix)
        return h

# Instantiate the GCN model
gcn = GCN(in_features=2, hidden_features=4, out_features=2)
node_features = torch.tensor([
    [1.0, 0.5],
    [0.8, 0.7],
    [0.9, 0.3]
], dtype=torch.float32)
adjacency_matrix = torch.tensor([
    [0, 1, 1],
    [1, 0, 1],
    [1, 1, 0]
], dtype=torch.float32)
output = gcn(node_features, adjacency_matrix)
print(output)
```

In this example, the 'GCNLayer' class implements a single graph convolutional layer that performs message passing and node update operations. The 'GCN' class stacks two such layers to form a simple GCN model. The degree matrix is used for normalization to ensure that the aggregated messages are appropriately scaled. The ReLU activation function introduces non-linearity, allowing the model to learn more complex patterns.

8.4.2 Graph Attention Networks (GATs)

Graph Attention Networks (GATs) introduce the concept of attention mechanisms to graph neural networks, enabling the model to learn the importance of neighboring nodes in a more flexible and data-driven manner. By leveraging attention, GATs can assign different weights to different nodes within the neighborhood, allowing for more expressive and powerful representations. Attention mechanisms have been successfully applied in various deep learning domains, particularly in natural language processing. The core idea of attention is to compute a set of weights that signify the importance of different elements in the input, allowing the model to focus on relevant parts of the input while ignoring less relevant parts. In the context of GNNs, attention mechanisms are used to determine the importance of neighboring nodes when aggregating their features. This enables the model to dynamically adjust the influence of each neighbor based on their relevance to the target node.

Consider a node v_i with a set of neighbors $\mathcal{N}(v_i)$. The goal is to compute an attention weight α_{ij} for each neighbor $v_j \in \mathcal{N}(v_i)$, indicating the importance of node v_j to node v_i. For each edge (v_i, v_j), compute an attention coefficient e_{ij} using the features of nodes v_i and v_j:

$$e_{ij} = \text{LeakyReLU}(\mathbf{a}^T [W\mathbf{h}_i \| W\mathbf{h}_j])$$

where W is a weight matrix, $\mathbf{a}$ is a learnable attention vector, $\mathbf{h}_i$ and $\mathbf{h}_j$ are the feature vectors of nodes v_i and v_j, and $\|$ denotes concatenation. Normalize the attention coefficients using the softmax function to obtain the attention weights α_{ij}:

$$\alpha_{ij} = \frac{\exp(e_{ij})}{\sum_{k \in \mathcal{N}(i)} \exp(e_{ik})}$$

Use the attention weights to compute a weighted sum of the neighbor features:

$$\mathbf{h}_i' = \sigma \left(\sum_{j \in \mathcal{N}(i)} \alpha_{ij} W\mathbf{h}_j \right)$$

where σ is a non-linear activation function (e.g., ReLU).

Graph Attention Layer A GAT layer can be formalized as follows:

1. **Linear Transformation**: Apply a linear transformation to the input features:

$$\mathbf{h}'_i = W\mathbf{h}_i$$

2. **Attention Coefficient Calculation**: Compute the unnormalized attention coefficients:

$$e_{ij} = \text{LeakyReLU}(\mathbf{a}^T[\mathbf{h}'_i \| \mathbf{h}'_j])$$

3. **Attention Weight Calculation**: Normalize the attention coefficients using the softmax function:

$$\alpha_{ij} = \frac{\exp(e_{ij})}{\sum_{k \in \mathcal{N}(i)} \exp(e_{ik})}$$

4. **Feature Aggregation**: Aggregate the neighbor features using the attention weights:

$$\mathbf{h}''_i = \sigma\left(\sum_{j \in \mathcal{N}(i)} \alpha_{ij}\mathbf{h}'_j\right)$$

Multi-head Attention GATs often employ multi-head attention to stabilize the learning process and improve performance. In multi-head attention, K independent attention mechanisms are used, and their outputs are concatenated or averaged:

$$\mathbf{h}''_i = \Big\|_{k=1}^{K} \sigma\left(\sum_{j \in \mathcal{N}(i)} \alpha_{ij}^{(k)}\mathbf{h}_j'^{(k)}\right)$$

where $\Big\|$ denotes concatenation, and $\alpha_{ij}^{(k)}$ and $\mathbf{h}_j'^{(k)}$ are the attention weights and transformed features from the k-th attention mechanism.

Example

```
import torch
import torch.nn as nn
import torch.nn.functional as F

class GraphAttentionLayer(nn.Module):
    def __init__(self, in_features, out_features, alpha, concat=True):
        super(GraphAttentionLayer, self).__init__()
        self.in_features = in_features
        self.out_features = out_features
        self.alpha = alpha
        self.concat = concat

        self.W = nn.Parameter(torch.zeros(size=(in_features,
                 out_features)))
        nn.init.xavier_uniform_(self.W.data, gain=1.414)
        self.a = nn.Parameter(torch.zeros(size=(2*out_features, 1)))
        nn.init.xavier_uniform_(self.a.data, gain=1.414)
```

```
        self.leakyrelu = nn.LeakyReLU(self.alpha)

    def forward(self, h, adj):
        Wh = torch.mm(h, self.W)
        a_input = self._prepare_attentional_mechanism_input(Wh)
        e = self.leakyrelu(torch.matmul(a_input, self.a).squeeze(2))

        zero_vec = -9e15 * torch.ones_like(e)
        attention = torch.where(adj > 0, e, zero_vec)
        attention = F.softmax(attention, dim=1)
        attention = F.dropout(attention, 0.6, training=self.training)
        h_prime = torch.matmul(attention, Wh)

        if self.concat:
            return F.elu(h_prime)
        else:
            return h_prime

    def _prepare_attentional_mechanism_input(self, Wh):
        N = Wh.size()[0]
        Wh_repeated_in_chunks = Wh.repeat_interleave(N, dim=0)
        Wh_repeated_alternating = Wh.repeat(N, 1)
        all_combinations_matrix = torch.cat([Wh_repeated_in_chunks,
                                   Wh_repeated_alternating], dim=1)
        return all_combinations_matrix.view(N, N, 2*self.out_features)

class GAT(nn.Module):
    def __init__(self, nfeat, nhid, nclass, dropout, alpha, nheads):
        super(GAT, self).__init__()
        self.dropout = dropout

        self.attentions = [GraphAttentionLayer(nfeat, nhid, alpha,
                           concat=True) for _ in range(nheads)]
        for i, attention in enumerate(self.attentions):
            self.add_module('attention_{}'.format(i), attention)

        self.out_att = GraphAttentionLayer(nhid * nheads, nclass,
                       alpha, concat=False)

    def forward(self, x, adj):
        x = F.dropout(x, self.dropout, training=self.training)
        x = torch.cat([att(x, adj) for att in self.attentions], dim=1)
        x = F.dropout(x, self.dropout, training=self.training)
        x = self.out_att(x, adj)
        return F.log_softmax(x, dim=1)

# Instantiate the GAT model
nfeat = node_features.shape[1]
nhid = 8
nclass = num_classes
dropout = 0.6
alpha = 0.2
nheads = 8

gat = GAT(nfeat, nhid, nclass, dropout, alpha, nheads)
output = gat(node_features, adjacency_matrix)
print(output)
```

The 'GraphAttentionLayer' class implements a single graph attention layer, computing the attention coefficients and aggregating neighbor features. The 'GAT' class stacks multiple attention heads in the first layer and combines them in the output layer to form a complete GAT model. The attention mechanism allows the model to dynamically weigh the importance of neighboring nodes, leading to more expressive and adaptive node representations.

8.4.3 GraphSAGE

GraphSAGE (Graph Sample and Aggregation) is a framework designed to efficiently generate node embeddings for large-scale graphs. Unlike traditional GNNs that operate on the entire neighborhood of a node, GraphSAGE employs a sampling-based approach to aggregate features from a fixed-size set of neighbors, enabling scalable learning on massive graphs. Traditional GNNs like GCNs and GATs aggregate features from all neighbors of a node, which can be computationally expensive and infeasible for large-scale graphs with millions of nodes and edges. GraphSAGE addresses this challenge by introducing a sampling strategy to aggregate information from a fixed-size set of neighbors, ensuring scalability and efficiency. In GraphSAGE, each node samples a fixed number of neighbors to aggregate features from, rather than using all neighbors. This approach reduces the computational complexity and memory requirements, making it suitable for large graphs.

We proceed as follows: For each node v_i, sample a fixed-size set of neighbors $\mathcal{N}_s(v_i)$ from its neighborhood $\mathcal{N}(v_i)$. The sampling can be performed uniformly or based on specific criteria (e.g., importance sampling). Aggregate the features from the sampled neighbors using an aggregation function. Common aggregation functions include mean, sum, and max-pooling. Combine the aggregated features with the node's own features using a combination function, such as concatenation followed by a linear transformation and non-linear activation. The core operation of GraphSAGE involves the following steps:

1. **Neighbor Sampling**: For each node v_i, sample a fixed-size set of neighbors $\mathcal{N}_s(v_i)$ from its neighborhood $\mathcal{N}(v_i)$. Let S denote the number of sampled neighbors.

$$\mathcal{N}_s(v_i) \subseteq \mathcal{N}(v_i), \quad |\mathcal{N}_s(v_i)| = S$$

2. **Feature Aggregation**: Compute the aggregated feature $m_i^{(t)}$ at layer t by applying an aggregation function AGG to the features of the sampled neighbors:

$$m_i^{(t)} = \text{AGG}\left(\{h_j^{(t-1)}, \forall v_j \in \mathcal{N}_s(v_i)\}\right)$$

Common aggregation functions include:
Mean Aggregation:

$$m_i^{(t)} = \frac{1}{S} \sum_{v_j \in \mathcal{N}_s(v_i)} h_j^{(t-1)}$$

Max Aggregation:

$$m_i^{(t)} = \max_{v_j \in \mathcal{N}_s(v_i)} h_j^{(t-1)}$$

LSTM Aggregation:

Use an LSTM to aggregate the features, capturing sequential dependencies among neighbors:

$$m_i^{(t)} = \text{LSTM}(\{h_j^{(t-1)}, \forall v_j \in \mathcal{N}_s(v_i)\})$$

3. **Combination**: Combine the aggregated feature $m_i^{(t)}$ with the node's own feature $h_i^{(t-1)}$ using a combination function, typically concatenation followed by a linear transformation and non-linear activation:

$$h_i^{(t)} = \sigma\left(W \cdot [h_i^{(t-1)} \| m_i^{(t)}] + b\right)$$

where: W is a weight matrix, b is a bias term, and σ is an activation function (e.g., ReLU).

Combining the sampling, aggregation, and combination steps, the operation of a GraphSAGE layer can be formalized as:

$$h_i^{(t)} = \sigma\left(W^{(t)} \cdot \left[h_i^{(t-1)} \| \text{AGG}\left(\{h_j^{(t-1)}, \forall v_j \in \mathcal{N}_s(v_i)\}\right)\right] + b^{(t)}\right)$$

Example

```
import torch
import torch.nn as nn
import torch.nn.functional as F

class GraphSAGELayer(nn.Module):
    def __init__(self, in_features, out_features, aggregator='mean'):
        super(GraphSAGELayer, self).__init__()
        self.aggregator = aggregator
        self.linear = nn.Linear(in_features * 2, out_features)

    def forward(self, node_features, adjacency_matrix):
        sampled_neighbors = self.sample_neighbors(adjacency_matrix)
        aggregated_features = self.aggregate(node_features,
                                  sampled_neighbors)
        combined_features = torch.cat([node_features,
                                aggregated_features], dim=1)
        return F.relu(self.linear(combined_features))

    def sample_neighbors(self, adjacency_matrix):
        # For simplicity, sample a fixed number of neighbors
        for each node
        num_nodes = adjacency_matrix.size(0)
        sampled_neighbors = []
        for i in range(num_nodes):
            neighbors = adjacency_matrix[i].nonzero(as_tuple=True)[0]
            sampled_neighbors.append(neighbors[:5])  # Sample up to
                                                      5 neighbors
        return sampled_neighbors

    def aggregate(self, node_features, sampled_neighbors):
        if self.aggregator == 'mean':
            aggregated_features = []
            for neighbors in sampled_neighbors:
                neighbor_features = node_features[neighbors]
```

```
                aggregated_features.append(neighbor_features
                .mean(dim=0))
            return torch.stack(aggregated_features)
        # Implement other aggregators (e.g., max, LSTM) as needed
        raise NotImplementedError('Aggregator not implemented')

class GraphSAGE(nn.Module):
    def __init__(self, in_features, hidden_features, out_features):
        super(GraphSAGE, self).__init__()
        self.sage1 = GraphSAGELayer(in_features, hidden_features)
        self.sage2 = GraphSAGELayer(hidden_features, out_features)

    def forward(self, node_features, adjacency_matrix):
        h = self.sage1(node_features, adjacency_matrix)
        h = self.sage2(h, adjacency_matrix)
        return h

# Instantiate the GraphSAGE model
sage = GraphSAGE(in_features=2, hidden_features=4, out_features=2)
node_features = torch.tensor([
    [1.0, 0.5],
    [0.8, 0.7],
    [0.9, 0.3]
], dtype=torch.float32)
adjacency_matrix = torch.tensor([
    [0, 1, 1],
    [1, 0, 1],
    [1, 1, 0]
], dtype=torch.float32)
output = sage(node_features, adjacency_matrix)
print(output)
```

In this example, the 'GraphSAGELayer' class implements a single GraphSAGE layer with a mean aggregator. The 'sample_neighbors' method samples a fixed number of neighbors for each node, while the 'aggregate' method aggregates the features of the sampled neighbors using the mean aggregator. The 'GraphSAGE' class stacks two such layers to form a complete GraphSAGE model. By employing sampling-based approaches, GraphSAGE significantly reduces the computational complexity and memory requirements for processing large-scale graphs. This enables scalable learning on massive graphs while maintaining the expressive power of the learned representations. The flexible aggregation and combination functions in GraphSAGE allow for capturing diverse patterns and dependencies within graph-structured data.

8.4.4 Other GNN Variants

GNNs have evolved significantly, resulting in various architectures designed to address specific types of graph data and learning tasks. Two notable variants are Message Passing Neural Networks (MPNNs) and Relational Graph Convolutional Networks (R-GCNs). These variants extend the basic principles of GNNs to handle more complex graph structures and relationships.

Message Passing Neural Networks (MPNNs) Message Passing Neural Networks are a general framework for designing GNNs that encompass a wide range of specific architectures. MPNNs operate by iteratively updating node representations through a message-passing mechanism, where nodes exchange information with their neighbors. MPNNs consist of two main phases for each layer: message aggregation and node update.

1. **Message Aggregation**: Each node v_i receives messages from its neighbors. The message m_{ij} from node v_j to node v_i is typically a function of the features of both nodes and the edge between them.

$$m_{ij}^{(t)} = \text{MessageFunction}(h_i^{(t-1)}, h_j^{(t-1)}, e_{ij})$$

where $h_i^{(t-1)}$ and $h_j^{(t-1)}$ are the features of nodes v_i and v_j at the $(t-1)$-th layer, and e_{ij} represents the edge features.
2. **Node Update**: The node v_i aggregates the messages from its neighbors and updates its feature using an update function, which is often a neural network.

$$h_i^{(t)} = \text{UpdateFunction}\left(h_i^{(t-1)}, \sum_{v_j \in \mathcal{N}(v_i)} m_{ij}^{(t)}\right)$$

The MPNN framework can be formalized as:

$$m_i^{(t)} = \sum_{v_j \in \mathcal{N}(v_i)} \text{MessageFunction}(h_i^{(t-1)}, h_j^{(t-1)}, e_{ij})$$

$$h_i^{(t)} = \text{UpdateFunction}(h_i^{(t-1)}, m_i^{(t)})$$

Example

```
import torch
import torch.nn as nn
import torch.nn.functional as F

class MPNNLayer(nn.Module):
    def __init__(self, in_features, out_features):
        super(MPNNLayer, self).__init__()
        self.message_function = nn.Linear(in_features * 2,
                                          out_features)
        self.update_function = nn.GRUCell(out_features,
                                         out_features)

    def forward(self, node_features, edge_features, adjacency_matrix):
        num_nodes = node_features.size(0)
        h = node_features
        for i in range(num_nodes):
            neighbors = adjacency_matrix[i].nonzero(as_tuple=True)[0]
            messages = []
            for j in neighbors:
                edge_feat = edge_features[i, j]
                message = F.relu(self.message_function(torch.cat([h[i],
                          h[j], edge_feat], dim=0)))
                messages.append(message)
            aggregated_message = torch.sum(torch.stack(messages),
                                           dim=0)
            h[i] = self.update_function(aggregated_message, h[i])
```

```
        return h

class MPNN(nn.Module):
    def __init__(self, in_features, hidden_features, out_features):
        super(MPNN, self).__init__()
        self.mpnn1 = MPNNLayer(in_features, hidden_features)
        self.mpnn2 = MPNNLayer(hidden_features, out_features)

    def forward(self, node_features, edge_features, adjacency_matrix):
        h = self.mpnn1(node_features, edge_features, adjacency_matrix)
        h = self.mpnn2(h, edge_features, adjacency_matrix)
        return h

# Example
node_features = torch.randn(3, 2)  # 3 nodes with 2 features each
edge_features = torch.randn(3, 3, 1)  # 3x3 adjacency matrix
                                        with 1 feature per edge
adjacency_matrix = torch.tensor([
    [0, 1, 1],
    [1, 0, 1],
    [1, 1, 0]
], dtype=torch.float32)

mpnn = MPNN(in_features=2, hidden_features=4, out_features=2)
output = mpnn(node_features, edge_features, adjacency_matrix)
print(output)
```

In this implementation, ‘MPNNLayer’ defines a single MPNN layer that performs message passing and node updates. The ‘MPNN’ class stacks two such layers to form a complete MPNN model. The message function and update function are implemented using neural network layers, allowing for flexible and expressive message aggregation and node updating.

Relational Graph Convolutional Networks (R-GCNs) Relational Graph Convolutional Networks extend the basic GCN framework to handle multi-relational graph data, where edges can have different types representing various relationships. R-GCNs are particularly useful for knowledge graphs and other applications where nodes are connected by multiple types of relationships. In R-GCNs, each relation type r is associated with a separate weight matrix W_r. The node feature update incorporates messages from all relation types, allowing the model to capture the diverse nature of the relationships.

1. **Message Passing**: For a node v_i, the message from a neighbor v_j through relation r is computed using the relation-specific weight matrix W_r:

$$m_{ij}^{(r)} = W_r h_j^{(t-1)}$$

2. **Node Update**: The aggregated message from all relations is used to update the node's feature:

$$h_i^{(t)} = \sigma\left(\sum_{r\in\mathcal{R}}\sum_{v_j\in\mathcal{N}_r(v_i)}\frac{1}{c_{i,r}}m_{ij}^{(r)} + W_0 h_i^{(t-1)}\right)$$

where $\mathcal{R}$ is the set of all relation types, $\mathcal{N}_r(v_i)$ is the set of neighbors of v_i under relation r, $c_{i,r}$ is a normalization constant, and W_0 is the weight matrix for self-loops.

Example

```
import torch
import torch.nn as nn
import torch.nn.functional as F

class RGCNLayer(nn.Module):
    def __init__(self, in_features, out_features, num_relations):
        super(RGCNLayer, self).__init__()
        self.num_relations = num_relations
        self.weight = nn.ParameterList([nn.Parameter(torch.randn
                      (in_features, out_features))
                      for _ in range(num_relations)])
        self.self_weight = nn.Parameter(torch.randn
                           (in_features, out_features))

    def forward(self, node_features, adjacency_matrices):
        out = torch.zeros_like(node_features)
        for r in range(self.num_relations):
            adj = adjacency_matrices[r]
            out += torch.matmul(adj, node_features)
                  .matmul(self.weight[r])
        out += node_features.matmul(self.self_weight)
        return F.relu(out)

class RGCN(nn.Module):
    def __init__(self, in_features, hidden_features,
    out_features, num_relations):
        super(RGCN, self).__init__()
        self.rgcn1 = RGCNLayer(in_features, hidden_features,
                     num_relations)
        self.rgcn2 = RGCNLayer(hidden_features, out_features,
                     num_relations)

    def forward(self, node_features, adjacency_matrices):
        h = self.rgcn1(node_features, adjacency_matrices)
        h = self.rgcn2(h, adjacency_matrices)
        return h

# Example
node_features = torch.randn(3, 2)  # 3 nodes with 2 features each
adjacency_matrices = [torch.tensor([
    [0, 1, 0],
    [0, 0, 1],
    [1, 0, 0]
], dtype=torch.float32) for _ in range(2)]  #Two types of relations

rgcn = RGCN(in_features=2, hidden_features=4, out_features=2,
       num_relations=2)
output = rgcn(node_features, adjacency_matrices)
print(output)
```

In this implementation, 'RGCNLayer' defines a single R-GCN layer that handles multiple types of relations. Each relation type has its own weight matrix, and the final node features are aggregated across all relations. The 'RGCN' class stacks two such layers to form a complete R-GCN model, demonstrating how R-GCNs can process multi-relational graph data.

Both MPNNs and R-GCNs extend the basic GNN framework to handle more complex graph structures and relationships. MPNNs provide a flexible message-passing framework that can be adapted to various tasks, while R-GCNs are specifically designed to handle multi-relational data, making them suitable for applications like knowledge graphs.

8.5 Training and Optimization of GNNs

Training GNNs involves optimizing the model's parameters to accurately capture the underlying patterns in graph-structured data. Various training techniques can be applied depending on the availability of labeled data and the specific learning task. This section delves into the different training paradigms and their implementations in GNNs.

8.5.1 *Training Techniques*

In supervised learning, GNNs are trained using labeled data, where each node (or edge) in the graph has an associated label. The goal is to learn a mapping from the node features to the labels by minimizing a loss function that measures the discrepancy between the predicted and true labels. For node classification, a common loss function is the cross-entropy loss, which is defined as:

$$\mathcal{L}_{\text{CE}} = -\frac{1}{N}\sum_{i=1}^{N}\sum_{c=1}^{C} y_i^c \log \hat{y}_i^c$$

where N is the number of nodes, C is the number of classes, y_i^c is the ground truth label for node i for class c, and $\hat{y}_i^c$ is the predicted probability for node i for class c.

Example: Consider a graph where each node represents a document, and the task is to classify documents into different categories. The GNN can be trained using labeled nodes with document categories as labels.

In unsupervised learning, GNNs are trained without labeled data. The objective is to learn meaningful node embeddings that capture the graph's structure and properties. Common unsupervised learning techniques for GNNs include node embedding methods like DeepWalk and GraphSAGE. A popular approach is to use reconstruction loss, where the goal is to reconstruct the graph's adjacency matrix or similarity matrix from the learned embeddings. The loss function can be defined as:

$$\mathcal{L}_{\text{recon}} = \sum_{(i,j)\in E} \|\mathbf{h}_i - \mathbf{h}_j\|_2^2 - \sum_{(i,j)\notin E} \|\mathbf{h}_i - \mathbf{h}_j\|_2^2$$

where $\mathbf{h}_i$ and $\mathbf{h}_j$ are the embeddings of nodes i and j, and E is the set of edges in the graph.

Example: Consider a social network graph where the goal is to learn node embeddings that capture the connections between users. These embeddings can be used for tasks like link prediction or clustering.

Semi-supervised learning is a hybrid approach where GNNs are trained using both labeled and unlabeled data. This paradigm leverages the structure of the graph to propagate label information from the labeled nodes to the unlabeled nodes, improving the model's performance. A key technique in semi-supervised learning is label propagation, where the labels of the labeled nodes are spread to their neighbors through the graph structure. This process iteratively updates the labels of the unlabeled nodes based on the labels of their neighbors. The loss function in semi-supervised learning combines the supervised loss on the labeled nodes and an unsupervised loss that encourages smoothness in the embeddings. A common formulation is:

$$\mathcal{L} = \mathcal{L}_{\text{sup}} + \lambda \mathcal{L}_{\text{unsup}}$$

where $\mathcal{L}_{\text{sup}}$ is the supervised loss on the labeled nodes, $\mathcal{L}_{\text{unsup}}$ is the unsupervised loss on the unlabeled nodes, and λ is a hyperparameter that balances the two losses.

GCNs are particularly well-suited for semi-supervised learning due to their ability to aggregate information from neighboring nodes. The forward propagation in a GCN can be interpreted as a form of label propagation, where the features and labels of the nodes are aggregated and transformed through the graph.

Example

```
import torch
import torch.nn as nn
import torch.nn.functional as F

class SemiSupervisedGCN(nn.Module):
    def __init__(self, in_features, hidden_features, out_features):
        super(SemiSupervisedGCN, self).__init__()
        self.gc1 = GraphConvolution(in_features, hidden_features)
        self.gc2 = GraphConvolution(hidden_features, out_features)

    def forward(self, x, adj):
        x = F.relu(self.gc1(x, adj))
        x = F.dropout(x, training=self.training)
        x = self.gc2(x, adj)
        return F.log_softmax(x, dim=1)

class GraphConvolution(nn.Module):
    def __init__(self, in_features, out_features):
        super(GraphConvolution, self).__init__()
        self.linear = nn.Linear(in_features, out_features)

    def forward(self, x, adj):
        support = torch.mm(adj, x)
        output = self.linear(support)
        return output

# Example
adjacency_matrix = torch.tensor([
    [1, 1, 0],
    [1, 1, 1],
```

```
    [0, 1, 1]
], dtype=torch.float32)
node_features = torch.tensor([
    [1.0, 0.5],
    [0.8, 0.7],
    [0.9, 0.3]
], dtype=torch.float32)
labels = torch.tensor([0, 1, 0])

model = SemiSupervisedGCN(in_features=2, hidden_features=4,
        out_features=2)
optimizer = torch.optim.Adam(model.parameters(), lr=0.01)

for epoch in range(100):
    model.train()
    optimizer.zero_grad()
    output = model(node_features, adjacency_matrix)
    loss = F.nll_loss(output, labels)
    loss.backward()
    optimizer.step()

    model.eval()
    _, predicted = output.max(1)
    correct = (predicted == labels).sum().item()
    print(f'Epoch {epoch+1}, Loss: {loss.item()},
          Accuracy: {correct / len(labels):.2f}')
```

In this implementation, the 'SemiSupervisedGCN' class defines a simple GCN for semi-supervised node classification. The adjacency matrix is used to propagate information through the graph, and the node features are transformed and aggregated at each layer. The model is trained using the negative log-likelihood loss on the labeled nodes, and the accuracy is evaluated after each epoch.

8.5.2 *Regularization Techniques*

Regularization is a critical aspect of training neural networks, including GNNs. It helps prevent overfitting and improves the model's generalization capabilities. In the context of GNNs, several regularization techniques are employed, such as dropout and graph-specific regularization methods.

Dropout is a widely used regularization technique that involves randomly dropping units (along with their connections) during training. This prevents the network from becoming too reliant on specific nodes, leading to better generalization. In the context of GNNs, dropout can be applied to both the node features and the edges in the graph. By randomly omitting parts of the graph, dropout forces the network to learn robust representations that do not rely on any particular subset of nodes or edges.

Consider a layer in a GNN where the input node features are represented by a matrix H. Applying dropout to the node features can be formulated as:

$$\tilde{H} = \text{Dropout}(H, p)$$

where p is the dropout probability, and $\tilde{H}$ is the resulting matrix after dropout. Each element in $\tilde{H}$ is independently set to zero with probability p or scaled by $\frac{1}{1-p}$ with probability $1 - p$. When applied to the adjacency matrix A, dropout can be represented as:

$$\tilde{A} = \text{Dropout}(A, p)$$

where $\tilde{A}$ is the adjacency matrix after applying dropout to the edges.

Implementation

```
import torch
import torch.nn as nn
import torch.nn.functional as F

class DropoutGCNLayer(nn.Module):
    def __init__(self, in_features, out_features, dropout_prob):
        super(DropoutGCNLayer, self).__init__()
        self.linear = nn.Linear(in_features, out_features)
        self.dropout_prob = dropout_prob

    def forward(self, x, adj):
        x = F.dropout(x, self.dropout_prob, training=self.training)
        adj = F.dropout(adj, self.dropout_prob,
              training=self.training)
        support = torch.mm(adj, x)
        output = self.linear(support)
        return F.relu(output)

class DropoutGCN(nn.Module):
    def __init__(self, in_features, hidden_features, out_features,
    dropout_prob):
        super(DropoutGCN, self).__init__()
        self.gcn1 = DropoutGCNLayer(in_features, hidden_features,
                    dropout_prob)
        self.gcn2 = DropoutGCNLayer(hidden_features, out_features,
                    dropout_prob)

    def forward(self, x, adj):
        x = self.gcn1(x, adj)
        x = self.gcn2(x, adj)
        return x

# Example
node_features = torch.randn(3, 2)  # 3 nodes with 2 features each
adjacency_matrix = torch.tensor([
    [1, 1, 0],
    [1, 1, 1],
    [0, 1, 1]
], dtype=torch.float32)
dropout_prob = 0.5

model = DropoutGCN(in_features=2, hidden_features=4, out_features=2,
        dropout_prob=dropout_prob)
output = model(node_features, adjacency_matrix)
print(output)
```

In this implementation, the 'DropoutGCNLayer' class defines a GCN layer with dropout applied to both node features and the adjacency matrix. The 'Dropout-GCN' class stacks two such layers to form a complete GCN model with dropout regularization. By applying dropout, the model is encouraged to learn more robust representations that generalize better to unseen data.

Graph-specific regularization methods leverage the unique structure of graph data to impose additional constraints and promote better generalization. These methods include node feature smoothness, edge weight regularization, and graph sparsification.

Node feature smoothness encourages the features of neighboring nodes to be similar, reflecting the homophily property of many real-world graphs. This can be enforced by adding a smoothness penalty to the loss function:

$$\mathcal{L}_{\text{smooth}} = \sum_{(i,j)\in E} \|\mathbf{h}_i - \mathbf{h}_j\|_2^2$$

where E is the set of edges, and $\mathbf{h}_i$ and $\mathbf{h}_j$ are the features of nodes i and j.

Edge weight regularization involves penalizing the model based on the weights of the edges. This can be particularly useful in applications where edge weights represent the strength of connections. A common approach is to use an L_2 regularization on the edge weights:

$$\mathcal{L}_{\text{edge}} = \lambda \sum_{(i,j)\in E} w_{ij}^2$$

where λ is a regularization coefficient, and w_{ij} is the weight of the edge between nodes i and j.

Graph sparsification aims to reduce the complexity of the graph by removing less important edges while preserving the overall structure and properties. This can be achieved through various techniques, such as thresholding the edge weights or using graph pruning algorithms.

Implementation

```
class GraphRegularizationGCN(nn.Module):
    def __init__(self, in_features, hidden_features, out_features,
    lambda_smooth, lambda_edge):
        super(GraphRegularizationGCN, self).__init__()
        self.gcn1 = GraphConvolution(in_features, hidden_features)
        self.gcn2 = GraphConvolution(hidden_features, out_features)
        self.lambda_smooth = lambda_smooth
        self.lambda_edge = lambda_edge

    def forward(self, x, adj):
        x = F.relu(self.gcn1(x, adj))
        x = F.dropout(x, training=self.training)
        x = self.gcn2(x, adj)
        return x
```

```
    def loss(self, output, labels, adj, node_features):
        ce_loss = F.nll_loss(output, labels)
        smoothness_loss = self.lambda_smooth*self.smoothness_penalty
                          (node_features, adj)
        edge_loss = self.lambda_edge * self.edge_weight_penalty(adj)
        return ce_loss + smoothness_loss + edge_loss

    def smoothness_penalty(self, node_features, adj):
        smoothness_loss = 0
        for i in range(adj.size(0)):
            neighbors = adj[i].nonzero(as_tuple=True)[0]
            for j in neighbors:
                smoothness_loss += torch.norm(node_features[i]
                                 - node_features[j]) ** 2
        return smoothness_loss

    def edge_weight_penalty(self, adj):
        return torch.norm(adj) ** 2

# Example
node_features = torch.randn(3, 2)  # 3 nodes with 2 features each
adjacency_matrix = torch.tensor([
    [1, 1, 0],
    [1, 1, 1],
    [0, 1, 1]
], dtype=torch.float32)
labels = torch.tensor([0, 1, 0])
lambda_smooth = 0.1
lambda_edge = 0.01

model = GraphRegularizationGCN(in_features=2, hidden_features=4,
        out_features=2, lambda_smooth=lambda_smooth,
         lambda_edge=lambda_edge)
optimizer = torch.optim.Adam(model.parameters(), lr=0.01)

for epoch in range(100):
    model.train()
    optimizer.zero_grad()
    output = model(node_features, adjacency_matrix)
    loss = model.loss(output, labels, adjacency_matrix, node_features)
    loss.backward()
    optimizer.step()

    model.eval()
    _, predicted = output.max(1)
    correct = (predicted == labels).sum().item()
    print(f'Epoch {epoch+1}, Loss: {loss.item()},
          Accuracy: {correct / len(labels):.2f}')
```

In this implementation, the 'GraphRegularizationGCN' class includes additional regularization terms in the loss function to enforce smoothness and penalize edge weights. The 'smoothness_penalty' method calculates the smoothness loss by summing the differences between node features for each edge, and the 'edge_weight_penalty' method calculates the L_2 norm of the adjacency matrix. These regularization techniques help improve the model's generalization by promoting smoother node features and controlling the influence of edge weights.

8.6 Practical Applications of GNNs

Graph Neural Networks (GNNs) have shown great potential in various domains due to their ability to capture the complex dependencies and relationships in graph-structured data. One prominent application area is social network analysis, where GNNs can be used for tasks such as node classification, community detection, link prediction, and influence analysis. This section explores these applications in detail, providing mathematical formulations and practical examples.

8.6.1 Social Network Analysis

Social networks are naturally represented as graphs, where nodes represent individuals, and edges represent relationships or interactions between them. Analyzing social networks involves understanding the structure and dynamics of these interactions. GNNs provide powerful tools for this analysis by leveraging the graph structure to extract meaningful patterns and insights.

Node Classification Node classification involves predicting the labels or categories of nodes in a graph. In a social network, this could mean categorizing users based on their interests, roles, or other attributes. GNNs excel at this task by aggregating information from a node's neighbors and learning representations that capture the node's context within the graph. Given a graph $G = (V, E)$ with node features X and an adjacency matrix A, the goal of node classification is to learn a function $f : V \rightarrow \mathbb{R}^C$ that maps each node to a probability distribution over C classes. This is typically achieved by training a GNN to minimize a cross-entropy loss:

$$\mathcal{L}_{\text{CE}} = -\frac{1}{N}\sum_{i=1}^{N}\sum_{c=1}^{C} y_i^c \log \hat{y}_i^c$$

where y_i^c is the true label, and $\hat{y}_i^c$ is the predicted probability for class c for node i.

Example

```
import torch
import torch.nn as nn
import torch.nn.functional as F

class NodeClassificationGCN(nn.Module):
    def __init__(self, in_features, hidden_features, out_features):
        super(NodeClassificationGCN, self).__init__()
        self.gc1 = GraphConvolution(in_features, hidden_features)
        self.gc2 = GraphConvolution(hidden_features, out_features)

    def forward(self, x, adj):
        x = F.relu(self.gc1(x, adj))
        x = F.dropout(x, training=self.training)
        x = self.gc2(x, adj)
        return F.log_softmax(x, dim=1)
```

```
# Example
node_features = torch.randn(5, 3)  # 5 nodes with 3 features each
adjacency_matrix = torch.tensor([
    [1, 1, 0, 0, 0],
    [1, 1, 1, 0, 0],
    [0, 1, 1, 1, 0],
    [0, 0, 1, 1, 1],
    [0, 0, 0, 1, 1]
], dtype=torch.float32)
labels = torch.tensor([0, 1, 0, 2, 2])

model = NodeClassificationGCN(in_features=3, hidden_features=4,
        out_features=3)
optimizer = torch.optim.Adam(model.parameters(), lr=0.01)

for epoch in range(100):
    model.train()
    optimizer.zero_grad()
    output = model(node_features, adjacency_matrix)
    loss = F.nll_loss(output, labels)
    loss.backward()
    optimizer.step()

    model.eval()
    _, predicted = output.max(1)
    correct = (predicted == labels).sum().item()
    print(f'Epoch {epoch+1}, Loss: {loss.item()},
         Accuracy: {correct / len(labels):.2f}')
```

Community Detection Community detection involves identifying clusters or groups of nodes that are more densely connected to each other than to the rest of the network. GNNs can be used to learn node embeddings that reveal community structures by capturing the connectivity patterns in the graph. Community detection can be framed as a clustering problem. One approach is to learn node embeddings H using a GNN and then apply a clustering algorithm like k-means:

$$\min_{C,\mu} \sum_{i=1}^{N} \|h_i - \mu_{c_i}\|^2$$

where μ_{c_i} is the centroid of cluster c_i, and h_i is the embedding of node i.

Example

```
from sklearn.cluster import KMeans

class CommunityDetectionGCN(nn.Module):
    def __init__(self, in_features, hidden_features, out_features):
        super(CommunityDetectionGCN, self).__init__()
        self.gc1 = GraphConvolution(in_features, hidden_features)
        self.gc2 = GraphConvolution(hidden_features, out_features)

    def forward(self, x, adj):
        x = F.relu(self.gc1(x, adj))
        x = F.dropout(x, training=self.training)
        x = self.gc2(x, adj)
        return x

# Example
node_features = torch.randn(5, 3)  # 5 nodes with 3 features each
adjacency_matrix = torch.tensor([
    [1, 1, 0, 0, 0],
```

```
    [1, 1, 1, 0, 0],
    [0, 1, 1, 1, 0],
    [0, 0, 1, 1, 1],
    [0, 0, 0, 1, 1]
], dtype=torch.float32)

model = CommunityDetectionGCN(in_features=3, hidden_features=4,
        out_features=2)
optimizer = torch.optim.Adam(model.parameters(), lr=0.01)

for epoch in range(100):
    model.train()
    optimizer.zero_grad()
    embeddings = model(node_features, adjacency_matrix)
    loss = F.mse_loss(embeddings,
           node_features)  # Dummy loss
                             for demonstration
    loss.backward()
    optimizer.step()

embeddings = embeddings.detach().numpy()
kmeans = KMeans(n_clusters=2).fit(embeddings)
print("Cluster assignments:", kmeans.labels_)
```

Link Prediction Link prediction involves predicting the existence of edges between nodes in a graph. In a social network, this could mean predicting potential friendships or connections between users. GNNs can be used to learn embeddings that encode the likelihood of edges between nodes. Link prediction can be approached by learning node embeddings H and then using a scoring function $f(u, v)$ to predict the existence of an edge between nodes u and v:

$$\hat{y}_{uv} = f(h_u, h_v)$$

A common choice for the scoring function is the dot product:

$$f(u, v) = h_u^\top h_v$$

Example

```
class LinkPredictionGCN(nn.Module):
    def __init__(self, in_features, hidden_features, out_features):
        super(LinkPredictionGCN, self).__init__()
        self.gc1 = GraphConvolution(in_features, hidden_features)
        self.gc2 = GraphConvolution(hidden_features, out_features)

    def forward(self, x, adj):
        x = F.relu(self.gc1(x, adj))
        x = F.dropout(x, training=self.training)
        x = self.gc2(x, adj)
        return x

# Example
node_features = torch.randn(5, 3)  # 5 nodes with 3 features each
adjacency_matrix = torch.tensor([
    [1, 1, 0, 0, 0],
    [1, 1, 1, 0, 0],
    [0, 1, 1, 1, 0],
    [0, 0, 1, 1, 1],
    [0, 0, 0, 1, 1]
], dtype=torch.float32)
```

```
model = LinkPredictionGCN(in_features=3, hidden_features=4,
        out_features=2)
optimizer = torch.optim.Adam(model.parameters(), lr=0.01)

for epoch in range(100):
    model.train()
    optimizer.zero_grad()
    embeddings = model(node_features, adjacency_matrix)
    # Dummy loss for demonstration (e.g., margin loss
    or binary cross-entropy)
    loss = F.mse_loss(embeddings, node_features)
    loss.backward()
    optimizer.step()

embeddings = embeddings.detach().numpy()
u, v = 0, 1  # Example node pair for link prediction
score = np.dot(embeddings[u], embeddings[v])
print("Link prediction score:", score)
```

Influence Analysis Influence analysis involves identifying influential nodes or edges in a social network, such as key opinion leaders or critical connections. GNNs can help quantify the influence of nodes based on their learned embeddings and the graph structure. One approach to influence analysis is to compute centrality measures based on node embeddings. For instance, the PageRank algorithm can be adapted to use node embeddings to measure influence:

$$PR(v_i) = \alpha \sum_{v_j \in \mathcal{N}(v_i)} \frac{PR(v_j)}{d_j} + (1 - \alpha)\frac{1}{N}$$

where $PR(v_i)$ is the PageRank score of node i, $\mathcal{N}(v_i)$ is the set of neighbors of i, d_j is the degree of node j, α is a damping factor, and N is the total number of nodes.

Example

```
import networkx as nx

# Create a NetworkX graph from adjacency matrix
G = nx.from_numpy_matrix(adjacency_matrix.numpy())

# Compute PageRank
pagerank_scores = nx.pagerank(G)
print("PageRank scores:", pagerank_scores)
```

In this example, the NetworkX library is used to compute PageRank scores for the nodes in the graph. These scores can be interpreted as measures of influence, helping identify the most influential nodes in the network.

8.6.2 Recommendation Systems

Recommendation systems are critical in various domains such as e-commerce, streaming services, and social media platforms. They help in suggesting relevant items to users based on their preferences and behavior. GNNs have shown significant potential in enhancing recommendation systems through collaborative filtering and personalized recommendations.

Collaborative Filtering Collaborative filtering is a popular technique used in recommendation systems. It leverages the interactions between users and items to predict the preferences of a user for items they haven't interacted with. Traditional collaborative filtering methods often use matrix factorization. However, GNNs provide a more powerful approach by capturing the complex relationships and interactions between users and items in a graph structure. Consider a bipartite graph $G = (U, I, E)$ where U represents the set of users, I represents the set of items, and E represents the interactions (edges) between users and items. The adjacency matrix A of this bipartite graph can be used to propagate information between users and items. The node features for users H_U and items H_I are updated using GNN layers:

$$H_U^{(t+1)} = \sigma \left(\sum_{i \in I} A_{ui} H_I^{(t)} W \right)$$

$$H_I^{(t+1)} = \sigma \left(\sum_{u \in U} A_{iu} H_U^{(t)} W \right)$$

where W is the weight matrix, and σ is an activation function such as ReLU.

Implementation

```
import torch
import torch.nn as nn
import torch.nn.functional as F

class GNNLayer(nn.Module):
    def __init__(self, in_features, out_features):
        super(GNNLayer, self).__init__()
        self.linear = nn.Linear(in_features, out_features)

    def forward(self, x, adj):
        out = torch.matmul(adj, x)
        out = self.linear(out)
        return F.relu(out)

class GNNRecommendation(nn.Module):
    def __init__(self, user_features, item_features, hidden_features):
        super(GNNRecommendation, self).__init__()
        self.user_gnn = GNNLayer(user_features, hidden_features)
        self.item_gnn = GNNLayer(item_features, hidden_features)
        self.prediction_layer = nn.Linear(hidden_features, 1)

    def forward(self, user_features, item_features, user_item_adj):
        user_hidden = self.user_gnn(user_features, user_item_adj)
        item_hidden = self.item_gnn(item_features, user_item_adj.T)
        scores = torch.matmul(user_hidden, item_hidden.T)
        return torch.sigmoid(scores)

# Example
user_features = torch.randn(5, 3)  # 5 users with 3 features each
item_features = torch.randn(4, 3)  # 4 items with 3 features each
```

```
user_item_adj = torch.tensor([
    [1, 1, 0, 0],
    [0, 1, 1, 0],
    [0, 0, 1, 1],
    [1, 0, 0, 1],
    [0, 1, 1, 0]
], dtype=torch.float32)

model = GNNRecommendation(user_features=3, item_features=3,
        hidden_features=4)
output = model(user_features, item_features, user_item_adj)
print(output)
```

In this implementation, the 'GNNRecommendation' class defines a simple GNN-based recommendation model. The user and item features are updated through GNN layers, and the final scores for user-item interactions are computed using a dot product followed by a sigmoid activation to produce probabilities.

Personalized Recommendations Personalized recommendations aim to suggest items that are tailored to individual users based on their unique preferences and behaviors. GNNs enhance personalized recommendations by learning from the user-item interaction graph and leveraging the rich relational information. The personalized recommendation can be formalized by learning a scoring function $f(u, i)$ that predicts the preference score of user u for item i:

$$\hat{y}_{ui} = f(h_u, h_i)$$

where h_u and h_i are the embeddings of user u and item i, respectively. The scoring function can be a simple dot product or a more complex neural network.

Implementation

```
class PersonalizedRecommendation(nn.Module):
    def __init__(self, user_features, item_features, hidden_features):
        super(PersonalizedRecommendation, self).__init__()
        self.user_gnn = GNNLayer(user_features, hidden_features)
        self.item_gnn = GNNLayer(item_features, hidden_features)
        self.prediction_layer = nn.Linear(hidden_features, 1)

    def forward(self, user_features, item_features, user_item_adj):
        user_hidden = self.user_gnn(user_features, user_item_adj)
        item_hidden = self.item_gnn(item_features, user_item_adj.T)
        user_item_score = torch.cat([user_hidden, item_hidden], dim=1)
        scores = self.prediction_layer(user_item_score)
        return torch.sigmoid(scores)

# Example
model = PersonalizedRecommendation(user_features=3, item_features=3,
        hidden_features=4)
output = model(user_features, item_features, user_item_adj)
print(output)
```

In this implementation, the 'PersonalizedRecommendation' class extends the basic GNN-based recommendation model by combining user and item embeddings through concatenation and passing them through an additional prediction layer. This allows for more complex interactions and better personalization.

8.6.3 Molecular Biology

Graph Neural Networks have found extensive applications in molecular biology, particularly in tasks such as molecule property prediction, drug discovery, and protein-protein interaction. Molecules and biological networks are naturally represented as graphs, making GNNs well-suited for these tasks.

Molecule Property Prediction Molecule property prediction involves predicting the properties of molecules, such as their solubility, toxicity, or biological activity. Molecules are represented as graphs where atoms are nodes, and chemical bonds are edges. Given a molecular graph $G = (V, E)$ with node features X representing atom types and edge features E representing bond types, the goal is to learn a function $f : G \rightarrow \mathbb{R}^P$ that maps the graph to a set of properties P.

Implementation

```
class MoleculeGCN(nn.Module):
    def __init__(self, in_features, hidden_features, out_features):
        super(MoleculeGCN, self).__init__()
        self.gc1 = GraphConvolution(in_features, hidden_features)
        self.gc2 = GraphConvolution(hidden_features, out_features)

    def forward(self, x, adj):
        x = F.relu(self.gc1(x, adj))
        x = F.dropout(x, training=self.training)
        x = self.gc2(x, adj)
        return x.mean(dim=0)

# Example
atom_features = torch.randn(5, 3)  # 5 atoms with 3 features each
bond_adj = torch.tensor([
    [1, 1, 0, 0, 0],
    [1, 1, 1, 0, 0],
    [0, 1, 1, 1, 0],
    [0, 0, 1, 1, 1],
    [0, 0, 0, 1, 1]
], dtype=torch.float32)

model = MoleculeGCN(in_features=3, hidden_features=4, out_features=1)
output = model(atom_features, bond_adj)
print(output)
```

In this implementation, the 'MoleculeGCN' class defines a GCN model for predicting molecular properties. The node features represent atom types, and the adjacency matrix represents chemical bonds. The model outputs a prediction for the molecule by averaging the node features in the final layer.

Drug Discovery and Protein-Protein Interaction In drug discovery, GNNs can be used to predict the interaction between drugs and their targets, identify potential drug candidates, and optimize lead compounds. For protein-protein interaction (PPI), GNNs can predict interactions between proteins based on their structural and functional properties. Given two molecular graphs $G_1 = (V_1, E_1)$ and $G_2 = (V_2, E_2)$, the goal is to predict the interaction I between them. This can be formulated as learning a function $f : (G_1, G_2) \rightarrow \mathbb{R}$.

Implementation

```
class DrugInteractionGCN(nn.Module):
    def __init__(self, in_features, hidden_features, out_features):
        super(DrugInteractionGCN, self).__init__()
        self.gcn1 = GraphConvolution(in_features, hidden_features)
        self.gcn2 = GraphConvolution(hidden_features, out_features)
        self.prediction_layer = nn.Linear(out_features * 2, 1)

    def forward(self, x1, adj1, x

2, adj2):
        h1 = F.relu(self.gcn1(x1, adj1))
        h1 = self.gcn2(h1, adj1)
        h2 = F.relu(self.gcn1(x2, adj2))
        h2 = self.gcn2(h2, adj2)
        h = torch.cat([h1.mean(dim=0), h2.mean(dim=0)], dim=1)
        score = self.prediction_layer(h)
        return torch.sigmoid(score)

# Example
drug_features1 = torch.randn(5, 3)  # 5 atoms with 3 features
                                      each for drug 1
bond_adj1 = torch.tensor([
    [1, 1, 0, 0, 0],
    [1, 1, 1, 0, 0],
    [0, 1, 1, 1, 0],
    [0, 0, 1, 1, 1],
    [0, 0, 0, 1, 1]
], dtype=torch.float32)
drug_features2 = torch.randn(5, 3)  # 5 atoms with 3 features
                                      each for drug 2
bond_adj2 = torch.tensor([
    [1, 0, 1, 0, 0],
    [0, 1, 0, 1, 0],
    [1, 0, 1, 0, 1],
    [0, 1, 0, 1, 0],
    [0, 0, 1, 0, 1]
], dtype=torch.float32)

model = DrugInteractionGCN(in_features=3, hidden_features=4,
        out_features=2)
output = model(drug_features1, bond_adj1, drug_features2,
         bond_adj2)
print(output)
```

In this implementation, the ‘DrugInteractionGCN’ class defines a GCN model for predicting drug interactions. The model processes two molecular graphs separately and concatenates their learned embeddings to predict the interaction score. This approach can be extended to other molecular interaction tasks, such as PPI prediction.

8.6.4 Other Applications

Graph Neural Networks (GNNs) have a wide range of applications beyond recommendation systems and molecular biology. This section explores their use in traffic and mobility analysis, and knowledge graphs and semantic networks, highlighting how GNNs can effectively model complex relationships and dynamics in these domains.

Traffic and Mobility Analysis Traffic and mobility analysis involves understanding and predicting the flow of vehicles or people through a transportation network. This is crucial for urban planning, traffic management, and improving transportation efficiency. The transportation network can be naturally represented as a graph, where nodes represent locations (e.g., intersections, bus stops) and edges represent roads or routes. Given a traffic network graph $G = (V, E)$, with node features X representing attributes like traffic volume, and edge features E representing road capacities, the goal is to learn functions for tasks such as traffic prediction, anomaly detection, and route optimization. One common task is traffic flow prediction, where we aim to predict the traffic volume at different locations based on historical data. This can be formulated as:

$$\hat{X}_t = f(X_{t-1}, X_{t-2}, \ldots, X_{t-n}, A)$$

where X_t is the traffic state at time t, and A is the adjacency matrix representing the traffic network.

Implementation

```
import torch
import torch.nn as nn
import torch.nn.functional as F

class TrafficGCN(nn.Module):
    def __init__(self, in_features, hidden_features, out_features):
        super(TrafficGCN, self).__init__()
        self.gc1 = GraphConvolution(in_features, hidden_features)
        self.gc2 = GraphConvolution(hidden_features, out_features)

    def forward(self, x, adj):
        x = F.relu(self.gc1(x, adj))
        x = F.dropout(x, training=self.training)
        x = self.gc2(x, adj)
        return x

# Example
traffic_features = torch.randn(10, 3)  # 10 locations with 3
                                         features each
adjacency_matrix = torch.tensor([
    [1, 1, 0, 0, 0, 0, 0, 0, 0, 0],
    [1, 1, 1, 0, 0, 0, 0, 0, 0, 0],
    [0, 1, 1, 1, 0, 0, 0, 0, 0, 0],
    [0, 0, 1, 1, 1, 0, 0, 0, 0, 0],
    [0, 0, 0, 1, 1, 1, 0, 0, 0, 0],
    [0, 0, 0, 0, 1, 1, 1, 0, 0, 0],
    [0, 0, 0, 0, 0, 1, 1, 1, 0, 0],
    [0, 0, 0, 0, 0, 0, 1, 1, 1, 0],
    [0, 0, 0, 0, 0, 0, 0, 1, 1, 1],
    [0, 0, 0, 0, 0, 0, 0, 0, 1, 1]
], dtype=torch.float32)

model = TrafficGCN(in_features=3, hidden_features=4, out_features=3)
output = model(traffic_features, adjacency_matrix)
print(output)
```

In this implementation, the ‘TrafficGCN’ class defines a GCN model for traffic flow prediction. The model updates the traffic features at each location using

GCN layers, leveraging the adjacency matrix to propagate information through the network. The output represents the predicted traffic state at the next time step.

Knowledge Graphs and Semantic Networks Knowledge graphs and semantic networks represent entities and their relationships in a structured form. They are widely used in information retrieval, natural language processing, and knowledge management systems. Nodes represent entities (e.g., people, places, concepts), and edges represent relationships (e.g.,"is a", "has a"). Given a knowledge graph $G = (V, E)$, with node features X representing entity attributes, and edge features E representing relationships, the goal is to learn representations for tasks such as entity classification, link prediction, and knowledge inference. One common task is link prediction, where we aim to predict the existence of a relationship between two entities. This can be formulated as learning a function $f(u, v)$ that predicts the likelihood of an edge between nodes u and v:

$$\hat{y}_{uv} = f(h_u, h_v)$$

where h_u and h_v are the embeddings of entities u and v.

Implementation

```
class KnowledgeGraphGCN(nn.Module):
    def __init__(self, in_features, hidden_features, out_features):
        super(KnowledgeGraphGCN, self).__init__()
        self.gc1 = GraphConvolution(in_features, hidden_features)
        self.gc2 = GraphConvolution(hidden_features, out_features)

    def forward(self, x, adj):
        x = F.relu(self.gc1(x, adj))
        x = F.dropout(x, training=self.training)
        x = self.gc2(x, adj)
        return x

# Example
entity_features = torch.randn(6, 3)  # 6 entities with
                                     3 features each
relationship_adj = torch.tensor([
    [1, 1, 0, 0, 0, 0],
    [1, 1, 1, 0, 0, 0],
    [0, 1, 1, 1, 0, 0],
    [0, 0, 1, 1, 1, 0],
    [0, 0, 0, 1, 1, 1],
    [0, 0, 0, 0, 1, 1]
], dtype=torch.float32)

model = KnowledgeGraphGCN(in_features=3, hidden_features=4,
        out_features=2)
output = model(entity_features, relationship_adj)
print(output)
```

In this implementation, the 'KnowledgeGraphGCN' class defines a GCN model for processing knowledge graphs. The model updates the entity features using GCN layers, leveraging the adjacency matrix to propagate relationship information. The output represents the updated entity embeddings, which can be used for various downstream tasks such as link prediction and entity classification.

8.7 Implementing GNNs with TensorFlow and PyTorch

Graph Neural Networks (GNNs) can be implemented using various deep learning frameworks. TensorFlow and PyTorch are two of the most popular frameworks that provide robust support for building and training GNNs. Here we provide a detailed guide on implementing GNNs with TensorFlow, covering step-by-step implementation and training and evaluation procedures.

8.7.1 Building GNNs with TensorFlow

Building GNNs in TensorFlow involves defining the graph convolutional layers, constructing the model, and writing the training loop.

1. Import Required Libraries:

```
import tensorflow as tf
from tensorflow.keras import layers, Model, optimizers, losses
import numpy as np
```

2. Define Graph Convolutional Layer: The core component of a GNN is the graph convolutional layer. In TensorFlow, we can define this layer by extending the 'tf.keras.layers.Layer' class.

```
class GraphConvolution(layers.Layer):
   def __init__(self, units):
       super(GraphConvolution, self).__init__()
       self.units = units

   def build(self, input_shape):
       self.weight = self.add_weight(shape=(input_shape[1][-1],
                                    self.units),
                                    initializer
                                    ='glorot_uniform',
                                    trainable=True)

   def call(self, inputs):
       features, adjacency_matrix = inputs
       support = tf.matmul(features, self.weight)
       output = tf.matmul(adjacency_matrix, support)
       return output
```

3. Define GNN Model: We can define a simple GNN model using the graph convolutional layer.

```
class GNN(Model):
    def __init__(self, hidden_units, output_units):
        super(GNN, self).__init__()
        self.gc1 = GraphConvolution(hidden_units)
        self.gc2 = GraphConvolution(output_units)

    def call(self, inputs):
        features, adjacency_matrix = inputs
        x = tf.nn.relu(self.gc1([features, adjacency_matrix]))
        x = self.gc2([x, adjacency_matrix])
        return x
```

4. Prepare Data: For demonstration purposes, let's create some synthetic data representing a small graph.

```
node_features = np.random.rand(5, 3)  # 5 nodes
                                        with 3 features each
adjacency_matrix = np.array([
    [1, 1, 0, 0, 0],
    [1, 1, 1, 0, 0],
    [0, 1, 1, 1, 0],
    [0, 0, 1, 1, 1],
    [0, 0, 0, 1, 1]
], dtype=np.float32)
labels = np.array([0, 1, 0, 2, 2])  # Node labels
                                      for classification
```

5. Training Loop: The training loop involves defining the loss function, optimizer, and the training step.

```
# Initialize model, loss, and optimizer
model = GNN(hidden_units=4, output_units=3)  # 3 output classes
loss_fn = losses.SparseCategoricalCrossentropy(from_logits=True)
optimizer = optimizers.Adam(learning_rate=0.01)

@tf.function
def train_step(features, adjacency_matrix, labels):
    with tf.GradientTape() as tape:
        logits = model([features, adjacency_matrix],
                 training=True)
        loss = loss_fn(labels, logits)
    gradients = tape.gradient(loss, model.trainable_variables)
    optimizer.apply_gradients(zip(gradients,
    model.trainable_variables))
    return loss

# Training loop
epochs = 100
for epoch in range(epochs):
    loss = train_step(node_features, adjacency_matrix, labels)
    if (epoch + 1) % 10 == 0:
        print(f'Epoch {epoch + 1}, Loss: {loss.numpy()}')
```

6. Evaluate the Model: After training, we evaluate the model's performance by computing accuracy or other relevant metrics.

```
# Prediction
logits = model([node_features, adjacency_matrix],
         training=False)
predicted_labels = tf.argmax(logits, axis=1)
accuracy = tf.reduce_mean(tf.cast(predicted_labels == labels,
           tf.float32))
print(f'Accuracy: {accuracy.numpy() * 100:.2f}%')
```

8.7.2 Building GNNs with PyTorch

Building GNNs with PyTorch involves defining the graph convolutional layers, constructing the model, and writing the training loop. PyTorch's dynamic computation graph and intuitive API make it a popular choice for developing GNNs.

1. Import Required Libraries:

```
import torch
import torch.nn as nn
import torch.nn.functional as F
import torch.optim as optim
```

2. Define Graph Convolutional Layer: The core component of a GNN is the graph convolutional layer. In PyTorch, we can define this layer by extending the 'nn.Module' class.

```
class GraphConvolution(nn.Module):
    def __init__(self, in_features, out_features):
        super(GraphConvolution, self).__init__()
        self.linear = nn.Linear(in_features, out_features)

    def forward(self, features, adj):
        support = self.linear(features)
        output = torch.matmul(adj, support)
        return output
```

3. Define GNN Model: We can define a simple GNN model using the graph convolutional layer.

```
class GNN(nn.Module):
    def __init__(self, input_dim, hidden_dim, output_dim):
        super(GNN, self).__init__()
        self.gc1 = GraphConvolution(input_dim, hidden_dim)
        self.gc2 = GraphConvolution(hidden_dim, output_dim)

    def forward(self, features, adj):
        x = F.relu(self.gc1(features, adj))
        x = self.gc2(x, adj)
        return x
```

4. Prepare Data: For demonstration purposes, let's create some synthetic data representing a small graph.

```
node_features = torch.rand(5, 3)  # 5 nodes with
                                    3 features each
adjacency_matrix = torch.tensor([
    [1, 1, 0, 0, 0],
    [1, 1, 1, 0, 0],
    [0, 1, 1, 1, 0],
    [0, 0, 1, 1, 1],
    [0, 0, 0, 1, 1]
], dtype=torch.float32)
labels = torch.tensor([0, 1, 0, 2, 2])  # Node labels
                                          for classification
```

5. Training Loop: The training loop involves defining the loss function, optimizer, and the training step.

```
# Initialize model, loss, and optimizer
model = GNN(input_dim=3, hidden_dim=4, output_dim=3)  # 3 output
                                                        classes
loss_fn = nn.CrossEntropyLoss()
optimizer = optim.Adam(model.parameters(), lr=0.01)

def train_step(features, adj, labels):
    model.train()
    optimizer.zero_grad()
```

```
        output = model(features, adj)
        loss = loss_fn(output, labels)
        loss.backward()
        optimizer.step()
        return loss.item()

    # Training loop
    epochs = 100
    for epoch in range(epochs):
        loss = train_step(node_features, adjacency_matrix, labels)
        if (epoch + 1) % 10 == 0:
            print(f'Epoch {epoch + 1}, Loss: {loss:.4f}')
```

6. Evaluate the Model: After training, we evaluate the model's performance by computing accuracy or other relevant metrics.

```
    model.eval()
    with torch.no_grad():
        logits = model(node_features, adjacency_matrix)
        predicted_labels = torch.argmax(logits, dim=1)
        accuracy = (predicted_labels == labels).float().mean()
        print(f'Accuracy: {accuracy.item() * 100:.2f}%')
```

In the 'GraphConvolution' class, the 'forward' method performs a linear transformation on the input features followed by a matrix multiplication with the adjacency matrix. This operation aggregates information from neighboring nodes. The 'GNN' class uses two graph convolutional layers. The first layer applies a ReLU activation function to introduce non-linearity, while the second layer produces the final output. The training loop involves iterating over the data for a specified number of epochs. In each epoch, we perform a forward pass to compute the model output, calculate the loss, perform a backward pass to compute gradients, and update the model parameters using the optimizer. After training, we set the model to evaluation mode using 'model.eval()'. We then perform a forward pass without computing gradients to obtain the predictions. The accuracy is calculated by comparing the predicted labels with the true labels.

8.8 Case Studies

The practical implementation of GNNs across various domains can be best understood through detailed case studies. Each case study below provides insights into how GNNs are applied to specific problems, highlighting their versatility and effectiveness.

Social Network Analysis This involves studying the structure of social networks to understand the relationships and influence between individuals or groups. GNNs can be used to analyze social networks for tasks such as community detection, node classification, and influence prediction. Consider a social network where nodes represent individuals and edges represent friendships. The task is to predict the community (e.g., groups of friends) to which each individual belongs. Given a graph $G = (V, E)$

where V represents individuals and E represents friendships, and node features X representing individual attributes (e.g., age, interests), we aim to learn a function $f : G \rightarrow C$ that maps each node to a community label C.

Implementation

```
import torch
import torch.nn as nn
import torch.optim as optim
import torch.nn.functional as F

class SocialNetworkGCN(nn.Module):
    def __init__(self, in_features, hidden_features, out_features):
        super(SocialNetworkGCN, self).__init__()
        self.gc1 = GraphConvolution(in_features, hidden_features)
        self.gc2 = GraphConvolution(hidden_features, out_features)

    def forward(self, x, adj):
        x = F.relu(self.gc1(x, adj))
        x = self.gc2(x, adj)
        return x

# Example
node_features = torch.rand(6, 3)  # 6 individuals with 3 features
                                    each
adjacency_matrix = torch.tensor([
    [1, 1, 0, 0, 0, 0],
    [1, 1, 1, 0, 0, 0],
    [0, 1, 1, 1, 0, 0],
    [0, 0, 1, 1, 1, 0],
    [0, 0, 0, 1, 1, 1],
    [0, 0, 0, 0, 1, 1]
], dtype=torch.float32)
labels = torch.tensor([0, 0, 1, 1, 2, 2])  # Community labels

model = SocialNetworkGCN(in_features=3, hidden_features=4,
        out_features=3)
optimizer = optim.Adam(model.parameters(), lr=0.01)
loss_fn = nn.CrossEntropyLoss()

def train_step(features, adj, labels):
    model.train()
    optimizer.zero_grad()
    output = model(features, adj)
    loss = loss_fn(output, labels)
    loss.backward()
    optimizer.step()
    return loss.item()

# Training loop
epochs = 100
for epoch in range(epochs):
    loss = train_step(node_features, adjacency_matrix, labels)
    if (epoch + 1) % 10 == 0:
        print(f'Epoch {epoch + 1}, Loss: {loss:.4f}')

# Evaluation
model.eval()
with torch.no_grad():
    logits = model(node_features, adjacency_matrix)
    predicted_labels = torch.argmax(logits, dim=1)
    accuracy = (predicted_labels == labels).float().mean()
    print(f'Accuracy: {accuracy.item() * 100:.2f}%')
```

This implementation demonstrates how a GNN can be used to predict community labels in a social network. The accuracy of the model provides an indication of its performance in correctly identifying communities.

Recommendation System The aim is to suggest relevant items (e.g., products, movies) to users based on their preferences and past interactions. GNNs enhance recommendation systems by capturing the complex relationships between users and items in a graph structure. Consider a bipartite graph where one set of nodes represents users and the other set represents items. The task is to predict user ratings for items they have not yet interacted with. Given a bipartite graph $G = (U, I, E)$ where U represents users, I represents items, and E represents interactions, the goal is to predict the interaction score $\hat{y}_{ui}$ for user u and item i.

Implementation

```
class RecommendationGCN(nn.Module):
    def __init__(self, in_features, hidden_features, out_features):
        super(RecommendationGCN, self).__init__()
        self.user_gc = GraphConvolution(in_features, hidden_features)
        self.item_gc = GraphConvolution(in_features, hidden_features)
        self.prediction_layer = nn.Linear(hidden_features,
                                          out_features)

    def forward(self, user_features, item_features, user_item_adj):
        user_hidden = F.relu(self.user_gc(user_features,
                                          user_item_adj))
        item_hidden = F.relu(self.item_gc(item_features,
                                          user_item_adj.T))
        scores = self.prediction_layer(user_hidden + item_hidden)
        return torch.sigmoid(scores)

# Example
user_features = torch.rand(4, 3)  # 4 users with 3 features each
item_features = torch.rand(5, 3)  # 5 items with 3 features each
user_item_adj = torch.tensor([
    [1, 1, 0, 0, 1],
    [0, 1, 1, 0, 0],
    [1, 0, 0, 1, 1],
    [0, 1, 1, 1, 0]
], dtype=torch.float32)
ratings = torch.tensor([
    [5.0, 3.0, 0.0, 0.0, 4.0],
    [0.0, 4.0, 2.0, 0.0, 0.0],
    [5.0, 0.0, 0.0, 4.0, 5.0],
    [0.0, 4.0, 3.0, 3.0, 0.0]
], dtype=torch.float32)

model = RecommendationGCN(in_features=3, hidden_features=4,
        out_features=1)
optimizer = optim.Adam(model.parameters(), lr=0.01)
loss_fn = nn.MSELoss()

def train_step(user_features, item_features, adj, ratings):
    model.train()
    optimizer.zero_grad()
    output = model(user_features, item_features, adj)
    loss = loss_fn(output, ratings)
    loss.backward()
    optimizer.step()
    return loss.item()

# Training loop
```

```
epochs = 100
for epoch in range(epochs):
    loss = train_step(user_features, item_features, user_item_adj,
          ratings)
    if (epoch + 1) % 10 == 0:
        print(f'Epoch {epoch + 1}, Loss: {loss:.4f}')

# Evaluation
model.eval()
with torch.no_grad():
    predicted_ratings = model(user_features, item_features,
                          user_item_adj)
    mse = loss_fn(predicted_ratings, ratings).item()
    print(f'Mean Squared Error: {mse:.4f}')
```

This implementation shows how a GNN can be used to predict user ratings for items. The mean squared error (MSE) of the model provides an indication of its performance in predicting ratings accurately.

Molecular Biology GNNs are powerful tools for molecular biology applications such as molecule property prediction, drug discovery, and protein-protein interaction. Molecules are represented as graphs where atoms are nodes and chemical bonds are edges. Consider a graph where nodes represent atoms and edges represent chemical bonds. The task is to predict a molecular property (e.g., solubility) based on the graph structure. Given a molecular graph $G = (V, E)$ with node features X representing atom types and edge features E representing bond types, the goal is to learn a function $f : G \to \mathbb{R}$ that maps the graph to a molecular property.

Implementation

```
class MoleculeGCN(nn.Module):
    def __init__(self, in_features, hidden_features, out_features):
        super(MoleculeGCN, self).__init__()
        self.gc1 = GraphConvolution(in_features, hidden_features)
        self.gc2 = GraphConvolution(hidden_features, out_features)

    def forward(self, x, adj):
        x = F.relu(self.gc1(x, adj))
        x = F.dropout(x, training=self.training)
        x = self.gc2(x, adj)
        return x.mean(dim=0)  # Global pooling

# Example
atom_features = torch.rand(6, 3)  # 6 atoms with 3 features each
bond_adj = torch.tensor([
    [1, 1, 0, 0, 0, 0],
    [1, 1, 1, 0, 0, 0],
    [0, 1, 1, 1, 0, 0],
    [0, 0, 1, 1, 1, 0],
    [0, 0, 0, 1, 1, 1],
    [0, 0, 0, 0, 1, 1]
], dtype=torch.float32)
property_label = torch.tensor([1.0])  # Example property value

model = MoleculeGCN(in_features=3, hidden_features=4, out_features=2)
optimizer = optim.Adam(model.parameters(), lr=0.01)
loss_fn = nn.MSELoss()

def train_step(features, adj, label):
    model.train()
    optimizer.zero_grad()
```

```
    output = model(features, adj)
    loss = loss_fn(output, label)
    loss.backward()
    optimizer.step()
    return loss.item()

# Training loop
epochs = 100
for epoch in range(epochs):
    loss = train_step(atom_features, bond_adj, property_label)
    if (epoch + 1) % 10 == 0:
        print(f'Epoch {epoch + 1}, Loss: {loss:.4f}')

# Evaluation
model.eval()
with torch.no_grad():
    predicted_property = model(atom_features, bond_adj)
    mse = loss_fn(predicted_property, property_label).item()
    print(f'Mean Squared Error: {mse:.4f}')
```

This implementation demonstrates how a GNN can be used to predict molecular properties. The mean squared error (MSE) of the model indicates its performance in predicting the property accurately.

Custom GNN Design Design a GNN for predicting the sentiment of user reviews in an e-commerce platform. Represent each review as a graph where nodes are words and edges are co-occurrences within a sliding window.

```
class SentimentGCN(nn.Module):
    def __init__(self, in_features, hidden_features, out_features):
        super(SentimentGCN, self).__init__()
        self.gc1 = GraphConvolution(in_features, hidden_features)
        self.gc2 = GraphConvolution(hidden_features, out_features)

    def forward(self, x, adj):
        x = F.relu(self.gc1(x, adj))
        x = F.dropout(x, training=self.training)
        x = self.gc2(x, adj)
        return x.mean(dim=0)  # Global pooling

# Example
word_features = torch.rand(10, 5)  # 10 words with 5 features each
co_occurrence_adj = torch.tensor([
    [1, 1, 0, 0, 0, 0, 0, 0, 0, 0],
    [1, 1, 1, 0, 0, 0, 0, 0, 0, 0],
    [0, 1, 1, 1, 0, 0, 0, 0, 0, 0],
    [0, 0, 1, 1, 1, 0, 0, 0, 0, 0],
    [0, 0, 0, 1, 1, 1, 0, 0, 0, 0],
    [0, 0, 0, 0, 1, 1, 1, 0, 0, 0],
    [0, 0, 0, 0, 0, 1, 1, 1, 0, 0],
    [0, 0, 0, 0, 0, 0, 1, 1, 1, 0],
    [0, 0, 0, 0, 0, 0, 0, 1, 1, 1],
    [0, 0, 0, 0, 0, 0, 0, 0, 1, 1]
], dtype=torch.float32)
sentiment_label = torch.tensor([1.0])  # Positive sentiment

model = SentimentGCN(in_features=5, hidden_features=6, out_features=2)
optimizer = optim.Adam(model.parameters(), lr=0.01)
loss_fn = nn.MSELoss()

def train_step(features, adj, label):
    model.train()
    optimizer.zero_grad()
    output = model(features, adj)
```

```
    loss = loss_fn(output, label)
    loss.backward()
    optimizer.step()
    return loss.item()

# Training loop
epochs = 100
for epoch in range(epochs):
    loss = train_step(word_features, co_occurrence_adj,
           sentiment_label)
    if (epoch + 1) % 10 == 0:
        print(f'Epoch {epoch + 1}, Loss: {loss:.4f}')

# Evaluation
model.eval()
with torch.no_grad():
    predicted_sentiment = model(word_features, co_occurrence_adj)
    mse = loss_fn(predicted_sentiment, sentiment_label).item()
    print(f'Mean Squared Error: {mse:.4f}')
```

8.9 Interpretable GNNs

Interpretable GNNs focus on making the decisions of these complex models understandable and transparent. This is crucial for gaining trust in the model's predictions, especially in sensitive applications such as healthcare and finance. Explainability in GNNs involves understanding how the model arrives at its predictions by identifying which parts of the input data (nodes, edges, features) are most influential. This is often achieved through techniques that highlight important components in the graph structure or node features. Let $G = (V, E)$ be a graph with nodes V and edges E, and let X be the node feature matrix. A GNN learns a function $f : G \to \mathbb{R}^C$ where C is the number of classes. To explain the model's decision for a node $v \in V$, we seek to identify a subgraph $G' \subset G$ and a subset of features $X' \subset X$ that significantly influence the prediction $f(G)$. For a node v, the importance score $s(v_i)$ for a neighboring node v_i can be defined based on the gradient of the prediction with respect to the input features:

$$s(v_i) = \left| \frac{\partial f(G)}{\partial X_{v_i}} \right|$$

Similarly, for an edge $e_{ij} \in E$:

$$s(e_{ij}) = \left| \frac{\partial f(G)}{\partial E_{ij}} \right|$$

Implementation Consider a GNN model implemented using PyTorch. We can compute importance scores for nodes and edges using gradient-based methods.

```
import torch
import torch.nn.functional as F

class ExplainableGNN(nn.Module):
    def __init__(self, in_features, hidden_features, out_features):
        super(ExplainableGNN, self).__init__()
        self.gc1 = GraphConvolution(in_features, hidden_features)
        self.gc2 = GraphConvolution(hidden_features, out_features)

    def forward(self, x, adj):
        x = F.relu(self.gc1(x, adj))
        x = self.gc2(x, adj)
        return x

def compute_node_importance(model, features, adj, target_node):
    model.eval()
    features.requires_grad_(True)
    output = model(features, adj)
    target = output[target_node]
    target.backward()
    importance_scores = features.grad.abs()
    return importance_scores

def compute_edge_importance(model, features, adj, target_node):
    model.eval()
    adj.requires_grad_(True)
    output = model(features, adj)
    target = output[target_node]
    target.backward()
    importance_scores = adj.grad.abs()
    return importance_scores

# Example
model = ExplainableGNN(in_features=3, hidden_features=4,
        out_features=2)
node_features = torch.rand(5, 3, requires_grad=True)
adjacency_matrix = torch.tensor([
    [1, 1, 0, 0, 0],
    [1, 1, 1, 0, 0],
    [0, 1, 1, 1, 0],
    [0, 0, 1, 1, 1],
    [0, 0, 0, 1, 1]
], dtype=torch.float32, requires_grad=True)

target_node = 2
node_importance = compute_node_importance(model, node_features,
                  adjacency_matrix, target_node)
edge_importance = compute_edge_importance(model, node_features,
                  adjacency_matrix, target_node)

print("Node Importance Scores:", node_importance)
print("Edge Importance Scores:", edge_importance)
```

In this implementation, the 'ExplainableGNN' class extends the basic GNN model. The 'compute_node_importance' and 'compute_edge_importance' functions calculate the importance scores for nodes and edges by computing the gradients of the model's output with respect to the input features and adjacency matrix. These scores indicate which nodes and edges are most influential in the model's decision for the target node.

Visualizing GNN decisions involves creating visual representations that highlight the important nodes and edges in the graph, as well as the relevant features that contribute to the model's predictions. This helps in interpreting the model's behavior and understanding its decision-making process. Given a graph $G = (V, E)$ and the importance scores $s(v_i)$ and $s(e_{ij})$ for nodes and edges, respectively, we aim to generate visualizations that highlight these important components. The visualization can be represented as a subgraph $G' \subset G$ where:

$$G' = (V', E') \quad \text{with} \quad V' = \{v_i \in V \mid s(v_i) > \theta_v\} \text{ and } E' = \{e_{ij} \in E \mid s(e_{ij}) > \theta_e\}$$

Here, θ_v and θ_e are thresholds for node and edge importance scores, respectively.

Implementation We can use libraries such as NetworkX and Matplotlib to visualize the important nodes and edges in the graph.

```
import networkx as nx
import matplotlib.pyplot as plt

def visualize_graph(graph, node_importance, edge_importance,
                    node_threshold=0.1, edge_threshold=0.1):
    G = nx.Graph()
    num_nodes = len(node_importance)
    for i in range(num_nodes):
        if node_importance[i].sum() > node_threshold:
            G.add_node(i)
    for i in range(num_nodes):
        for j in range(num_nodes):
            if edge_importance[i, j] > edge_threshold:
                G.add_edge(i, j)

    pos = nx.spring_layout(G)
    nx.draw(G, pos, with_labels=True, node_size=700,
           node_color='lightblue', edge_color='gray')
    plt.show()

# Example
node_threshold = 0.1
edge_threshold = 0.1
visualize_graph(adjacency_matrix, node_importance,
                edge_importance, node_threshold,
                edge_threshold)
```

In this implementation, the 'visualize_graph' function uses NetworkX to create a graph visualization. Nodes and edges with importance scores above the specified thresholds are included in the visualization. The 'nx.spring_layout' function is used to position the nodes, and the graph is drawn with labels, node colors, and edge colors for clarity.

8.10 Hybrid GNN Models

Hybrid GNN models combine the strengths of GNNs with other neural network architectures to leverage the unique advantages of each for improved performance on complex tasks. These hybrid models can integrate CNNs, RNNs, or Transformer architectures to handle diverse data types and tasks. Hybrid models integrate GNNs with other neural networks to enhance their ability to process and learn from both graph-structured data and other data types. For example, CNNs are excellent for image data, RNNs for sequential data, and Transformers for capturing long-range dependencies. By combining these with GNNs, hybrid models can effectively handle multi-modal and complex data scenarios. Let $G = (V, E)$ be a graph with node features X and adjacency matrix A. The hybrid model can be expressed as:

$$H = f_{\text{hybrid}}(X, A, \theta)$$

where f_{hybrid} is a function representing the hybrid model and θ are the parameters. This function can be decomposed into components representing different neural networks:

$$H = f_{\text{GNN}}(f_{\text{other}}(X, A, \theta_{\text{other}}), A, \theta_{\text{GNN}})$$

Here, f_{other} represents another neural network (e.g., CNN, RNN) that processes input features X and outputs intermediate representations which are then fed into the GNN component f_{GNN}.

Implementation Consider a hybrid model combining a GNN with a CNN for image classification tasks where the image data is represented as a graph of superpixels.

1. Define Graph Convolutional Layer:

```
import torch
import torch.nn as nn
import torch.nn.functional as F

class GraphConvolution(nn.Module):
    def __init__(self, in_features, out_features):
        super(GraphConvolution, self).__init__()
        self.linear = nn.Linear(in_features, out_features)

    def forward(self, features, adj):
        support = self.linear(features)
        output = torch.matmul(adj, support)
        return output
```

2. Define CNN Component:

```
class CNN(nn.Module):
    def __init__(self):
        super(CNN, self).__init__()
        self.conv1 = nn.Conv2d(in_channels=3, out_channels=16,
                               kernel_size=3, stride=1, padding=1)
        self.conv2 = nn.Conv2d(in_channels=16, out_channels=32,
                               kernel_size=3, stride=1, padding=1)
        self.pool = nn.MaxPool2d(kernel_size=2, stride=2,
                              padding=0)
        self.fc1 = nn.Linear(32 * 8 * 8, 64)

    def forward(self, x):
        x = self.pool(F.relu(self.conv1(x)))
        x = self.pool(F.relu(self.conv2(x)))
```

```
        x = x.view(-1, 32 * 8 * 8)
        x = F.relu(self.fc1(x))
        return x
```

3. Define Hybrid Model:

```
class HybridGNN(nn.Module):
    def __init__(self, in_features, hidden_features,
                 out_features):
        super(HybridGNN, self).__init__()
        self.cnn = CNN()
        self.gc1 = GraphConvolution(in_features,
                   hidden_features)
        self.gc2 = GraphConvolution(hidden_features,
                   out_features)

    def forward(self, x, adj):
        x_cnn = self.cnn(x)
        x_gnn = F.relu(self.gc1(x_cnn, adj))
        x_gnn = self.gc2(x_gnn, adj)
        return x_gnn
```

4. Prepare Data:

```
# Example data: 5 images with 3 channels (RGB), each 32x32
  pixels
images = torch.rand(5, 3, 32, 32)
adjacency_matrix = torch.tensor([
    [1, 1, 0, 0, 0],
    [1, 1, 1, 0, 0],
    [0, 1, 1, 1, 0],
    [0, 0, 1, 1, 1],
    [0, 0, 0, 1, 1]
], dtype=torch.float32)
labels = torch.tensor([0, 1, 0, 2, 2]) # Class labels
                                       for classification
```

5. Training Loop:

```
model = HybridGNN(in_features=64, hidden_features=32,
        out_features=3)
optimizer = torch.optim.Adam(model.parameters(), lr=0.01)
loss_fn = nn.CrossEntropyLoss()

def train_step(images, adj, labels):
    model.train()
    optimizer.zero_grad()
    output = model(images, adj)
    loss = loss_fn(output, labels)
    loss.backward()
    optimizer.step()
    return loss.item()

# Training loop
epochs = 100
for epoch in range(epochs):
    loss = train_step(images, adjacency_matrix, labels)
    if (epoch + 1) % 10 == 0:
        print(f'Epoch {epoch + 1}, Loss: {loss:.4f}')

# Evaluation
model.eval()
with torch.no_grad():
    logits = model(images, adjacency_matrix)
    predicted_labels = torch.argmax(logits, dim=1)
```

```
accuracy = (predicted_labels == labels).float().mean()
print(f'Accuracy: {accuracy.item() * 100:.2f}%')
```

In this implementation, the 'HybridGNN' class integrates a CNN for image feature extraction and a GNN for graph-based learning. The model processes image data through the CNN, and the output features are then used as input for the GNN. This approach leverages the strengths of both CNNs and GNNs, making it suitable for tasks that involve both image and graph data.

Hybrid models have numerous applications across various domains where multi-modal data is prevalent. In image classification tasks, hybrid models can combine CNNs for feature extraction with GNNs for capturing relationships between different parts of the image. For example, superpixels or regions of interest in an image can be represented as nodes in a graph, and their spatial relationships can be modeled using GNNs. In social network analysis, hybrid models can integrate RNNs for sequential data (e.g., user interactions over time) with GNNs for structural data (e.g., network connections). This allows for a comprehensive analysis that captures both temporal and relational patterns. In molecular biology, hybrid models can combine GNNs for graph-based molecule representation with other neural networks (e.g., CNNs or RNNs) to incorporate additional data types such as 3D molecular structures or sequential protein data. This approach enhances the predictive power for tasks such as drug discovery and protein-protein interaction prediction. Hybrid models in recommendation systems can integrate GNNs for capturing user-item interactions with other models (e.g., collaborative filtering methods) to improve recommendation accuracy. GNNs can model complex relationships in user-item graphs, while traditional methods can leverage user and item embeddings.

8.11 Exercises

1. Given the following adjacency matrix $\mathbf{A}$ and feature matrix $\mathbf{X}$ for a simple graph with 4 nodes:

$$\mathbf{A} = \begin{pmatrix} 0\,1\,0\,1 \\ 1\,0\,1\,0 \\ 0\,1\,0\,1 \\ 1\,0\,1\,0 \end{pmatrix}, \quad \mathbf{X} = \begin{pmatrix} 1\,0 \\ 0\,1 \\ 1\,1 \\ 0\,0 \end{pmatrix}$$

and the weight matrix $\mathbf{W} = \begin{pmatrix} 0.2\;0.3 \\ 0.4\;0.5 \end{pmatrix}$.

   (a) Compute the normalized adjacency matrix $\hat{\mathbf{A}} = \mathbf{D}^{-1/2}\mathbf{A}\mathbf{D}^{-1/2}$, where $\mathbf{D}$ is the degree matrix.
   (b) Perform the first layer of graph convolution using the equation $\mathbf{H}^{(1)} = \hat{\mathbf{A}}\mathbf{X}\mathbf{W}$.
   (c) Apply a non-linear activation function (ReLU) to the result.

2. Consider a GAT with 3 nodes and the following adjacency matrix $\mathbf{A}$:

$$\mathbf{A} = \begin{pmatrix} 0 & 1 & 1 \\ 1 & 0 & 1 \\ 1 & 1 & 0 \end{pmatrix}$$

Feature matrix $\mathbf{X} = \begin{pmatrix} 1 & 0 \\ 0 & 1 \\ 1 & 1 \end{pmatrix}$, and attention mechanism with parameters:

$$\mathbf{a} = \begin{pmatrix} 0.1 \\ 0.2 \end{pmatrix}, \quad \mathbf{W} = \begin{pmatrix} 0.3 & 0.5 \\ 0.7 & 0.9 \end{pmatrix}$$

   (a) Compute the linear transformation of the features $\mathbf{XW}$.
   (b) Calculate the attention coefficients α_{ij} for all pairs of nodes.
   (c) Apply the softmax function to normalize the attention coefficients.
   (d) Compute the new node representations using the attention mechanism.

3. Given a small graph with the following adjacency list:

Node 1: [2, 3], Node 2: [1, 3, 4], Node 3: [1, 2], Node 4: [2]

Feature matrix $\mathbf{X} = \begin{pmatrix} 1 & 0 \\ 0 & 1 \\ 1 & 1 \\ 0 & 0 \end{pmatrix}$.

   (a) Implement the GraphSAGE mean aggregation for the nodes.
   (b) Compute the new node embeddings after one iteration of aggregation.
   (c) Use the updated embeddings to classify nodes based on their features.

4. Given the feature matrix $\mathbf{X} = \begin{pmatrix} 1 & 2 \\ 2 & 3 \\ 3 & 4 \end{pmatrix}$ and adjacency matrix $\mathbf{A} = \begin{pmatrix} 0 & 1 & 0 \\ 1 & 0 & 1 \\ 0 & 1 & 0 \end{pmatrix}$.

   (a) Implement L2 regularization in the loss function for a GCN.
   (b) Calculate the regularized loss for the given graph.
   (c) Analyze how regularization affects the model training and convergence.

5. Consider a graph G with 4 nodes and the adjacency matrix $\mathbf{A} \in \mathbb{R}^{4\times4}$:

$$\mathbf{A} = \begin{pmatrix} 0 & 1 & 0 & 1 \\ 1 & 0 & 1 & 0 \\ 0 & 1 & 0 & 1 \\ 1 & 0 & 1 & 0 \end{pmatrix}$$

The initial node features are given by $\mathbf{X} \in \mathbb{R}^{4\times 3}$:

$$\mathbf{X} = \begin{pmatrix} 1 & 2 & 3 \\ 4 & 5 & 6 \\ 7 & 8 & 9 \\ 10 & 11 & 12 \end{pmatrix}$$

The weight matrix for the convolution layer is $\mathbf{W} \in \mathbb{R}^{3\times 2}$:

$$\mathbf{W} = \begin{pmatrix} 0.1 & 0.2 \\ 0.3 & 0.4 \\ 0.5 & 0.6 \end{pmatrix}$$

(a) Compute the normalized adjacency matrix $\hat{\mathbf{A}}$ using $\hat{\mathbf{A}} = \mathbf{D}^{-1/2}\mathbf{A}\mathbf{D}^{-1/2}$, where $\mathbf{D}$ is the degree matrix.

(b) Perform the message aggregation step and update the node features using the formula $\mathbf{H} = \hat{\mathbf{A}}\mathbf{X}\mathbf{W}$.

6. Consider the same graph G and initial node features $\mathbf{X}$ as in above problem. The attention coefficients are computed using a shared attention mechanism, where the attention weight matrix is $\mathbf{a} \in \mathbb{R}^{1\times 6}$:

$$\mathbf{a} = \begin{pmatrix} 0.1 & 0.2 & 0.3 & 0.4 & 0.5 & 0.6 \end{pmatrix}$$

(a) Compute the attention scores for each edge using the formula:

$$e_{ij} = \mathbf{a}\left[\mathbf{W}\mathbf{x}_i \,\|\, \mathbf{W}\mathbf{x}_j\right]$$

where $\|$ denotes concatenation and $\mathbf{W}$ is the identity matrix for simplicity.

(b) Apply the softmax function to obtain the normalized attention coefficients and compute the updated node features using the attention mechanism.

7. For the same graph G and initial node features $\mathbf{X}$, GraphSAGE uses a mean aggregator. The weight matrix for the node update is $\mathbf{W} \in \mathbb{R}^{3\times 2}$ as given in Exercise 1.

(a) Perform the mean aggregation of neighboring node features for each node.

(b) Update the node features using the aggregated messages and the weight matrix $\mathbf{W}$, followed by ReLU activation function.

8. Consider a graph G with 3 nodes and the following adjacency matrix $\mathbf{A} \in \mathbb{R}^{3\times3}$:

$$\mathbf{A} = \begin{pmatrix} 0 & 1 & 1 \\ 1 & 0 & 1 \\ 1 & 1 & 0 \end{pmatrix}$$

The initial node features are given by $\mathbf{X} \in \mathbb{R}^{3\times2}$:

$$\mathbf{X} = \begin{pmatrix} 1 & 0 \\ 0 & 1 \\ 1 & 1 \end{pmatrix}$$

The message function is defined as $m_{ij} = \mathrm{ReLU}(\mathbf{W}_m \mathbf{x}_j)$ with $\mathbf{W}_m \in \mathbb{R}^{2\times2}$:

$$\mathbf{W}_m = \begin{pmatrix} 0.2 & 0.3 \\ 0.4 & 0.5 \end{pmatrix}$$

(a) Compute the messages m_{ij} for each edge in the graph.

(b) Aggregate the messages and update the node features using the update function $\mathbf{h}_i' = \mathrm{ReLU}(\mathbf{W}_u(\mathbf{x}_i + \sum_{j\in\mathcal{N}(i)} m_{ij}))$, where $\mathbf{W}_u = \mathbf{W}_m$.

9. Consider a graph G with 3 nodes and the following adjacency matrix $\mathbf{A} \in \mathbb{R}^{3\times3}$:

$$\mathbf{A} = \begin{pmatrix} 0 & 1 & 0 \\ 1 & 0 & 1 \\ 0 & 1 & 0 \end{pmatrix}$$

The degree matrix $\mathbf{D}$ and the graph Laplacian $\mathbf{L}$ are defined as:

$$\mathbf{D} = \begin{pmatrix} 1 & 0 & 0 \\ 0 & 2 & 0 \\ 0 & 0 & 1 \end{pmatrix}, \quad \mathbf{L} = \mathbf{D} - \mathbf{A}$$

(a) Compute the eigenvalues and eigenvectors of the graph Laplacian $\mathbf{L}$.

(b) Using the Chebyshev polynomial approximation of order 1, compute the spectral convolution operation on the node feature matrix $\mathbf{X} \in \mathbb{R}^{3\times2}$:

$$\mathbf{X} = \begin{pmatrix} 1 & 2 \\ 3 & 4 \\ 5 & 6 \end{pmatrix}$$

with filter coefficients $\mathbf{c} = \begin{pmatrix} 0.5 & 0.5 \end{pmatrix}$.

10. Given a graph G with 4 nodes and adjacency matrix $\mathbf{A} \in \mathbb{R}^{4\times4}$:

$$\mathbf{A} = \begin{pmatrix} 0 & 1 & 0 & 1 \\ 1 & 0 & 1 & 0 \\ 0 & 1 & 0 & 1 \\ 1 & 0 & 1 & 0 \end{pmatrix}$$

and node features $\mathbf{X} \in \mathbb{R}^{4\times3}$:

$$\mathbf{X} = \begin{pmatrix} 1 & 0 & 2 \\ 0 & 1 & 3 \\ 1 & 1 & 1 \\ 2 & 0 & 0 \end{pmatrix}$$

(a) Compute the normalized adjacency matrix $\hat{\mathbf{A}} = \mathbf{D}^{-1/2}\mathbf{A}\mathbf{D}^{-1/2}$, where $\mathbf{D}$ is the degree matrix.

(b) Perform a single-layer graph convolution using the weight matrix $\mathbf{W} \in \mathbb{R}^{3\times2}$:

$$\mathbf{W} = \begin{pmatrix} 0.2 & 0.5 \\ 0.3 & 0.4 \\ 0.6 & 0.1 \end{pmatrix}$$

and compute the updated node features $\mathbf{H} = \hat{\mathbf{A}}\mathbf{X}\mathbf{W}$.

11. Consider a graph G with 3 nodes and the adjacency matrix $\mathbf{A} \in \mathbb{R}^{3\times3}$ given by:

$$\mathbf{A} = \begin{pmatrix} 0 & 1 & 0 \\ 1 & 0 & 1 \\ 0 & 1 & 0 \end{pmatrix}$$

The initial node features are $\mathbf{X} \in \mathbb{R}^{3\times2}$:

$$\mathbf{X} = \begin{pmatrix} 1 & 2 \\ 3 & 4 \\ 5 & 6 \end{pmatrix}$$

(a) Perform the first layer graph convolution using the weight matrix $\mathbf{W}_1 \in \mathbb{R}^{2\times2}$:

$$\mathbf{W}_1 = \begin{pmatrix} 0.1 & 0.3 \\ 0.2 & 0.4 \end{pmatrix}$$

Compute the output $\mathbf{H}_1$.

(b) Apply a second layer of graph convolution using the weight matrix $\mathbf{W}_2 \in \mathbb{R}^{2\times1}$:

$$\mathbf{W}_2 = \begin{pmatrix} 0.5 \\ 0.6 \end{pmatrix}$$

Compute the final output $\mathbf{H}_2$.

12. Given a graph G with 3 nodes and adjacency matrix $\mathbf{A} \in \mathbb{R}^{3\times3}$:

$$\mathbf{A} = \begin{pmatrix} 0 & 1 & 1 \\ 1 & 0 & 1 \\ 1 & 1 & 0 \end{pmatrix}$$

The edge features $\mathbf{E} \in \mathbb{R}^{3\times3}$ are defined as:

$$\mathbf{E}_{ij} = \begin{cases} 1 & \text{if } (i, j) \text{ is an edge} \\ 0 & \text{otherwise} \end{cases}$$

(a) Define the combined node and edge feature matrix $\mathbf{X} \in \mathbb{R}^{3\times3}$:

$$\mathbf{X} = \begin{pmatrix} 1 & 2 & 3 \\ 4 & 5 & 6 \\ 7 & 8 & 9 \end{pmatrix}$$

(b) Perform a graph convolution considering the edge features, using the weight matrix $\mathbf{W} \in \mathbb{R}^{3\times2}$:

$$\mathbf{W} = \begin{pmatrix} 0.1 & 0.2 \\ 0.3 & 0.4 \\ 0.5 & 0.6 \end{pmatrix}$$

and compute the updated node features. Incorporate the edge features by modifying the aggregation step to include $\mathbf{E}$.

13. Consider a graph G with 4 nodes and the adjacency matrix $\mathbf{A} \in \mathbb{R}^{4\times4}$:

$$\mathbf{A} = \begin{pmatrix} 0 & 1 & 0 & 1 \\ 1 & 0 & 1 & 0 \\ 0 & 1 & 0 & 1 \\ 1 & 0 & 1 & 0 \end{pmatrix}$$

The initial node features are $\mathbf{X} \in \mathbb{R}^{4\times3}$:

$$\mathbf{X} = \begin{pmatrix} 1 & 0 & 2 \\ 0 & 1 & 3 \\ 1 & 1 & 1 \\ 2 & 0 & 0 \end{pmatrix}$$

The attention mechanism uses the weight matrix $\mathbf{W} \in \mathbb{R}^{3\times 2}$:

$$\mathbf{W} = \begin{pmatrix} 0.2 & 0.5 \\ 0.3 & 0.4 \\ 0.6 & 0.1 \end{pmatrix}$$

and the attention weight vector $\mathbf{a} \in \mathbb{R}^4$:

$$\mathbf{a} = \begin{pmatrix} 0.1 & 0.2 & 0.3 & 0.4 \end{pmatrix}$$

(a) Compute the transformed node features $\mathbf{H} = \mathbf{X}\mathbf{W}$.

(b) Calculate the attention coefficients e_{ij} using:

$$e_{ij} = \mathbf{a}^T[\mathbf{h}_i \| \mathbf{h}_j]$$

where $\|$ denotes concatenation.

(c) Apply the softmax function to the attention coefficients and compute the final node features using the attention mechanism.

14. Consider the same graph G and node features $\mathbf{X}$ as in above exercise. Assume the graph attention network has 2 attention heads with weight matrices $\mathbf{W}^{(1)}$ and $\mathbf{W}^{(2)}$:

$$\mathbf{W}^{(1)} = \begin{pmatrix} 0.1 & 0.3 \\ 0.4 & 0.5 \\ 0.6 & 0.2 \end{pmatrix}, \quad \mathbf{W}^{(2)} = \begin{pmatrix} 0.3 & 0.4 \\ 0.2 & 0.6 \\ 0.1 & 0.5 \end{pmatrix}$$

(a) Compute the transformed node features for each head.

(b) Calculate the attention coefficients for each head using the shared attention weight vector:

$$\mathbf{a} = \begin{pmatrix} 0.1 & 0.2 \end{pmatrix}$$

(c) Apply the softmax function to obtain the normalized attention coefficients and compute the final concatenated node features from the two heads.

15. Consider the same graph G and node features $\mathbf{X}$ as in above exercise. The positional encodings for the nodes are given by $\mathbf{P} \in \mathbb{R}^{4\times 3}$:

$$\mathbf{P} = \begin{pmatrix} 0.1 & 0.2 & 0.3 \\ 0.2 & 0.3 & 0.1 \\ 0.3 & 0.1 & 0.2 \\ 0.1 & 0.3 & 0.2 \end{pmatrix}$$

(a) Incorporate the positional encodings into the node features $\mathbf{X}$ to obtain $\mathbf{X}' = \mathbf{X} + \mathbf{P}$.

(b) Perform the graph attention mechanism using the transformed node features $\mathbf{X}'$ and compute the attention coefficients and updated node features.

16. Consider a graph G with 3 nodes and the adjacency matrix $\mathbf{A} \in \mathbb{R}^{3\times3}$:

$$\mathbf{A} = \begin{pmatrix} 0 & 2 & 1 \\ 2 & 0 & 1 \\ 1 & 1 & 0 \end{pmatrix}$$

The initial node features are $\mathbf{X} \in \mathbb{R}^{3\times2}$:

$$\mathbf{X} = \begin{pmatrix} 1 & 2 \\ 3 & 4 \\ 5 & 6 \end{pmatrix}$$

The attention mechanism uses the weight matrix $\mathbf{W} \in \mathbb{R}^{2\times2}$:

$$\mathbf{W} = \begin{pmatrix} 0.2 & 0.3 \\ 0.4 & 0.5 \end{pmatrix}$$

and the attention weight vector $\mathbf{a} \in \mathbb{R}^4$:

$$\mathbf{a} = \begin{pmatrix} 0.1 & 0.2 & 0.3 & 0.4 \end{pmatrix}$$

(a) Compute the transformed node features $\mathbf{H} = \mathbf{XW}$.

(b) Calculate the attention coefficients e_{ij} considering the edge weights A_{ij}:

$$e_{ij} = \mathbf{a}^T [\mathbf{h}_i \| \mathbf{h}_j] + A_{ij}$$

(c) Apply the softmax function to the attention coefficients and compute the final node features using the attention mechanism.

Chapter 9
Self-supervised and Unsupervised Learning in Deep Learning

AI is a fundamental risk to the existence of human civilization.

—Elon Musk, Interview with CNBC

9.1 Introduction

Self-supervised and unsupervised learning are two fundamental paradigms in deep learning that enable models to learn from data without explicit human-provided labels. Unlike supervised learning, where models are trained on labeled datasets, self-supervised and unsupervised methods leverage the intrinsic structure of data to extract meaningful representations and patterns. Self-supervised learning (SSL) involves creating pseudo-labels from the data itself. This is done by designing pretext tasks that force the model to learn useful features that can be later fine-tuned for downstream tasks. For example, predicting the rotation angle of an image, inpainting missing parts of an image, or predicting the next word in a sentence are common pretext tasks that help the model learn useful representations. Unsupervised learning, on the other hand, focuses on learning from unlabeled data by discovering underlying structures and patterns. Techniques such as clustering, dimensionality reduction, and generative models fall under this category. For example, clustering algorithms group similar data points together, while dimensionality reduction techniques like Principal Component Analysis (PCA) and t-SNE reduce the data's dimensionality to highlight important features. Mathematically, both paradigms aim to model the underlying data distribution $p(x)$ or the conditional distribution $p(x \mid z)$, where z represents latent variables or pseudo-labels. These methods are crucial for leveraging vast amounts of unlabeled data, which is often more abundant and easier to obtain than labeled data.

P. Singh and B. Raman, *Deep Learning Through the Prism of Tensors*, Studies in Big Data 162, https://doi.org/10.1007/978-981-97-8019-8_9

Historical Context and Development

The concepts of self-supervised and unsupervised learning have evolved significantly over the years. In the early days, unsupervised learning methods like clustering and PCA were widely used for data analysis. However, with the advent of deep learning, more sophisticated methods have been developed. In the context of deep learning, self-supervised learning gained prominence with the introduction of models like Word2Vec (Mikolov et al. 2013, which learns word embeddings by predicting the context of words in sentences. This idea was further extended to images with models like RotNet (Gidaris et al. 2018), which learns image representations by predicting image rotations. Unsupervised learning has also seen substantial advancements with the development of deep generative models. Variational Autoencoders (VAEs) (Kingma and Welling 2014) and Generative Adversarial Networks (GANs) (Goodfellow et al. 2014) are prime examples that learn to generate new data samples similar to the training data. These models have not only advanced the field of generative modeling but also contributed to tasks like anomaly detection and data augmentation. Recent developments in SSL have focused on contrastive learning methods, where models learn by contrasting positive pairs (similar samples) with negative pairs (dissimilar samples). Methods like SimCLR (Chen et al. 2020) and MoCo (He et al. 2019) have demonstrated impressive results by learning high-quality representations from large-scale unlabeled data.

Applications in Modern Deep Learning Self-supervised and unsupervised learning have a wide range of applications across various domains in modern deep learning. Some key applications include:

1. Natural Language Processing (NLP):
   Word Embeddings: Techniques like Word2Vec, GloVe, and BERT utilize self-supervised learning to learn word representations from large text corpora.
   Pre-training Language Models: Models like GPT-3 and BERT are pretrained on massive text datasets using self-supervised tasks and then fine-tuned for specific NLP tasks such as text classification, translation, and question answering.
2. Computer Vision:
   Representation Learning: Models like SimCLR and BYOL learn visual representations from unlabeled images, which can be fine-tuned for tasks like image classification, object detection, and segmentation.
   Anomaly Detection: Generative models like VAEs and GANs can detect anomalies by learning the normal data distribution and identifying deviations from it.
3. Recommender Systems:
   Collaborative Filtering: Unsupervised methods like matrix factorization and graph-based approaches are used to learn user-item interaction patterns for making recommendations.
   Representation Learning: Self-supervised learning techniques are employed to learn user and item embeddings from user behavior data.

4. Healthcare:
   Medical Imaging: Self-supervised learning is used to learn representations from medical images, which can be fine-tuned for tasks like disease diagnosis and segmentation.
   Electronic Health Records (EHRs): Unsupervised learning techniques are applied to EHR data to discover patient subgroups and predict health outcomes.
5. Autonomous Systems:
   Sensor Data Analysis: Unsupervised learning methods are used to analyze sensor data from autonomous vehicles, enabling tasks like anomaly detection and predictive maintenance.
   Self-supervised Robotics: Self-supervised learning is applied to robotics for learning policies from raw sensory inputs without manual labeling.

Mathematically, the objective functions for self-supervised and unsupervised learning can be expressed as:

Self-supervised Learning:

$$\mathcal{L}_{\text{SSL}} = \sum_{i=1}^{N} \mathcal{L}_{\text{pretext}}(f(x_i), y_i^*)$$

where $\mathcal{L}_{\text{pretext}}$ is the loss for the pretext task, $f(x_i)$ is the model's prediction, and y_i^* are the pseudo-labels generated from the data.

Unsupervised Learning:

$$\mathcal{L}_{\text{UL}} = \sum_{i=1}^{N} \mathcal{L}_{\text{reconstruction}}(f(x_i), x_i)$$

where $\mathcal{L}_{\text{reconstruction}}$ is the loss function that measures the discrepancy between the input x_i and its reconstruction $f(x_i)$.

These loss functions guide the model to learn meaningful representations from the data without relying on labeled examples.

9.2 Fundamentals of Unsupervised Learning

9.2.1 Clustering-Based Approaches

Clustering is a fundamental technique in unsupervised learning that involves grouping a set of objects in such a way that objects in the same group (or cluster) are more similar to each other than to those in other groups. Clustering methods are widely used for exploratory data analysis, pattern recognition, and data compression. In this section, we explore three primary clustering techniques: K-Means Clustering, Gaussian Mixture Models (GMM), and Hierarchical Clustering.

K-Means Clustering K-Means Clustering aims to partition n observations into k clusters in which each observation belongs to the cluster with the nearest mean. The goal is to minimize the within-cluster sum of squares (WCSS), which measures the variance within each cluster. Given a dataset $\mathbf{X} = \{x_1, x_2, \ldots, x_n\}$, where each x_i is a data point, K-Means clustering seeks to minimize the following objective function:

$$\mathcal{L} = \sum_{i=1}^{k} \sum_{x \in C_i} \|x - \mu_i\|^2$$

where C_i represents the i-th cluster, and μ_i is the mean (centroid) of the i-th cluster. The algorithm iteratively updates the centroids and the assignments of points to clusters until convergence.

Algorithm:

1. Initialization: Randomly select k initial centroids.
2. Assignment Step: Assign each data point to the nearest centroid.
3. Update Step: Recalculate the centroids as the mean of all points assigned to each cluster.
4. Convergence Check: Repeat steps 2 and 3 until the centroids no longer change significantly.

Example: Consider a dataset with two-dimensional points that we want to cluster into three groups. The algorithm proceeds as follows: Initialize centroids μ_1, μ_2, μ_3 randomly. Assign each point to the nearest centroid based on Euclidean distance. Update the centroids by calculating the mean of the points assigned to each cluster. Iterate the assignment and update steps until the centroids stabilize.
Python Implementation (see Fig. 9.1):

```
import numpy as np
import matplotlib.pyplot as plt
from sklearn.datasets import make_blobs

# Generate synthetic data
X, y = make_blobs(n_samples=300, centers=3, random_state=42)

# K-Means algorithm
def kmeans(X, k, max_iters=100):
    n_samples, n_features = X.shape
    centroids = X[np.random.choice(n_samples, k, replace=False)]

    for _ in range(max_iters):
        distances = np.linalg.norm(X[:, np.newaxis] - centroids, axis=2)
        cluster_assignments = np.argmin(distances, axis=1)
```

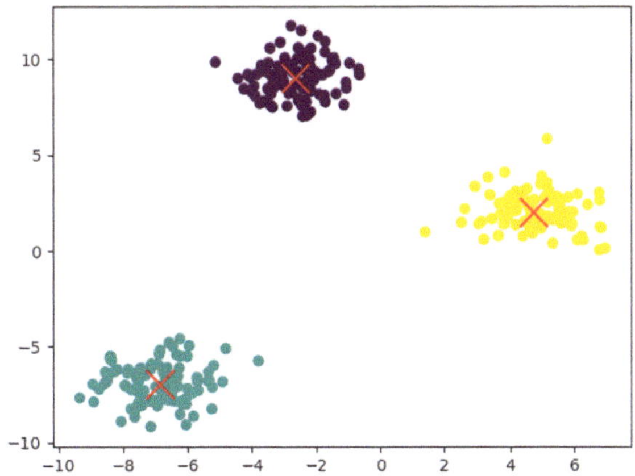

Fig. 9.1 An illustration of K-means clusters

```
        new_centroids = np.array([X[cluster_assignments == i].mean(axis=0) for i in
range(k)])

        if np.all(centroids == new_centroids):
            break
        centroids = new_centroids

    return centroids, cluster_assignments

# Apply K-Means
centroids, cluster_assignments = kmeans(X, k=3)

# Plot results
plt.scatter(X[:, 0], X[:, 1], c=cluster_assignments, cmap='viridis')
plt.scatter(centroids[:, 0], centroids[:, 1], s=300, c='red', marker='x')
plt.show()
```

Gaussian Mixture Models (GMM) Gaussian Mixture Models (GMM) are probabilistic models that assume the data is generated from a mixture of several Gaussian distributions with unknown parameters. GMM provides a flexible way to model the distribution of data and can capture more complex cluster shapes compared to K-Means. A GMM can be expressed as a weighted sum of k Gaussian components:

$$p(x) = \sum_{i=1}^{k} \pi_i \mathcal{N}(x \mid \mu_i, \Sigma_i)$$

where π_i are the mixture weights, μ_i are the mean vectors, and Σ_i are the covariance matrices of the Gaussian components. The parameters are estimated using the Expectation-Maximization (EM) algorithm.

Algorithm:

1. Initialization: Initialize the parameters π_i, μ_i, Σ_i.
2. E-Step (Expectation): Calculate the responsibility $\gamma_i(x)$ for each data point:

$$\gamma_i(x) = \frac{\pi_i \mathcal{N}(x \mid \mu_i, \Sigma_i)}{\sum_{j=1}^{k} \pi_j \mathcal{N}(x \mid \mu_j, \Sigma_j)}$$

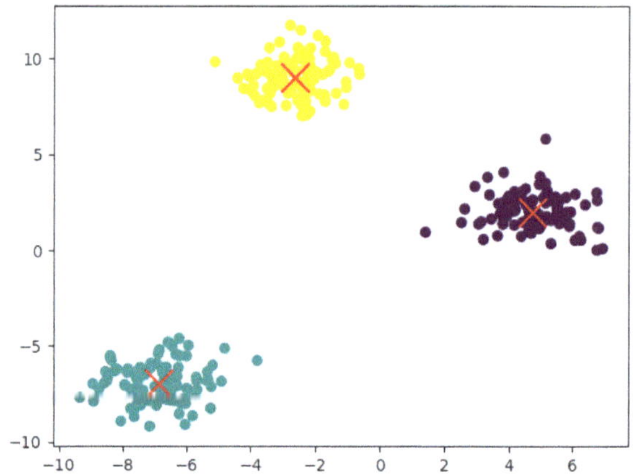

Fig. 9.2 An illustration of Gaussian mixtures

3. M-Step (Maximization): Update the parameters using the responsibilities:

$$\pi_i = \frac{1}{n}\sum_{j=1}^{n}\gamma_i(x_j)$$

$$\mu_i = \frac{\sum_{j=1}^{n}\gamma_i(x_j)x_j}{\sum_{j=1}^{n}\gamma_i(x_j)}$$

$$\Sigma_i = \frac{\sum_{j=1}^{n}\gamma_i(x_j)(x_j - \mu_i)(x_j - \mu_i)^T}{\sum_{j=1}^{n}\gamma_i(x_j)}$$

4. Convergence Check: Repeat the E-Step and M-Step until convergence.

Example: Consider the same dataset with two-dimensional points. We fit a GMM to identify the underlying Gaussian distributions.

Python Implementation (see Fig. 9.2):

```
from sklearn.mixture import GaussianMixture

# Generate synthetic data
X, y = make_blobs(n_samples=300, centers=3, random_state=42)

# Fit GMM
gmm = GaussianMixture(n_components=3, random_state=42)
gmm.fit(X)
labels = gmm.predict(X)

# Plot results
plt.scatter(X[:, 0], X[:, 1], c=labels, cmap='viridis')
plt.scatter(gmm.means_[:, 0], gmm.means_[:, 1], s=300, c='red', marker='x')
plt.show()
```

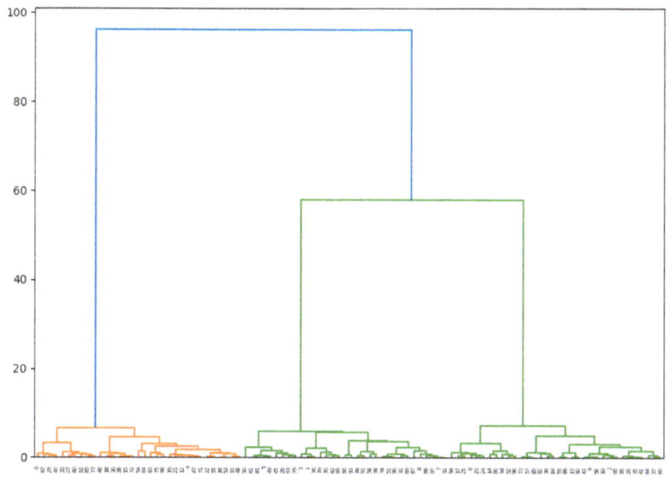

Fig. 9.3 An illustration of a dendogram

Hierarchical Clustering Hierarchical clustering builds a hierarchy of clusters either in a bottom-up approach (agglomerative) or a top-down approach (divisive). The result is a tree-like structure called a dendrogram that represents the nested grouping of data points. In agglomerative clustering, each data point starts as its own cluster. Pairs of clusters are iteratively merged based on a distance metric until a single cluster remains. Common distance metrics include:

Single Linkage: Minimum distance between points in different clusters.
Complete Linkage: Maximum distance between points in different clusters.
Average Linkage: Average distance between points in different clusters.

Algorithm:

1. Initialization: Start with each data point as its own cluster.
2. Merge Clusters: At each step, merge the two closest clusters based on the chosen linkage criterion.
3. Repeat: Continue merging until all points belong to a single cluster.

Example: Consider a dataset with two-dimensional points. We apply agglomerative hierarchical clustering to build the cluster hierarchy.

Python Implementation (see Fig. 9.3):

```
from scipy.cluster.hierarchy import dendrogram, linkage
from sklearn.datasets import make_blobs

# Generate synthetic data
X, y = make_blobs(n_samples=100, centers=3, random_state=42)

# Perform hierarchical clustering
Z = linkage(X, 'ward')

# Plot dendrogram
```

```
plt.figure(figsize=(10, 7))
dendrogram(Z)
plt.show()
```

Clustering-based approaches are fundamental in unsupervised learning, providing insights into the structure of data. Each method has its strengths and weaknesses, making them suitable for different types of data and applications. K-Means is simple and efficient for spherical clusters, GMM is flexible for capturing complex distributions, and hierarchical clustering provides a detailed hierarchy of clusters.

9.2.2 Dimensionality Reduction Techniques

Dimensionality reduction techniques are essential in unsupervised learning to reduce the number of random variables under consideration by obtaining a set of principal variables. These techniques help in visualizing high-dimensional data, improving computational efficiency, and mitigating the curse of dimensionality. We will explore three key dimensionality reduction methods: Principal Component Analysis (PCA), t-Distributed Stochastic Neighbor Embedding (t-SNE), and Uniform Manifold Approximation and Projection (UMAP).

Principal Component Analysis (PCA) Principal Component Analysis is a linear dimensionality reduction technique that transforms the data into a new coordinate system such that the greatest variance by any projection of the data lies on the first coordinate (called the first principal component), the second greatest variance on the second coordinate, and so on. PCA is widely used for reducing the dimensionality of large datasets while preserving as much variance as possible. Given a dataset $\mathbf{X}$ with n samples and d features, PCA seeks to find a set of orthogonal vectors (principal components) that capture the maximum variance in the data. The steps involved are:

1. Standardize the Data: Center the data by subtracting the mean of each feature.

$$\mathbf{X}_{\text{centered}} = \mathbf{X} - \mu$$

2. Compute the Covariance Matrix: Calculate the covariance matrix of the centered data.

$$\mathbf{C} = \frac{1}{n-1}\mathbf{X}_{\text{centered}}^{\top}\mathbf{X}_{\text{centered}}$$

3. Eigen Decomposition: Perform eigen decomposition on the covariance matrix to obtain eigenvalues and eigenvectors.

$$\mathbf{CV} = \mathbf{V}\Lambda$$

where Λ is the diagonal matrix of eigenvalues, and $\mathbf{V}$ is the matrix of eigenvectors.

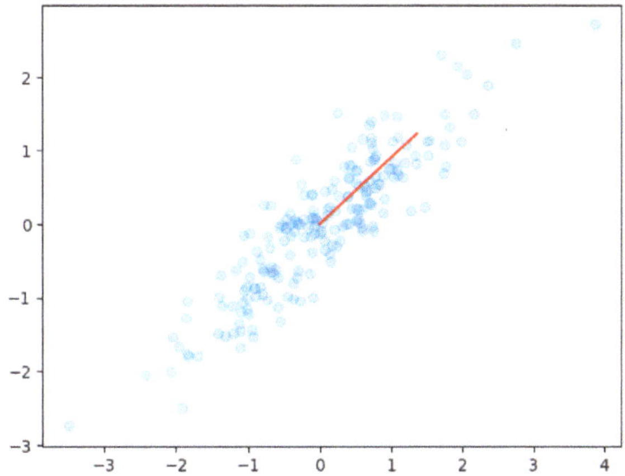

Fig. 9.4 An illustration of PCA

4. Principal Components: Select the top k eigenvectors corresponding to the largest eigenvalues to form the projection matrix.

$$\mathbf{X}_{\text{reduced}} = \mathbf{X}_{\text{centered}}\mathbf{V}_k$$

Example: Consider a dataset with two-dimensional points that we want to project onto one dimension using PCA.

Python Implementation (see Fig. 9.4):

```
import numpy as np
import matplotlib.pyplot as plt
from sklearn.decomposition import PCA

# Generate synthetic data
np.random.seed(42)
X = np.dot(np.random.rand(2, 2), np.random.randn(2, 200)).T

# Apply PCA
pca = PCA(n_components=1)
X_reduced = pca.fit_transform(X)

# Plot results
plt.scatter(X[:, 0], X[:, 1], alpha=0.2)
plt.plot([0, pca.components_[0, 0] * pca.explained_variance_[0]],
         [0, pca.components_[0, 1] * pca.explained_variance_[0]], color='red')
plt.title('PCA')
plt.show()
```

t-Distributed Stochastic Neighbor Embedding (t-SNE) t-Distributed Stochastic Neighbor Embedding (t-SNE) is a non-linear dimensionality reduction technique particularly well-suited for embedding high-dimensional data into a two- or three-dimensional space for visualization. t-SNE is designed to capture local structure and is effective in revealing clusters in the data. t-SNE minimizes the divergence between two distributions: one that measures pairwise similarities of the input objects in the high-dimensional space and one that measures pairwise similarities of the corresponding low-dimensional points.

1. Compute Pairwise Similarities in High-Dimensional Space: Compute the joint probabilities p_{ij} using a Gaussian distribution.

$$p_{ij} = \frac{\exp(-\|x_i - x_j\|^2/2\sigma_i^2)}{\sum_{k \neq l} \exp(-\|x_k - x_l\|^2/2\sigma_k^2)}$$

2. Compute Pairwise Similarities in Low-Dimensional Space: Compute the joint probabilities q_{ij} using a Student t-distribution.

$$q_{ij} = \frac{(1 + \|y_i - y_j\|^2)^{-1}}{\sum_{k \neq l} (1 + \|y_k - y_l\|^2)^{-1}}$$

3. Minimize KL Divergence: Minimize the Kullback-Leibler (KL) divergence between the distributions P and Q.

$$\mathcal{L} = \sum_{i \neq j} p_{ij} \log \frac{p_{ij}}{q_{ij}}$$

Example: Apply t-SNE to visualize a high-dimensional dataset in two dimensions.

Python Implementation (see Fig. 9.5):

```
from sklearn.datasets import load_digits
from sklearn.manifold import TSNE

# Load dataset
digits = load_digits()
X = digits.data
y = digits.target

# Apply t-SNE
tsne = TSNE(n_components=2, random_state=42)
X_embedded = tsne.fit_transform(X)

# Plot results
plt.scatter(X_embedded[:, 0], X_embedded[:, 1], c=y, cmap='tab10')
plt.colorbar()
plt.title('t-SNE')
plt.show()
```

Uniform Manifold Approximation and Projection (UMAP) Uniform Manifold Approximation and Projection (UMAP) is a non-linear dimensionality reduction technique that aims to preserve the global structure of data while capturing local relationships. UMAP is highly scalable and can handle large datasets effectively. UMAP constructs a high-dimensional graph representation of the data and then optimizes a low-dimensional graph that maintains the same topological structure.

1. Construct High-Dimensional Graph: Compute the nearest neighbors for each data point. Construct a weighted k-nearest neighbor graph based on these neighbors.

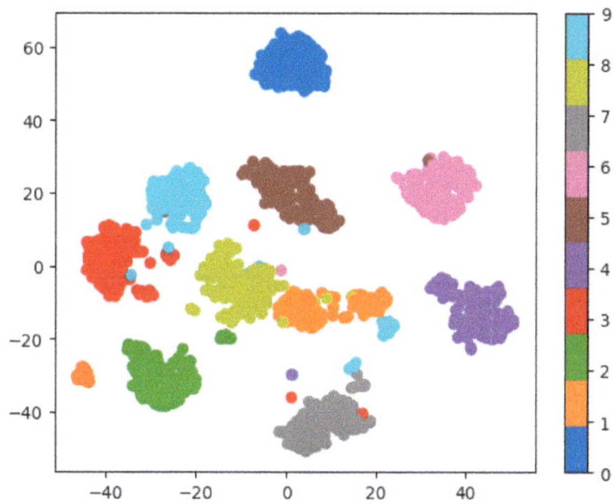

Fig. 9.5 An illustration of t-SNE

2. Optimize Low-Dimensional Embedding: Initialize the low-dimensional representation. Minimize the cross-entropy between the high-dimensional and low-dimensional representations.

$$\mathcal{L} = \sum_{i \neq j} w_{ij} \left(d_{ij}^2 + (1 - w_{ij})(1 + d_{ij}^2)^{-1}\right)$$

where w_{ij} are the edge weights in the high-dimensional graph and d_{ij} are the distances in the low-dimensional space.

Example: Apply UMAP to visualize a high-dimensional dataset in two dimensions.

Python Implementation (see Fig. 9.6):

```
import umap
from sklearn.datasets import load_digits

# Load dataset
digits = load_digits()
X = digits.data
y = digits.target

# Apply UMAP
reducer = umap.UMAP(n_components=2, random_state=42)
X_embedded = reducer.fit_transform(X)

# Plot results
plt.scatter(X_embedded[:, 0], X_embedded[:, 1], c=y, cmap='tab10')
plt.colorbar()
plt.title('UMAP')
plt.show()
```

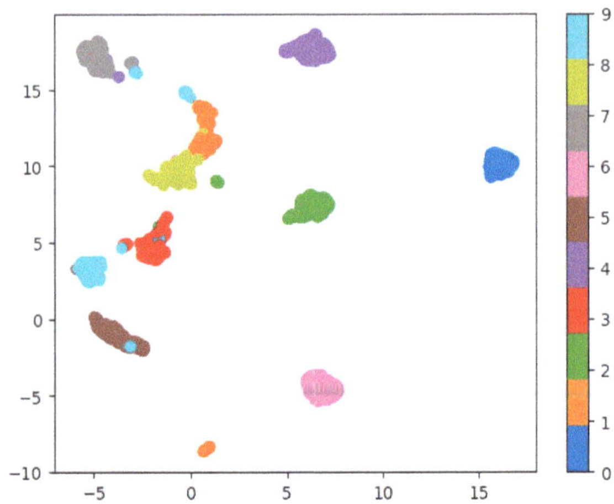

Fig. 9.6 An illustration of UMAP

Dimensionality reduction techniques like PCA, t-SNE, and UMAP are powerful tools for visualizing and understanding high-dimensional data. Each method has its unique strengths: PCA is simple and computationally efficient, t-SNE excels at revealing local structures, and UMAP balances local and global structures while being scalable.

9.3 Self-supervised Learning

Self-supervised learning (SSL) is a type of unsupervised learning where the system learns to predict part of its input from other parts of its input, effectively creating its own supervisory signal from the data itself. The primary idea is to use the inherent structure of the data to generate labels, known as pseudo-labels, which are then used to train the model. SSL leverages large amounts of unlabeled data to learn useful representations that can be transferred to downstream tasks. Let $\mathbf{X} = \{x_1, x_2, \ldots, x_n\}$ represent the dataset, where each x_i is a data point. The objective of SSL is to define a pretext task T such that a model f_θ can be trained to predict the output y_i^* generated from x_i itself. The loss function for SSL can be written as:

$$\mathcal{L}_{\text{SSL}} = \sum_{i=1}^{n} \mathcal{L}(f_\theta(T(x_i)), y_i^*)$$

where $\mathcal{L}$ is the loss function, f_θ is the model parameterized by θ, and y_i^* is the pseudo-label generated from x_i.

Key Concepts:

1. Pretext Task: A task designed to create pseudo-labels from the input data. Examples include predicting the rotation angle of an image, filling in missing parts of an image, or predicting the next word in a sentence.
2. Pseudo-Labels: Labels generated from the data itself, used to train the model.
3. Representation Learning: The process of learning useful features from the data that can be transferred to downstream tasks.

Example: A common example of a pretext task in SSL is the rotation prediction task for images. The model is trained to predict the degree of rotation (0°, 90°, 180°, or 270°) applied to an image. This task forces the model to learn meaningful representations of the image content.

```
import tensorflow as tf
from tensorflow.keras.layers import Conv2D, Dense, Flatten
from tensorflow.keras.models import Sequential
import numpy as np

# Example dataset
images = np.random.rand(1000, 28, 28, 1)
rotations = np.random.choice([0, 90, 180, 270], 1000)
rotated_images = np.array([np.rot90(img, k=rot//90) for img, rot in zip(images,
 rotations)])

# Simple SSL model for rotation prediction
model = Sequential([
    Conv2D(32, (3, 3), activation='relu', input_shape=(28, 28, 1)),
    Flatten(),
    Dense(128, activation='relu'),
    Dense(4, activation='softmax')
])

model.compile(optimizer='adam', loss='sparse_categorical_crossentropy', metrics=
['accuracy'])
model.fit(rotated_images, rotations, epochs=10)
```

Supervised versus Unsupervised versus Self-supervised Learning

In supervised learning, the model is trained on labeled data, where each input data point x is paired with a corresponding label y. The objective is to learn a mapping $f : X \to Y$ that can accurately predict labels for new, unseen data.

$$\mathcal{L}_{\text{sup}} = \sum_{i=1}^{n} \mathcal{L}(f_\theta(x_i), y_i)$$

Unsupervised learning deals with unlabeled data. The goal is to discover hidden patterns or structures in the data. Clustering and dimensionality reduction are common tasks in unsupervised learning.

$$\mathcal{L}_{\text{unsup}} = \sum_{i=1}^{n} \mathcal{L}(x_i)$$

Self-supervised learning bridges the gap between supervised and unsupervised learning. It uses the data itself to generate pseudo-labels, creating a supervised learning problem out of an unsupervised one. SSL aims to learn representations that can be useful for various downstream tasks without relying on manually labeled data.

$$\mathcal{L}_{\text{SSL}} = \sum_{i=1}^{n} \mathcal{L}(f_\theta(T(x_i)), y_i^*)$$

Benefits:

1. Utilization of Unlabeled Data: SSL can leverage vast amounts of unlabeled data, which is often more abundant and easier to obtain than labeled data.
2. Improved Representations: SSL pretext tasks encourage the model to learn rich and meaningful representations that are transferable to various downstream tasks.
3. Reduced Labeling Effort: By generating labels from the data itself, SSL reduces the need for extensive manual labeling, saving time and resources.

Challenges:

1. Designing Effective Pretext Tasks: Creating pretext tasks that lead to useful representations can be challenging and often requires domain knowledge.
2. Computational Complexity: SSL methods can be computationally intensive, especially when dealing with large-scale datasets and complex pretext tasks.
3. Evaluation Metrics: Evaluating the quality of learned representations without labeled data can be difficult. SSL models are often assessed based on their performance on downstream tasks.

Example Pretext Tasks:

Image Rotation Prediction: Predict the rotation angle applied to an image (0°, 90°, 180°, 270°).
Context Prediction: Predict the relative positions of patches in an image.
Contrastive Learning: Maximize agreement between different views of the same data point while minimizing agreement between different data points.

9.3.1 Contrastive Learning

Contrastive learning is a popular self-supervised learning approach that focuses on learning representations by comparing positive pairs (similar data points) and negative pairs (dissimilar data points). The objective is to make the representations of positive pairs closer in the latent space while pushing the representations of negative pairs apart. This technique has shown impressive results in various domains, particularly in visual representation learning. Let $\mathbf{X} = \{x_1, x_2, \ldots, x_n\}$ be a dataset of n samples. Contrastive learning involves creating augmented views of the data, such that x_i and x_i^+ are positive pairs (augmented versions of the same data point) and x_i and x_j^- are negative pairs (augmented versions of different data points). The goal is to learn an embedding function f_θ that minimizes the distance between positive pairs and maximizes the distance between negative pairs. The contrastive loss function, such as the InfoNCE loss, is commonly used:

$$\mathcal{L}_{\text{contrastive}} = -\sum_{i=1}^{n} \log \frac{\exp(\text{sim}(f_\theta(x_i), f_\theta(x_i^+))/\tau)}{\sum_{j=1}^{n} \exp(\text{sim}(f_\theta(x_i), f_\theta(x_j^-))/\tau)}$$

where sim is a similarity measure (e.g., cosine similarity), and τ is a temperature parameter.

SimCLR: A Simple Framework for Contrastive Learning SimCLR (Simple Framework for Contrastive Learning of Visual Representations) is a self-supervised learning method that employs contrastive learning to learn visual representations. SimCLR uses a straightforward approach to generate positive pairs through data augmentation and relies on a large batch size to include many negative samples in each training iteration.

Key Components:

1. Data Augmentation: SimCLR generates positive pairs by applying random data augmentations to the same input image. Common augmentations include random cropping, color distortion, and Gaussian blur.
2. Embedding Function: The base encoder f_θ (e.g., a ResNet) maps the augmented images to a latent space. A projection head g_ϕ maps the embeddings to a lower-dimensional space where the contrastive loss is applied.
3. Contrastive Loss: The InfoNCE loss is used to train the model, encouraging the embeddings of positive pairs to be close and those of negative pairs to be distant.

Given an input image x, two augmented views x_i and x_i^+ are generated. The model learns representations $h_i = f_\theta(x_i)$ and $h_i^+ = f_\theta(x_i^+)$, which are then projected to a lower-dimensional space $z_i = g_\phi(h_i)$ and $z_i^+ = g_\phi(h_i^+)$. The contrastive loss is applied to these projections. The loss for a positive pair (z_i, z_i^+) is:

$$\ell_i = -\log \frac{\exp(\text{sim}(z_i, z_i^+)/\tau)}{\sum_{k=1}^{2N} \mathbb{1}_{[k \neq i]} \exp(\text{sim}(z_i, z_k)/\tau)}$$

where N is the batch size and $\mathbb{1}_{[k \neq i]}$ is an indicator function that excludes the positive pair from the denominator.

Example: Implementing SimCLR with TensorFlow:

```
import tensorflow as tf
import tensorflow_addons as tfa
from tensorflow.keras import layers

# Data augmentation function
def augment_image(image):
```

```
    image = tf.image.random_crop(image, size=[28, 28, 3])
    image = tf.image.random_flip_left_right(image)
    image = tf.image.random_brightness(image, max_delta=0.5)
    return image

# Encoder network (e.g., ResNet)
base_model = tf.keras.applications.ResNet50(include_top=False, input_shape=(28, 28, 3),
 pooling='avg')
base_model.trainable = True

# Projection head
projection_head = tf.keras.Sequential([
    layers.Dense(128, activation='relu'),
    layers.Dense(64)
])

# Contrastive learning model
inputs = tf.keras.Input(shape=(28, 28, 3))
augmented = augment_image(inputs)
features = base_model(augmented, training=True)
projections = projection_head(features)
model = tf.keras.Model(inputs=inputs, outputs=projections)

# Contrastive loss
def contrastive_loss(z_i, z_j, temperature=0.1):
    z_i = tf.math.l2_normalize(z_i, axis=1)
    z_j = tf.math.l2_normalize(z_j, axis=1)
    similarities = tf.matmul(z_i, z_j, transpose_b=True) / temperature
    labels = tf.range(tf.shape(z_i)[0])
    return tf.reduce_mean(tf.nn.sparse_softmax_cross_entropy_with_logits(labels,
 similarities))

# Optimizer and training loop
optimizer = tf.keras.optimizers.Adam()

@tf.function
def train_step(images):
    with tf.GradientTape() as tape:
        augmented_images_1 = tf.map_fn(augment_image, images)
        augmented_images_2 = tf.map_fn(augment_image, images)
        z_i = model(augmented_images_1, training=True)
        z_j = model(augmented_images_2, training=True)
        loss = contrastive_loss(z_i, z_j)
    gradients = tape.gradient(loss, model.trainable_variables)
    optimizer.apply_gradients(zip(gradients, model.trainable_variables))
    return loss

# Training loop
for epoch in range(10):
    for images in dataset:  # Assume 'dataset' is a preloaded dataset of images
        loss = train_step(images)
    print(f'Epoch {epoch+1}, Loss: {loss.numpy()}')
```

Momentum Contrast (MoCo) Momentum Contrast (MoCo) is a self-supervised learning framework designed to address the need for large batch sizes in contrastive learning methods like SimCLR. MoCo maintains a dynamic dictionary with a queue of encoded representations and uses a momentum encoder to provide consistent and stable keys for contrastive learning.

Key Components:

1. Encoder Networks: MoCo uses two encoders: a query encoder f_q and a key encoder f_k. The query encoder is updated with standard backpropagation, while the key encoder is updated with a momentum mechanism.

2. Momentum Update: The key encoder is updated as an exponential moving average of the query encoder to ensure consistency:

$$\theta_k \leftarrow m\theta_k + (1 - m)\theta_q$$

   where θ_q and θ_k are the parameters of the query and key encoders, respectively, and m is the momentum coefficient.
3. Dynamic Dictionary: A queue stores the encoded keys from previous batches, allowing MoCo to use a large number of negative samples without requiring a large batch size.

Given a query image x_q and a key image x_k, MoCo aims to minimize the InfoNCE loss:

$$\mathcal{L}_{\text{MoCo}} = -\log \frac{\exp(\text{sim}(f_q(x_q), f_k(x_k^+))/\tau)}{\sum_{i=0}^{K} \exp(\text{sim}(f_q(x_q), k_i)/\tau)}$$

where k_i are the keys stored in the queue.

Example: Implementing MoCo with TensorFlow:

```
class MoCo(tf.keras.Model):
    def __init__(self, base_encoder, projection_head, queue_size=4096, momentum=0.999):
        super(MoCo, self).__init__()
        self.query_encoder = tf.keras.Sequential([base_encoder, projection_head])
        self.key_encoder = tf.keras.Sequential([tf.keras.models.clone_model
(base_encoder), projection_head])
        self.key_encoder.trainable = False
        self.queue = tf.Variable(tf.random.normal([queue_size,
projection_head.layers[-1].units]), trainable=False)
        self.queue_ptr = tf.Variable(0, trainable=False)
        self.momentum = momentum

    def update_key_encoder(self):
        for query_weights, key_weights in zip(self.query_encoder.weights, self.key_
encoder.weights):
            key_weights.assign(self.momentum * key_weights + (1 - self.momentum) * query_
weights)

    def enqueue_and_dequeue(self, keys):
        batch_size = tf.shape(keys)[0]
        self.queue.scatter_update(tf.IndexedSlices(keys, tf.range(self.queue_ptr,
 self.queue_ptr + batch_size) % tf.shape(self.queue)[0]))
        self.queue_ptr.assign((self.queue_ptr + batch_size) % tf.shape(self.queue)[0])

    def call(self, images_query, images_key):
        queries = self.query_encoder(images_query, training=True)
        keys = self.key_encoder(images_key, training=True)
        keys = tf.stop_gradient(keys)
        self.update_key_encoder()
        self.enqueue_and_dequeue(keys)
        return queries, keys, self.queue

# Instantiate MoCo
base_encoder = tf.keras.applications.ResNet50(include_top=False, input_shape=(224, 224,
 3), pooling='avg')
projection_head = tf.keras.Sequential([layers.Dense(128, activation='relu'), layers.
```

```
Dense(64)])
moco = MoCo(base_encoder, projection_head)

# Training loop
optimizer = tf.keras.optimizers.Adam()

@tf.function
def train_step(images_query, images_key):
    with tf.GradientTape() as tape:
        queries, keys, queue = moco(images_query, images_key)
        loss = contrastive_loss(queries, tf.concat([keys, queue], axis=0))
    gradients = tape.gradient(loss, moco.trainable_variables)
    optimizer.apply_gradients(zip(gradients, moco.trainable_variables))
    return loss

# Training loop
for epoch in range(10):
    for images_query, images_key in dataset:  # Assume 'dataset' provides query and key
 image pairs
        loss = train_step(images_query, images_key)
    print(f'Epoch {epoch+1}, Loss: {loss.numpy()}')
```

Contrastive learning techniques like SimCLR and MoCo have significantly advanced the field of self-supervised learning, particularly in visual representation learning. They leverage large amounts of unlabeled data to learn useful representations that can be fine-tuned for various downstream tasks. SimCLR's simplicity and MoCo's efficient use of negative samples provide robust frameworks for training high-quality models in a self-supervised manner.

9.3.2 Clustering-Based Self-supervised Learning

Clustering-based self-supervised learning leverages clustering algorithms to generate pseudo-labels for unlabeled data, thereby learning useful representations. Two notable methods in this category are DeepCluster and SwAV.

DeepCluster It is a method that iteratively applies k-means clustering on the learned representations to create pseudo-labels and then uses these pseudo-labels to update the model. The core idea is to cluster the features extracted by a neural network and use these cluster assignments as targets to train the network.

1. Feature Extraction: Let $\mathbf{X} = \{x_1, x_2, \ldots, x_n\}$ be the dataset, and f_θ be the feature extractor. Extract features $\mathbf{F} = f_\theta(\mathbf{X})$.
2. Clustering: Apply k-means clustering on $\mathbf{F}$ to obtain cluster assignments $\mathbf{C} = \{c_1, c_2, \ldots, c_n\}$.

$$\mathbf{C} = \arg\min_{\mathbf{C}} \sum_{i=1}^{n} \| f_\theta(x_i) - \mu_{c_i} \|^2$$

   where μ_{c_i} is the centroid of cluster c_i.
3. Model Update: Use the cluster assignments as pseudo-labels to update the network by minimizing the cross-entropy loss:

$$\mathcal{L}_{\text{DeepCluster}} = -\sum_{i=1}^{n} \log P(c_i | x_i; \theta)$$

4. Iterative Process: Repeat the feature extraction and clustering steps iteratively to refine the representations.

Example: Implementing DeepCluster with PyTorch:

```
import torch
import torch.nn as nn
import torch.optim as optim
from sklearn.cluster import KMeans

# Define a simple CNN model
class SimpleCNN(nn.Module):
    def __init__(self):
        super(SimpleCNN, self).__init__()
        self.conv1 = nn.Conv2d(1, 32, kernel_size=3)
        self.pool = nn.MaxPool2d(kernel_size=2)
        self.fc1 = nn.Linear(32 * 13 * 13, 128)
        self.fc2 = nn.Linear(128, 10)  # Assume 10 clusters

    def forward(self, x):
        x = self.pool(F.relu(self.conv1(x)))
        x = x.view(-1, 32 * 13 * 13)
        x = F.relu(self.fc1(x))
        x = self.fc2(x)
        return x

model = SimpleCNN()
optimizer = optim.Adam(model.parameters(), lr=0.001)
criterion = nn.CrossEntropyLoss()

# Assume 'data_loader' is a PyTorch DataLoader that provides the training data
for epoch in range(10):
    features = []
    for images, _ in data_loader:
        outputs = model(images)
        features.append(outputs.detach().cpu().numpy())

    features = np.concatenate(features)
    kmeans = KMeans(n_clusters=10).fit(features)
    pseudo_labels = kmeans.predict(features)

    for i, (images, _) in enumerate(data_loader):
        outputs = model(images)
        loss = criterion(outputs, torch.tensor(pseudo_labels[i * batch_size: (i + 1) *
batch_size], dtype=torch.long))
        optimizer.zero_grad()
        loss.backward()
        optimizer.step()
    print(f'Epoch {epoch + 1}, Loss: {loss.item()}')
```

SwAV: Swapping Assignments between Views SwAV is a clustering-based self-supervised method that combines the advantages of contrastive learning and clustering. SwAV learns representations by simultaneously clustering the data and enforcing consistency between cluster assignments of different augmentations of the same image.

1. Multi-Crop Augmentation: For each image x_i, generate multiple augmented views $\{x_i^1, x_i^2, \ldots, x_i^k\}$.

2. Feature Extraction: Extract features for each augmented view using a shared encoder f_θ.

$$\mathbf{Z}_i^j = f_\theta(x_i^j)$$

3. Swapped Assignment Problem: Compute cluster assignments for each view and enforce consistency between them. Let $\mathbf{Q}_i^j$ be the assignment of $\mathbf{Z}_i^j$ to clusters. The SwAV loss is formulated as:

$$\mathcal{L}_{\text{SwAV}} = \sum_{i=1}^{n}\sum_{j=1}^{k}\sum_{l=1}^{k} \mathbb{1}_{[j\neq l]} D(\mathbf{Q}_i^j, \mathbf{Q}_i^l)$$

where D is a divergence measure (e.g., cosine similarity) that ensures consistency between cluster assignments of different views.
4. Prototypes: Learn a set of prototypes $\mathbf{P}$ that represent the clusters. The assignment is computed using the Sinkhorn-Knopp algorithm to balance the clusters.

Example: Implementing SwAV with PyTorch:

```
import torch
import torch.nn as nn
import torch.optim as optim
from torch.nn.functional import normalize

class SwAV(nn.Module):
    def __init__(self, base_encoder, num_prototypes):
        super(SwAV, self).__init__()
        self.base_encoder = base_encoder
        self.prototypes = nn.Linear(base_encoder.fc.in_features, num_prototypes,
 bias=False)
        self.base_encoder.fc = nn.Identity()

    def forward(self, x):
        features = self.base_encoder(x)
        return features, self.prototypes(features)

def sinkhorn(Q, epsilon=0.05, iterations=3):
    Q = torch.exp(Q / epsilon).t()
    Q /= Q.sum()
    K, B = Q.shape

    u = torch.zeros(K).to(Q.device)
    r = torch.ones(K).to(Q.device) / K
    c = torch.ones(B).to(Q.device) / B

    for _ in range(iterations):
        u = Q.sum(dim=1)
        Q *= (r / u).unsqueeze(1)
        Q *= (c / Q.sum(dim=0)).unsqueeze(0)

    return (Q / Q.sum(dim=0, keepdim=True)).t()

def swav_loss(embeddings, prototypes):
    assignments = sinkhorn(prototypes @ embeddings.t())
    loss = -torch.mean(torch.sum(assignments * torch.log_softmax(prototypes @ embeddings.
t(), dim=0), dim=0))
    return loss
```

```
base_encoder = torch.hub.load('pytorch/vision:v0.10.0', 'resnet50', pretrained=False)
swav_model = SwAV(base_encoder, num_prototypes=3000)
optimizer = optim.Adam(swav_model.parameters(), lr=0.001)

for epoch in range(10):
    for images, _ in data_loader:
        crops = [images, random_crop(images), random_crop(images)]
        embeddings, prototypes = swav_model(torch.cat(crops, dim=0))
        loss = swav_loss(embeddings, prototypes)
        optimizer.zero_grad()
        loss.backward()
        optimizer.apply_gradients()
    print(f'Epoch {epoch + 1}, Loss: {loss.item()}')
```

9.3.3 Predictive Coding and Masked Modeling

Predictive coding and masked modeling are powerful techniques in self-supervised learning that focus on predicting missing parts of the input data. These methods are widely used in natural language processing and computer vision.

Masked Language Models Masked Language Models (MLMs) predict missing or masked tokens in a sequence. BERT (Bidirectional Encoder Representations from Transformers) is a prominent example that trains on masked tokens to learn deep bidirectional representations. Given a sequence of tokens $\mathbf{X} = \{x_1, x_2, \ldots, x_T\}$, randomly mask a subset of tokens $\mathbf{X}_M = \{x_{m_1}, x_{m_2}, \ldots, x_{m_k}\}$. The objective is to predict the original tokens from the masked sequence using the BERT model f_θ:

$$\mathcal{L}_{\text{MLM}} = -\sum_{i \in M} \log P(x_i | \mathbf{X}_{\setminus i}; \theta)$$

where $\mathbf{X}_{\setminus i}$ is the sequence with the i-th token masked.

Example: Implementing BERT with Hugging Face Transformers:

```
from transformers import BertTokenizer, BertForMaskedLM, AdamW
import torch

tokenizer = BertTokenizer.from_pretrained('bert-base-uncased')
model = BertForMaskedLM.from_pretrained('bert-base-uncased')
optimizer = AdamW(model.parameters(), lr=0.001)

inputs = tokenizer("The quick brown fox jumps over the lazy dog", return_tensors='pt')
inputs['labels'] = inputs.input_ids.detach().clone()
inputs['input_ids'][0, 3] = tokenizer.mask_token_id  # Mask 'brown'

outputs = model(inputs)
loss = outputs.loss
loss.backward()
optimizer.step()
print(f'Loss: {loss.item()}')
```

Masked Image Modeling Masked Image Modeling extends the idea of masked language modeling to images. It involves masking parts of an image and training the model to predict the missing regions. This approach helps the model learn contextual information and robust representations. Given an image $\mathbf{X}$, mask a subset of pixels $\mathbf{X}_M$. The objective is to predict the masked pixels using the model f_θ:

$$\mathcal{L}_{\text{MIM}} = \sum_{i \in M} \|x_i - f_\theta(\mathbf{X}_{\setminus i})\|^2$$

where $\mathbf{X}_{\setminus i}$ is the image with the i-th pixel masked.

Example: Implementing masked image modeling with PyTorch:

```
import torch
import torch.nn as nn
import torch.optim as optim
import torchvision.transforms as transforms
from torchvision.datasets import CIFAR10
from torch.utils.data import DataLoader

# Define a simple CNN model
class SimpleCNN(nn.Module):
    def __init__(self):
        super(SimpleCNN, self).__init__()
        self.conv1 = nn.Conv2d(3, 64, kernel_size=3, padding=1)
        self.conv2 = nn.Conv2d(64, 128, kernel_size=3, padding=1)
        self.fc1 = nn.Linear(128 * 8 * 8, 256)
        self.fc2 = nn.Linear(256, 3 * 32 * 32)  # Output the full image

    def forward(self, x):
        x = F.relu(self.conv1(x))
        x = F.max_pool2d(x, 2)
        x = F.relu(self.conv2(x))
        x = F.max_pool2d(x, 2)
        x = x.view(-1, 128 * 8 * 8)
        x = F.relu(self.fc1(x))
        x = self.fc2(x)
        x = x.view(-1, 3, 32, 32)
        return x

transform = transforms.Compose([transforms.ToTensor(), transforms.RandomErasing(p=0.5,
 scale=(0.02, 0.33), ratio=(0.3, 3.3))])
train_dataset = CIFAR10(root='./data', train=True, download=True, transform=transform)
train_loader = DataLoader(train_dataset, batch_size=64, shuffle=True)

model = SimpleCNN()
optimizer = optim.Adam(model.parameters(), lr=0.001)
criterion = nn.MSELoss()

# Training loop
for epoch in range(10):
    for images, _ in train_loader:
        masked_images = transform(images)
        outputs = model(masked_images)
        loss = criterion(outputs, images)
        optimizer.zero_grad()
        loss.backward()
        optimizer.step()
    print(f'Epoch {epoch + 1}, Loss: {loss.item()}')
```

9.4 Recent Advancements in Self-supervised Learning

9.4.1 *SimCLR (Simple Framework for Contrastive Learning of Visual Representations)*

SimCLR is a pioneering self-supervised learning framework that has significantly advanced the field of visual representation learning. By leveraging contrastive learning principles, SimCLR learns robust and transferable representations from unlabeled data through a combination of data augmentation, neural network encoders, and a contrastive loss function (Chen et al. 2020). This section delves into the architecture, training strategy, performance, and applications of SimCLR. The architecture of SimCLR comprises several key components, including data augmentation, an encoder network, a projection head, and a contrastive loss function. These components work synergistically to learn meaningful representations from images.

1. Data Augmentation: SimCLR heavily relies on data augmentation to create different views of the same image. Each input image x is transformed into two different augmented views, $\tilde{x}_i$ and $\tilde{x}_j$, using a series of stochastic augmentations such as random cropping, color distortions, and Gaussian blur.
2. Encoder Network: The encoder network f_θ maps the augmented views to a latent space. Common choices for the encoder include convolutional neural networks like ResNet. The output of the encoder is a high-dimensional feature vector $h = f_\theta(\tilde{x})$.
3. Projection Head: Following the encoder, a projection head g_ϕ is used to map the feature vector h to a lower-dimensional space where the contrastive loss is applied. The projection head typically consists of a few fully connected layers. The output of the projection head is $z = g_\phi(h)$.
4. Contrastive Loss Function: SimCLR employs the normalized temperature-scaled cross-entropy loss (NT-Xent loss) to maximize agreement between positive pairs and minimize agreement between negative pairs. The loss function for a positive pair (i, j) is defined as:

$$\ell_{i,j} = -\log \frac{\exp(\text{sim}(z_i, z_j)/\tau)}{\sum_{k=1}^{2N} \mathbb{1}_{[k \neq i]} \exp(\text{sim}(z_i, z_k)/\tau)}$$

Here, $\text{sim}(z_i, z_j)$ denotes the cosine similarity between z_i and z_j, and τ is a temperature parameter. The denominator sums over all positive and negative pairs in the batch, excluding i.

Training Strategy:

1. Data Augmentation: For each image x in the batch, generate two augmented views $\tilde{x}_i$ and $\tilde{x}_j$.
2. Feature Extraction: Pass the augmented views through the encoder f_θ to obtain feature vectors $h_i = f_\theta(\tilde{x}_i)$ and $h_j = f_\theta(\tilde{x}_j)$.

3. Projection: Use the projection head g_ϕ to map the feature vectors to the projection space: $z_i = g_\phi(h_i)$ and $z_j = g_\phi(h_j)$.
4. Contrastive Loss Calculation: Compute the NT-Xent loss for the batch, considering all positive and negative pairs.
5. Backpropagation and Optimization: Update the parameters θ and ϕ of the encoder and projection head using gradient descent.

Example Implementation:

```
import tensorflow as tf
from tensorflow.keras import layers, models, optimizers

# Define data augmentation pipeline
data_augmentation = tf.keras.Sequential([
    layers.RandomResizedCrop(height=32, width=32),
    layers.RandomFlip("horizontal"),
    layers.RandomBrightness(factor=0.5),
    layers.RandomContrast(factor=0.5),
])

# Define encoder (ResNet) and projection head
base_encoder = tf.keras.applications.ResNet50(include_top=False, input_shape=(32, 32, 3),
 pooling='avg')
projection_head = models.Sequential([
    layers.Dense(128, activation='relu'),
    layers.Dense(64)
])

# SimCLR model
class SimCLR(tf.keras.Model):
    def __init__(self, encoder, projection_head):
        super(SimCLR, self).__init__()
        self.encoder = encoder
        self.projection_head = projection_head

    def call(self, x):
        h = self.encoder(x)
        z = self.projection_head(h)
        return z

simclr_model = SimCLR(base_encoder, projection_head)

# NT-Xent loss function
def nt_xent_loss(z_i, z_j, temperature=0.1):
    z_i = tf.math.l2_normalize(z_i, axis=1)
    z_j = tf.math.l2_normalize(z_j, axis=1)
    similarities = tf.matmul(z_i, z_j, transpose_b=True) / temperature
    labels = tf.range(tf.shape(z_i)[0])
    return tf.reduce_mean(tf.nn.sparse_softmax_cross_entropy_with_logits(labels,
 similarities))

# Training step
optimizer = optimizers.Adam()

@tf.function
def train_step(images):
    with tf.GradientTape() as tape:
        aug_1 = data_augmentation(images)
        aug_2 = data_augmentation(images)
        z_i = simclr_model(aug_1, training=True)
        z_j = simclr_model(aug_2, training=True)
        loss = nt_xent_loss(z_i, z_j)
    gradients = tape.gradient(loss, simclr_model.trainable_variables)
```

```
    optimizer.apply_gradients(zip(gradients, simclr_model.trainable_variables))
    return loss

# Training loop
for epoch in range(10):
    for images in dataset:  # Assume 'dataset' is a preloaded dataset of images
        loss = train_step(images)
    print(f'Epoch {epoch+1}, Loss: {loss.numpy()}')
```

SimCLR has demonstrated state-of-the-art performance in several visual representation learning benchmarks. Its effectiveness is largely attributed to the powerful combination of data augmentation, contrastive loss, and the ability to leverage large batches during training. SimCLR's performance improves with larger batch sizes, more aggressive data augmentation, and longer training times.

1. Transfer Learning: The representations learned by SimCLR can be fine-tuned for various downstream tasks such as image classification, object detection, and segmentation. Fine-tuning typically involves replacing the projection head with a task-specific head and training on a labeled dataset.
2. Linear Evaluation Protocol: A common evaluation method for self-supervised models is the linear evaluation protocol. In this setup, a linear classifier is trained on top of the frozen encoder representations to measure their quality. SimCLR has consistently outperformed previous methods in this protocol.

Applications:

1. Image Classification: SimCLR can be used to pre-train models on large unlabeled datasets, which can then be fine-tuned on smaller labeled datasets for image classification tasks. This approach has shown significant improvements in accuracy compared to models trained from scratch.
2. Object Detection and Segmentation: The rich feature representations learned by SimCLR can be transferred to object detection and segmentation tasks. Pre-trained SimCLR models have been fine-tuned for tasks such as COCO object detection and segmentation, demonstrating competitive performance.
3. Medical Imaging: In medical imaging, where labeled data is often scarce and expensive to obtain, SimCLR can be used to pre-train models on large amounts of unlabeled medical images. These pre-trained models can then be fine-tuned for specific tasks like disease detection and segmentation.
4. Unsupervised Learning: Beyond supervised tasks, SimCLR's learned representations can be used for unsupervised clustering and anomaly detection. The high-quality features extracted by the encoder can facilitate better clustering and identification of outliers.

9.4.2 BYOL (Bootstrap Your Own Latent)

BYOL (Bootstrap Your Own Latent) is a self-supervised learning method that learns representations without requiring negative samples or explicit comparison between different data points (Grill et al. 2020). It leverages a dual network architecture and a momentum update mechanism to learn robust representations from unlabeled data. BYOL consists of two neural networks: an online network and a target network. Both networks share the same architecture but have different parameters.

1. Online Network: The online network f_θ consists of an encoder, a projection head, and a prediction head. Encoder f_θ: Extracts features from the input images. Projection head g_θ: Maps the features to a latent space. Prediction head q_θ: Predicts the target network's output.
2. Target Network: The target network f_ξ also consists of an encoder and a projection head but without the prediction head. Encoder f_ξ: Extracts features from the input images. Projection head g_ξ: Maps the features to a latent space.

Training Strategy:

1. Data Augmentation: Generate two augmented views $\tilde{x}_i$ and $\tilde{x}_j$ for each input image x.
2. Forward Pass: Pass $\tilde{x}_i$ through the online network to obtain the prediction $q_\theta(g_\theta(f_\theta(\tilde{x}_i)))$. Pass $\tilde{x}_j$ through the target network to obtain the target representation $g_\xi(f_\xi(\tilde{x}_j))$.
3. Loss Calculation: Minimize the mean squared error between the prediction and the target representation:

$$\mathcal{L}_{\text{BYOL}} = \|q_\theta(g_\theta(f_\theta(\tilde{x}_i))) - g_\xi(f_\xi(\tilde{x}_j))\|^2$$

4. Momentum Update: Update the target network parameters ξ as an exponential moving average of the online network parameters θ:

$$\xi \leftarrow \tau\xi + (1-\tau)\theta$$

where τ is the momentum coefficient.

Example: Implementing BYOL with PyTorch:

```
import torch
import torch.nn as nn
import torch.optim as optim

class BYOL(nn.Module):
    def __init__(self, base_encoder, projection_dim=256, hidden_dim=4096):
        super(BYOL, self).__init__()
        self.online_encoder = base_encoder
        self.online_projection = nn.Sequential(
            nn.Linear(base_encoder.output_dim, hidden_dim),
            nn.BatchNorm1d(hidden_dim),
            nn.ReLU(),
            nn.Linear(hidden_dim, projection_dim)
```

```
        )
        self.online_prediction = nn.Sequential(
            nn.Linear(projection_dim, hidden_dim),
            nn.BatchNorm1d(hidden_dim),
            nn.ReLU(),
            nn.Linear(hidden_dim, projection_dim)
        )
        self.target_encoder = base_encoder
        self.target_projection = nn.Sequential(
            nn.Linear(base_encoder.output_dim, hidden_dim),
            nn.BatchNorm1d(hidden_dim),
            nn.ReLU(),
            nn.Linear(hidden_dim, projection_dim)
        )
        for param in self.target_encoder.parameters():
            param.requires_grad = False
        for param in self.target_projection.parameters():
            param.requires_grad = False

    def forward(self, x1, x2):
        online_z1 = self.online_projection(self.online_encoder(x1))
        online_z2 = self.online_projection(self.online_encoder(x2))
        online_p1 = self.online_prediction(online_z1)
        online_p2 = self.online_prediction(online_z2)
        with torch.no_grad():
            target_z1 = self.target_projection(self.target_encoder(x1))
            target_z2 = self.target_projection(self.target_encoder(x2))
        return online_p1, online_p2, target_z1, target_z2

def byol_loss(p1, p2, z1, z2):
    loss1 = torch.nn.functional.mse_loss(p1, z2.detach())
    loss2 = torch.nn.functional.mse_loss(p2, z1.detach())
    return (loss1 + loss2) / 2

# Training loop
base_encoder = ...  # Define base encoder model
byol_model = BYOL(base_encoder)
optimizer = optim.Adam(byol_model.parameters(), lr=0.001)
momentum = 0.99

for epoch in range(10):
    for x1, x2 in dataloader:  # Assume 'dataloader' provides pairs of augmented images
        p1, p2, z1, z2 = byol_model(x1, x2)
        loss = byol_loss(p1, p2, z1, z2)
        optimizer.zero_grad()
        loss.backward()
        optimizer.step()
        # Update target network parameters
        for param_q, param_k in zip(byol_model.online_encoder.parameters(), byol_model.
target_encoder.parameters()):
            param_k.data = momentum * param_k.data + (1 - momentum) * param_q.data
        for param_q, param_k in zip(byol_model.online_projection.parameters(), byol_model.
target_projection.parameters()):
            param_k.data = momentum * param_k.data + (1 - momentum) * param_q.data
    print(f'Epoch {epoch+1}, Loss: {loss.item()}')
```

BYOL operates without negative samples, relying solely on positive pairs (augmented views of the same image) for learning representations. The momentum update mechanism stabilizes the training process and prevents the network from collapsing into trivial solutions. The dual network architecture ensures that the target representations provide a stable reference for the online network to learn from. BYOL achieves state-of-the-art performance on several benchmarks for self-supervised learning. It demonstrates that negative samples are not strictly necessary for contrastive learning, provided that the training process is stabilized by mechanisms such as momentum updates and dual networks.

1. Transfer Learning: BYOL's learned representations can be transferred to various downstream tasks, including image classification, object detection, and segmentation, often outperforming previous methods.
2. Linear Evaluation Protocol: BYOL's performance in the linear evaluation protocol is comparable to or better than contrastive learning methods that use negative samples.

Applications:

1. Image Classification: Pre-trained BYOL models can be fine-tuned on labeled datasets for image classification, achieving high accuracy and robustness.
2. Medical Imaging: BYOL's self-supervised approach can be applied to large unlabeled medical image datasets, learning representations that can be fine-tuned for specific tasks like disease detection.
3. Unsupervised Learning: BYOL's robust representations can be used for unsupervised clustering and anomaly detection in various domains.

9.4.3 MoCo (Momentum Contrast)

MoCo (Momentum Contrast) is another self-supervised learning framework that maintains a dynamic dictionary with a queue of encoded representations and employs a momentum encoder to provide consistent and stable keys for contrastive learning (He et al. 2020).

Architecture:

1. Query and Key Encoders: MoCo uses two encoders: a query encoder f_q and a key encoder f_k. The query encoder is updated with standard backpropagation, while the key encoder is updated with a momentum mechanism.
2. Momentum Update: The key encoder f_k is updated as an exponential moving average of the query encoder f_q:

$$\theta_k \leftarrow m\theta_k + (1 - m)\theta_q$$

   where θ_q and θ_k are the parameters of the query and key encoders, respectively, and m is the momentum coefficient.
3. Dynamic Dictionary: A queue stores the encoded keys from previous batches, allowing MoCo to use a large number of negative samples without requiring a large batch size.

Training Strategy:

1. Data Augmentation: Generate two augmented views $\tilde{x}_q$ and $\tilde{x}_k$ for each input image x.
2. Forward Pass: Pass $\tilde{x}_q$ through the query encoder f_q to obtain the query representation $q = f_q(\tilde{x}_q)$. Pass $\tilde{x}_k$ through the key encoder f_k to obtain the key representation $k = f_k(\tilde{x}_k)$.
3. Contrastive Loss Calculation: Compute the InfoNCE loss between the query and key representations, using the keys in the queue as negative samples:

$$\mathcal{L}_{\text{MoCo}} = -\log \frac{\exp(\text{sim}(q, k^+))/\tau}{\sum_{i=0}^{K} \exp(\text{sim}(q, k_i))/\tau}$$

   where k^+ is the positive key and k_i are the keys in the queue.
4. Momentum Update and Queue Management: Update the key encoder parameters with momentum. Enqueue the current batch of keys and dequeue the oldest keys to maintain a fixed queue size.

Example: Implementing MoCo with PyTorch:

```
import torch
import torch.nn as nn
import torch.optim as optim

class MoCo(nn.Module):
    def __init__(self, base_encoder, dim=128, K=65536, m=0.999, T=0.07):
        super(MoCo, self).__init__()
        self.K = K
        self.m = m
        self.T = T
        self.encoder_q = base_encoder
        self.encoder_k = base_encoder
        self.register_buffer("queue", torch.randn(dim, K))
        self.queue = nn.functional.normalize(self.queue, dim=0)
        self.register_buffer("queue_ptr", torch.zeros(1, dtype=torch.long))

    @torch.no_grad()
    def _momentum_update_key_encoder(self):
        for param_q, param_k in zip(self.encoder_q.parameters(), self.encoder_k.
parameters()):
            param_k.data = param_k.data * self.m + param_q.data * (1. - self.m)

    @torch.no_grad()
    def _dequeue_and_enqueue(self, keys):
        batch_size = keys.shape[0]
        ptr = int(self.queue_ptr)
        self.queue[:, ptr:ptr + batch_size] = keys.T
        ptr = (ptr + batch_size) % self.K
        self.queue_ptr[0] = ptr

    def forward(self, im_q, im_k):
        q = self.encoder_q(im_q)
        q = nn.functional.normalize(q, dim=1)
        with torch.no_grad():
            self._momentum_update_key_encoder()
            k = self.encoder_k(im_k)
            k = nn.functional.normalize(k, dim=1)
        return q, k
```

```
def moco_loss(q, k, queue, T=0.07):
    N = q.shape[0]
    l_pos = torch.einsum('nc,nc->n', [q, k]).unsqueeze(-1)
    l_neg = torch.einsum('nc,ck->nk', [q, queue.clone().detach()])
    logits = torch.cat([l_pos, l_neg], dim=1)
    logits /= T
    labels = torch.zeros(N, dtype=torch.long).cuda()
    return nn.CrossEntropyLoss()(logits, labels)

# Training loop
base_encoder = ...  # Define base encoder model
moco_model = MoCo(base_encoder)
optimizer = optim.Adam(moco_model.parameters(), lr=0.001)

for epoch in range(10):
    for im_q, im_k in dataloader:  # Assume 'dataloader' provides pairs of augmented
images
        q, k = moco_model(im_q, im_k)
        loss = moco_loss(q, k, moco_model.queue)
        optimizer.zero_grad()
        loss.backward()
        optimizer.step()
        moco_model._dequeue_and_enqueue(k)
    print(f'Epoch {epoch+1}, Loss: {loss.item()}')
```

MoCo has shown remarkable performance on several self-supervised learning benchmarks, demonstrating the effectiveness of maintaining a dynamic dictionary and using a momentum encoder.

1. Transfer Learning: MoCo's learned representations can be fine-tuned for various downstream tasks, achieving competitive performance on image classification and object detection.
2. Linear Evaluation Protocol: MoCo performs well in the linear evaluation protocol, indicating the quality of its learned representations.

Applications:

1. Image Classification: Pre-trained MoCo models can be fine-tuned on labeled datasets, resulting in high accuracy for image classification tasks.
2. Object Detection: MoCo's robust feature representations can be transferred to object detection tasks, improving performance and generalization.
3. Medical Imaging: Similar to other self-supervised methods, MoCo can be applied to large unlabeled medical image datasets, aiding in disease detection and segmentation tasks.

9.5 Applications in Various Domains

9.5.1 Computer Vision

In computer vision, self-supervised and unsupervised learning techniques have significantly enhanced image classification and object detection tasks. These methods enable models to learn robust features from large amounts of unlabeled data, which can then be fine-tuned on smaller labeled datasets for specific tasks.

Image Classification Self-supervised learning methods like SimCLR and BYOL have demonstrated that models can learn meaningful representations by predicting transformations applied to images. These representations can be used to classify images into predefined categories with high accuracy. Let x_i be an input image, and $\tilde{x}_i$ be its augmented version. The self-supervised loss function, such as the contrastive loss used in SimCLR, can be formulated as:

$$\mathcal{L}_{\text{contrastive}} = -\log \frac{\exp(\text{sim}(z_i, z_j)/\tau)}{\sum_{k=1}^{2N} \mathbb{1}_{[k \neq i]} \exp(\text{sim}(z_i, z_k)/\tau)}$$

where z_i and z_j are the embeddings of x_i and $\tilde{x}_i$, respectively, and τ is the temperature parameter. This loss encourages similar images to have similar embeddings.

Example: Fine-tuning a pre-trained SimCLR model for image classification:

```
import tensorflow as tf
from tensorflow.keras import layers, models

# Load pre-trained SimCLR model
simclr_model = ...  # Pre-trained SimCLR model
base_model = tf.keras.Model(inputs=simclr_model.input, outputs=simclr_model.layers[-3].
output)  # Extract encoder part

# Add classification head
classification_model = models.Sequential([
    base_model,
    layers.GlobalAveragePooling2D(),
    layers.Dense(256, activation='relu'),
    layers.Dense(num_classes, activation='softmax')
])

classification_model.compile(optimizer='adam', loss='sparse_categorical_crossentropy',
metrics=['accuracy'])
classification_model.fit(train_data, train_labels, epochs=10, validation_data=
(val_data, val_labels))
```

Object Detection For object detection, models need to localize and classify multiple objects within an image. Self-supervised learning methods, such as those based on contrastive learning, can pre-train models to learn useful features that improve detection accuracy. The detection task involves predicting bounding boxes $B = \{b_1, b_2, \ldots, b_n\}$ and class labels $C = \{c_1, c_2, \ldots, c_n\}$ for each image x. The loss function typically combines classification and localization losses:

$$\mathcal{L}_{\text{detection}} = \mathcal{L}_{\text{cls}} + \lambda \mathcal{L}_{\text{loc}}$$

where $\mathcal{L}_{\text{cls}}$ is the classification loss (e.g., cross-entropy), and $\mathcal{L}_{\text{loc}}$ is the localization loss (e.g., smooth L1 loss).

Example: Fine-tuning a pre-trained backbone for object detection using an architecture like Faster R-CNN:

```
import torchvision
from torchvision.models.detection import FasterRCNN
```

```
from torchvision.models.detection.rpn import AnchorGenerator

# Load pre-trained backbone
backbone = torchvision.models.resnet50(pretrained=True)
backbone = torch.nn.Sequential(*(list(backbone.children())[:-2]))

# Define RPN and ROI head
rpn_anchor_generator = AnchorGenerator(sizes=((32, 64, 128, 256, 512),), aspect_ratios=
((0.5, 1.0, 2.0),))
roi_pooler = torchvision.ops.MultiScaleRoIAlign(featmap_names=['0'], output_size=7,
sampling_ratio=2)

# Create Faster R-CNN model
model = FasterRCNN(backbone, num_classes=num_classes, rpn_anchor_generator=rpn_anchor_
generator, box_roi_pool=roi_pooler)
model.train()

# Training loop
optimizer = torch.optim.Adam(model.parameters(), lr=0.001)
for epoch in range(num_epochs):
    for images, targets in dataloader:
        loss_dict = model(images, targets)
        losses = sum(loss for loss in loss_dict.values())
        optimizer.zero_grad()
        losses.backward()
        optimizer.step()
    print(f'Epoch {epoch+1}, Loss: {losses.item()}')
```

Self-supervised and unsupervised learning techniques have also been applied to image segmentation and reconstruction tasks, where the goal is to partition an image into meaningful regions or to reconstruct the original image from corrupted inputs.

Image Segmentation In image segmentation, the model assigns a class label to each pixel in the image. Self-supervised learning methods, such as those using contrastive learning or clustering, can pre-train segmentation models to learn spatially coherent features. The segmentation task can be formulated as predicting a mask M for each input image x, where M has the same spatial dimensions as x but with a class label for each pixel. The loss function often used is the cross-entropy loss applied pixel-wise:

$$\mathcal{L}_{\text{segmentation}} = -\sum_{i=1}^{H}\sum_{j=1}^{W}\sum_{c=1}^{C} y_{ijc}\log(p_{ijc})$$

where H and W are the height and width of the image, C is the number of classes, y_{ijc} is the ground truth label, and p_{ijc} is the predicted probability for class c at pixel (i, j).

Example: Fine-tuning a pre-trained model for semantic segmentation using U-Net:

```
import tensorflow as tf
from tensorflow.keras import layers, models

# Load pre-trained encoder (e.g., ResNet)
base_model = tf.keras.applications.ResNet50(include_top=False, input_shape=(128, 128,
3))

# Define U-Net decoder
inputs = tf.keras.Input(shape=(128, 128, 3))
```

```
encoder_outputs = base_model(inputs)
x = layers.Conv2D(256, (3, 3), activation='relu', padding='same')(encoder_outputs)
x = layers.UpSampling2D((2, 2))(x)
x = layers.Conv2D(128, (3, 3), activation='relu', padding='same')(x)
x = layers.UpSampling2D((2, 2))(x)
x = layers.Conv2D(64, (3, 3), activation='relu', padding='same')(x)
x = layers.UpSampling2D((2, 2))(x)
outputs = layers.Conv2D(num_classes, (1, 1), activation='softmax')(x)

segmentation_model = models.Model(inputs, outputs)
segmentation_model.compile(optimizer='adam', loss='sparse_categorical_crossentropy',
metrics=['accuracy'])
segmentation_model.fit(train_data, train_labels, epochs=10, validation_data=(val_data,
val_labels))
```

Image Reconstruction This task involves recovering the original image from its degraded version. Self-supervised learning methods, such as autoencoders or masked image modeling, are effective for this purpose. The reconstruction task can be formulated as minimizing the difference between the original image x and its reconstructed version $\hat{x}$. The loss function often used is the mean squared error (MSE):

$$\mathcal{L}_{\text{reconstruction}} = \|x - \hat{x}\|^2$$

Example: Training a denoising autoencoder for image reconstruction:

```
import tensorflow as tf
from tensorflow.keras import layers, models

# Define autoencoder model
inputs = tf.keras.Input(shape=(128, 128, 3))
x = layers.Conv2D(64, (3, 3), activation='relu', padding='same')(inputs)
x = layers.MaxPooling2D((2, 2), padding='same')(x)
x = layers.Conv2D(128, (3, 3), activation='relu', padding='same')(x)
encoded = layers.MaxPooling2D((2, 2), padding='same')(x)

x = layers.Conv2D(128, (3, 3), activation='relu', padding='same')(encoded)
x = layers.UpSampling2D((2, 2))(x)
x = layers.Conv2D(64, (3, 3), activation='relu', padding='same')(x)
x = layers.UpSampling2D((2, 2))(x)
decoded = layers.Conv2D(3, (3, 3), activation='sigmoid', padding='same')(x)

autoencoder = models.Model(inputs, decoded)
autoencoder.compile(optimizer='adam', loss='mse')
autoencoder.fit(noisy_train_data, train_data, epochs=10, validation_data=(noisy_val_
data, val_data))
```

9.5.2 Natural Language Processing

In natural language processing (NLP), self-supervised learning methods have revolutionized the way models learn text representations. Techniques such as masked language modeling (MLM) have enabled models to learn contextual embeddings from large corpora of unlabeled text data.

Masked Language Modeling (MLM) The objective of MLM is to predict masked words within a sentence, allowing the model to learn bidirectional context. The loss function for MLM is the cross-entropy loss over the masked tokens:

$$\mathcal{L}_{\text{MLM}} = -\sum_{i=1}^{N} y_i \log(p_i)$$

where N is the number of masked tokens, y_i is the true label, and p_i is the predicted probability.

Example: Pre-training BERT with MLM:

```
from transformers import BertForMaskedLM, BertTokenizer, Trainer, TrainingArguments

model = BertForMaskedLM.from_pretrained('bert-base-uncased')
tokenizer = BertTokenizer.from_pretrained('bert-base-uncased')

# Prepare dataset for MLM
def tokenize_function(examples):
    return tokenizer(examples['text'], truncation=True, padding='max_length',
max_length=128)

dataset = dataset.map(tokenize_function, batched=True)
dataset = dataset.map(mask_tokens, batched=True)  # Function to mask tokens for MLM

training_args = TrainingArguments(output_dir='./results', num_train_epochs=3,
per_device_train_batch_size=32)
trainer = Trainer(model=model, args=training_args, train_dataset=dataset)
trainer.train()
```

Self-supervised learning methods have also been applied to text generation and summarization tasks, where models generate coherent text or summarize long documents into concise summaries.

Text Generation Autoregressive models like GPT-3 generate text by predicting the next word in a sequence, trained using the language modeling objective:

$$\mathcal{L}_{\text{LM}} = -\sum_{i=1}^{N} y_i \log(p(y_i | y_{<i}))$$

where N is the sequence length, y_i is the true label, and $p(y_i | y_{<i})$ is the predicted probability given the previous tokens.

Example: Fine-tuning GPT-3 for text generation:

```
from transformers import GPT2LMHeadModel, GPT2Tokenizer, Trainer, TrainingArguments

model = GPT2LMHeadModel.from_pretrained('gpt2')
tokenizer = GPT2Tokenizer.from_pretrained('gpt2')

# Prepare dataset for language modeling
def tokenize_function(examples):
    return tokenizer(examples['text'], truncation=True, padding='max_length',
max_length=128)

dataset = dataset.map(tokenize_function, batched=True)

training_args = TrainingArguments(output_dir='./results', num_train_epochs=3,
```

```
per_device_train_batch_size=32)
  trainer = Trainer(model=model, args=training_args, train_dataset=dataset)
  trainer.train()
```

Text Summarization Summarization models can be trained using sequence-to-sequence objectives. The loss function for summarization typically combines the cross-entropy loss over the target sequence:

$$\mathcal{L}_{\text{summarization}} = -\sum_{i=1}^{N} y_i \log(p(y_i|y_{<i}, x))$$

where N is the length of the summary, y_i is the target token, $p(y_i|y_{<i}, x)$ is the predicted probability given the input text x and previous tokens.

Example: Fine-tuning BART for summarization:

```
from transformers import BartForConditionalGeneration, BartTokenizer, Trainer,
TrainingArguments

model = BartForConditionalGeneration.from_pretrained('facebook/bart-large-cnn')
tokenizer = BartTokenizer.from_pretrained('facebook/bart-large-cnn')

# Prepare dataset for summarization
def tokenize_function(examples):
    inputs = tokenizer(examples['text'], max_length=512, truncation=True, padding='max_
length')
    targets = tokenizer(examples['summary'], max_length=128, truncation=True, padding=
'max_length')
    return {'input_ids': inputs.input_ids, 'attention_mask': inputs.attention_mask,
'labels': targets.input_ids}

dataset = dataset.map(tokenize_function, batched=True)

training_args = TrainingArguments(output_dir='./results', num_train_epochs=3, per_
device_train_batch_size=8)
trainer = Trainer(model=model, args=training_args, train_dataset=dataset)
trainer.train()
```

9.5.3 *Anomaly Detection*

Detecting Anomalies in Images and Videos Anomaly detection in images and videos involves identifying patterns that deviate from the norm. Self-supervised learning techniques can significantly enhance this task by pre-training models to understand normal patterns and detect deviations during inference. Anomaly detection can be approached by training a model to minimize reconstruction error or maximize consistency between augmented views. For reconstruction-based methods, the mean-squared error (MSE) is commonly used:

$$\mathcal{L}_{\text{reconstruction}} = \|x - \hat{x}\|^2$$

Here, x is the input image, and $\hat{x}$ is the reconstructed image.

Example: Training a convolutional autoencoder for anomaly detection:

```
import tensorflow as tf
from tensorflow.keras import layers, models

# Define convolutional autoencoder
inputs = tf.keras.Input(shape=(128, 128, 3))
x = layers.Conv2D(64, (3, 3), activation='relu', padding='same')(inputs)
x = layers.MaxPooling2D((2, 2), padding='same')(x)
x = layers.Conv2D(128, (3, 3), activation='relu', padding='same')(x)
encoded = layers.MaxPooling2D((2, 2), padding='same')(x)

x = layers.Conv2D(128, (3, 3), activation='relu', padding='same')(encoded)
x = layers.UpSampling2D((2, 2))(x)
x = layers.Conv2D(64, (3, 3), activation='relu', padding='same')(x)
x = layers.UpSampling2D((2, 2))(x)
decoded = layers.Conv2D(3, (3, 3), activation='sigmoid', padding='same')(x)

autoencoder = models.Model(inputs, decoded)
autoencoder.compile(optimizer='adam', loss='mse')
autoencoder.fit(normal_data, normal_data, epochs=10, validation_data=(val_data, val_data))

# Detect anomalies
reconstructions = autoencoder.predict(test_data)
reconstruction_errors = np.mean((test_data - reconstructions) ** 2, axis=(1, 2, 3))
anomalies = reconstruction_errors > threshold
```

This approach ensures that the model learns to reconstruct normal images accurately, while anomalies will have higher reconstruction errors.

Detecting Anomalies in Time Series Data Anomaly detection in time series data involves identifying unusual patterns in sequential data. Self-supervised learning methods can pre-train models to recognize normal temporal patterns, which aids in detecting anomalies. For time series data, an LSTM autoencoder can be used. The goal is to minimize the forecasting or reconstruction error. The mean squared error (MSE) is a typical choice for the loss function:

$$\mathcal{L}_{\text{forecasting}} = \|y_t - \hat{y}_t\|^2$$

where y_t is the true value at time t and $\hat{y}_t$ is the predicted value.

Example: Training an LSTM autoencoder for anomaly detection in time series:

```
import tensorflow as tf
from tensorflow.keras import layers, models

# Define LSTM autoencoder
timesteps, features = 100, 1  # Example shape
inputs = tf.keras.Input(shape=(timesteps, features))
encoded = layers.LSTM(128, activation='relu', return_sequences=True)(inputs)
encoded = layers.LSTM(64, activation='relu')(encoded)

decoded = layers.RepeatVector(timesteps)(encoded)
decoded = layers.LSTM(64, activation='relu', return_sequences=True)(decoded)
decoded = layers.LSTM(128, activation='relu', return_sequences=True)(decoded)
outputs = layers.TimeDistributed(layers.Dense(features))(decoded)

autoencoder = models.Model(inputs, outputs)
autoencoder.compile(optimizer='adam', loss='mse')
autoencoder.fit(normal_data, normal_data, epochs=10, validation_data=(val_data, val_data))

# Detect anomalies
```

```
reconstructions = autoencoder.predict(test_data)
reconstruction_errors = np.mean((test_data - reconstructions) ** 2, axis=1)
anomalies = reconstruction_errors > threshold
```

9.6 Implementing Self-supervised and Unsupervised Learning with TensorFlow and PyTorch

9.6.1 Self-supervised Learning with TensorFlow

Self-supervised learning can be implemented in TensorFlow using contrastive learning frameworks like SimCLR or BYOL. Here, we focus on implementing SimCLR, which uses contrastive loss to learn representations. The contrastive loss function used in SimCLR can be written as:

$$\mathcal{L}_{\text{contrastive}} = -\log \frac{\exp(\text{sim}(z_i, z_j)/\tau)}{\sum_{k=1}^{2N} \mathbb{1}_{[k \neq i]} \exp(\text{sim}(z_i, z_k)/\tau)}$$

where z_i and z_j are the embeddings of the input x_i and its augmented view $\tilde{x}_i$, and τ is the temperature parameter.

Example: Implementing SimCLR with TensorFlow:

```
import tensorflow as tf
from tensorflow.keras import layers, models

# Data augmentation
def data_augmentation(x):
    x = tf.image.random_flip_left_right(x)
    x = tf.image.random_crop(x, size=[128, 128, 3])
    return x

# SimCLR model
base_model = tf.keras.applications.ResNet50(include_top=False, input_shape=(128, 128, 3),
 pooling='avg')
inputs = tf.keras.Input(shape=(128, 128, 3))
augmented = layers.Lambda(data_augmentation)(inputs)
features = base_model(augmented, training=True)
projection_head = layers.Dense(128)(features)
projection_head = layers.BatchNormalization()(projection_head)
projection_head = layers.ReLU()(projection_head)
outputs = layers.Dense(128)(projection_head)

simclr_model = models.Model(inputs, outputs)
simclr_model.compile(optimizer='adam', loss='contrastive_loss')
simclr_model.fit(train_data, epochs=10, validation_data=val_data)

# Contrastive loss function
def contrastive_loss(y_true, y_pred, temperature=0.07):
    y_true = tf.cast(y_true, tf.int32)
    logits = tf.matmul(y_pred, y_pred, transpose_b=True) / temperature
    labels = tf.eye(tf.shape(logits)[0], dtype=tf.int32)
    return tf.losses.softmax_cross_entropy(labels, logits)

# Training loop
for epoch in range(num_epochs):
```

```
    for images in dataloader:
        with tf.GradientTape() as tape:
            augmented_images = data_augmentation(images)
            representations = simclr_model(augmented_images)
            loss = contrastive_loss(representations)
        gradients = tape.gradient(loss, simclr_model.trainable_variables)
        optimizer.apply_gradients(zip(gradients, simclr_model.trainable_variables))
    print(f'Epoch {epoch+1}, Loss: {loss.numpy()}')
```

9.6.2 Self-supervised Learning with PyTorch

Self-supervised learning in PyTorch can be effectively implemented using contrastive learning frameworks like SimCLR or BYOL. Here, we demonstrate the implementation of SimCLR, which employs a contrastive loss to learn useful representations from unlabeled data. The contrastive loss used in SimCLR aims to maximize the agreement between different augmented views of the same image while minimizing the agreement between views of different images. The loss function can be formulated as follows:

$$\mathcal{L}_{\text{contrastive}} = -\log \frac{\exp(\text{sim}(z_i, z_j)/\tau)}{\sum_{k=1}^{2N} \mathbb{1}_{[k \neq i]} \exp(\text{sim}(z_i, z_k)/\tau)}$$

Here, z_i and z_j are the embeddings of the original image x_i and its augmented version $\tilde{x}_i$, respectively. τ is a temperature parameter, and $\text{sim}(\cdot, \cdot)$ denotes the cosine similarity between two vectors.

Implementation:

```
import torch
import torch.nn as nn
import torch.optim as optim
from torchvision import datasets, transforms, models

# Data augmentation
train_transform = transforms.Compose([
    transforms.RandomResizedCrop(128),
    transforms.RandomHorizontalFlip(),
    transforms.ToTensor()
])

train_dataset = datasets.CIFAR10(root='./data', train=True, download=True,
 transform=train_transform)
train_loader = torch.utils.data.DataLoader(train_dataset, batch_size=256, shuffle=True,
 num_workers=4)

# SimCLR model
class SimCLR(nn.Module):
    def __init__(self, base_model, projection_dim):
        super(SimCLR, self).__init__()
        self.encoder = base_model
        self.encoder.fc = nn.Identity()  # Remove original FC layer
        self.projector = nn.Sequential(
            nn.Linear(self.encoder.fc.in_features, 2048),
            nn.ReLU(),
            nn.Linear(2048, projection_dim)
        )
```

```
    def forward(self, x):
        h = self.encoder(x)
        z = self.projector(h)
        return z

base_model = models.resnet50(pretrained=True)
model = SimCLR(base_model, projection_dim=128).cuda()

# Contrastive loss function
def contrastive_loss(out_1, out_2, temperature=0.5):
    batch_size = out_1.shape[0]
    out = torch.cat([out_1, out_2], dim=0)
    sim_matrix = nn.functional.cosine_similarity(out.unsqueeze(1), out.unsqueeze(0),
 dim=2)
    sim_matrix = sim_matrix / temperature
    sim_matrix.fill_diagonal_(-float('inf'))
    labels = torch.cat([torch.arange(batch_size) for _ in range(2)], dim=0).cuda()
    loss = nn.CrossEntropyLoss()(sim_matrix, labels)
    return loss

# Optimizer
optimizer = optim.Adam(model.parameters(), lr=0.001)

# Training loop
model.train()
for epoch in range(10):
    total_loss = 0
    for (images, _) in train_loader:
        images = torch.cat([images, images], dim=0).cuda()  # Duplicate images for
 augmentation
        augmented_images = torch.cat([train_transform(img.unsqueeze(0)).cuda() for img in
 images], dim=0)

        optimizer.zero_grad()
        representations = model(augmented_images)
        out_1, out_2 = torch.split(representations, len(representations) // 2, dim=0)
        loss = contrastive_loss(out_1, out_2)
        loss.backward()
        optimizer.step()

        total_loss += loss.item()

    print(f'Epoch [{epoch+1}/10], Loss: {total_loss/len(train_loader):.4f}')
```

In this implementation: We use data augmentation to create two different views of each image. A ResNet-50 model is used as the encoder, with a projection head added for the contrastive learning task. The contrastive loss is computed to maximize the agreement between augmented views of the same image while minimizing the agreement between views of different images. The model is trained using the Adam optimizer.

Training self-supervised models involves careful handling of the training data, transformations, and ensuring the model learns robust representations. Evaluation often requires transferring the learned representations to downstream tasks, such as image classification, to assess their quality.

Training Procedure:

1. Data Augmentation: Apply various augmentations to generate different views of the input images.

2. Forward Pass: Pass the augmented images through the encoder and projection head to obtain embeddings.
3. Compute Loss: Calculate the contrastive loss between embeddings of different views.
4. Backpropagation: Update the model weights using backpropagation.

After training the self-supervised model, we typically evaluate it by transferring the learned representations to a downstream task, such as image classification. This involves freezing the encoder and training a classifier on top of the learned representations.

Example: Evaluating the learned representations on an image classification task:

```
# Define a simple classifier
class Classifier(nn.Module):
    def __init__(self, encoder, num_classes):
        super(Classifier, self).__init__()
        self.encoder = encoder
        self.encoder.fc = nn.Identity()  # Remove projection head
        self.classifier = nn.Linear(2048, num_classes)

    def forward(self, x):
        with torch.no_grad():
            features = self.encoder(x)
        out = self.classifier(features)
        return out

classifier_model = Classifier(model.encoder, num_classes=10).cuda()
criterion = nn.CrossEntropyLoss()
optimizer = optim.Adam(classifier_model.parameters(), lr=0.001)

# Training loop for classifier
classifier_model.train()
for epoch in range(10):
    total_loss = 0
    for (images, labels) in train_loader:
        images, labels = images.cuda(), labels.cuda()
        optimizer.zero_grad()
        outputs = classifier_model(images)
        loss = criterion(outputs, labels)
        loss.backward()
        optimizer.step()
        total_loss += loss.item()
    print(f'Epoch [{epoch+1}/10], Loss: {total_loss/len(train_loader):.4f}')
```

In this evaluation step: We define a simple classifier that uses the frozen encoder from the self-supervised model. The classifier is trained on the downstream task (e.g., CIFAR-10 image classification). The training loop updates only the classifier weights, leveraging the representations learned during self-supervised training.

9.7 Case Studies

9.7.1 *Image Representation Learning Case Study*

In this case study, we explore the application of self-supervised learning to image representation learning. We will implement and analyze the performance of SimCLR, a popular self-supervised learning framework, on the CIFAR-10 dataset. The objective is to learn robust image representations that can be transferred to downstream tasks such as image classification. We use the CIFAR-10 dataset, which consists of 60,000 32×32 color images in 10 classes, with 6,000 images per class. The contrastive loss function for SimCLR is formulated as:

$$\mathcal{L}_{\text{contrastive}} = -\log \frac{\exp(\text{sim}(z_i, z_j)/\tau)}{\sum_{k=1}^{2N} \mathbb{1}_{[k \neq i]} \exp(\text{sim}(z_i, z_k)/\tau)}$$

where z_i and z_j are the embeddings of the original image and its augmented view, respectively, and τ is the temperature parameter.

Implementation:

```
import torch
import torch.nn as nn
import torch.optim as optim
from torchvision import datasets, transforms, models

# Data augmentation for self-supervised learning
train_transform = transforms.Compose([
    transforms.RandomResizedCrop(32),
    transforms.RandomHorizontalFlip(),
    transforms.ToTensor()
])

train_dataset = datasets.CIFAR10(root='./data', train=True, download=True,
 transform=train_transform)
train_loader = torch.utils.data.DataLoader(train_dataset, batch_size=256, shuffle=True,
 num_workers=4)

# SimCLR model definition
class SimCLR(nn.Module):
    def __init__(self, base_model, projection_dim):
        super(SimCLR, self).__init__()
        self.encoder = base_model
        self.encoder.fc = nn.Identity()
        self.projector = nn.Sequential(
            nn.Linear(self.encoder.fc.in_features, 2048),
            nn.ReLU(),
            nn.Linear(2048, projection_dim)
        )

    def forward(self, x):
        h = self.encoder(x)
        z = self.projector(h)
        return z

base_model = models.resnet50(pretrained=False)
model = SimCLR(base_model, projection_dim=128).cuda()

# Contrastive loss function
```

```
def contrastive_loss(out_1, out_2, temperature=0.5):
    batch_size = out_1.shape[0]
    out = torch.cat([out_1, out_2], dim=0)
    sim_matrix = nn.functional.cosine_similarity(out.unsqueeze(1), out.unsqueeze(0), dim=2)
    sim_matrix = sim_matrix / temperature
    sim_matrix.fill_diagonal_(-float('inf'))
    labels = torch.cat([torch.arange(batch_size) for _ in range(2)], dim=0).cuda()
    loss = nn.CrossEntropyLoss()(sim_matrix, labels)
    return loss

# Optimizer
optimizer = optim.Adam(model.parameters(), lr=0.001)

# Training loop
model.train()
for epoch in range(10):
    total_loss = 0
    for (images, _) in train_loader:
        images = images.cuda()
        optimizer.zero_grad()
        representations = model(images)
        out_1, out_2 = torch.split(representations, len(representations) // 2, dim=0)
        loss = contrastive_loss(out_1, out_2)
        loss.backward()
        optimizer.step()
        total_loss += loss.item()
    print(f'Epoch [{epoch+1}/10], Loss: {total_loss/len(train_loader):.4f}')
```

To evaluate the learned representations, we train a linear classifier on top of the frozen encoder.

```
class Classifier(nn.Module):
    def __init__(self, encoder, num_classes):
        super(Classifier, self).__init__()
        self.encoder = encoder
        self.encoder.fc = nn.Identity()
        self.classifier = nn.Linear(2048, num_classes)

    def forward(self, x):
        with torch.no_grad():
            features = self.encoder(x)
        out = self.classifier(features)
        return out

classifier_model = Classifier(model.encoder, num_classes=10).cuda()
criterion = nn.CrossEntropyLoss()
optimizer = optim.Adam(classifier_model.parameters(), lr=0.001)

# Training loop for classifier
classifier_model.train()
for epoch in range(10):
    total_loss = 0
    for (images, labels) in train_loader:
        images, labels = images.cuda(), labels.cuda()
        optimizer.zero_grad()
        outputs = classifier_model(images)
        loss = criterion(outputs, labels)
        loss.backward()
        optimizer.step()
        total_loss += loss.item()
    print(f'Epoch [{epoch+1}/10], Loss: {total_loss/len(train_loader):.4f}')
```

9.7.2 Text Representation Learning Case Study

In this case study, we apply self-supervised learning to text representation using masked language modeling (MLM) with the BERT architecture. We use a large corpus of text, such as Wikipedia or a collection of news articles, to train the BERT model. The masked language modeling loss function is:

$$\mathcal{L}_{\text{MLM}} = -\sum_{i=1}^{N} y_i \log(p(y_i))$$

where y_i is the true label for the masked token and $p(y_i)$ is the predicted probability.

Implementation:

```
from transformers import BertForMaskedLM, BertTokenizer, Trainer, TrainingArguments

model = BertForMaskedLM.from_pretrained('bert-base-uncased')
tokenizer = BertTokenizer.from_pretrained('bert-base-uncased')

# Prepare dataset for MLM
def tokenize_function(examples):
    return tokenizer(examples['text'], truncation=True, padding='max_length',
 max_length=128)

dataset = dataset.map(tokenize_function, batched=True)
dataset = dataset.map(mask_tokens, batched=True)  # Function to mask tokens for MLM

training_args = TrainingArguments(output_dir='./results', num_train_epochs=3,
per_device_train_batch_size=32)
trainer = Trainer(model=model, args=training_args, train_dataset=dataset)
trainer.train()
```

Evaluate the learned text representations by fine-tuning the BERT model on a downstream task, such as sentiment analysis.

```
from transformers import BertForSequenceClassification, Trainer, TrainingArguments

model = BertForSequenceClassification.from_pretrained('bert-base-uncased', num_labels=2)
training_args = TrainingArguments(output_dir='./results', num_train_epochs=3,
 per_device_train_batch_size=32)
trainer = Trainer(model=model, args=training_args, train_dataset=sentiment_dataset)
trainer.train()
```

9.7.3 Anomaly Detection Case Study

This case study focuses on using self-supervised learning for anomaly detection in time series data. We will train an LSTM autoencoder to identify anomalies in a dataset of sensor readings. We use a dataset of time series data, such as sensor readings from an industrial machine, to train the model. The reconstruction error is used to detect anomalies:

$$\mathcal{L}_{\text{anomaly}} = \|x - \hat{x}\|^2$$

Implementation:

```
import torch
import torch.nn as nn
import torch.optim as optim

class LSTMAutoencoder(nn.Module):
    def __init__(self, input_dim, hidden_dim, num_layers):
        super(LSTMAutoencoder, self).__init__()
        self.encoder = nn.LSTM(input_dim, hidden_dim, num_layers, batch_first=True)
        self.decoder = nn.LSTM(hidden_dim, input_dim, num_layers, batch_first=True)

    def forward(self, x):
        _, (hidden, _) = self.encoder(x)
        decoded, _ = self.decoder(hidden.repeat(x.size(1), 1, 1))
        return decoded

model = LSTMAutoencoder(input_dim=1, hidden_dim=128, num_layers=2).cuda()
criterion = nn.MSELoss()
optimizer = optim.Adam(model.parameters(), lr=0.001)

# Training loop
for epoch in range(10):
    total_loss = 0
    for batch in train_loader:
        inputs = batch['data'].cuda()
        outputs = model(inputs)
        loss = criterion(outputs, inputs)
        optimizer.zero_grad()
        loss.backward()
        optimizer.step()
        total_loss += loss.item()
    print(f'Epoch [{epoch+1}/10], Loss: {total_loss/len(train_loader):.4f}')

# Detect anomalies
model.eval()
with torch.no_grad():
    for batch in test_loader:
        inputs = batch['data'].cuda()
        outputs = model(inputs)
        reconstruction_error = torch.mean((inputs - outputs) ** 2, dim=1)
        anomalies = reconstruction_error > threshold
```

9.8 Advanced Topics and Research Directions

9.8.1 Scalability and Efficiency

As self-supervised learning (SSL) continues to evolve, scalability and efficiency are becoming paramount. Efficient training techniques and scalable architectures are essential to handle large datasets and complex models without compromising performance. Training self-supervised models on massive datasets can be computationally expensive and time-consuming. To address this, several efficient training techniques have been developed:

1. Mixed Precision Training: Mixed precision training uses both 16-bit and 32-bit floating point operations to reduce memory usage and increase computational

speed without sacrificing model accuracy. This technique leverages the capabilities of modern GPUs to perform faster computations while maintaining the precision required for training deep neural networks.

$$\text{loss} = \sum_{i=1}^{N} \frac{1}{N} \left(y_i - \hat{y}_i\right)^2$$

The above formula can be computed using mixed precision, where y_i and $\hat{y}_i$ are represented in 16-bit floating point, and the final sum is accumulated in 32-bit precision.

```
import torch
from torch.cuda.amp import autocast, GradScaler

scaler = GradScaler()

for epoch in range(num_epochs):
    for inputs, _ in dataloader:
        optimizer.zero_grad()
        with autocast():
            outputs = model(inputs)
            loss = criterion(outputs, inputs)
        scaler.scale(loss).backward()
        scaler.step(optimizer)
        scaler.update()
```

2. Distributed Training: Distributed training involves splitting the training data across multiple GPUs or machines, allowing parallel processing. This approach significantly reduces training time for large-scale SSL models.

$$\text{loss} = \sum_{i=1}^{N} \frac{1}{N} \left(y_i - \hat{y}_i\right)^2$$

The above loss can be computed in a distributed manner where each GPU computes the loss for a subset of data, and the results are aggregated.

```
import torch.distributed as dist

dist.init_process_group(backend='nccl')

model = torch.nn.parallel.DistributedDataParallel(model, device_ids=[local_rank])

for epoch in range(num_epochs):
    for inputs, _ in dataloader:
        optimizer.zero_grad()
        outputs = model(inputs)
        loss = criterion(outputs, inputs)
        loss.backward()
        optimizer.step()
```

9.8.2 Scalable Architectures

Scalable architectures are designed to handle increasing amounts of data and model complexity without significant performance degradation. Some approaches to achieving scalability include:

1. Hierarchical Architectures: Hierarchical models, such as hierarchical attention networks, process data at multiple levels of granularity. This approach allows the model to scale effectively by breaking down complex tasks into simpler sub-tasks.

$$h_i = \text{Attention}(h_{i-1}, h_{i-2}, \ldots, h_0)$$

In hierarchical attention networks, each level of the hierarchy processes a different level of abstraction, allowing the model to capture fine-grained details and high-level features.

```
class HierarchicalAttention(nn.Module):
    def __init__(self):
        super(HierarchicalAttention, self).__init__()
        self.word_attention = AttentionLayer()
        self.sentence_attention = AttentionLayer()

    def forward(self, x):
        word_embeddings = self.word_attention(x)
        sentence_embeddings = self.sentence_attention(word_embeddings)
        return sentence_embeddings
```

2. Sparse Architectures: Sparse architectures, such as sparse attention mechanisms, reduce the computational complexity by focusing on the most relevant parts of the data. This approach is particularly useful in models like transformers, where full attention can be prohibitively expensive.

$$\text{Attention}(Q, K, V) = \text{softmax}\left(\frac{QK^T}{\sqrt{d_k}}\right)V$$

Sparse attention mechanisms selectively compute the attention weights for a subset of the input sequence, reducing the number of operations required.

```
class SparseAttention(nn.Module):
    def __init__(self, d_model):
        super(SparseAttention, self).__init__()
        self.query = nn.Linear(d_model, d_model)
        self.key = nn.Linear(d_model, d_model)
        self.value = nn.Linear(d_model, d_model)

    def forward(self, x):
        Q = self.query(x)
        K = self.key(x)
        V = self.value(x)
        attention_scores = torch.matmul(Q, K.transpose(-2, -1)) / math.sqrt
(Q.size(-1))
        sparse_mask = (attention_scores > threshold).float()
        attention_scores = attention_scores * sparse_mask
        attention_weights = nn.Softmax(dim=-1)(attention_scores)
```

```
    output = torch.matmul(attention_weights, V)
    return output
```

Interpretable Self-supervised Learning Interpretability in self-supervised learning is crucial for understanding the representations learned by the model and ensuring transparency in decision-making processes. Two key aspects of interpretable self-supervised learning are understanding learned representations and visualization techniques.

1. Feature Analysis: Analyzing the features learned by the model helps in understanding what the model has captured during training. This can be done by visualizing the feature space using techniques like t-SNE or UMAP.

$$\text{t-SNE:} \quad y_i = f(x_i)$$

where y_i represents the low-dimensional embedding of the high-dimensional feature vector x_i.

```
from sklearn.manifold import TSNE
import matplotlib.pyplot as plt

features = model.encoder(images).detach().cpu().numpy()
tsne = TSNE(n_components=2)
tsne_results = tsne.fit_transform(features)

plt.scatter(tsne_results[:, 0], tsne_results[:, 1], c=labels)
plt.show()
```

2. Attention Maps: For models that use attention mechanisms, visualizing attention maps can provide insights into which parts of the input the model focuses on. This is particularly useful for understanding decision-making in tasks like image classification and natural language processing.

$$\alpha_{ij} = \frac{\exp(e_{ij})}{\sum_{k=1}^{n} \exp(e_{ik})}$$

where α_{ij} is the attention weight and e_{ij} is the attention score.

```
def visualize_attention(model, image):
    image = image.unsqueeze(0).cuda()
    attention_weights = model.get_attention_weights(image)
    attention_map = attention_weights.squeeze().detach().cpu().numpy()
    plt.imshow(attention_map, cmap='viridis')
    plt.show()
```

Visualization Techniques

1. Saliency Maps: Saliency maps highlight the regions of the input that are most influential in the model's prediction. These maps can be computed by taking the gradient of the output with respect to the input.

$$S = \left| \frac{\partial y}{\partial x} \right|$$

where S is the saliency map, y is the model output, and x is the input.

```
def compute_saliency(model, image):
    image = image.requires_grad_().cuda()
    output = model(image)
    output_idx = output.argmax()
    output_max = output[0, output_idx]
    output_max.backward()
    saliency, _ = torch.max(image.grad.data.abs(), dim=1)
    return saliency

saliency = compute_saliency(model, image)
plt.imshow(saliency.squeeze().cpu().numpy(), cmap='hot')
plt.show()
```

2. Gradient-weighted Class Activation Mapping (Grad-CAM): Grad-CAM visualizes the regions of the input that are most important for a particular class prediction. It uses the gradients of the target class with respect to the feature maps of the convolutional layer.

$$\text{Grad-CAM:} \quad L^c_{\text{Grad-CAM}} = \sum_k \alpha^c_k A^k$$

where α^c_k are the weights computed from the gradients and A^k are the activation maps.

```
def grad_cam(model, image, target_layer, target_class):
    image = image.requires_grad_().cuda()
    model.eval()

    features = []
    def hook_fn(module, input, output):
        features.append(output)

    handle = target_layer.register_forward_hook(hook_fn)
    output = model(image)
    handle.remove()

    one_hot = torch.zeros(output.size()).cuda()
    one_hot[0, target_class] = 1
    model.zero_grad()
    output.backward(gradient=one_hot, retain_graph=True)

    gradients = model.get_activations_gradient()
    pooled_gradients = torch.mean(gradients, dim=[0, 2, 3])
    activations = features[0].detach()

    for i in range(activations.size(1)):
        activations[:, i, :, :] *= pooled_gradients[i]

    heatmap = torch.mean(activations, dim=1).squeeze()
    heatmap = np.maximum(heatmap.cpu(), 0)
    heatmap /= torch.max(heatmap)
    return heatmap

heatmap = grad_cam(model, image, model.layer4[1].conv2, target_class)
plt.matshow(heatmap.squeeze().numpy(), cmap='viridis')
plt.show()
```

9.8.3 Hybrid Models

Combining Self-supervised and Supervised Learning Hybrid models that combine self-supervised learning (SSL) with supervised learning have emerged as powerful tools for leveraging unlabeled data to improve performance on labeled tasks. These models exploit the strengths of both paradigms: the ability of SSL to learn rich representations from vast amounts of unlabeled data and the precision of supervised learning to fine-tune these representations for specific tasks.

1. Pre-training Phase (Self-supervised Learning): In the pre-training phase, a model learns to solve a self-supervised task, such as predicting masked parts of the input or contrastive learning. This phase aims to learn a robust feature representation.

$$\mathcal{L}_{\text{SSL}} = \sum_{i=1}^{N} \ell(f_\theta(x_i), y_i^{\text{pseudo}})$$

where f_θ is the model with parameters θ, x_i is the input, and y_i^{pseudo} are the pseudo-labels generated by the self-supervised task.
For example, in contrastive learning (SimCLR), the loss function is:

$$\mathcal{L}_{\text{contrastive}} = -\log \frac{\exp(\text{sim}(z_i, z_j)/\tau)}{\sum_{k=1}^{2N} \mathbb{1}_{[k \neq i]} \exp(\text{sim}(z_i, z_k)/\tau)}$$

where z_i and z_j are representations of two augmented views of the same input, and τ is a temperature hyperparameter.
2. Fine-Tuning Phase (Supervised Learning): After pre-training, the model is fine-tuned on a labeled dataset to adapt the learned representations to the specific task at hand.

$$\mathcal{L}_{\text{supervised}} = \sum_{i=1}^{M} \ell(g_\phi(f_\theta(x_i)), y_i)$$

where g_ϕ is the task-specific head with parameters ϕ, and y_i are the true labels for the supervised task.

Implementation Example:

```
import torch
import torch.nn as nn
import torch.optim as optim
from torchvision import datasets, transforms, models

# Self-supervised pre-training (SimCLR)
class SimCLR(nn.Module):
    def __init__(self, base_model, projection_dim):
        super(SimCLR, self).__init__()
        self.encoder = base_model
```

```
        self.encoder.fc = nn.Identity()
        self.projector = nn.Sequential(
            nn.Linear(self.encoder.fc.in_features, 2048),
            nn.ReLU(),
            nn.Linear(2048, projection_dim)
        )

    def forward(self, x):
        h = self.encoder(x)
        z = self.projector(h)
        return z

base_model = models.resnet50(pretrained=False)
model = SimCLR(base_model, projection_dim=128).cuda()

# Pre-training optimizer and loss function
optimizer = optim.Adam(model.parameters(), lr=0.001)
criterion = nn.CrossEntropyLoss()

# Pre-training loop (self-supervised)
for epoch in range(10):
    for inputs, _ in train_loader:
        inputs = inputs.cuda()
        optimizer.zero_grad()
        representations = model(inputs)
        loss = contrastive_loss(representations)
        loss.backward()
        optimizer.step()

# Fine-tuning phase (supervised)
class SupervisedHead(nn.Module):
    def __init__(self, base_model, num_classes):
        super(SupervisedHead, self).__init__()
        self.encoder = base_model
        self.classifier = nn.Linear(2048, num_classes)

    def forward(self, x):
        with torch.no_grad():
            features = self.encoder(x)
        out = self.classifier(features)
        return out

supervised_model = SupervisedHead(model.encoder, num_classes=10).cuda()
optimizer = optim.Adam(supervised_model.parameters(), lr=0.001)
criterion = nn.CrossEntropyLoss()

# Fine-tuning loop (supervised)
for epoch in range(10):
    for inputs, labels in train_loader:
        inputs, labels = inputs.cuda(), labels.cuda()
        optimizer.zero_grad()
        outputs = supervised_model(inputs)
        loss = criterion(outputs, labels)
        loss.backward()
        optimizer.step()
```

Applications of Hybrid Approaches Hybrid models have demonstrated significant improvements in various applications by effectively combining self-supervised and supervised learning. Here are a few notable examples:

1. Image Classification: In image classification, hybrid models pre-trained with self-supervised techniques like SimCLR or MoCo can achieve superior performance when fine-tuned on labeled datasets like ImageNet. The self-supervised

pre-training phase enables the model to learn robust visual features that transfer well to the classification task.

$$\mathcal{L}_{\text{total}} = \mathcal{L}_{\text{SSL}} + \lambda \mathcal{L}_{\text{supervised}}$$

where λ is a balancing hyperparameter.

2. Natural Language Processing: In NLP, models like BERT use a hybrid approach where masked language modeling (MLM) serves as the self-supervised pre-training task, followed by fine-tuning on specific downstream tasks such as sentiment analysis, question answering, and named entity recognition. This hybrid strategy significantly boosts performance due to the rich contextual representations learned during pre-training.

$$\mathcal{L}_{\text{total}} = \mathcal{L}_{\text{MLM}} + \lambda \mathcal{L}_{\text{task}}$$

3. Medical Imaging: Hybrid models are also effective in medical imaging, where labeled data is often scarce. Self-supervised techniques can learn meaningful representations from large unlabeled medical image datasets. These representations can then be fine-tuned for tasks such as disease diagnosis or segmentation, resulting in improved accuracy and robustness.

$$\mathcal{L}_{\text{total}} = \mathcal{L}_{\text{reconstruction}} + \lambda \mathcal{L}_{\text{diagnosis}}$$

4. Time Series Forecasting: In time series forecasting, self-supervised learning can be used to pre-train models on tasks like predicting the next value in a sequence or reconstructing corrupted segments. These pre-trained models can then be fine-tuned on labeled datasets for specific forecasting tasks, enhancing the model's ability to capture temporal patterns.

$$\mathcal{L}_{\text{total}} = \mathcal{L}_{\text{forecast}} + \lambda \mathcal{L}_{\text{task}}$$

9.9 Exercises

1. Given the following dataset of points:

$$\mathbf{X} = \begin{pmatrix} 1.0 & 2.0 \\ 1.5 & 1.8 \\ 5.0 & 8.0 \\ 8.0 & 8.0 \\ 1.0 & 0.6 \\ 9.0 & 11.0 \end{pmatrix}$$

   (a) Perform K-means clustering with $k = 2$.

(b) Initialize the centroids randomly and compute the assignments of the points to clusters.
(c) Update the centroids and repeat the process until convergence.
(d) Calculate the final cluster centers and the total within-cluster sum of squares.

2. Given the following dataset:

$$\mathbf{X} = \begin{pmatrix} 2.5 & 2.4 \\ 0.5 & 0.7 \\ 2.2 & 2.9 \\ 1.9 & 2.2 \\ 3.1 & 3.0 \\ 2.3 & 2.7 \\ 2.0 & 1.6 \\ 1.0 & 1.1 \\ 1.5 & 1.6 \\ 1.1 & 0.9 \end{pmatrix}$$

(a) Standardize the dataset.
(b) Compute the covariance matrix.
(c) Calculate the eigenvalues and eigenvectors of the covariance matrix.
(d) Transform the original dataset to the new PCA basis and compute the first two principal components.

3. Given a batch of image embeddings:

$$\mathbf{H} = \begin{pmatrix} 0.1 & 0.3 & 0.7 \\ 0.4 & 0.6 & 0.8 \\ 0.5 & 0.9 & 0.2 \\ 0.7 & 0.5 & 0.4 \end{pmatrix}$$

and the corresponding positive pairs:

$$\text{Positive Pairs:} \quad (1, 2), (3, 4)$$

(a) Compute the normalized embeddings.
(b) Calculate the cosine similarities between all pairs of embeddings.
(c) Compute the contrastive loss for the given batch using the NT-Xent loss function.

4. Given a sequence of data points and a predictive coding model with the following parameters:

$$\mathbf{W}_{\text{enc}} = \begin{pmatrix} 0.2 & 0.4 \\ 0.6 & 0.8 \end{pmatrix}, \quad \mathbf{W}_{\text{dec}} = \begin{pmatrix} 0.1 & 0.3 \\ 0.5 & 0.7 \end{pmatrix}$$

Input sequence:

$$\mathbf{X}_t = \begin{pmatrix} 0.5 \\ 0.9 \end{pmatrix}, \quad \mathbf{X}_{t+1} = \begin{pmatrix} 0.6 \\ 0.8 \end{pmatrix}$$

(a) Encode the input sequence using the encoder.
(b) Predict the next data point using the decoder.
(c) Calculate the prediction error and update the model parameters using gradient descent.

5. Given a batch of image embeddings from the online and target networks:
Online network embeddings:

$$\mathbf{Z}_{\text{online}} = \begin{pmatrix} 0.2 & 0.4 \\ 0.3 & 0.6 \\ 0.5 & 0.7 \end{pmatrix}$$

Target network embeddings:

$$\mathbf{Z}_{\text{target}} = \begin{pmatrix} 0.1 & 0.3 \\ 0.4 & 0.8 \\ 0.6 & 0.9 \end{pmatrix}$$

(a) Compute the L2-normalized embeddings for both networks.
(b) Calculate the cosine similarity between the embeddings from the online and target networks.
(c) Compute the mean squared error loss for the BYOL objective.

6. Given a batch of query and key embeddings:
Query embeddings:

$$\mathbf{Q} = \begin{pmatrix} 0.5 & 0.7 \\ 0.3 & 0.8 \\ 0.6 & 0.9 \end{pmatrix}$$

Key embeddings:

$$\mathbf{K} = \begin{pmatrix} 0.2 & 0.4 \\ 0.1 & 0.3 \\ 0.7 & 0.6 \end{pmatrix}$$

(a) Compute the dot-product similarities between query and key embeddings.
(b) Apply the softmax function to obtain the similarity scores.
(c) Calculate the contrastive loss for MoCo using the similarity scores.

7. Consider a dataset of 1000 images, each represented as a vector in $\mathbb{R}^{128}$. You are tasked with implementing a contrastive learning framework. Define the positive

pairs as augmentations of the same image and the negative pairs as different images.

(a) Write down the contrastive loss function using the cosine similarity measure.
(b) Suppose you have a mini-batch of 5 images with their corresponding embeddings after augmentation:

$$\mathbf{E} = \begin{pmatrix} 0.1 & 0.2 & 0.3 & 0.4 & 0.5 & 0.6 & 0.7 & 0.8 & 0.9 & 1.0 \\ 0.9 & 0.8 & 0.7 & 0.6 & 0.5 & 0.4 & 0.3 & 0.2 & 0.1 & 0.0 \\ 0.1 & 0.1 & 0.1 & 0.1 & 0.1 & 0.1 & 0.1 & 0.1 & 0.1 & 0.1 \\ 0.2 & 0.3 & 0.4 & 0.5 & 0.6 & 0.7 & 0.8 & 0.9 & 1.0 & 1.1 \\ 1.1 & 1.0 & 0.9 & 0.8 & 0.7 & 0.6 & 0.5 & 0.4 & 0.3 & 0.2 \end{pmatrix}$$

Compute the contrastive loss for this mini-batch.

8. In the context of natural language processing, you are given a sequence of tokens represented by their embeddings $\mathbf{X} \in \mathbb{R}^{10\times 50}$ where each row corresponds to a token in a sequence of length 10, and each token is represented by a 50-dimensional vector.

   (a) Define the objective of a masked language model (MLM) and explain the masking strategy.
   (b) Given a specific example where tokens at positions 3 and 7 are masked, outline the steps to predict these masked tokens using a transformer-based model.

9. Assume you are using a self-supervised learning approach to pre-train a neural network on a dataset of 5000 images. One popular method is to maximize the mutual information between different views of the same image.

   (a) Describe the process of generating different views of the same image.
   (b) Given an encoder $\mathbf{f}$ and two augmented views $\mathbf{v}_1$ and $\mathbf{v}_2$ of an image, define the mutual information maximization objective.
   (c) Assume the following embeddings for a batch of 3 images and their augmented views:

$$\mathbf{Z}_1 = \begin{pmatrix} 0.2 & 0.4 & 0.6 \\ 0.1 & 0.3 & 0.5 \\ 0.0 & 0.2 & 0.4 \end{pmatrix}, \quad \mathbf{Z}_2 = \begin{pmatrix} 0.3 & 0.5 & 0.7 \\ 0.2 & 0.4 & 0.6 \\ 0.1 & 0.3 & 0.5 \end{pmatrix}$$

   Compute the mutual information maximization objective for this batch.

10. One common approach in self-supervised learning is to use pretext tasks, such as predicting the rotation angle of an image.

   (a) Explain the concept of pretext tasks and provide examples of different pretext tasks that can be used for self-supervised learning.

(b) Given a dataset of images, describe the steps to create a self-supervised learning pipeline using the task of predicting rotation angles (0°, 90°, 180°, 270°).

(c) Assume you have a set of 4 images, each rotated by one of the four angles. Their embeddings after passing through the network are:

$$\mathbf{E}_0 = \begin{pmatrix} 0.1 & 0.3 & 0.5 \\ 0.2 & 0.4 & 0.6 \\ 0.3 & 0.5 & 0.7 \\ 0.4 & 0.6 & 0.8 \end{pmatrix}, \quad \mathbf{E}_{90} = \begin{pmatrix} 0.2 & 0.4 & 0.6 \\ 0.3 & 0.5 & 0.7 \\ 0.4 & 0.6 & 0.8 \\ 0.5 & 0.7 & 0.9 \end{pmatrix},$$

$$\mathbf{E}_{180} = \begin{pmatrix} 0.3 & 0.5 & 0.7 \\ 0.4 & 0.6 & 0.8 \\ 0.5 & 0.7 & 0.9 \\ 0.6 & 0.8 & 1.0 \end{pmatrix}, \quad \mathbf{E}_{270} = \begin{pmatrix} 0.4 & 0.6 & 0.8 \\ 0.5 & 0.7 & 0.9 \\ 0.6 & 0.8 & 1.0 \\ 0.7 & 0.9 & 1.1 \end{pmatrix}$$

Describe how you would train a classifier to predict the rotation angle using these embeddings.

11. Given a dataset represented as a matrix $\mathbf{X} \in \mathbb{R}^{6\times 3}$, where each row is a data point with three features:

$$\mathbf{X} = \begin{pmatrix} 2.5 & 0.5 & 2.2 \\ 1.0 & 1.5 & 2.9 \\ 3.1 & 3.3 & 1.1 \\ 4.0 & 2.0 & 0.1 \\ 0.5 & 4.5 & 3.2 \\ 3.8 & 2.7 & 1.5 \end{pmatrix}$$

(a) Center the dataset $\mathbf{X}$ by subtracting the mean of each column.

(b) Compute the covariance matrix of the centered dataset.

(c) Find the eigenvalues and eigenvectors of the covariance matrix.

(d) Project the data onto the first two principal components to obtain a 2D representation of the dataset.

12. Given a 2D dataset represented as a matrix $\mathbf{Z} \subset \mathbb{R}^{8\times 2}$:

$$\mathbf{Z} = \begin{pmatrix} 0.1 & 0.5 \\ 0.2 & 0.6 \\ 0.3 & 0.7 \\ 0.4 & 0.8 \\ 0.5 & 0.9 \\ 0.6 & 1.0 \\ 0.7 & 1.1 \\ 0.8 & 1.2 \end{pmatrix}$$

(a) Describe the steps involved in applying UMAP to reduce the dimensionality of $\mathbf{Z}$ to 1D.

(b) Perform the UMAP transformation manually for the first two data points using the simplified version of the algorithm. Write down the new coordinates of these points in the 1D space.

13. Assume you have a noisy dataset represented as a matrix $\mathbf{D} \in \mathbb{R}^{10\times 3}$:

$$\mathbf{D} = \begin{pmatrix} 2.1 & 1.1 & 0.9 \\ 1.9 & 0.8 & 1.2 \\ 3.2 & 2.2 & 1.5 \\ 2.9 & 1.7 & 0.9 \\ 1.2 & 0.5 & 0.8 \\ 2.8 & 2.1 & 1.3 \\ 2.4 & 1.3 & 1.0 \\ 2.6 & 1.8 & 1.1 \\ 3.0 & 2.0 & 1.2 \\ 1.5 & 0.7 & 0.9 \end{pmatrix}$$

(a) Perform PCA on $\mathbf{D}$ and retain only the first principal component. Write down the new dataset.
(b) Reconstruct the original dataset from the first principal component and compare it with the original noisy dataset. Discuss the effect of noise reduction achieved through PCA.

14. Given a dataset $\mathbf{X} \in \mathbb{R}^{8\times 2}$ where each row represents a data point in a 2-dimensional space:

$$\mathbf{X} = \begin{pmatrix} 1.0 & 2.0 \\ 1.5 & 1.8 \\ 5.0 & 8.0 \\ 8.0 & 8.0 \\ 1.0 & 0.6 \\ 9.0 & 11.0 \\ 8.0 & 2.0 \\ 10.0 & 2.0 \end{pmatrix}$$

(a) Initialize the cluster centroids for $k = 2$ randomly from the dataset.
(b) Perform one iteration of the k-means algorithm. Assign each data point to the nearest centroid and compute the new centroids.
(c) Repeat the iteration until convergence and provide the final cluster assignments and centroids.

15. Consider the same dataset $\mathbf{X}$ from above exercise.

(a) Calculate the pairwise Euclidean distance matrix for $\mathbf{X}$.
(b) Using the single-linkage method, construct the dendrogram step by step until all data points are clustered into one cluster. Show the sequence of merges and the distance at each step.

(c) Determine the clusters when cutting the dendrogram at a height that results in 2 clusters. Provide the final cluster assignments.

16. Given the dataset $\mathbf{Y} \in \mathbb{R}^{6\times 2}$:

$$\mathbf{Y} = \begin{pmatrix} 1.1 & 2.2 \\ 1.3 & 2.4 \\ 5.5 & 8.4 \\ 8.1 & 8.2 \\ 1.0 & 0.5 \\ 9.2 & 11.1 \end{pmatrix}$$

(a) Assume an initial guess for the parameters of a GMM with 2 components: means $\boldsymbol{\mu}_1 = (1.0, 2.0), \boldsymbol{\mu}_2 = (8.0, 9.0)$; covariances $\Sigma_1 = \Sigma_2 = \begin{pmatrix} 1 & 0 \\ 0 & 1 \end{pmatrix}$; and equal mixing coefficients.
(b) Perform one Expectation-Maximization (EM) iteration to update the parameters of the GMM. Compute the responsibilities for each data point and update the means, covariances, and mixing coefficients.
(c) Repeat the EM steps until convergence and provide the final parameter estimates.

17. Consider a dataset $\mathbf{Z} \in \mathbb{R}^{5\times 2}$ where each row is a data point:

$$\mathbf{Z} = \begin{pmatrix} 1.0 & 2.0 \\ 2.0 & 3.0 \\ 3.0 & 3.0 \\ 8.0 & 7.0 \\ 8.5 & 8.0 \end{pmatrix}$$

(a) Construct the similarity matrix $\mathbf{S}$ using a Gaussian (RBF) kernel with $\sigma = 1$:

$$S_{ij} = \exp\left(-\frac{\|\mathbf{z}_i - \mathbf{z}_j\|^2}{2\sigma^2}\right)$$

(b) Compute the unnormalized Laplacian matrix $\mathbf{L}$ and its eigenvalues and eigenvectors.
(c) Perform spectral clustering by using the first two eigenvectors of $\mathbf{L}$ to cluster the data points into 2 clusters. Provide the cluster assignments.

18. Given a dataset $\mathbf{W} \in \mathbb{R}^{10\times 2}$:

$$\mathbf{W} = \begin{pmatrix} 1.0 & 2.0 \\ 1.2 & 1.9 \\ 2.0 & 2.1 \\ 8.0 & 8.0 \\ 8.2 & 8.1 \\ 8.1 & 8.2 \\ 25.0 & 30.0 \\ 25.1 & 30.1 \\ 25.2 & 29.9 \\ 25.0 & 30.2 \end{pmatrix}$$

(a) Define the parameters ϵ (eps) and (minPts) for DBSCAN.
(b) Perform DBSCAN clustering on the dataset with $\epsilon = 1.5$ and minPts $= 2$. Identify core points, border points, and noise points.
(c) Provide the final cluster assignments, including noise points.

Chapter 10
Learning Representations via Autoencoders and Generative Models

Information is the resolution of uncertainty.

—Claude Shannon, A Mathematical Theory of Communication

10.1 Introduction

Autoencoders and generative models are crucial components in the field of unsupervised learning, enabling the discovery of meaningful representations from unlabeled data. They are designed to learn compressed, latent representations of input data, facilitating various tasks such as data compression, reconstruction, and generation.

Autoencoders are a type of neural network that aims to learn a representation (encoding) for a set of data, typically for the purpose of dimensionality reduction or feature learning. The network consists of an encoder that maps the input to a latent space, and a decoder that reconstructs the input from the latent representation (see Fig. 10.1). The learning objective is to minimize the reconstruction error, which is often formulated as the mean squared error between the input and its reconstruction:

$$\mathcal{L} = \frac{1}{N} \sum_{i=1}^{N} \left\| \mathbf{x}_i - \hat{\mathbf{x}}_i \right\|^2$$

where $\mathbf{x}_i$ is the input data point and $\hat{\mathbf{x}}_i$ is its reconstruction.

Generative models, on the other hand, are focused on modeling the distribution of data to generate new samples. These models can capture complex data distributions and generate new data points that resemble the original data. Prominent generative models include Generative Adversarial Networks (GANs) and Variational Autoen-

P. Singh and B. Raman, *Deep Learning Through the Prism of Tensors*, Studies in Big Data 162, https://doi.org/10.1007/978-981-97-8019-8_10

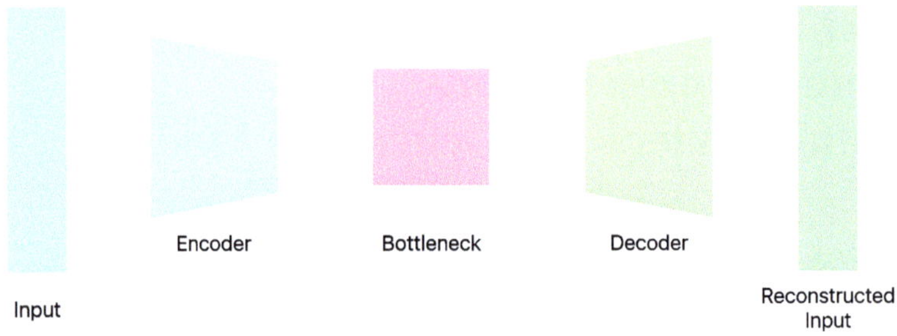

Fig. 10.1 An illustration of an autoencoder

coders (VAEs). GANs consist of a generator and a discriminator network that are trained adversarially, while VAEs employ a probabilistic approach to model the latent space and generate new data.

10.1.1 Historical Development

The development of autoencoders and generative models has been marked by several key milestones. Autoencoders were introduced in the 1980s as a neural network-based method for dimensionality reduction and feature learning (Rumelhart et al. 1986). Early autoencoders suffered from limitations such as poor scalability and inability to learn complex representations, but advancements in deep learning architectures and training techniques have significantly improved their performance. The introduction of deep autoencoders by Hinton and Salakhutdinov (2006) demonstrated the potential of deep networks in learning hierarchical representations. This work laid the foundation for further advancements in unsupervised learning and representation learning. GANs, introduced by Goodfellow et al. (2014), marked a significant breakthrough in generative modeling. GANs leverage the adversarial training paradigm, where the generator aims to produce realistic samples, and the discriminator attempts to distinguish between real and generated samples. The adversarial nature of GANs has led to impressive results in various applications, including image synthesis, style transfer, and data augmentation. VAEs, introduced by Kingma and Welling (2014), provide a probabilistic framework for learning latent representations and generating new data. VAEs combine the principles of autoencoders with variational inference, allowing the model to capture uncertainty in the latent space and generate diverse samples. VAEs have found applications in image generation, anomaly detection, and data imputation.

10.1.2 Applications in Data Compression, Reconstruction, and Generation

Autoencoders and generative models have diverse applications across various domains, leveraging their ability to learn compact and meaningful representations of data.

Data Compression: Autoencoders can be used for data compression by learning a compact representation of the input data. The encoder compresses the input into a lower-dimensional latent space, and the decoder reconstructs the input from this representation. This process can be particularly useful for compressing images, audio, and other high-dimensional data. The reconstruction error indicates the quality of compression, and techniques such as sparse autoencoders and contractive autoencoders have been developed to enhance the robustness and interpretability of the compressed representations. For an autoencoder with an encoder function $f_\theta(\cdot)$ and a decoder function $g_\phi(\cdot)$, the objective is to minimize the reconstruction loss:

$$\mathcal{L}(\theta, \phi) = \frac{1}{N} \sum_{i=1}^{N} \left\| \mathbf{x}_i - g_\phi(f_\theta(\mathbf{x}_i)) \right\|^2$$

where θ and ϕ are the parameters of the encoder and decoder networks, respectively.

Data Reconstruction: In data reconstruction, the goal is to reconstruct the input data as accurately as possible. Autoencoders excel in this task by learning to capture the underlying structure of the data. Applications include image denoising, where the autoencoder learns to remove noise from images, and missing data imputation, where the model reconstructs missing values in a dataset. Consider a noisy input $\tilde{\mathbf{x}}_i$ and the clean target $\mathbf{x}_i$. The autoencoder is trained to minimize the reconstruction loss:

$$\mathcal{L}(\theta, \phi) = \frac{1}{N} \sum_{i=1}^{N} \left\| \mathbf{x}_i - g_\phi(f_\theta(\tilde{\mathbf{x}}_i)) \right\|^2$$

Data Generation: Generative models such as GANs and VAEs are designed to generate new data samples that resemble the training data. These models have been used in various creative and practical applications, including image synthesis, text generation, and drug discovery. GANs, for example, can generate realistic images of faces, animals, and even artwork, while VAEs can generate diverse samples by sampling from the learned latent space. In VAEs, the generative process involves sampling from the latent space and decoding the samples to generate new data. The objective combines the reconstruction loss and a regularization term that ensures the latent space follows a prior distribution, typically a Gaussian:

$$\mathcal{L}(\theta, \phi) = \mathbb{E}_{q_\phi(\mathbf{z}|\mathbf{x})} \left[\log p_\theta(\mathbf{x} \mid \mathbf{z}) \right] - \mathrm{KL}(q_\phi(\mathbf{z} \mid \mathbf{x}) \| p(\mathbf{z}))$$

where $q_\phi(\mathbf{z} \mid \mathbf{x})$ is the encoder's approximate posterior distribution, $p_\theta(\mathbf{x} \mid \mathbf{z})$ is the decoder's likelihood, and KL denotes the Kullback-Leibler divergence.

10.2 Basic Architecture of Autoencoders

Autoencoders are neural networks designed to learn efficient representations of data by compressing input into a latent space and then reconstructing the input from this compressed representation. This process involves two main components: the encoder and the decoder networks, which work together to learn meaningful features of the input data. The architecture of autoencoders is fundamental to various applications in data compression, reconstruction, and generation.

Components of Autoencoders

Autoencoders consist of three main components: the encoder network, the decoder network, and the latent space representation. Each component plays a critical role in the functioning of the autoencoder, enabling it to learn and utilize compressed representations of data.

Encoder Network

The encoder network is responsible for mapping the input data to a lower-dimensional latent space. It compresses the high-dimensional input $\mathbf{x}$ into a latent representation $\mathbf{z}$, capturing the most important features of the data. Mathematically, the encoder can be represented as a function f_θ parameterized by θ:

$$\mathbf{z} = f_\theta(\mathbf{x})$$

where $\mathbf{z}$ is the latent representation, and θ denotes the parameters of the encoder network, such as weights and biases. The architecture of the encoder typically consists of multiple layers of neurons, where each layer applies a linear transformation followed by a non-linear activation function. The choice of activation function, such as ReLU, sigmoid, or tanh, influences the properties of the learned representation.

Example: Consider an encoder with three layers. The input $\mathbf{x}$ is transformed through successive layers as follows:

$$\mathbf{h}_1 = \sigma(W_1\mathbf{x} + \mathbf{b}_1)$$

$$\mathbf{h}_2 = \sigma(W_2\mathbf{h}_1 + \mathbf{b}_2)$$

$$\mathbf{z} = W_3\mathbf{h}_2 + \mathbf{b}_3$$

where σ denotes the activation function, W_i are the weight matrices, and $\mathbf{b}_i$ are the bias vectors.

Decoder Network

The decoder network is responsible for reconstructing the input data from the latent representation. It maps the latent code $\mathbf{z}$ back to the original data space, producing a reconstruction $\hat{\mathbf{x}}$. Mathematically, the decoder can be represented as a function g_ϕ parameterized by ϕ:

$$\hat{\mathbf{x}} = g_\phi(\mathbf{z})$$

where $\hat{\mathbf{x}}$ is the reconstructed output, and ϕ denotes the parameters of the decoder network. Similar to the encoder, the decoder typically consists of multiple layers of neurons, each applying a linear transformation followed by a non-linear activation function. The final layer of the decoder outputs the reconstructed data, often using an activation function appropriate for the data type (e.g., sigmoid for binary data, linear for continuous data).

Example: Consider a decoder with three layers. The latent representation $\mathbf{z}$ is transformed through successive layers as follows:

$$\mathbf{h}_3 = \sigma(W_4\mathbf{z} + \mathbf{b}_4)$$

$$\mathbf{h}_4 = \sigma(W_5\mathbf{h}_3 + \mathbf{b}_5)$$

$$\hat{\mathbf{x}} = \sigma(W_6\mathbf{h}_4 + \mathbf{b}_6)$$

where σ denotes the activation function, W_i are the weight matrices, and $\mathbf{b}_i$ are the bias vectors.

Latent Space Representation

The latent space representation, also known as the bottleneck or code, is the intermediate, lower-dimensional representation of the input data learned by the encoder. This representation captures the essential features of the data, enabling efficient compression and reconstruction. The dimensionality of the latent space is a crucial hyperparameter in the design of autoencoders. A smaller latent space forces the model to learn more compact and meaningful representations, but may also lead to loss of information if the compression is too severe. Conversely, a larger latent space may capture more information but at the cost of redundancy. The quality of the latent representation is often evaluated by the reconstruction error, which measures how well the decoder can reconstruct the input data from the latent code. The reconstruction error is typically defined as the mean squared error (MSE) between the input $\mathbf{x}$ and the reconstruction $\hat{\mathbf{x}}$:

$$\mathcal{L}(\mathbf{x}, \hat{\mathbf{x}}) = \frac{1}{N}\sum_{i=1}^{N} \|\mathbf{x}_i - \hat{\mathbf{x}}_i\|^2$$

where N is the number of data points.

Example: In an image compression task, the encoder maps high-dimensional image data to a lower-dimensional latent code. The decoder then reconstructs the image from this latent code. The latent space can be visualized to understand the learned features and the quality of the representation.

10.3 Types of Autoencoders

Autoencoders come in various types, each designed to address specific tasks and challenges. One important variant is the denoising autoencoder, which is designed to handle noisy input data.

10.3.1 Denoising Autoencoders

Denoising autoencoders (DAEs) are a type of autoencoder specifically designed to remove noise from input data. They learn to reconstruct clean data from noisy inputs by training on pairs of noisy and clean data.

Architecture and Objective

The architecture of a denoising autoencoder is similar to that of a basic autoencoder, consisting of an encoder and a decoder network. However, the training process involves adding noise to the input data and training the model to reconstruct the original, clean data. Let $\mathbf{x}$ be the original clean data and $\tilde{\mathbf{x}}$ be the noisy version of the data, generated by adding noise $\mathbf{n}$:

$$\tilde{\mathbf{x}} = \mathbf{x} + \mathbf{n}$$

The objective of the denoising autoencoder is to minimize the reconstruction error between the original data $\mathbf{x}$ and the reconstructed output $\hat{\mathbf{x}}$:

$$\mathcal{L}(\mathbf{x}, \hat{\mathbf{x}}) = \frac{1}{N} \sum_{i=1}^{N} \left\| \mathbf{x}_i - g_\phi(f_\theta(\tilde{\mathbf{x}}_i)) \right\|^2$$

Here, f_θ is the encoder function, g_ϕ is the decoder function, and θ and ϕ are the parameters of the encoder and decoder networks, respectively.

Training Process:

1. Generate Noisy Data: Add noise to the input data to create noisy versions of the data points.
2. Forward Pass: Pass the noisy data $\tilde{\mathbf{x}}$ through the encoder to obtain the latent representation $\mathbf{z}$:

$$\mathbf{z} = f_\theta(\tilde{\mathbf{x}})$$

Pass the latent representation through the decoder to obtain the reconstruction $\hat{\mathbf{x}}$:

$$\hat{\mathbf{x}} = g_\phi(\mathbf{z})$$

3. Loss Computation: Compute the reconstruction loss between the original clean data $\mathbf{x}$ and the reconstruction $\hat{\mathbf{x}}$:

$$\mathcal{L}(\mathbf{x}, \hat{\mathbf{x}})$$

4. Backward Pass and Parameter Update: Compute the gradients of the loss function with respect to the parameters θ and ϕ, and update the parameters using an optimization algorithm.

Example: Consider training a denoising autoencoder to remove Gaussian noise from images. The input images are corrupted by adding Gaussian noise, and the model is trained to reconstruct the clean images from the noisy inputs.

```
import torch
import torch.nn as nn
import torch.optim as optim

class DenoisingAutoencoder(nn.Module):
    def __init__(self, input_dim, hidden_dim, output_dim):
        super(DenoisingAutoencoder, self).__init__()
        self.encoder = nn.Sequential(
            nn.Linear(input_dim, hidden_dim),
            nn.ReLU()
        )
        self.decoder = nn.Sequential(
            nn.Linear(hidden_dim, output_dim),
            nn.Sigmoid()
        )

    def forward(self, x):
        z = self.encoder(x)
        x_hat = self.decoder(z)
        return x_hat

# Training loop
model = DenoisingAutoencoder(input_dim=784, hidden_dim=256, output_dim=784)
criterion = nn.MSELoss()
optimizer = optim.Adam(model.parameters(), lr=0.001)

for epoch in range(10):
    for data in dataloader:
        noisy_data = data + 0.1 * torch.randn_like(data)  # Adding Gaussian noise
        optimizer.zero_grad()
        output = model(noisy_data)
        loss = criterion(output, data)
        loss.backward()
        optimizer.step()
    print(f'Epoch {epoch+1}, Loss: {loss.item()}')
```

In this example, the denoising autoencoder is trained to reconstruct clean images from noisy inputs. The training loop involves adding Gaussian noise to the input images, passing the noisy images through the autoencoder, computing the reconstruction loss, and updating the model parameters.

Applications in Noise Reduction

Denoising autoencoders are widely used in applications requiring noise reduction and signal restoration. Some key applications include:

1. Image Denoising: Denoising autoencoders are used to remove noise from images, enhancing image quality and facilitating better visual analysis. This is particularly useful in medical imaging, where noise reduction can improve the accuracy of diagnoses.
2. Audio Denoising: In audio processing, denoising autoencoders can be employed to remove background noise from audio signals, improving the clarity and quality of speech recordings.
3. Data Preprocessing: Denoising autoencoders can be used as a preprocessing step to clean noisy data before further analysis or modeling. This can enhance the performance of subsequent machine learning models by providing cleaner input data.
4. Anomaly Detection: By learning to reconstruct normal data, denoising autoencoders can be used to detect anomalies. Data points that deviate significantly from the normal reconstruction pattern can be identified as anomalies.

10.3.2 Sparse Autoencoders

Sparse autoencoders are designed to learn efficient representations by imposing sparsity constraints on the hidden units during training. This encourages the model to activate only a small number of neurons for any given input, leading to more interpretable and robust feature representations.

Architecture and Sparsity Constraints

The architecture of a sparse autoencoder is similar to that of a basic autoencoder, consisting of an encoder and a decoder. However, the key difference lies in the training process, where an additional sparsity constraint is applied to the activation of the hidden units. Let $\mathbf{h}$ be the hidden layer activations of the autoencoder. Sparsity can be enforced by adding a regularization term to the loss function, which penalizes the deviation of the average activation of the hidden units from a desired sparsity level ρ. One common approach is to use the Kullback-Leibler (KL) divergence as the regularization term:

$$\mathrm{KL}(\rho \| \hat{\rho}) = \sum_{j=1}^{m} \left(\rho \log \frac{\rho}{\hat{\rho}_j} + (1 - \rho) \log \frac{1 - \rho}{1 - \hat{\rho}_j} \right)$$

where $\hat{\rho}_j$ is the average activation of hidden unit j over the training examples, and m is the number of hidden units. The overall loss function for the sparse autoencoder then becomes:

$$\mathcal{L}(\mathbf{x}, \hat{\mathbf{x}}) = \frac{1}{N} \sum_{i=1}^{N} \left\| \mathbf{x}_i - \hat{\mathbf{x}}_i \right\|^2 + \beta \sum_{j=1}^{m} \mathrm{KL}(\rho \| \hat{\rho}_j)$$

where β is a hyperparameter that controls the importance of the sparsity penalty.

Training Process:

1. Forward Pass:
   Pass the input data $\mathbf{x}$ through the encoder to obtain the hidden representation $\mathbf{h}$:

   $$\mathbf{h} = f_\theta(\mathbf{x})$$

   Pass the hidden representation through the decoder to obtain the reconstruction $\hat{\mathbf{x}}$:

   $$\hat{\mathbf{x}} = g_\phi(\mathbf{h})$$

2. Loss Computation:
   Compute the reconstruction loss and the sparsity penalty:

   $$\mathcal{L}(\mathbf{x}, \hat{\mathbf{x}}) + \beta \sum_{j=1}^{m} \mathrm{KL}(\rho \| \hat{\rho}_j)$$

3. Backward Pass and Parameter Update:
   Compute the gradients of the loss function with respect to the parameters θ and ϕ, and update the parameters using an optimization algorithm.

Example: Consider training a sparse autoencoder on image data. The hidden layer activations are constrained to maintain sparsity, leading to more interpretable features that capture essential structures in the images.

```
import torch
import torch.nn as nn
import torch.optim as optim

class SparseAutoencoder(nn.Module):
    def __init__(self, input_dim, hidden_dim, output_dim, sparsity_param=0.05, beta=3):
        super(SparseAutoencoder, self).__init__()
        self.encoder = nn.Sequential(
            nn.Linear(input_dim, hidden_dim),
            nn.ReLU()
        )
```

```
        self.decoder = nn.Sequential(
            nn.Linear(hidden_dim, output_dim),
            nn.Sigmoid()
        )
        self.sparsity_param = sparsity_param
        self.beta = beta

    def forward(self, x):
        h = self.encoder(x)
        x_hat = self.decoder(h)
        return x_hat, h

    def sparsity_loss(self, h):
        rho_hat = torch.mean(h, dim=0)
        kl_div = self.sparsity_param * torch.log(self.sparsity_param / rho_hat) +
(1 - self.sparsity_param) * torch.log((1 - self.sparsity_param) / (1 - rho_hat))
        return torch.sum(kl_div)

# Training loop
model = SparseAutoencoder(input_dim=784, hidden_dim=256, output_dim=784)
criterion = nn.MSELoss()
optimizer = optim.Adam(model.parameters(), lr=0.001)

for epoch in range(10):
    for data in dataloader:
        optimizer.zero_grad()
        output, hidden = model(data)
        loss = criterion(output, data) + model.beta * model.sparsity_loss(hidden)
        loss.backward()
        optimizer.step()
    print(f'Epoch {epoch+1}, Loss: {loss.item()}')
```

Applications in Feature Extraction

Sparse autoencoders are particularly effective in feature extraction due to their ability to learn sparse and meaningful representations. Key applications include:

1. Image Classification: Sparse autoencoders can learn features that are useful for image classification tasks. By extracting sparse features, the model can achieve better generalization and improved classification performance.
2. Anomaly Detection: The sparse features learned by the autoencoder can be used to detect anomalies in data. Data points that deviate significantly from the normal pattern of sparse activations can be flagged as anomalies.
3. Data Visualization: Sparse autoencoders can be used for dimensionality reduction and data visualization. The learned sparse representations can be projected into lower-dimensional spaces, facilitating visualization and interpretation of high-dimensional data.
4. Pretraining for Deep Networks: Sparse autoencoders can be used to pretrain deep neural networks. The learned sparse features provide a good initialization for subsequent layers, leading to faster convergence and better performance in supervised learning tasks.

10.3.3 *Variational Autoencoders (VAEs)*

Variational autoencoders (VAEs) are a type of generative model that combines the principles of autoencoders with probabilistic graphical models. VAEs aim to learn a probabilistic latent space from which new data samples can be generated.

Architecture and Variational Inference

The architecture of a VAE consists of an encoder, a decoder, and a latent space. However, unlike traditional autoencoders, VAEs introduce a probabilistic approach to the latent space representation. The encoder outputs parameters of a probability distribution, typically a Gaussian, from which the latent variables are sampled. The decoder then generates data samples from these latent variables. The encoder network outputs the mean μ and the log-variance $\log \sigma^2$ of the latent variables:

$$\mu, \log \sigma^2 = f_\theta(\mathbf{x})$$

The latent variables $\mathbf{z}$ are then sampled from the Gaussian distribution parameterized by μ and σ:

$$\mathbf{z} \sim \mathcal{N}(\mu, \sigma^2)$$

The decoder network generates the reconstruction $\hat{\mathbf{x}}$ from the latent variables $\mathbf{z}$:

$$\hat{\mathbf{x}} = g_\phi(\mathbf{z})$$

The objective of the VAE is to maximize the evidence lower bound (ELBO), which consists of two terms: the reconstruction loss and the KL divergence between the approximate posterior and the prior distribution:

$$\mathcal{L}(\mathbf{x}, \hat{\mathbf{x}}) = \mathbb{E}_{q_\phi(\mathbf{z}|\mathbf{x})}\left[\log p_\theta(\mathbf{x} \mid \mathbf{z})\right] - \mathrm{KL}(q_\phi(\mathbf{z} \mid \mathbf{x}) \| p(\mathbf{z}))$$

Here, $q_\phi(\mathbf{z} \mid \mathbf{x})$ is the approximate posterior distribution, $p_\theta(\mathbf{x} \mid \mathbf{z})$ is the likelihood of the data given the latent variables, and $p(\mathbf{z})$ is the prior distribution, typically a standard Gaussian.

The loss function for a VAE combines the reconstruction loss and the KL divergence. The reconstruction loss measures how well the decoder can reconstruct the input data from the latent variables, while the KL divergence ensures that the approximate posterior distribution is close to the prior distribution. The reconstruction loss is typically the negative log-likelihood of the data given the latent variables. For continuous data, this can be the mean squared error:

$$\mathcal{L}_{\text{recon}} = \mathbb{E}_{q_\phi(\mathbf{z}|\mathbf{x})}\left[\log p_\theta(\mathbf{x} \mid \mathbf{z})\right]$$

The KL divergence term regularizes the latent space to follow the prior distribution:

$$\mathcal{L}_{\text{KL}} = \text{KL}(q_\phi(\mathbf{z} \mid \mathbf{x}) \| p(\mathbf{z})) = -\frac{1}{2} \sum_{j=1}^{d} \left(1 + \log \sigma_j^2 - \mu_j^2 - \sigma_j^2\right)$$

The overall loss function for the VAE is the sum of the reconstruction loss and the KL divergence:

$$\mathcal{L}(\mathbf{x}, \hat{\mathbf{x}}) = \mathcal{L}_{\text{recon}} + \mathcal{L}_{\text{KL}}$$

Example: Consider training a VAE on image data. The encoder maps the input images to the parameters of the Gaussian distribution in the latent space, and the decoder generates images from the sampled latent variables.

```
import torch
import torch.nn as nn
import torch.optim as optim

class VAE(nn.Module):
    def __init__(self, input_dim, hidden_dim, latent_dim):
        super(VAE, self).__init__()
        self.encoder = nn.Sequential(
            nn.Linear(input_dim, hidden_dim),
            nn.ReLU()
        )
        self.mu_layer = nn.Linear(hidden_dim, latent_dim)
        self.logvar_layer = nn.Linear(hidden_dim, latent_dim)
        self.decoder = nn.Sequential(
            nn.Linear(latent_dim, hidden_dim),
            nn.ReLU(),
            nn.Linear(hidden_dim, input_dim),
            nn.Sigmoid()
        )

    def encode(self, x):
        h = self.encoder(x)
        mu = self.mu_layer(h)
        logvar = self.logvar_layer(h)
        return mu, logvar

    def reparameterize(self, mu, logvar):
        std = torch.exp(0.5 * logvar)
        eps = torch.randn_like(std)
        return mu + eps * std

    def decode(self, z):
        return self.decoder(z)

    def forward(self, x):
        mu, logvar = self.encode(x)
        z = self.reparameterize(mu, logvar)
        x_hat = self.decode(z)
        return x_hat, mu, logvar

# Training loop
model = VAE(input_dim=784, hidden_dim=256, latent_dim=20)
criterion = nn.MSELoss()
optimizer = optim.Adam(model.parameters(), lr=0.001)
```

```
def vae_loss(x, x_hat, mu, logvar):
    recon_loss = criterion(x_hat, x)
    kl_loss = -0.5 * torch.sum(1 + logvar - mu.pow(2) - logvar.exp())
    return recon_loss + kl_loss

for epoch in range(10):
    for data in dataloader:
        optimizer.zero_grad()
        output, mu, logvar = model(data)
        loss = vae_loss(data, output, mu, logvar)
        loss.backward()
        optimizer.step()
    print(f'Epoch {epoch+1}, Loss: {loss.item()}')
```

Applications in Data Generation

VAEs have wide-ranging applications in data generation and have been used to generate new data points that resemble the training data. Key applications include:

1. Image Generation: VAEs can generate realistic images by sampling from the learned latent space. This capability is useful in various creative and practical applications, such as generating artworks, synthesizing faces, and augmenting datasets.
2. Text Generation: VAEs can be used for generating coherent and contextually relevant text. By learning the latent representations of text data, VAEs can generate new sentences and paragraphs that mimic the training data.
3. Drug Discovery: In the field of drug discovery, VAEs can be used to generate novel molecular structures with desired properties. The generative capability of VAEs enables the exploration of new chemical compounds and accelerates the drug development process.
4. Anomaly Detection: VAEs can be employed for anomaly detection by learning the distribution of normal data. Data points that significantly deviate from this distribution can be flagged as anomalies, making VAEs useful for detecting outliers in various domains.

10.4 Advanced Autoencoder Architectures

Autoencoders have evolved to address more complex data types and tasks, leading to the development of advanced architectures such as convolutional autoencoders and recurrent autoencoders. These architectures leverage the specific structures of the data, such as spatial relationships in images and temporal dependencies in sequential data, to improve performance and applicability.

10.4.1 Convolutional Autoencoders

Convolutional autoencoders (CAEs) are designed to handle image data by incorporating convolutional layers into the encoder and decoder networks. These layers are adept at capturing spatial hierarchies and local features, making CAEs particularly suitable for image reconstruction and related tasks.

Architecture for Image Data

The architecture of a convolutional autoencoder consists of convolutional layers in both the encoder and decoder networks. The encoder network uses convolutional layers to extract features from the input images, reducing their spatial dimensions through pooling operations. The latent representation retains the essential features in a compressed form. The decoder network then reconstructs the images by progressively upsampling the latent representation, typically using transposed convolutions.

1. Encoder: The encoder applies a series of convolutional operations to the input image $\mathbf{x}$:

$$\mathbf{h}_l = \sigma(\mathbf{W}_l * \mathbf{h}_{l-1} + \mathbf{b}_l)$$

   Here, $\mathbf{W}_l$ and $\mathbf{b}_l$ are the weights and biases of the l-th convolutional layer, $*$ denotes the convolution operation, and σ is a non-linear activation function such as ReLU. Pooling layers may be used to reduce the spatial dimensions.
2. Latent Representation: The final output of the encoder is a low-dimensional latent representation $\mathbf{z}$:

$$\mathbf{z} = f_\theta(\mathbf{x})$$

3. Decoder: The decoder reconstructs the image by applying transposed convolution operations:

$$\hat{\mathbf{h}}_{l-1} = \sigma(\mathbf{W}_l^T * \hat{\mathbf{h}}_l + \mathbf{b}_l)$$

   The final output is the reconstructed image $\hat{\mathbf{x}}$:

$$\hat{\mathbf{x}} = g_\phi(\mathbf{z})$$

Example: Consider a convolutional autoencoder for image denoising, where the goal is to reconstruct clean images from noisy inputs.

```
import torch
import torch.nn as nn
import torch.optim as optim

class ConvAutoencoder(nn.Module):
    def __init__(self):
        super(ConvAutoencoder, self).__init__()
        # Encoder
```

```
        self.encoder = nn.Sequential(
            nn.Conv2d(1, 16, 3, stride=2, padding=1),  # (batch, 1, 28, 28) ->
(batch, 16, 14, 14)
            nn.ReLU(),
            nn.Conv2d(16, 32, 3, stride=2, padding=1),  # (batch, 16, 14, 14) ->
(batch, 32, 7, 7)
            nn.ReLU()
        )
        # Decoder
        self.decoder = nn.Sequential(
            nn.ConvTranspose2d(32, 16, 3, stride=2, padding=1, output_padding=1),  #
(batch, 32, 7, 7) -> (batch, 16, 14, 14)
            nn.ReLU(),
            nn.ConvTranspose2d(16, 1, 3, stride=2, padding=1, output_padding=1),  #
(batch, 16, 14, 14) -> (batch, 1, 28, 28)
            nn.Sigmoid()
        )

    def forward(self, x):
        z = self.encoder(x)
        x_hat = self.decoder(z)
        return x_hat

# Training loop
model = ConvAutoencoder()
criterion = nn.MSELoss()
optimizer = optim.Adam(model.parameters(), lr=0.001)

for epoch in range(10):
    for data in dataloader:
        noisy_data = data + 0.1 * torch.randn_like(data)  # Adding Gaussian noise
        optimizer.zero_grad()
        output = model(noisy_data)
        loss = criterion(output, data)
        loss.backward()
        optimizer.step()
    print(f'Epoch {epoch+1}, Loss: {loss.item()}')
```

Applications in Image Reconstruction

Convolutional autoencoders are widely used in applications requiring image reconstruction, including:

1. Image Denoising: CAEs are effective at removing noise from images, improving the quality of visual data. They are commonly used in medical imaging to enhance the clarity of diagnostic images.
2. Image Inpainting: CAEs can be used to fill in missing parts of an image, a technique known as image inpainting. This is useful in applications such as restoring damaged photographs or removing unwanted objects from images.
3. Super-Resolution: CAEs can be trained to generate high-resolution images from low-resolution inputs. This application is valuable in fields like satellite imaging and video enhancement.
4. Feature Extraction: The features learned by the encoder of a CAE can be used for downstream tasks such as image classification and object detection, providing a rich representation of the input data.

10.4.2 Recurrent Autoencoders

Recurrent autoencoders (RAEs) are designed to handle sequential data, leveraging the power of RNNs to capture temporal dependencies in the data. These architectures are particularly suited for time-series data and sequential text data.

Architecture for Sequential Data

The architecture of a recurrent autoencoder consists of an encoder and a decoder, both built using recurrent layers such as LSTMs or GRUs. The encoder processes the input sequence to produce a hidden state that captures the temporal dependencies, while the decoder reconstructs the sequence from this hidden state.

1. Encoder: The encoder processes the input sequence $\mathbf{x} = (\mathbf{x}_1, \mathbf{x}_2, \ldots, \mathbf{x}_T)$ using recurrent layers to produce a sequence of hidden states $\mathbf{h} = (\mathbf{h}_1, \mathbf{h}_2, \ldots, \mathbf{h}_T)$:

$$\mathbf{h}_t = f_\theta(\mathbf{x}_t, \mathbf{h}_{t-1})$$

   The final hidden state $\mathbf{h}_T$ serves as the latent representation $\mathbf{z}$.
2. Decoder: The decoder reconstructs the sequence from the latent representation $\mathbf{z}$ using recurrent layers:

$$\hat{\mathbf{x}}_t = g_\phi(\hat{\mathbf{x}}_{t-1}, \mathbf{h}_t)$$

   The initial input to the decoder is typically a special start-of-sequence token, and the hidden state $\mathbf{h}_t$ is initialized with the latent representation $\mathbf{z}$.

Example: Consider a recurrent autoencoder for sequence-to-sequence modeling of time-series data.

```
import torch
import torch.nn as nn
import torch.optim as optim

class RecurrentAutoencoder(nn.Module):
    def __init__(self, input_dim, hidden_dim, num_layers):
        super(RecurrentAutoencoder, self).__init__()
        self.encoder = nn.LSTM(input_dim, hidden_dim, num_layers, batch_first=True)
        self.decoder = nn.LSTM(hidden_dim, input_dim, num_layers, batch_first=True)
        self.hidden_dim = hidden_dim
        self.num_layers = num_layers

    def forward(self, x):
        # Encode
        _, (h_n, _) = self.encoder(x)
        h_n = h_n[-1]  # Get the hidden state of the last layer
        h_n = h_n.unsqueeze(0).repeat(self.num_layers, 1, 1)
        c_n = torch.zeros_like(h_n)  # Initialize cell state
        # Decode
        outputs, _ = self.decoder(x, (h_n, c_n))
        return outputs

# Training loop
model = RecurrentAutoencoder(input_dim=1, hidden_dim=50, num_layers=2)
criterion = nn.MSELoss()
```

```
optimizer = optim.Adam(model.parameters(), lr=0.001)

for epoch in range(10):
    for data in dataloader:
        optimizer.zero_grad()
        output = model(data)
        loss = criterion(output, data)
        loss.backward()
        optimizer.step()
    print(f'Epoch {epoch+1}, Loss: {loss.item()}')
```

Applications in Time Series and Text Data

Recurrent autoencoders are widely used in applications involving sequential data, including:

1. Time Series Forecasting: RAEs can be used for time series forecasting by learning the underlying patterns in the data and predicting future values. This is valuable in applications such as stock price prediction, weather forecasting, and demand forecasting.
2. Anomaly Detection: RAEs can detect anomalies in time series data by identifying deviations from the normal sequence patterns. This is useful in applications such as fault detection in machinery, fraud detection in financial transactions, and monitoring of critical systems.
3. Text Generation: RAEs can generate coherent text by learning the structure and dependencies in the training data. This is useful in applications such as automated content creation, chatbots, and language translation.
4. Sequence-to-Sequence Learning: RAEs can be used for sequence-to-sequence learning tasks, where the goal is to map an input sequence to an output sequence. This includes applications such as machine translation, text summarization, and speech recognition.

10.4.3 Hybrid Autoencoders

Hybrid autoencoders combine the strengths of different types of autoencoders to handle complex and multimodal data. By integrating various architectures, such as convolutional and recurrent layers, hybrid autoencoders can effectively process data with both spatial and temporal dependencies. These models are particularly useful in applications where the data exhibits multiple characteristics, requiring a more flexible and comprehensive approach.

Combining Different Autoencoder Types

Hybrid autoencoders leverage the advantages of various autoencoder architectures to address specific challenges presented by complex datasets. For instance, combining convolutional autoencoders (CAEs) with recurrent autoencoders (RAEs) allows the model to capture both spatial and temporal features in multimodal data, such as video sequences or sensor data streams.

1. Encoder: The encoder in a hybrid autoencoder typically starts with convolutional layers to extract spatial features from the input data, followed by recurrent layers to capture temporal dependencies. For an input sequence of images $\mathbf{X} = (\mathbf{x}_1, \mathbf{x}_2, \ldots, \mathbf{x}_T)$, the convolutional layers process each image $\mathbf{x}_t$ independently:

$$\mathbf{h}_t^{(1)} = \sigma(\mathbf{W}_1 * \mathbf{x}_t + \mathbf{b}_1)$$

$$\mathbf{h}_t^{(2)} = \sigma(\mathbf{W}_2 * \mathbf{h}_t^{(1)} + \mathbf{b}_2)$$

   The output of the convolutional layers is then fed into a recurrent layer to capture the temporal dependencies:

$$\mathbf{z}_t = f_\theta(\mathbf{h}_t^{(2)}, \mathbf{z}_{t-1})$$

   The final hidden state $\mathbf{z}_T$ serves as the latent representation.
2. Decoder: The decoder reconstructs the sequence by first generating latent variables through a recurrent layer and then using transposed convolutional layers to produce the output images:

$$\hat{\mathbf{h}}_t^{(2)} = g_\phi(\hat{\mathbf{h}}_{t+1}^{(2)}, \mathbf{z}_t)$$

$$\hat{\mathbf{h}}_t^{(1)} = \sigma(\mathbf{W}_3^T * \hat{\mathbf{h}}_t^{(2)} + \mathbf{b}_3)$$

$$\hat{\mathbf{x}}_t = \sigma(\mathbf{W}_4^T * \hat{\mathbf{h}}_t^{(1)} + \mathbf{b}_4)$$

Example: Consider a hybrid autoencoder designed for video sequence reconstruction, where the goal is to reconstruct a sequence of frames from a noisy video input.

```
import torch
import torch.nn as nn
import torch.optim as optim

class HybridAutoencoder(nn.Module):
    def __init__(self, input_channels, conv_dim, hidden_dim, num_layers):
        super(HybridAutoencoder, self).__init__()
        # Convolutional encoder
        self.conv_encoder = nn.Sequential(
            nn.Conv2d(input_channels, conv_dim, kernel_size=3, stride=2, padding=1),
            nn.ReLU(),
            nn.Conv2d(conv_dim, conv_dim * 2, kernel_size=3, stride=2, padding=1),
            nn.ReLU()
        )
        # Recurrent encoder
        self.rnn_encoder = nn.LSTM(conv_dim * 2 * 7 * 7, hidden_dim, num_layers,
 batch_first=True)
        # Recurrent decoder
        self.rnn_decoder = nn.LSTM(hidden_dim, conv_dim * 2 * 7 * 7, num_layers,
batch_first=True)
```

```
        # Convolutional decoder
        self.conv_decoder = nn.Sequential(
            nn.ConvTranspose2d(conv_dim * 2, conv_dim, kernel_size=3, stride=2,
padding=1, output_padding=1),
            nn.ReLU(),
            nn.ConvTranspose2d(conv_dim, input_channels, kernel_size=3, stride=2,
padding=1, output_padding=1),
            nn.Sigmoid()
        )

    def forward(self, x):
        # Encode
        batch_size, seq_len, c, h, w = x.size()
        x = x.view(batch_size * seq_len, c, h, w)
        x = self.conv_encoder(x)
        x = x.view(batch_size, seq_len, -1)
        _, (h_n, _) = self.rnn_encoder(x)
        h_n = h_n[-1]  # Get the last hidden state
        # Decode
        h_n = h_n.unsqueeze(0).repeat(seq_len, 1, 1)
        x, _ = self.rnn_decoder(h_n)
        x = x.view(batch_size * seq_len, -1, 7, 7)
        x = self.conv_decoder(x)
        x = x.view(batch_size, seq_len, c, h, w)
        return x

# Training loop
model = HybridAutoencoder(input_channels=3, conv_dim=64, hidden_dim=256, num_layers=2)
criterion = nn.MSELoss()
optimizer = optim.Adam(model.parameters(), lr=0.001)

for epoch in range(10):
    for data in dataloader:
        optimizer.zero_grad()
        output = model(data)
        loss = criterion(output, data)
        loss.backward()
        optimizer.step()
    print(f'Epoch {epoch+1}, Loss: {loss.item()}')
```

Applications in Multimodal Data Processing

Hybrid autoencoders are particularly effective in processing multimodal data, where the data exhibits multiple types of dependencies, such as spatial, temporal, and contextual. Key applications include:

1. Video Analysis: Hybrid autoencoders can be used for video denoising, inpainting, and super-resolution by leveraging the spatial information from individual frames and the temporal coherence across frames.
2. Multimodal Sensor Data: In applications such as autonomous driving or smart homes, data from various sensors (e.g., cameras, LIDAR, and microphones) can be processed using hybrid autoencoders to fuse different types of information and enhance perception and decision-making.
3. Speech and Audio Processing: Hybrid autoencoders can process speech signals by combining convolutional layers to capture local patterns in the audio spectrograms and recurrent layers to model the temporal dynamics of speech.

4. Healthcare Data: In healthcare, hybrid autoencoders can integrate various types of patient data, such as medical images, time-series vital signs, and clinical notes, to improve diagnostics, treatment planning, and outcome prediction.

Example: Consider a hybrid autoencoder for multimodal sensor data in an autonomous vehicle. The model processes images from cameras and time-series data from LIDAR sensors to detect obstacles and navigate the environment.

10.5 Introduction to Generative Models

Generative models are a class of machine learning models that aim to generate new data samples that resemble the training data. These models learn the underlying distribution of the data and can be used to produce new instances that share the same properties as the original data. Generative models are essential for tasks such as data augmentation, image synthesis, and creative applications.

10.5.1 Overview of Generative Models

Generative models are designed to model the distribution $p(\mathbf{x})$ of the data, allowing them to generate new samples $\mathbf{x}$ that are similar to those in the training set. Unlike discriminative models, which focus on learning the decision boundary between classes, generative models capture the joint probability distribution of the features. The goal of a generative model is to learn the distribution $p(\mathbf{x})$ from a set of training samples $\{\mathbf{x}_i\}_{i=1}^N$. Once the model has learned this distribution, it can generate new samples by drawing from $p(\mathbf{x})$.

Let θ represent the parameters of the generative model. The model is trained to maximize the likelihood of the training data:

$$\mathcal{L}(\theta) = \sum_{i=1}^{N} \log p(\mathbf{x}_i; \theta)$$

In practice, the exact form of $p(\mathbf{x})$ and the optimization process depend on the specific type of generative model used.

10.5.2 Types of Generative Models

Generative models can be broadly classified into two categories: explicit density models and implicit density models. Each category has its own set of techniques and approaches for modeling data distributions.

Explicit Density Models

Explicit density models explicitly define and estimate the probability density function $p(\mathbf{x})$. These models can be further divided into two types: parametric and non-parametric models.

1. Parametric Models: Parametric models assume a specific form for the probability distribution and estimate the parameters of this distribution. Examples include Gaussian Mixture Models (GMMs) and Hidden Markov Models (HMMs).
   Gaussian Mixture Models (GMMs): GMMs model the data as a mixture of multiple Gaussian distributions. The probability density function is given by:

   $$p(\mathbf{x}) = \sum_{k=1}^{K} \pi_k \mathcal{N}(\mathbf{x} \mid \mu_k, \Sigma_k)$$

   where π_k are the mixture weights, μ_k are the means, and Σ_k are the covariance matrices of the Gaussian components.
2. Non-Parametric Models: Non-parametric models do not assume a specific form for the probability distribution. Instead, they use the data to estimate the density function. Examples include Kernel Density Estimation (KDE) and k-Nearest Neighbors (k-NN).

Kernel Density Estimation (KDE): KDE estimates the probability density function by summing kernel functions centered at each data point:

$$p(\mathbf{x}) = \frac{1}{Nh^d} \sum_{i=1}^{N} K\left(\frac{\mathbf{x} - \mathbf{x}_i}{h}\right)$$

where K is the kernel function, h is the bandwidth parameter, and d is the dimensionality of the data.

Implicit Density Models

Implicit density models do not explicitly define the probability density function $p(\mathbf{x})$. Instead, they define a process for generating samples from the distribution. Examples include Generative Adversarial Networks (GANs) and Variational Autoencoders (VAEs).

1. Generative Adversarial Networks (GANs): GANs consist of two neural networks: a generator G and a discriminator D. The generator produces synthetic samples, while the discriminator distinguishes between real and synthetic samples. The training process involves a minimax game:

$$\min_G \max_D \mathbb{E}_{\mathbf{x}\sim p_{\text{data}}}[\log D(\mathbf{x})] + \mathbb{E}_{\mathbf{z}\sim p_{\mathbf{z}}}[\log(1 - D(G(\mathbf{z})))]$$

where $\mathbf{z}$ is a random noise vector.

2. Variational Autoencoders (VAEs):
   VAEs introduce a probabilistic approach to autoencoders, modeling the latent space with a probability distribution. The objective is to maximize the evidence lower bound (ELBO):

$$\mathcal{L}(\theta, \phi) = \mathbb{E}_{q_\phi(\mathbf{z}|\mathbf{x})}[\log p_\theta(\mathbf{x} \mid \mathbf{z})] - \text{KL}(q_\phi(\mathbf{z} \mid \mathbf{x}) \| p(\mathbf{z}))$$

Here, $q_\phi(\mathbf{z} \mid \mathbf{x})$ is the approximate posterior distribution, and $p(\mathbf{z})$ is the prior distribution.

10.5.3 Applications in Data Generation and Augmentation

Generative models have numerous applications in data generation and augmentation, providing valuable tools for enhancing datasets and creating new content.

1. Data Augmentation: Generative models can augment training datasets by generating new samples that resemble the original data. This is particularly useful in scenarios with limited data, helping to improve the performance and generalization of machine learning models.
2. Image Synthesis: Generative models can create realistic images from scratch. Applications include generating high-resolution images, creating artistic content, and producing images for virtual environments.
3. Text Generation: Generative models can produce coherent and contextually relevant text. This has applications in automated content creation, dialogue systems, and language translation.
4. Drug Discovery: In drug discovery, generative models can generate novel molecular structures with desired properties. This accelerates the exploration of new chemical compounds and aids in the development of new drugs.
5. Anomaly Detection: Generative models can be used for anomaly detection by modeling the distribution of normal data. Data points that deviate significantly from this distribution can be identified as anomalies.
6. Creative Applications: Generative models are used in creative fields to generate art, music, and other forms of content. They enable artists and designers to explore new styles and create unique pieces.

Example: Consider training a GAN to generate synthetic images of handwritten digits. The generator network produces images from random noise vectors, while the discriminator network distinguishes between real and synthetic images. The training process involves optimizing both networks to improve the quality of the generated images.

```
import torch
import torch.nn as nn
import torch.optim as optim

class Generator(nn.Module):
    def __init__(self, input_dim, output_dim):
        super(Generator, self).__init__()
        self.model = nn.Sequential(
            nn.Linear(input_dim, 128),
            nn.ReLU(),
            nn.Linear(128, 256),
            nn.ReLU(),
            nn.Linear(256, output_dim),
            nn.Tanh()
        )

    def forward(self, x):
        return self.model(x)

class Discriminator(nn.Module):
    def __init__(self, input_dim):
        super(Discriminator, self).__init__()
        self.model = nn.Sequential(
            nn.Linear(input_dim, 256),
            nn.LeakyReLU(0.2),
            nn.Linear(256, 128),
            nn.LeakyReLU(0.2),
            nn.Linear(128, 1),
            nn.Sigmoid()
        )

    def forward(self, x):
        return self.model(x)

# Training loop
generator = Generator(input_dim=100, output_dim=784)
discriminator = Discriminator(input_dim=784)
criterion = nn.BCELoss()
optimizer_G = optim.Adam(generator.parameters(), lr=0.0002)
optimizer_D = optim.Adam(discriminator.parameters(), lr=0.0002)

for epoch in range(10):
    for data in dataloader:
        # Train Discriminator
        optimizer_D.zero_grad()
        real_data = data
        fake_data = generator(torch.randn(data.size(0), 100))
        real_labels = torch.ones(data.size(0), 1)
        fake_labels = torch.zeros(data.size(0), 1)
        loss_real = criterion(discriminator(real_data), real_labels)
        loss_fake = criterion(discriminator(fake_data.detach()), fake_labels)
        loss_D = loss_real + loss_fake
        loss_D.backward()
        optimizer_D.step()

        # Train Generator
        optimizer_G.zero_grad()
        loss_G = criterion(discriminator(fake_data), real_labels)
        loss_G.backward()
        optimizer_G.step()

    print(f'Epoch {epoch+1}, Loss D: {loss_D.item()}, Loss G: {loss_G.item()}')
```

10.6 Generative Adversarial Networks (GANs)

Generative Adversarial Networks (GANs) are a class of generative models that have gained significant attention for their ability to produce highly realistic synthetic data. GANs consist of two neural networks, a generator and a discriminator, that compete against each other in a minimax game. This section delves into the architecture of GANs and the intricacies of adversarial training.

10.6.1 Basic Architecture of GANs

The architecture of GANs involves two main components: the generator network and the discriminator network. These networks are trained simultaneously with opposing objectives, leading to a dynamic and competitive learning process.

Generator Network

The generator network G aims to produce synthetic data that resembles the real data distribution. It takes a random noise vector $\mathbf{z}$ as input and transforms it into a data sample $G(\mathbf{z})$. The objective of the generator is to generate data that the discriminator cannot distinguish from real data. Let $\mathbf{z}$ be a noise vector drawn from a prior distribution $p_{\mathbf{z}}$, typically a standard normal distribution $\mathcal{N}(0, I)$. The generator network G maps $\mathbf{z}$ to the data space:

$$G(\mathbf{z}; \theta_G)$$

where θ_G represents the parameters of the generator. The goal of the generator is to maximize the probability that the discriminator classifies the generated samples as real. This can be formulated as minimizing the following loss function:

$$\mathcal{L}_G = \mathbb{E}_{\mathbf{z}\sim p_{\mathbf{z}}}[\log(1 - D(G(\mathbf{z})))]$$

However, for better gradient properties, this is often reformulated to:

$$\mathcal{L}_G = -\mathbb{E}_{\mathbf{z}\sim p_{\mathbf{z}}}[\log D(G(\mathbf{z}))]$$

Example: Consider a generator network designed to generate 28x28 grayscale images from a 100-dimensional noise vector:

```
import torch.nn as nn

class Generator(nn.Module):
    def __init__(self, input_dim, output_dim):
        super(Generator, self).__init__()
        self.model = nn.Sequential(
            nn.Linear(input_dim, 128),
            nn.ReLU(),
            nn.Linear(128, 256),
```

```
            nn.ReLU(),
            nn.Linear(256, 512),
            nn.ReLU(),
            nn.Linear(512, output_dim),
            nn.Tanh()
        )

    def forward(self, x):
        return self.model(x)
```

Discriminator Network

The discriminator network D aims to distinguish between real data samples and synthetic data generated by the generator. It takes a data sample $\mathbf{x}$ as input and outputs a probability $D(\mathbf{x})$ indicating the likelihood that $\mathbf{x}$ is real. The discriminator network D maps a data sample $\mathbf{x}$ to a probability score:

$$D(\mathbf{x}; \theta_D)$$

where θ_D represents the parameters of the discriminator. The goal of the discriminator is to maximize the probability of correctly classifying real and fake samples. This can be formulated as maximizing the following loss function:

$$\mathcal{L}_D = \mathbb{E}_{\mathbf{x} \sim p_{\text{data}}}[\log D(\mathbf{x})] + \mathbb{E}_{\mathbf{z} \sim p_{\mathbf{z}}}[\log(1 - D(G(\mathbf{z})))]$$

Example: Consider a discriminator network designed to classify 28x28 grayscale images as real or fake:

```
import torch.nn as nn

class Discriminator(nn.Module):
    def __init__(self, input_dim):
        super(Discriminator, self).__init__()
        self.model = nn.Sequential(
            nn.Linear(input_dim, 512),
            nn.LeakyReLU(0.2),
            nn.Linear(512, 256),
            nn.LeakyReLU(0.2),
            nn.Linear(256, 1),
            nn.Sigmoid()
        )

    def forward(self, x):
        return self.model(x)
```

10.6.2 Adversarial Training

Adversarial training involves the simultaneous training of the generator and discriminator networks through a minimax game. The generator seeks to fool the discriminator, while the discriminator strives to correctly classify real and synthetic samples.

Minimax Game and Objective Functions

The training process of GANs can be viewed as a minimax game with the following objective function:

$$\min_G \max_D \mathbb{E}_{\mathbf{x}\sim p_{\text{data}}}[\log D(\mathbf{x})] + \mathbb{E}_{\mathbf{z}\sim p_{\mathbf{z}}}[\log(1 - D(G(\mathbf{z})))]$$

In this game, the discriminator D tries to maximize the objective by correctly classifying real and fake samples, while the generator G tries to minimize the objective by producing realistic samples that can deceive the discriminator.

Optimization Process:

1. Update Discriminator: For each batch of real data $\mathbf{x}$ and noise vectors $\mathbf{z}$: Generate fake data: $\mathbf{z} \sim p_{\mathbf{z}}$, Compute discriminator loss: $\mathcal{L}_D$, Backpropagate and update θ_D

$$\theta_D \leftarrow \theta_D + \eta \nabla_{\theta_D} \mathcal{L}_D$$

2. Update Generator: For each batch of noise vectors $\mathbf{z}$: Generate fake data: $\mathbf{z} \sim p_{\mathbf{z}}$, Compute generator loss: $\mathcal{L}_G$, Backpropagate and update θ_G

$$\theta_G \leftarrow \theta_G - \eta \nabla_{\theta_G} \mathcal{L}_G$$

Training Challenges and Solutions

Training GANs can be challenging due to issues such as instability, mode collapse, and convergence problems. Several techniques have been proposed to address these challenges:

1. Instability: GAN training can be unstable, leading to oscillations and divergence. Techniques such as feature matching, mini-batch discrimination, and one-sided label smoothing can help stabilize the training process.
2. Mode Collapse: Mode collapse occurs when the generator produces limited diversity in its outputs, focusing on a few modes of the data distribution. Techniques such as unrolled GANs, gradient penalties, and multi-generator architectures can mitigate mode collapse.
3. Convergence Issues: GANs can have difficulty converging due to the adversarial nature of the training process. Approaches like progressive growing of GANs, spectral normalization, and using more sophisticated loss functions (e.g., Wasserstein loss) can improve convergence.

Example: Consider the training loop for a GAN:

```
import torch.optim as optim

# Initialize models and optimizers
generator = Generator(input_dim=100, output_dim=784)
discriminator = Discriminator(input_dim=784)
optimizer_G = optim.Adam(generator.parameters(), lr=0.0002)
optimizer_D = optim.Adam(discriminator.parameters(), lr=0.0002)
criterion = nn.BCELoss()

for epoch in range(num_epochs):
    for real_data in dataloader:
        # Train Discriminator
        optimizer_D.zero_grad()
        batch_size = real_data.size(0)
        real_labels = torch.ones(batch_size, 1)
        fake_labels = torch.zeros(batch_size, 1)
        noise = torch.randn(batch_size, 100)
        fake_data = generator(noise)
        loss_real = criterion(discriminator(real_data), real_labels)
        loss_fake = criterion(discriminator(fake_data.detach()), fake_labels)
        loss_D = loss_real + loss_fake
        loss_D.backward()
        optimizer_D.step()

        # Train Generator
        optimizer_G.zero_grad()
        noise = torch.randn(batch_size, 100)
        fake_data = generator(noise)
        loss_G = criterion(discriminator(fake_data), real_labels)
        loss_G.backward()
        optimizer_G.step()

    print(f'Epoch {epoch+1}, Loss D: {loss_D.item()}, Loss G: {loss_G.item()}')
```

10.6.3 Variants of GANs

GANs have seen numerous variations since their inception, each tailored to address specific challenges or expand their capabilities. Among the most influential variants are Conditional GANs (cGANs), CycleGANs, and StyleGANs. These models extend the basic GAN architecture to handle additional input conditions, unpaired data transformations, and high-resolution image synthesis, respectively.

Conditional GANs

Conditional GANs (cGANs) extend the GAN framework by incorporating additional information into both the generator and discriminator, enabling the generation of data conditioned on specific attributes. This conditioning can be in the form of class labels, text descriptions, or other modalities, allowing for more controlled and targeted data generation. In a cGAN, both the generator G and discriminator D are conditioned on an auxiliary variable y. The objective function is modified to include this conditioning:

$$\min_G \max_D \mathbb{E}_{\mathbf{x}\sim p_{\text{data}}(\mathbf{x})}[\log D(\mathbf{x} \mid y)] + \mathbb{E}_{\mathbf{z}\sim p_{\mathbf{z}}(\mathbf{z})}[\log(1 - D(G(\mathbf{z} \mid y) \mid y))]$$

Here, y could be a class label or any other conditional information.

Example: Consider a cGAN that generates images conditioned on class labels:

```
import torch
import torch.nn as nn

class ConditionalGenerator(nn.Module):
    def __init__(self, noise_dim, class_dim, output_dim):
        super(ConditionalGenerator, self).__init__()
        self.model = nn.Sequential(
            nn.Linear(noise_dim + class_dim, 256),
            nn.ReLU(),
            nn.Linear(256, 512),
            nn.ReLU(),
            nn.Linear(512, output_dim),
            nn.Tanh()
        )

    def forward(self, noise, labels):
        input = torch.cat((noise, labels), dim=1)
        return self.model(input)

class ConditionalDiscriminator(nn.Module):
    def __init__(self, input_dim, class_dim):
        super(ConditionalDiscriminator, self).__init__()
        self.model = nn.Sequential(
            nn.Linear(input_dim + class_dim, 512),
            nn.LeakyReLU(0.2),
            nn.Linear(512, 256),
            nn.LeakyReLU(0.2),
            nn.Linear(256, 1),
            nn.Sigmoid()
        )

    def forward(self, data, labels):
        input = torch.cat((data, labels), dim=1)
        return self.model(input)

# Training loop
generator = ConditionalGenerator(noise_dim=100, class_dim=10, output_dim=784)
discriminator = ConditionalDiscriminator(input_dim=784, class_dim=10)
# ... (same training loop structure as basic GAN, but with label inputs)
```

Applications:

Image Synthesis: Generating images conditioned on specific classes or attributes (e.g., generating faces with specific features).

Text-to-Image: Generating images from textual descriptions by conditioning on the text embeddings.

CycleGANs

CycleGANs are designed to learn mappings between two domains without requiring paired training examples. This is particularly useful for tasks such as image-to-image translation where paired datasets are scarce. CycleGANs consist of two generators G

and F and two discriminators D_X and D_Y. The goal is to learn mappings $G : X \to Y$ and $F : Y \to X$ such that the translations are cycle-consistent:

$$\mathcal{L}_{\text{cycle}}(G, F) = \mathbb{E}_{\mathbf{x}\sim p_{\text{data}}(X)}[\|F(G(\mathbf{x})) - \mathbf{x}\|_1] + \mathbb{E}_{\mathbf{y}\sim p_{\text{data}}(Y)}[\|G(F(\mathbf{y})) - \mathbf{y}\|_1]$$

The adversarial losses for the two mappings are:

$$\mathcal{L}_{\text{GAN}}(G, D_Y, X, Y) = \mathbb{E}_{\mathbf{y}\sim p_{\text{data}}(Y)}[\log D_Y(\mathbf{y})] + \mathbb{E}_{\mathbf{x}\sim p_{\text{data}}(X)}[\log(1 - D_Y(G(\mathbf{x})))]$$

$$\mathcal{L}_{\text{GAN}}(F, D_X, Y, X) = \mathbb{E}_{\mathbf{x}\sim p_{\text{data}}(X)}[\log D_X(\mathbf{x})] + \mathbb{E}_{\mathbf{y}\sim p_{\text{data}}(Y)}[\log(1 - D_X(F(\mathbf{y})))]$$

Total Loss:

$$\mathcal{L}(G, F, D_X, D_Y) = \mathcal{L}_{\text{GAN}}(G, D_Y, X, Y) + \mathcal{L}_{\text{GAN}}(F, D_X, Y, X) + \lambda\mathcal{L}_{\text{cycle}}(G, F)$$

Example: CycleGAN implementation for unpaired image-to-image translation:

```
import torch.nn as nn

class ResidualBlock(nn.Module):
    def __init__(self, dim):
        super(ResidualBlock, self).__init__()
        self.block = nn.Sequential(
            nn.Conv2d(dim, dim, kernel_size=3, padding=1),
            nn.InstanceNorm2d(dim),
            nn.ReLU(inplace=True),
            nn.Conv2d(dim, dim, kernel_size=3, padding=1),
            nn.InstanceNorm2d(dim)
        )

    def forward(self, x):
        return x + self.block(x)

class Generator(nn.Module):
    def __init__(self, input_nc, output_nc, n_res_blocks=9):
        super(Generator, self).__init__()
        layers = [
            nn.Conv2d(input_nc, 64, kernel_size=7, stride=1, padding=3),
            nn.InstanceNorm2d(64),
            nn.ReLU(inplace=True)
        ]
        for _ in range(n_res_blocks):
            layers += [ResidualBlock(64)]
        layers += [
            nn.Conv2d(64, output_nc, kernel_size=7, stride=1, padding=3),
            nn.Tanh()
        ]
        self.model = nn.Sequential(*layers)

    def forward(self, x):
        return self.model(x)
```

```
class Discriminator(nn.Module):
    def __init__(self, input_nc):
        super(Discriminator, self).__init__()
        self.model = nn.Sequential(
            nn.Conv2d(input_nc, 64, kernel_size=4, stride=2, padding=1),
            nn.LeakyReLU(0.2, inplace=True),
            nn.Conv2d(64, 128, kernel_size=4, stride=2, padding=1),
            nn.InstanceNorm2d(128),
            nn.LeakyReLU(0.2, inplace=True),
            nn.Conv2d(128, 1, kernel_size=4, stride=1, padding=1)
        )

    def forward(self, x):
        return self.model(x)
```

Applications:

Image-to-Image Translation: Tasks like converting images from one style to another (e.g., summer to winter, day to night).

Artistic Style Transfer: Applying artistic styles to photographs without needing paired datasets.

StyleGANs

StyleGANs introduce a novel approach to generating high-resolution images with unprecedented quality. They allow control over the generated image through style transfer and manipulation at different levels of detail. StyleGANs use a mapping network f to transform the latent code $\mathbf{z}$ into an intermediate latent space $\mathbf{w}$. This intermediate representation controls the synthesis network g at various stages, enabling detailed style manipulation. The synthesis network generates images by progressively increasing resolution, with noise inputs at each layer to enhance stochastic variations:

$$\mathbf{w} = f(\mathbf{z})$$

$$\mathbf{x} = g(\mathbf{w}, \mathbf{n})$$

Example: StyleGAN implementation for high-resolution image generation:

```
import torch.nn as nn

class MappingNetwork(nn.Module):
    def __init__(self, latent_dim, dlatent_dim, num_layers=8):
        super(MappingNetwork, self).__init__()
        layers = []
        for _ in range(num_layers):
            layers.append(nn.Linear(latent_dim, dlatent_dim))
            layers.append(nn.LeakyReLU(0.2))
        self.model = nn.Sequential(*layers)

    def forward(self, z):
        return self.model(z)

class SynthesisNetwork(nn.Module):
    def __init__(self, dlatent_dim, channels):
```

```
        super(SynthesisNetwork, self).__init__()
        self.initial_block = nn.Sequential(
            nn.Linear(dlatent_dim, channels * 4 * 4),
            nn.LeakyReLU(0.2)
        )
        self.blocks = nn.ModuleList([
            nn.Sequential(
                nn.Conv2d(channels, channels, kernel_size=3, padding=1),
                nn.LeakyReLU(0.2)
            ) for _ in range(4)
        ])
        self.to_rgb = nn.Conv2d(channels, 3, kernel_size=1)

    def forward(self, w, noise):
        x = self.initial_block(w).view(-1, 512, 4, 4)
        for block in self.blocks:
            x = block(x + noise)
        return self.to_rgb(x)

class StyleGAN(nn.Module):
    def __init__(self, latent_dim, dlatent_dim, channels):
        super(StyleGAN, self).__init__()
        self.mapping = MappingNetwork(latent_dim, dlatent_dim)
        self.synthesis = SynthesisNetwork(dlatent_dim, channels)

    def forward(self, z, noise):
        w = self.mapping(z)
        return self.synthesis(w, noise)
```

Applications:

High-Resolution Image Synthesis: Generating photorealistic images with fine-grained control over details.

Style Transfer: Manipulating specific styles (e.g., texture, color) of generated images for creative applications.

10.7 Advanced Topics in Generative Models

Generative models, particularly GANs, have achieved remarkable success, but training them can be challenging due to issues like instability and mode collapse. This section explores advanced techniques designed to improve GAN training, making the models more stable and reliable.

10.7.1 Improving GAN Training

Several advanced techniques have been developed to address the challenges inherent in training GANs. Among these, Wasserstein GANs, spectral normalization, and progressive growing of GANs are notable for their effectiveness in enhancing training stability and quality.

Wasserstein GANs

Wasserstein GANs (WGANs) reformulate the GAN objective to address the issues of instability and mode collapse by leveraging the Wasserstein distance (also known as the Earth Mover's distance) instead of the Jensen-Shannon divergence. The Wasserstein distance $W(p_{\text{data}}, p_G)$ between the real data distribution p_{data} and the generator's distribution p_G is defined as:

$$W(p_{\text{data}}, p_G) = \inf_{\gamma \in \Pi(p_{\text{data}}, p_G)} \mathbb{E}_{(\mathbf{x},\mathbf{y})\sim\gamma}[\|\mathbf{x} - \mathbf{y}\|]$$

where $\Pi(p_{\text{data}}, p_G)$ denotes the set of all joint distributions $\gamma(\mathbf{x}, \mathbf{y})$ whose marginals are p_{data} and p_G. To make this distance computationally feasible, WGAN employs the Kantorovich-Rubinstein duality, which reformulates the Wasserstein distance as:

$$W(p_{\text{data}}, p_G) = \frac{1}{K} \sup_{\|f\|_L \leq K} \mathbb{E}_{\mathbf{x}\sim p_{\text{data}}}[f(\mathbf{x})] - \mathbb{E}_{\mathbf{z}\sim p_{\mathbf{z}}}[f(G(\mathbf{z}))]$$

Here, f is a K-Lipschitz function. In practice, this supremum is approximated using a neural network (the critic network), and the K-Lipschitz constraint is enforced by weight clipping.

Objective Function: The WGAN objective for the critic network C and the generator G is:

$$\mathcal{L}_C = -\mathbb{E}_{\mathbf{x}\sim p_{\text{data}}}[C(\mathbf{x})] + \mathbb{E}_{\mathbf{z}\sim p_{\mathbf{z}}}[C(G(\mathbf{z}))]$$

$$\mathcal{L}_G = -\mathbb{E}_{\mathbf{z}\sim p_{\mathbf{z}}}[C(G(\mathbf{z}))]$$

The critic is trained to maximize $\mathcal{L}_C$, while the generator is trained to minimize $\mathcal{L}_G$.

Spectral Normalization

Spectral normalization is a technique designed to stabilize the training of GANs by normalizing the spectral norm of the weight matrices in the discriminator. This helps to control the Lipschitz constant of the discriminator, which is crucial for maintaining the balance between the generator and the discriminator. For a weight matrix W in the discriminator, the spectral norm $\sigma(W)$ is defined as the largest singular value of W:

$$\sigma(W) = \sup_{\|\mathbf{h}\|_2 \leq 1} \|W\mathbf{h}\|_2$$

Spectral normalization normalizes W by dividing it by its spectral norm:

$$\tilde{W} = \frac{W}{\sigma(W)}$$

The normalized weight matrix $\tilde{W}$ is then used in place of W in the discriminator, ensuring that the Lipschitz constant of the discriminator is controlled, which is vital for stable GAN training. Spectral normalization is applied to each weight matrix W in the discriminator network. The process involves computing the spectral norm using power iteration and normalizing the weight matrix accordingly.

Progressive Growing of GANs

Progressive growing of GANs is a technique introduced to enhance the training of GANs, particularly for high-resolution image generation. The idea is to start training with a low-resolution version of the images and progressively increase the resolution by adding layers to the generator and discriminator.

Training Process:

1. Start with Low Resolution: Training begins with low-resolution images (e.g., 4x4 or 8x8). Both the generator and discriminator networks are initially designed to operate at this resolution.
2. Progressive Layer Addition: New layers are gradually added to both the generator and discriminator networks. Each new layer increases the resolution of the images (e.g., 16x16, 32x32, and so on). During the transition, a blending technique is used to smoothly integrate the new layers.
3. Blending Technique: When transitioning to a higher resolution, the output of the generator is a weighted sum of the outputs from the old and new layers. Similarly, the discriminator processes both the high-resolution images and downsampled versions, smoothly transitioning between resolutions.

For a transition from resolution R to $2R$, the generator's output G_{new} is blended as:

$$G_{\text{blend}}(\mathbf{z}) = \alpha G_{\text{old}}(\mathbf{z}) + (1 - \alpha) G_{\text{new}}(\mathbf{z})$$

where α transitions from 0 to 1 over the course of training.

Advantages:

Stability: Starting with low-resolution images allows the networks to learn stable features before increasing the complexity.

Efficiency: Progressive growing reduces computational requirements initially, as training at lower resolutions is less demanding.

Quality: This method improves the quality of generated high-resolution images by gradually refining details.

10.7.2 Other Generative Models

In addition to GANs and their variants, several other types of generative models have been developed, each with unique approaches to modeling data distributions. These include flow-based models, autoregressive models, and energy-based models. This section explores these models, focusing on their mathematical foundations and applications.

Flow-based Models

Flow-based models are a class of generative models that provide an exact and tractable likelihood for the data by transforming a simple distribution (usually Gaussian) into a complex distribution using a series of invertible transformations. Let $\mathbf{x}$ be the data and $\mathbf{z}$ be the latent variable. A flow-based model defines a series of bijective transformations f that map $\mathbf{x}$ to $\mathbf{z}$:

$$\mathbf{z} = f(\mathbf{x})$$

Given the invertibility of f, the likelihood of $\mathbf{x}$ can be computed using the change of variables formula:

$$p(\mathbf{x}) = p_{\mathbf{z}}(f(\mathbf{x})) \left| \det \left(\frac{\partial f(\mathbf{x})}{\partial \mathbf{x}} \right) \right|$$

where $\det \left(\frac{\partial f(\mathbf{x})}{\partial \mathbf{x}} \right)$ is the determinant of the Jacobian of f at $\mathbf{x}$. For example, consider the RealNVP (Real-valued Non-Volume Preserving) model, which uses affine coupling layers for the transformation:

For $\mathbf{x} = (\mathbf{x}_1, \mathbf{x}_2)$:

$$\mathbf{y}_1 = \mathbf{x}_1$$

$$\mathbf{y}_2 = \mathbf{x}_2 \odot \exp(s(\mathbf{y}_1)) + t(\mathbf{y}_1)$$

where $\odot$ denotes element-wise multiplication, and s and t are scaling and translation functions. The log-likelihood is:

$$\log p(\mathbf{x}) = \log p_{\mathbf{z}}(\mathbf{y}) - \sum_{i=1}^{d/2} s_i(\mathbf{y}_1)$$

Applications:

Density Estimation: Flow-based models provide exact density estimation, making them suitable for tasks requiring precise likelihoods.

Image Generation: Flow-based models can generate high-quality images with exact likelihoods, useful for applications in computer vision.

Autoregressive Models

Autoregressive models generate data one element at a time, conditioning each new element on the previously generated elements. This sequential approach allows for exact likelihood computation and is particularly effective for sequence data. Let $\mathbf{x} = (x_1, x_2, \ldots, x_d)$ be a data vector. An autoregressive model factorizes the joint distribution $p(\mathbf{x})$ as a product of conditional distributions:

$$p(\mathbf{x}) = \prod_{i=1}^{d} p(x_i \mid x_{1:i-1})$$

where $x_{1:i-1}$ denotes the preceding elements in the sequence. For example, consider the PixelRNN model for image generation, where each pixel value is conditioned on all previous pixels:

$$p(\mathbf{x}) = \prod_{i=1}^{d} p(x_i \mid x_{1:i-1})$$

In a PixelRNN, the conditional distribution $p(x_i \mid x_{1:i-1})$ is modeled using a recurrent neural network.

Applications:

Image Generation: PixelRNN and PixelCNN generate images pixel-by-pixel, achieving high-quality results.

Language Modeling: Models like GPT-3 generate text word-by-word or token-by-token, capturing complex language patterns.

Energy-Based Models

Energy-based models (EBMs) define a probability distribution over data using an energy function. The energy function assigns low values to high-probability regions and high values to low-probability regions. These models are flexible and can capture complex dependencies in data. Let $E(\mathbf{x})$ be an energy function that assigns an energy value to each data point $\mathbf{x}$. The probability distribution is defined as:

$$p(\mathbf{x}) = \frac{\exp(-E(\mathbf{x}))}{Z}$$

where Z is the partition function, ensuring that the distribution sums to one:

$$Z = \int \exp(-E(\mathbf{x}))\, d\mathbf{x}$$

Example: Consider the Restricted Boltzmann Machine (RBM), an EBM with an energy function defined as:

$$E(\mathbf{v}, \mathbf{h}) = -\mathbf{v}^\top \mathbf{W}\mathbf{h} - \mathbf{b}^\top \mathbf{v} - \mathbf{c}^\top \mathbf{h}$$

where $\mathbf{v}$ and $\mathbf{h}$ are visible and hidden units, respectively, and $\mathbf{W}$, $\mathbf{b}$, and $\mathbf{c}$ are parameters. The joint distribution is:

$$p(\mathbf{v}, \mathbf{h}) = \frac{\exp(-E(\mathbf{v}, \mathbf{h}))}{Z}$$

Applications:

Density Estimation: EBMs provide a flexible framework for modeling complex data distributions.

Feature Learning: RBMs and Deep Belief Networks (DBNs) are used for unsupervised feature learning.

10.8 Practical Applications of Autoencoders and Generative Models

10.8.1 Style Transfer

Style transfer is a technique that allows the separation and recombination of content and style from different images. By leveraging deep neural networks, particularly convolutional neural networks (CNNs), this process enables the creation of visually appealing images that combine the content of one image with the style of another.

Neural Style Transfer Techniques

Neural style transfer (NST) involves using neural networks to apply the artistic style of one image (the style image) to another image (the content image) while preserving the original content. Neural style transfer typically involves minimizing a loss function that combines a content loss and a style loss.

1. Content Loss: The content loss measures the difference between the feature representations of the content image and the generated image. Let $\mathbf{p}$ be the

content image, **x** the generated image, and P^l and F^l the feature representations at layer l for **p** and **x**, respectively. The content loss is defined as:

$$\mathcal{L}_{\text{content}}^l = \frac{1}{2}\sum_{i,j}(F_{ij}^l - P_{ij}^l)^2$$

2. Style Loss: The style loss measures the difference between the Gram matrices of the style image and the generated image. Let **a** be the style image and A^l and G^l the Gram matrices at layer l for **a** and **x**, respectively. The style loss is defined as:

$$\mathcal{L}_{\text{style}}^l = \frac{1}{4N_l^2 M_l^2}\sum_{i,j}(G_{ij}^l - A_{ij}^l)^2$$

   where N_l and M_l are the dimensions of the feature map at layer l.
3. Total Loss: The total loss is a weighted combination of the content and style losses:

$$\mathcal{L}_{\text{total}} = \alpha\mathcal{L}_{\text{content}} + \beta\sum_{l}\mathcal{L}_{\text{style}}^l$$

   where α and β are the weights for the content and style losses, respectively.

Example: Given a content image of a landscape and a style image of a painting, the neural style transfer process generates a new image that maintains the landscape's structure while adopting the painting's style.

Applications:

Artistic Creation: Artists can use NST to create new artworks by blending different styles with their original content.

Image Enhancement: NST can be used to enhance photographs with artistic styles for commercial or personal use.

Applications and Variants

Various applications and variants of neural style transfer have been developed to enhance its versatility and effectiveness.

Real-Time Style Transfer To make style transfer feasible for real-time applications, models like Fast Style Transfer use feedforward networks to apply the style in a single pass. Instead of iterative optimization, a pre-trained network generates styled images directly:

$$\mathbf{y} = f(\mathbf{x}; \theta)$$

where f is a feedforward network parameterized by θ, which is trained to minimize the style transfer loss.

Multi-style Transfer Some models are designed to handle multiple styles, allowing the user to switch between styles dynamically. For example, a multi-style transfer model can generate images with different artistic styles, such as Picasso, Van Gogh, and Monet, from a single model by conditioning on style codes.

Video Style Transfer Extending style transfer to videos involves maintaining temporal consistency to avoid flickering and artifacts. Temporal consistency loss $\mathcal{L}_{\text{temporal}}$ is added to ensure smooth transitions between frames:

$$\mathcal{L}_{\text{temporal}} = \sum_t \|f(\mathbf{x}_t;\theta) - w \cdot f(\mathbf{x}_{t-1};\theta)\|_2$$

where w is a weighting factor for previous frames.

10.8.2 Data Augmentation

Data augmentation is a technique used to artificially increase the size and diversity of a dataset by generating new data samples. This is particularly useful in deep learning to improve the robustness and generalization of models.

Generating New Data Samples

Generative models can be employed to create new data samples that resemble the training data, thereby augmenting the dataset. Example: GANs can generate realistic images to augment datasets for training deep learning models. Let G be a generator trained on a dataset $\mathbf{X}$. New data samples $\mathbf{x}'$ can be generated by sampling from the latent space $\mathbf{z}$:

$$\mathbf{x}' = G(\mathbf{z}), \quad \mathbf{z} \sim p(\mathbf{z})$$

Applications:

Image Classification: Augmented images can enhance the performance of image classifiers by providing additional training examples.

Medical Imaging: Synthetic medical images can help in training models where annotated data is scarce.

Improving Model Robustness

Data augmentation techniques improve the robustness of models by exposing them to a wider variety of data during training. Example: Training with augmented data can help models generalize better to unseen data, reducing overfitting. Consider an augmentation function T that applies transformations to the data. The training objective with data augmentation becomes:

$$\mathcal{L}_{\text{augmented}} = \mathbb{E}_{\mathbf{x} \sim \mathbf{X}} \mathbb{E}_{T \sim \mathcal{T}} \left[\mathcal{L}(f(T(\mathbf{x})), y) \right]$$

where $\mathcal{L}$ is the loss function, f is the model, $\mathcal{T}$ is the set of transformations, and y is the label.

Types of Data Augmentation:

1. Geometric Transformations: Rotations, translations, and scaling to create variations in spatial configuration.
2. Color Jittering: Adjusting brightness, contrast, and saturation to create variations in color properties.
3. Noise Addition: Adding random noise to images to improve robustness against noisy data.

Applications:

Robust Classification Models: Data augmentation improves the performance of classifiers by making them invariant to transformations.

Domain Adaptation: Augmented data can help models adapt to different domains by simulating various environmental conditions.

10.8.3 Other Applications

Generative models have a wide range of applications beyond image synthesis and data augmentation. This section explores some of the other prominent applications, including anomaly detection, super-resolution imaging, and voice synthesis.

Anomaly Detection

Anomaly detection involves identifying unusual patterns that do not conform to expected behavior. Generative models, particularly autoencoders, are well-suited for this task due to their ability to learn representations of normal data and detect deviations from this norm. Autoencoders learn a compact representation (encoding) of input data and then reconstruct the input from this representation. The reconstruction error is used as an anomaly score:

$$\mathcal{L}_{\text{reconstruction}} = \|\mathbf{x} - \hat{\mathbf{x}}\|^2$$

where $\mathbf{x}$ is the input data and $\hat{\mathbf{x}}$ is the reconstructed data.

Anomaly Score: For an input $\mathbf{x}$, the anomaly score $s(\mathbf{x})$ is the reconstruction error:

$$s(\mathbf{x}) = \|\mathbf{x} - \hat{\mathbf{x}}\|^2$$

Higher reconstruction errors indicate anomalies. For example, in an industrial setting, autoencoders can be trained on normal operational data. During operation, new data is fed through the autoencoder, and high reconstruction errors signal potential anomalies in the system.

Applications:

Industrial Monitoring: Detecting faults in machinery by identifying deviations from normal operational data.

Cybersecurity: Identifying unusual patterns in network traffic that may indicate security breaches.

Super-Resolution Imaging

Super-resolution imaging aims to enhance the resolution of an image by predicting high-resolution details from low-resolution inputs. Generative models, especially GANs, have been highly effective in this domain. Super-resolution can be framed as learning a mapping from low-resolution images $\mathbf{I}_{\text{LR}}$ to high-resolution images $\mathbf{I}_{\text{HR}}$:

$$\mathbf{I}_{\text{HR}} = G(\mathbf{I}_{\text{LR}})$$

where G is a generator network trained to minimize the difference between generated high-resolution images and ground truth high-resolution images.

Loss Function: The loss function for super-resolution typically combines pixel-wise loss and perceptual loss:

$$\mathcal{L}_{\text{SR}} = \mathcal{L}_{\text{pixel}} + \lambda\mathcal{L}_{\text{perceptual}}$$

where $\mathcal{L}_{\text{pixel}} = \|\mathbf{I}_{\text{HR}} - \hat{\mathbf{I}}_{\text{HR}}\|^2$ is the pixel-wise mean squared error, $\mathcal{L}_{\text{perceptual}}$ measures the difference in high-level feature representations obtained from a pre-trained network. Example: GAN-based models like SRGAN (Super-Resolution GAN) use a generator to produce high-resolution images and a discriminator to ensure that the generated images are indistinguishable from real high-resolution images.

Applications:

Medical Imaging: Enhancing the resolution of medical images to reveal finer details.

Satellite Imaging: Improving the resolution of satellite images for better analysis and interpretation.

Voice Synthesis

Voice synthesis involves generating natural-sounding speech from text or other inputs. Generative models, such as WaveNet and Tacotron, have significantly advanced this field by producing high-quality, realistic speech. Voice synthesis models learn a mapping from text sequences $\mathbf{T}$ to audio waveforms $\mathbf{A}$:

$$\mathbf{A} = G(\mathbf{T})$$

where G is a generative model trained on pairs of text and corresponding audio data.

WaveNet Example: WaveNet models the distribution of audio samples directly as a conditional distribution:

$$p(\mathbf{A}) = \prod_{t=1}^{T} p(a_t \mid a_{1:t-1}, \mathbf{T})$$

where a_t is the audio sample at time t and $\mathbf{T}$ is the text input.

Tacotron Example: Tacotron converts text sequences to spectrograms, which are then converted to waveforms using a vocoder:

1. Text-to-Spectrogram:

$$\mathbf{S} = G_{\text{T2S}}(\mathbf{T})$$

   where $\mathbf{S}$ is the spectrogram and G_{T2S} is the text-to-spectrogram model.
2. Spectrogram-to-Waveform:

$$\mathbf{A} = G_{\text{S2W}}(\mathbf{S})$$

   where G_{S2W} is the spectrogram-to-waveform vocoder.

Applications:

Text-to-Speech (TTS): Converting written text into natural-sounding speech for applications like virtual assistants and audiobooks.

Voice Cloning: Creating personalized synthetic voices by training models on specific individuals' voice data.

10.9 Exercises

1. Given a denoising autoencoder with the following architecture and parameters:

$$\text{Encoder:} \quad \mathbf{W}_1 = \begin{pmatrix} 0.5 & 0.2 \\ 0.3 & 0.7 \end{pmatrix}, \quad \mathbf{b}_1 = \begin{pmatrix} 0.1 \\ 0.2 \end{pmatrix}$$

$$\text{Decoder:}\quad \mathbf{W}_2 = \begin{pmatrix} 0.6 & 0.4 \\ 0.5 & 0.8 \end{pmatrix}, \quad \mathbf{b}_2 = \begin{pmatrix} 0.3 \\ 0.4 \end{pmatrix}$$

Noisy input:

$$\mathbf{x} = \begin{pmatrix} 1.0 \\ 0.5 \end{pmatrix}, \quad \text{Noisy input:}\quad \tilde{\mathbf{x}} = \begin{pmatrix} 1.1 \\ 0.6 \end{pmatrix}$$

(a) Compute the latent representation using the encoder.
(b) Reconstruct the denoised output using the decoder.
(c) Calculate the reconstruction loss using mean squared error.

2. Consider a sparse autoencoder with the following architecture and parameters:

$$\text{Encoder:}\quad \mathbf{W}_1 = \begin{pmatrix} 0.4 & 0.1 \\ 0.2 & 0.6 \end{pmatrix}, \quad \mathbf{b}_1 = \begin{pmatrix} 0.2 \\ 0.3 \end{pmatrix}$$

$$\text{Decoder:}\quad \mathbf{W}_2 = \begin{pmatrix} 0.7 & 0.3 \\ 0.5 & 0.9 \end{pmatrix}, \quad \mathbf{b}_2 = \begin{pmatrix} 0.4 \\ 0.5 \end{pmatrix}$$

Input:

$$\mathbf{x} = \begin{pmatrix} 1.0 \\ 0.8 \end{pmatrix}$$

(a) Compute the latent representation using the encoder with a sparsity constraint.
(b) Reconstruct the output using the decoder.
(c) Calculate the reconstruction loss and sparsity penalty using Kullback-Leibler divergence.

3. Given the following parameters for a variational autoencoder:

$$\text{Encoder:}\quad \mathbf{W}_\mu = \begin{pmatrix} 0.3 & 0.2 \\ 0.5 & 0.7 \end{pmatrix}, \quad \mathbf{W}_\sigma = \begin{pmatrix} 0.1 & 0.4 \\ 0.6 & 0.8 \end{pmatrix}$$

$$\text{Decoder:}\quad \mathbf{W}_d = \begin{pmatrix} 0.2 & 0.6 \\ 0.3 & 0.9 \end{pmatrix}, \quad \mathbf{b}_d = \begin{pmatrix} 0.2 \\ 0.5 \end{pmatrix}$$

Input:

$$\mathbf{x} = \begin{pmatrix} 0.5 \\ 0.9 \end{pmatrix}$$

(a) Compute the mean μ and standard deviation σ using the encoder.
(b) Sample the latent variable $\mathbf{z}$ using the reparameterization trick.

(c) Reconstruct the output using the decoder.
(d) Calculate the reconstruction loss and the KL divergence between the approximate posterior and the prior.

4. Consider a recurrent autoencoder with the following architecture and parameters:

$$\text{Encoder:} \quad \mathbf{W}_{enc} = \begin{pmatrix} 0.1 & 0.3 \\ 0.5 & 0.7 \end{pmatrix}, \quad \mathbf{b}_{enc} = \begin{pmatrix} 0.2 \\ 0.4 \end{pmatrix}$$

$$\text{Decoder:} \quad \mathbf{W}_{dec} = \begin{pmatrix} 0.3 & 0.6 \\ 0.4 & 0.8 \end{pmatrix}, \quad \mathbf{b}_{dec} = \begin{pmatrix} 0.1 \\ 0.5 \end{pmatrix}$$

Input sequence:

$$\mathbf{X}_t = \begin{pmatrix} 0.1 & 0.2 \\ 0.3 & 0.4 \end{pmatrix}$$

(a) Compute the hidden state for each time step using the encoder.
(b) Reconstruct the sequence using the decoder.
(c) Calculate the sequence reconstruction loss using mean squared error.

5. Consider an autoencoder with an input layer of size 4, a hidden layer of size 2, and an output layer of size 4. The input vector $\mathbf{x} \in \mathbb{R}^4$ is given by:

$$\mathbf{x} = \begin{pmatrix} 1 \\ 2 \\ 3 \\ 4 \end{pmatrix}$$

The weight matrices for the encoder and decoder are $\mathbf{W}_e \in \mathbb{R}^{2\times 4}$ and $\mathbf{W}_d \in \mathbb{R}^{4\times 2}$:

$$\mathbf{W}_e = \begin{pmatrix} 0.1 & 0.2 & 0.3 & 0.4 \\ 0.5 & 0.6 & 0.7 & 0.8 \end{pmatrix}, \quad \mathbf{W}_d = \begin{pmatrix} 0.2 & 0.1 \\ 0.3 & 0.4 \\ 0.5 & 0.6 \\ 0.7 & 0.8 \end{pmatrix}$$

(a) Compute the encoded representation $\mathbf{z}$ using a ReLU activation function.
(b) Compute the reconstructed output $\hat{\mathbf{x}}$ using a sigmoid activation function.

6. Using the same autoencoder structure as in the previous question, consider the true input $\mathbf{x}$ and the reconstructed output $\hat{\mathbf{x}}$. The loss function is the mean squared error (MSE):

$$\text{MSE}(\mathbf{x}, \hat{\mathbf{x}}) = \frac{1}{4}\sum_{i=1}^{4}(x_i - \hat{x}_i)^2$$

(a) Compute the MSE loss for the input $\mathbf{x}$ and the reconstructed output $\hat{\mathbf{x}}$.

(b) Derive the gradients of the loss with respect to the weight matrices $\mathbf{W}_e$ and $\mathbf{W}_d$.

7. Consider a denoising autoencoder where the input vector $\mathbf{x} \in \mathbb{R}^4$ is corrupted by adding Gaussian noise, resulting in $\tilde{\mathbf{x}} \in \mathbb{R}^4$:

$$\mathbf{x} = \begin{pmatrix} 1 \\ 2 \\ 3 \\ 4 \end{pmatrix}, \quad \tilde{\mathbf{x}} = \begin{pmatrix} 1.1 \\ 1.9 \\ 3.2 \\ 3.8 \end{pmatrix}$$

Using the same weight matrices $\mathbf{W}_e$ and $\mathbf{W}_d$ as in the first question:

(a) Compute the encoded representation $\mathbf{z}$ for the corrupted input $\tilde{\mathbf{x}}$.
(b) Compute the reconstructed output $\hat{\mathbf{x}}$ from the encoded representation $\mathbf{z}$ and compare it to the original input $\mathbf{x}$.

8. Consider a variational autoencoder with an input layer of size 3 and a latent space of size 2. The encoder produces a mean vector $\mu \in \mathbb{R}^2$ and a standard deviation vector $\sigma \in \mathbb{R}^2$. Given the input vector $\mathbf{x} \in \mathbb{R}^3$:

$$\mathbf{x} = \begin{pmatrix} 0.5 \\ 1.0 \\ 1.5 \end{pmatrix}$$

The parameters of the encoder are:

$$\mu = \begin{pmatrix} 0.2 \\ 0.5 \end{pmatrix}, \quad \sigma = \begin{pmatrix} 0.1 \\ 0.3 \end{pmatrix}$$

(a) Sample a latent vector $\mathbf{z}$ from the latent space using the reparameterization trick: $\mathbf{z} = \mu + \sigma \circ \epsilon$, where $\epsilon \sim \mathcal{N}(0, \mathbf{I})$.
(b) Using a decoder with weight matrix $\mathbf{W}_d \in \mathbb{R}^{3\times 2}$:

$$\mathbf{W}_d = \begin{pmatrix} 0.3 & 0.2 \\ 0.5 & 0.4 \\ 0.7 & 0.6 \end{pmatrix}$$

Compute the reconstructed output $\hat{\mathbf{x}}$ from the sampled latent vector $\mathbf{z}$.

9. Consider a simple GAN where the generator G and discriminator D are represented by linear transformations. The input noise vector $\mathbf{z} \in \mathbb{R}^2$ is:

$$\mathbf{z} = \begin{pmatrix} 0.1 \\ 0.2 \end{pmatrix}$$

The weight matrices for the generator and discriminator are $\mathbf{W}_G \in \mathbb{R}^{3\times 2}$ and $\mathbf{W}_D \in \mathbb{R}^{1\times 3}$:

$$\mathbf{W}_G = \begin{pmatrix} 0.5 & 0.6 \\ 0.7 & 0.8 \\ 0.9 & 1.0 \end{pmatrix}, \quad \mathbf{W}_D = \begin{pmatrix} 0.2 & 0.4 & 0.6 \end{pmatrix}$$

(a) Compute the generated data $\mathbf{x}_{\text{fake}} = G(\mathbf{z})$.
(b) Compute the discriminator output for the generated data, $D(\mathbf{x}_{\text{fake}})$.

10. Using the same GAN structure as in the previous question, consider a real data sample $\mathbf{x}_{\text{real}} \in \mathbb{R}^3$:

$$\mathbf{x}_{\text{real}} = \begin{pmatrix} 1.0 \\ 1.5 \\ 2.0 \end{pmatrix}$$

(a) Compute the discriminator output for the real data, $D(\mathbf{x}_{\text{real}})$.
(b) Using the binary cross-entropy loss, calculate the loss for the discriminator L_D and the loss for the generator L_G. Assume the labels for real data and fake data are 1 and 0, respectively.

11. With the given generator and discriminator weight matrices from the first question, and using the real and fake data samples $\mathbf{x}_{\text{real}}$ and $\mathbf{x}_{\text{fake}}$:

(a) Compute the gradients of the discriminator loss L_D with respect to the discriminator weight matrix $\mathbf{W}_D$.
(b) Compute the gradients of the generator loss L_G with respect to the generator weight matrix $\mathbf{W}_G$.

12. Consider a mini-batch of 2 noise vectors $\mathbf{z}_1$ and $\mathbf{z}_2$:

$$\mathbf{z}_1 = \begin{pmatrix} 0.1 \\ 0.2 \end{pmatrix}, \quad \mathbf{z}_2 = \begin{pmatrix} 0.3 \\ 0.4 \end{pmatrix}$$

(a) Compute the generated data $\mathbf{x}_{\text{fake},1} = G(\mathbf{z}_1)$ and $\mathbf{x}_{\text{fake},2} = G(\mathbf{z}_2)$.
(b) Update the discriminator weights $\mathbf{W}_D$ using the gradients from the previous question and a learning rate of $\alpha = 0.01$.
(c) Update the generator weights $\mathbf{W}_G$ using the gradients from the previous question and the same learning rate $\alpha = 0.01$.

Chapter 11
Recent Advances and Future Perspectives

We can only see a short distance ahead, but we can see plenty there that needs to be done.
—Alan Turing, Computing Machinery and Intelligence

11.1 Introduction

The field of deep learning has witnessed remarkable advancements in recent years, significantly impacting both industry and academic research. These developments have been driven by breakthroughs in model architectures, training techniques, and computational resources. This section provides an overview of these recent developments, their impact, and emerging trends that are shaping the future of deep learning. Recent advancements in deep learning have been fueled by innovations across various aspects of the field, from model architectures and optimization techniques to data augmentation and hardware acceleration. Techniques such as mixed precision training, gradient checkpointing, and distributed training have significantly reduced the training time and computational cost of deep learning models. Mixed precision training uses both 16-bit and 32-bit floating point representations to balance computational efficiency and model accuracy. The loss function $\mathcal{L}$ and gradients are computed in 32-bit precision, while weights and activations are stored in 16-bit precision, i.e., $\mathcal{L}_{\text{16-bit}} \approx \mathcal{L}_{\text{32-bit}}$.

Pre-trained models like BERT, GPT-3, and ResNet have become foundational tools in deep learning. These models are trained on large datasets and fine-tuned for specific tasks, dramatically reducing the data and computational requirements for new applications. Transfer learning involves adapting a pre-trained model $M_{\text{pretrained}}$ to a new task by fine-tuning its parameters on a smaller dataset $\mathcal{D}_{\text{new}}$:

P. Singh and B. Raman, *Deep Learning Through the Prism of Tensors*, Studies in Big Data 162, https://doi.org/10.1007/978-981-97-8019-8_11

$$\theta_{\text{fine-tuned}} = \arg\min_{\theta} \mathbb{E}_{(x,y)\sim\mathcal{D}_{\text{new}}} \left[\mathcal{L}(M_{\text{pretrained}}(x;\theta), y)\right]$$

Advanced data augmentation techniques like AutoAugment and Mixup have enhanced model robustness and generalization by generating diverse training samples. Regularization methods such as dropout, weight decay, and batch normalization continue to play a crucial role in preventing overfitting. Mixup creates new training samples by linearly interpolating between pairs of examples:

$$\tilde{x} = \lambda x_i + (1 - \lambda)x_j$$

$$\tilde{y} = \lambda y_i + (1 - \lambda)y_j$$

where (x_i, y_i) and (x_j, y_j) are pairs of training examples and $\lambda \in [0, 1]$ is sampled from a beta distribution.

11.1.1 Impact on Industry and Research

The advancements in deep learning have had profound impacts across various industries and research domains.

1. Industry Impact:

Healthcare: Deep learning models are used for medical image analysis, drug discovery, and personalized medicine. They assist in diagnosing diseases with high accuracy, often surpassing human experts.

Finance: Algorithms predict market trends, assess credit risk, and detect fraudulent activities. Automated trading systems leverage deep learning for decision-making.

Autonomous Vehicles: Self-driving cars use deep learning for object detection, path planning, and navigation. These models enable vehicles to perceive and interact with their environment safely.

Natural Language Processing (NLP): Chatbots, virtual assistants, and language translation services rely on deep learning to understand and generate human language. Models like BERT and GPT-3 have set new benchmarks in NLP tasks.

2. Research Impact:

Scientific Discovery: Deep learning accelerates research in fields like physics, chemistry, and biology by modeling complex phenomena and predicting experimental outcomes.

Climate Science: Models predict weather patterns, climate changes, and natural disasters, aiding in environmental protection and disaster preparedness.

Education: Personalized learning systems adapt to students' needs, providing customized educational content and feedback to enhance learning outcomes.

11.1.2 Emerging Trends in Deep Learning

Several emerging trends are shaping the future of deep learning, promising further advancements and broader applications.

1. Federated Learning: Federated learning enables decentralized training of models across multiple devices while preserving data privacy. It allows models to learn from data distributed across various sources without transferring raw data. In federated learning, local models θ_i are trained on separate datasets $\mathcal{D}_i$, and their updates are aggregated to form a global model θ:

$$\theta = \frac{1}{n} \sum_{i=1}^{n} \theta_i$$

2. Self-supervised Learning: Self-supervised learning leverages unlabeled data to pre-train models by solving auxiliary tasks, such as predicting parts of the input from other parts. This approach reduces the dependence on large labeled datasets. Self-supervised learning involves defining a pretext task $\mathcal{T}_{\text{pretext}}$ with a loss function $\mathcal{L}_{\text{pretext}}$ and optimizing the model parameters θ:

$$\theta = \arg\min_{\theta} \mathbb{E}_{(x, \mathcal{T}_{\text{pretext}}(x)) \sim \mathcal{D}_{\text{unlabeled}}} \left[\mathcal{L}_{\text{pretext}}(f(x; \theta), \mathcal{T}_{\text{pretext}}(x)) \right]$$

3. Explainable AI (XAI): Explainable AI focuses on making deep learning models interpretable and transparent. Techniques such as saliency maps, LIME (Local Interpretable Model-agnostic Explanations), and SHAP (SHapley Additive exPlanations) provide insights into model predictions. Saliency maps highlight the most influential parts of the input for a given prediction:

$$\text{Saliency}(\mathbf{x}) = \left| \frac{\partial f(\mathbf{x})}{\partial \mathbf{x}} \right|$$

4. Quantum Machine Learning: Quantum machine learning explores the integration of quantum computing with machine learning. Quantum algorithms have the potential to solve certain problems more efficiently than classical algorithms. Quantum machine learning leverages quantum states and operations, with quantum circuits representing data and computations:

$$|\psi\rangle = U|0\rangle$$

where $|\psi\rangle$ is a quantum state and U is a unitary transformation.

11.2 Emerging Models and Architectures

The continuous evolution of deep learning has led to the development of novel models and architectures that push the boundaries of what is possible. This section explores some of these emerging models and architectures, focusing on self-supervised learning, reinforcement learning enhancements, and hybrid models.

11.2.1 Self-supervised Learning Models

Self-supervised learning has emerged as a powerful paradigm, enabling models to learn useful representations from unlabeled data by solving auxiliary tasks. This approach significantly reduces the dependency on large annotated datasets.

Predictive Coding
Predictive coding is a framework for self-supervised learning inspired by theories of brain function. It posits that the brain continuously generates predictions about sensory inputs and updates these predictions based on the actual inputs. In predictive coding, a hierarchical model is trained to predict future inputs based on past inputs. Let $\mathbf{x}_t$ be the input at time t, and $\mathbf{h}_t$ be the hidden state representing the model's prediction. The model minimizes the prediction error:

$$\mathcal{L}_{\text{pred}} = \sum_{t=1}^{T} \|\mathbf{x}_t - \hat{\mathbf{x}}_t\|^2$$

where $\hat{\mathbf{x}}_t$ is the predicted input.

Example: Consider a video sequence where the model predicts future frames based on past frames. The model learns to capture temporal dependencies and motion patterns by minimizing the prediction error.

Applications:

Video Prediction: Enhancing the ability of models to predict future frames in a video sequence.

Anomaly Detection: Identifying deviations from normal patterns by analyzing prediction errors.

11.2.2 Reinforcement Learning Enhancements

Reinforcement learning (RL) has seen significant advancements, particularly with the integration of deep learning techniques. These enhancements have enabled RL to solve complex tasks that were previously infeasible.

Deep Reinforcement Learning Algorithms

Deep reinforcement learning combines neural networks with reinforcement learning algorithms, enabling agents to learn from high-dimensional sensory inputs like images and text. In deep reinforcement learning, an agent interacts with an environment $\mathcal{E}$ and learns a policy $\pi(a \mid s)$ that maximizes the expected cumulative reward R:

$$\pi^* = \arg\max_{\pi} \mathbb{E}\left[\sum_{t=0}^{T} \gamma^t r_t\right]$$

where γ is the discount factor, r_t is the reward at time t, s is the state, and a is the action.

Example: The Deep Q-Network (DQN) algorithm uses a neural network to approximate the Q-function, which estimates the expected reward for each action in a given state:

$$Q(s, a; \theta) \approx \mathbb{E}\left[r + \gamma \max_{a'} Q(s', a'; \theta') \mid s, a\right]$$

where θ are the parameters of the neural network.

Applications:

Game Playing: Achieving superhuman performance in games like Go, chess, and Atari.

Robotics: Enabling robots to learn complex tasks through trial and error.

Applications and Future Directions

Deep reinforcement learning has broad applications across various domains and continues to evolve with new research directions.

Applications:

Autonomous Vehicles: Training self-driving cars to navigate complex environments.

Healthcare: Personalizing treatment plans and optimizing drug discovery processes.

Finance: Developing trading strategies and portfolio management systems.

Future Directions:

Hierarchical Reinforcement Learning: Decomposing tasks into hierarchies of subtasks to improve learning efficiency.

Meta-Reinforcement Learning: Enabling agents to learn how to learn, adapting quickly to new tasks.

Multi-Agent Reinforcement Learning: Training multiple agents to cooperate or compete, leading to more complex behaviors and strategies.

11.2.3 Hybrid Models

Hybrid models combine the strengths of different approaches to address the limitations of individual models. One promising direction is the integration of neural networks with symbolic AI, leading to neuro-symbolic systems.

Combining Neural Networks with Symbolic AI
Symbolic AI, which involves the manipulation of symbols and rules, has strengths in reasoning and interpretability. Combining it with neural networks, which excel in pattern recognition, results in powerful hybrid models. A hybrid model can be represented as a combination of a neural network f_{NN} and a symbolic component f_{sym}:

$$y = f_{\text{sym}}(f_{\text{NN}}(x))$$

where x is the input and y is the output. For example, a hybrid model for visual question answering (VQA) might use a neural network to extract features from images and a symbolic reasoning engine to answer questions based on these features.

Applications:

Explainable AI: Enhancing the interpretability of neural network decisions through symbolic reasoning.

Knowledge Graphs: Integrating neural networks with knowledge graphs to improve information retrieval and reasoning.

Neuro-symbolic Integration
Neuro-symbolic integration aims to create systems that can leverage both neural and symbolic representations, combining the flexibility of neural networks with the structure and interpretability of symbolic reasoning. Neuro-symbolic systems can be formalized as:

$$y = g(f_{\text{NN}}(x), \mathcal{R})$$

where g is a function that combines neural outputs $f_{\text{NN}}(x)$ with a set of symbolic rules $\mathcal{R}$. For example, in a neuro-symbolic system for natural language understanding, a neural network might encode the semantics of a sentence, while symbolic rules are used to perform logical reasoning on the encoded representations.

Applications:

Robotic Systems: Combining perception and reasoning for autonomous robots capable of complex decision-making.

Healthcare Diagnostics: Integrating patient data analysis with medical knowledge bases to improve diagnostic accuracy.

11.3 Improvements in Tensor Computation

Tensor computation forms the backbone of deep learning, enabling the efficient processing and manipulation of multi-dimensional arrays. Recent advancements in tensor computation have been driven by both software optimizations in tensor libraries and the development of specialized hardware accelerators. This section delves into these improvements, highlighting key updates and their implications for deep learning.

11.3.1 Optimizations in Tensor Libraries

Tensor libraries such as TensorFlow and PyTorch have undergone significant enhancements to improve performance, usability, and functionality. These updates have made it easier for researchers and practitioners to develop, train, and deploy deep learning models.

TensorFlow Updates
TensorFlow, developed by Google, has introduced several updates aimed at improving efficiency and flexibility. TensorFlow 2.x, in particular, has emphasized ease of use with the introduction of the Keras API as the default high-level API.

Key Enhancements:

1. Eager Execution: Eager execution provides an imperative programming environment that evaluates operations immediately, making it easier to debug and develop models dynamically. Traditional TensorFlow uses a static computation graph. In eager execution, operations are evaluated as they are called:

$$y = f(x) = \text{operation}(x)$$

This allows for immediate feedback and interactive model development.

2. tf.data API: The 'tf.data' API enables efficient data loading and preprocessing, supporting complex input pipelines that can handle large datasets and perform transformations on the fly.
3. TensorFlow Lite: TensorFlow Lite optimizes models for mobile and edge devices, allowing for efficient inference with reduced latency and power consumption.

PyTorch Updates
PyTorch, developed by Facebook, has also seen major updates, focusing on dynamic computation graphs and improved support for research and production.

Key Enhancements:

1. JIT Compilation: PyTorch's Just-In-Time (JIT) compiler allows for the conversion of models to TorchScript, enabling optimizations and deployment on various platforms. JIT compilation converts dynamic graphs to static graphs for optimization while maintaining flexibility:

$$\text{scripted\_model} = \text{torch.jit.script}(model)$$

2. Distributed Training: PyTorch supports distributed training through 'torch.distributed', allowing models to be trained on multiple GPUs or nodes, accelerating training times. Example: Setting up distributed training with 'torch.distributed':

$$\text{torch.distributed.init\_process\_group("nccl")}$$

$$\text{model} = \text{torch.nn.parallel.DistributedDataParallel}(model)$$

3. PyTorch Lightning: PyTorch Lightning is a lightweight framework that simplifies the research workflow, promoting best practices and reducing boilerplate code.

Performance Enhancements and New Features
Both TensorFlow and PyTorch have incorporated numerous performance enhancements and new features that streamline model development and deployment.

1. Mixed Precision Training: Mixed precision training uses both 16-bit and 32-bit floating-point types to accelerate training while maintaining accuracy. This is achieved through hardware support such as NVIDIA's Tensor Cores. Mixed precision training involves scaling the loss to prevent underflow in 16-bit precision and then scaling back during backpropagation:

$$\mathcal{L}_{\text{scaled}} = \mathcal{L} \times \text{scale\_factor}$$

$$\frac{\partial \mathcal{L}_{\text{scaled}}}{\partial \theta} = \frac{\partial (\mathcal{L} \times \text{scale\_factor})}{\partial \theta}$$

2. Accelerated Linear Algebra Libraries: Both libraries leverage highly optimized linear algebra libraries such as cuBLAS, cuDNN, and MKL-DNN to perform tensor operations efficiently.

11.3.2 Efficient Training Techniques

The growing complexity of deep learning models and the increasing size of datasets have necessitated the development of efficient training techniques. These techniques aim to reduce computational costs, accelerate training times, and make models more suitable for deployment on resource-constrained devices.

Model Pruning and Quantization
Model pruning and quantization are two prominent techniques used to optimize neural networks by reducing their size and computational requirements without significantly sacrificing accuracy.

Model Pruning Model pruning involves removing less important parameters (weights) from the neural network. This process reduces the number of parameters and computational overhead, making the model more efficient. Pruning can be formalized as the optimization problem:

$$\min_{\mathbf{W}} \mathcal{L}(\mathbf{W}; \mathbf{X}, \mathbf{Y}) + \lambda \|\mathbf{W}\|_0$$

where $\mathcal{L}$ is the loss function, $\mathbf{W}$ are the model weights, $\mathbf{X}$ and $\mathbf{Y}$ are the input data and labels, respectively, and $\|\mathbf{W}\|_0$ is the ℓ_0-norm representing the number of non-zero weights. The regularization parameter λ controls the trade-off between loss minimization and sparsity. For example, a convolutional neural network (CNN) can be pruned by iteratively removing filters with small ℓ_1-norms, which contribute less to the final output:

$$\mathbf{W}_i = \mathbf{W}_i \cdot \mathbf{m}_i$$

where $\mathbf{m}_i$ is a binary mask indicating whether the ith filter is kept (1) or pruned (0).

Quantization Quantization reduces the precision of the model parameters from floating-point to fixed-point representations, such as 8-bit integers. This technique significantly decreases the model size and speeds up inference, especially on hardware with limited precision support. Quantization maps floating-point weights w to lower precision values w_q:

$$w_q = \text{round}\left(\frac{w - \min(w)}{\max(w) - \min(w)} \cdot (2^b - 1)\right)$$

where b is the number of bits used for quantization. For example, in an 8-bit quantization, weights are scaled and rounded to fit within the range [0, 255], allowing for efficient storage and computation on specialized hardware.

11.4 Integration with Other Fields

The integration of deep learning with other fields, such as quantum computing, has opened up new avenues for research and application. These interdisciplinary approaches promise to enhance the capabilities of deep learning and address some of its current limitations.

11.4.1 Quantum Computing and Deep Learning

Quantum computing leverages the principles of quantum mechanics to perform computations that are infeasible for classical computers. The integration of quantum computing with deep learning aims to harness the unique properties of quantum systems to enhance machine learning algorithms.

Quantum Machine Learning Algorithms
Quantum machine learning algorithms exploit quantum parallelism and entanglement to perform computations more efficiently than classical algorithms.

Quantum Circuit Model: A quantum circuit consists of qubits and quantum gates. A qubit is a quantum analog of a classical bit, representing a superposition of $|0\rangle$ and $|1\rangle$:

$$|\psi\rangle = \alpha|0\rangle + \beta|1\rangle$$

where α and β are complex amplitudes. Quantum gates manipulate the state of qubits, enabling complex computations.

Quantum Algorithm for Machine Learning: Quantum algorithms, such as the Quantum Support Vector Machine (QSVM) and Quantum Principal Component Analysis (QPCA), leverage quantum mechanics for efficient data processing.

Example: In QSVM, a quantum computer computes the kernel matrix exponentially faster than a classical computer, enabling the solution of large-scale classification problems.

Potential and Challenges
The integration of quantum computing and deep learning holds great promise but also presents several challenges.

Potential:

Speed: Quantum computers can solve certain problems exponentially faster than classical computers, potentially accelerating deep learning training and inference.

Complexity: Quantum algorithms can process complex data structures more efficiently, enabling the development of more sophisticated models.

Challenges:

Scalability: Current quantum computers are limited in the number of qubits and coherence time, restricting the size of problems that can be tackled.

Noise: Quantum computations are prone to errors due to environmental interference, requiring robust error correction methods.

Integration: Developing efficient algorithms that leverage both quantum and classical computing resources remains a significant challenge.

11.4.2 Neuroscience-Inspired Models

Neuroscience-inspired models draw inspiration from the structure and function of the human brain, aiming to replicate its computational efficiency and adaptability.

Two key approaches in this domain are neuromorphic computing and brain-inspired algorithms.

Neuromorphic Computing

Neuromorphic computing involves designing hardware that mimics the neural architecture and processing mechanisms of the brain. These systems aim to achieve higher efficiency and lower power consumption compared to traditional computing architectures. Neuromorphic hardware typically employs spiking neural networks (SNNs), where neurons communicate via discrete spikes or events rather than continuous signals. The dynamics of an SNN can be described by differential equations governing membrane potential V:

$$\tau \frac{\mathrm{d}V_i}{\mathrm{d}t} = -V_i + \sum_j w_{ij} S_j(t)$$

where τ is the membrane time constant, V_i is the membrane potential of neuron i, w_{ij} is the synaptic weight from neuron j to neuron i, and $S_j(t)$ is the spike train of neuron j. IBM's TrueNorth and Intel's Loihi are examples of neuromorphic chips that implement SNNs, demonstrating significant energy efficiency and parallelism for tasks such as pattern recognition and sensory processing.

Applications:

Real-Time Processing: Neuromorphic systems excel in real-time processing tasks, such as autonomous driving and robotics, where low latency and high efficiency are crucial.

Edge Computing: These systems are ideal for edge devices that require energy-efficient computation, enabling complex processing capabilities in portable and wearable devices.

Brain-Inspired Algorithms

Brain-inspired algorithms aim to emulate the learning and adaptation mechanisms of the brain. These algorithms leverage insights from neuroscience to develop models that can learn more efficiently and robustly. One example of a brain-inspired algorithm is Hebbian learning, which adjusts synaptic weights based on the correlation between the activity of pre- and post-synaptic neurons. The weight update rule can be expressed as:

$$\Delta w_{ij} = \eta x_i y_j$$

where η is the learning rate, x_i is the activity of the pre-synaptic neuron, and y_j is the activity of the post-synaptic neuron. For example, deep learning models incorporating mechanisms such as synaptic plasticity and metaplasticity can enhance learning efficiency and robustness, particularly in dynamic and noisy environments.

Applications:

Adaptive Systems: Brain-inspired algorithms enable the development of adaptive systems capable of continuous learning and self-optimization.

Robust Learning: These algorithms improve the robustness of models against adversarial attacks and data distribution shifts, making them more reliable in real-world applications.

11.4.3 Deep Learning in Healthcare and Biotech

Deep learning has transformative potential in healthcare and biotechnology, where it can enhance diagnostics, drug discovery, and personalized medicine. This section explores key applications in drug discovery, genomics, medical imaging, and diagnostics.

Drug Discovery and Genomics

Deep learning models can accelerate the drug discovery process by predicting the interactions between drugs and targets, optimizing drug formulations, and analyzing genomic data. In drug discovery, deep learning models such as graph neural networks (GNNs) are used to predict molecular properties. A molecular graph G with nodes V (atoms) and edges E (bonds) can be represented as:

$$h_v^{(k+1)} = \text{AGGREGATE}^{(k)} \left(\{h_u^{(k)} : u \in \mathcal{N}(v)\} \right)$$

$$h_v^{(k+1)} = \text{COMBINE}^{(k)} \left(h_v^{(k)}, h_v^{(k+1)} \right)$$

where $h_v^{(k)}$ is the feature vector of node v at iteration k, and $\mathcal{N}(v)$ represents the neighbors of v. For example, DeepChem, an open-source toolkit, leverages deep learning for computational chemistry and drug discovery, enabling the prediction of molecular properties and biological activities.

Applications:

Virtual Screening: Identifying potential drug candidates by predicting their interactions with biological targets.

Genomic Analysis: Analyzing genomic data to identify genetic variations associated with diseases and potential therapeutic targets.

Medical Imaging and Diagnostics

Deep learning has revolutionized medical imaging by enhancing image analysis, segmentation, and interpretation, leading to more accurate and timely diagnoses. CNNs are commonly used in medical imaging for tasks such as image classification and segmentation.

For example, deep learning models such as U-Net and V-Net are designed for medical image segmentation, enabling precise delineation of anatomical structures and pathological regions.

Applications:

Radiology: Enhancing the detection and diagnosis of conditions such as tumors, fractures, and cardiovascular diseases through advanced image analysis.

Pathology: Automating the analysis of histopathological slides to identify cellular abnormalities and diagnose diseases at an early stage.

11.5 Ethical Considerations in Deep Learning

The rapid advancements in deep learning have brought about significant benefits across various domains. However, they have also raised important ethical concerns, particularly regarding bias, fairness, transparency, and interpretability. This section delves into these ethical considerations, discussing methods to identify and mitigate bias, ensure fairness, and improve the transparency and interpretability of deep learning models.

11.5.1 Bias and Fairness in AI

Bias in AI systems can lead to unfair treatment of individuals or groups, perpetuating and even amplifying existing societal inequalities. Ensuring fairness in AI involves identifying and mitigating these biases to create more equitable systems.

Identifying and Mitigating Bias
Identifying Bias: Bias can arise at various stages of the machine learning pipeline, including data collection, model training, and deployment. To identify bias, it is crucial to analyze the data and model outputs for disparities across different demographic groups. Consider a classifier f that predicts an outcome $\hat{y}$ based on input features $\mathbf{x}$. Bias can be quantified by measuring the difference in performance metrics (e.g., accuracy, false positive rate) across different groups A and B:

$$\text{Bias}_{\text{accuracy}} = |\text{Accuracy}_A - \text{Accuracy}_B|$$

Example: In a facial recognition system, bias can be identified by comparing the false positive and false negative rates across different gender and racial groups. A significant disparity indicates the presence of bias.

Several techniques can be employed to mitigate bias, including re-weighting, re-sampling, and adversarial debiasing.

Re-weighting: Assign different weights to samples from underrepresented groups to ensure balanced representation during training.

Let w_i be the weight assigned to sample i. The weighted loss function $\mathcal{L}$ is:

$$\mathcal{L} = \sum_{i=1}^{n} w_i \cdot \mathcal{L}_i$$

where $\mathcal{L}_i$ is the individual loss for sample i.

Adversarial Debiasing: Train the model with an adversarial network that aims to identify and remove bias-related features from the representations learned by the model. Let f be the main model and g be the adversary. The adversarial training objective is:

$$\min_f \max_g \left(\mathcal{L}_{\text{main}}(f) - \lambda \mathcal{L}_{\text{adv}}(g, f)\right)$$

where $\mathcal{L}_{\text{main}}$ is the primary task loss and $\mathcal{L}_{\text{adv}}$ is the adversarial loss.

Fairness-Aware Machine Learning

Fairness-aware machine learning focuses on developing algorithms that ensure equitable treatment of all individuals, regardless of their demographic characteristics.

Fairness Metrics Several metrics can be used to assess fairness, including demographic parity, equalized odds, and disparate impact.

Demographic Parity: The probability of a positive prediction should be the same for all groups:

$$P(\hat{y} = 1 \mid A) = P(\hat{y} = 1 \mid B)$$

Equalized Odds: The true positive and false positive rates should be equal across groups:

$$P(\hat{y} = 1 \mid y = 1, A) = P(\hat{y} = 1 \mid y = 1, B)$$

$$P(\hat{y} = 1 \mid y = 0, A) = P(\hat{y} = 1 \mid y = 0, B)$$

Algorithmic Approaches: Fairness-aware algorithms incorporate fairness constraints into the optimization process to ensure equitable outcomes.

Example: In a hiring algorithm, fairness constraints can be added to ensure that candidates from different demographic groups have equal probabilities of being selected based on their qualifications.

11.5.2 Transparency and Interpretability

Transparency and interpretability are critical for understanding and trusting deep learning models. Explainable AI (XAI) techniques and interpretability methods provide insights into how models make decisions, enabling stakeholders to assess their reliability and fairness.

Explainable AI (XAI) Techniques

Explainable AI aims to make model predictions understandable to humans, enhancing trust and facilitating decision-making. XAI techniques often involve approximating complex models with simpler, interpretable models or highlighting the most important features contributing to a prediction.

Local Interpretable Model-agnostic Explanations (LIME) LIME approximates the behavior of a complex model f locally around a specific prediction $\hat{y} = f(\mathbf{x})$ using a simpler model g:

$$g(\mathbf{x}') \approx f(\mathbf{x}') \text{ for } \mathbf{x}' \text{ near } \mathbf{x}$$

Example: LIME generates explanations by perturbing the input data and observing the changes in predictions, allowing users to understand which features are most influential for a specific prediction.

Model Interpretability Methods
Interpretability methods provide insights into the internal workings of a model and its decision-making process.

Feature Importance Feature importance methods identify the most influential features for model predictions. Techniques like permutation importance and SHapley Additive exPlanations (SHAP) quantify the contribution of each feature.

Permutation Importance: Measure the change in model performance when a feature is permuted:

$$\text{Importance}(x_j) = \mathbb{E}_{\mathbf{x}}\left[\mathcal{L}(f(\mathbf{x})) - \mathcal{L}(f(\mathbf{x}_{\text{perm}(j)}))\right]$$

where $\mathbf{x}_{\text{perm}(j)}$ is the input with feature j permuted.

SHAP Values: Calculate the contribution of each feature based on cooperative game theory:

$$\phi_j = \sum_{S \subseteq \{1,\ldots,p\} \setminus \{j\}} \frac{|S|!(p-|S|-1)!}{p!}[f(S \cup \{j\}) - f(S)]$$

where ϕ_j is the SHAP value for feature j and S is a subset of features. For example, SHAP values provide a unified measure of feature importance, allowing users to understand the impact of each feature on individual predictions and overall model behavior.

11.5.3 Privacy and Security

With the increasing deployment of deep learning models in various applications, concerns regarding privacy and security have become paramount. This section discusses the challenges and strategies related to data privacy and the defenses against adversarial attacks.

Data Privacy in Deep Learning
Data privacy is crucial, especially when dealing with sensitive information such as medical records, financial data, and personal identifiers. Ensuring privacy involves techniques that allow model training without compromising individual data privacy.

Federated Learning Federated learning allows multiple devices to collaboratively train a model while keeping the data localized. Each device computes local updates, which are aggregated to form a global model. Let $\mathbf{w}_i$ be the local model parameters on device i. The global model is updated by aggregating these local parameters:

$$\mathbf{w} = \frac{1}{N}\sum_{i=1}^{N}\mathbf{w}_i$$

where N is the number of devices.

Differential Privacy Differential privacy adds noise to the data or model gradients to protect individual data points. It ensures that the inclusion or exclusion of a single data point does not significantly affect the model's output. A mechanism $\mathcal{M}$ is (ϵ, δ)-differentially private if, for all datasets D and D' differing by one element, and for all $S \subseteq \text{Range}(\mathcal{M})$:

$$P(\mathcal{M}(D) \in S) \leq e^{\epsilon} P(\mathcal{M}(D') \in S) + \delta$$

Example: In a differentially private stochastic gradient descent (DP-SGD), noise is added to the gradients:

$$\mathbf{g} = \frac{1}{N}\sum_{i=1}^{N}\mathbf{g}_i + \mathcal{N}(0, \sigma^2)$$

where $\mathbf{g}_i$ are the individual gradients and $\mathcal{N}(0, \sigma^2)$ is Gaussian noise.

Adversarial Attacks and Defenses

Adversarial attacks involve subtly modifying input data to deceive a deep learning model, leading to incorrect predictions. Defending against these attacks is critical for the robustness of AI systems.

Adversarial Attacks Adversarial examples are inputs altered by small perturbations that cause the model to make mistakes. These perturbations are often imperceptible to humans but significantly affect the model's output. Given an input $\mathbf{x}$ and a model f, an adversarial example $\mathbf{x}^*$ is generated by adding a perturbation δ such that:

$$\mathbf{x}^* = \mathbf{x} + \delta$$

$$\text{where} \quad \|\delta\| \leq \epsilon$$

and $f(\mathbf{x}^*) \neq f(\mathbf{x})$.

Adversarial Training One effective defense against adversarial attacks is adversarial training, where the model is trained on both original and adversarial examples. The adversarial training objective modifies the loss function to include adversarial examples:

$$\mathcal{L}_{\text{adv}} = \mathbb{E}_{(\mathbf{x},y)\sim\mathcal{D}}\left[\max_{\|\delta\|\leq\epsilon} \mathcal{L}(f(\mathbf{x}+\delta), y)\right]$$

Defensive Techniques Other defensive techniques include: Obscuring the model's gradients to make it harder for attackers to compute effective adversarial perturbations. Applying transformations to the input data, such as randomization or denoising, to remove adversarial perturbations.

11.6 Ongoing Research and Future Developments

The field of deep learning is rapidly evolving, with continuous advancements and breakthroughs. This section explores cutting-edge research areas and potential future developments that promise to shape the next generation of AI systems.

11.6.1 Cutting-Edge Research Areas

Researchers are exploring various innovative approaches to push the boundaries of deep learning. Two notable areas of research are meta-learning and few-shot learning, and generative models and creative AI.

Meta-Learning and Few-Shot Learning

Meta-learning, or "learning to learn," aims to develop models that can adapt quickly to new tasks with minimal data. Few-shot learning focuses on training models that can generalize well from only a few examples. Meta-learning involves optimizing a model's learning algorithm across multiple tasks. Let $\mathcal{T}$ be a distribution over tasks, and $\mathcal{L}_{\mathcal{T}}(f)$ be the loss on task $\mathcal{T}$. The meta-learning objective is:

$$\min_{\theta} \mathbb{E}_{\mathcal{T} \sim p(\mathcal{T})} \left[\mathcal{L}_{\mathcal{T}}(f_{\theta}) \right]$$

where θ are the meta-parameters.

Example: Model-Agnostic Meta-Learning (MAML) optimizes for initial parameters that can quickly adapt to new tasks with a few gradient steps:

$$\theta \leftarrow \theta - \alpha \nabla_{\theta} \sum_{\mathcal{T}_i} \mathcal{L}_{\mathcal{T}_i}(f_{\theta - \beta \nabla_{\theta} \mathcal{L}_{\mathcal{T}_i}(f_{\theta})})$$

Applications:

Personalized Models: Quickly adapting models to individual user preferences and behaviors.

Robotics: Enabling robots to learn new tasks from a few demonstrations.

Generative Models and Creative AI

Generative models, such as GANs and VAEs, are being used to create new data, artworks, and designs, pushing the limits of creative AI. Generative models learn to approximate the data distribution $p(\mathbf{x})$. In GANs, this involves a generator G and a discriminator D:

$$\min_G \max_D \mathbb{E}_{\mathbf{x}\sim p_{\text{data}}(\mathbf{x})}\left[\log D(\mathbf{x})\right] + \mathbb{E}_{\mathbf{z}\sim p_{\mathbf{z}}(\mathbf{z})}\left[\log(1 - D(G(\mathbf{z})))\right]$$

Example: StyleGAN generates high-quality images by controlling the style at different levels of the synthesis process, leading to impressive visual results.

Applications:

Art and Design: Creating artworks, music, and fashion designs.

Data Augmentation: Generating synthetic data to improve the training of machine learning models.

11.6.2 Potential Future Breakthroughs

As deep learning continues to evolve, several potential breakthroughs could redefine its capabilities and applications. This section explores the possibilities in AI and human-computer interaction, and autonomous systems and robotics.

AI and Human-Computer Interaction
Advancements in AI are poised to revolutionize human-computer interaction (HCI), making it more intuitive, efficient, and personalized.

Natural Language Understanding Improving natural language understanding (NLU) will enable more sophisticated and context-aware interactions between humans and machines. Transformer models, such as BERT and GPT, have significantly advanced NLU by leveraging self-attention mechanisms. For example, virtual assistants like Google Assistant and Siri use advanced NLU models to understand and respond to user queries in a more human-like manner.

Applications:

Conversational AI: Enhancing chatbots and virtual assistants for more natural and effective communication.

Personalized Interfaces: Developing adaptive interfaces that respond to individual user preferences and behaviors.

Autonomous Systems and Robotics
Autonomous systems and robotics are set to benefit immensely from deep learning, enabling more intelligent and capable machines. Reinforcement learning (RL) is crucial for training autonomous agents that can navigate complex environments and perform tasks autonomously. In RL, an agent learns a policy $\pi(a \mid s)$ that maximizes the expected cumulative reward R:

$$\pi^* = \arg\max_\pi \mathbb{E}_\pi \left[\sum_{t=0}^{\infty} \gamma^t r_t\right]$$

where γ is the discount factor, r_t is the reward at time step t, and s and a are the states and actions, respectively. For example, Deep Q-Networks (DQNs) combine Q-learning with deep neural networks to learn policies for playing video games, achieving superhuman performance.

Applications:

Self-Driving Cars: Developing autonomous vehicles that can navigate safely and efficiently in various traffic conditions.

Service Robots: Creating robots that can assist with tasks in homes, hospitals, and workplaces.

Bibliography

D. Bahdanau, K. Cho, Y. Bengio, Neural machine translation by jointly learning to align and translate. CoRR, abs/1409.0473 (2014)

J. Bruna, W. Zaremba, A. Szlam, Y. LeCun, Spectral networks and locally connected networks on graphs, in *International Conference on Learning Representations (ICLR)* (2014)

I. Chavel, *Riemannian Geometry: A Modern Introduction* (Cambridge University Press, 2006)

T. Chen, S. Kornblith, M. Norouzi, G. Hinton, A simple framework for contrastive learning of visual representations, in *Proceedings of the 37th International Conference on Machine Learning* (2020)

K. Cho, B. Van Merriënboer, C. Gulcehre, D. Bahdanau, F. Bougares, H. Schwenk, Y. Bengio, Learning phrase representations using RNN encoder-decoder for statistical machine translation. arXiv preprint arXiv:1406.1078 (2014)

J. Deng, W. Dong, R. Socher, L.-J. Li, K. Li, L. Fei-Fei, Imagenet: a large-scale hierarchical image database, in *2009 IEEE Conference on Computer Vision and Pattern Recognition* (2009), pp. 248–255

J. Devlin, M.-W. Chang, K. Lee, K. Toutanova, Bert: pre-training of deep bidirectional transformers for language understanding, in *North American Chapter of the Association for Computational Linguistics* (2019)

R. Diestel, *Graph Theory* (Springer, 2010)

M.P. do Carmo, *Riemannian Geometry* (Birkhäuser, 1992)

C. Dong, C.C. Loy, K. He, X. Tang, Image super-resolution using deep convolutional networks. IEEE Trans. Pattern Anal. Mach. Intell. **38**, 295–307 (2014)

A. Dosovitskiy, L. Beyer, A. Kolesnikov, D. Weissenborn, X. Zhai, T. Unterthiner, M. Dehghani, M. Minderer, G. Heigold, S. Gelly, J. Uszkoreit, N. Houlsby, An image is worth 16x16 words: transformers for image recognition at scale. arXiv, abs/2010.11929 (2020)

J.C. Duchi, E. Hazan, Y. Singer, Adaptive subgradient methods for online learning and stochastic optimization. J. Mach. Learn. Res. **12**, 2121–2159 (2011)

K. Fukushima, Neocognitron: a self-organizing neural network model for a mechanism of pattern recognition unaffected by shift in position. Biol. Cybern. **36**(4), 193–202 (1980)

S. Gidaris, P. Singh, N. Komodakis, Unsupervised representation learning by predicting image rotations. arXiv, abs/1803.07728 (2018)

J. Gilmer, S.S. Schoenholz, P.F. Riley, O. Vinyals, G.E. Dahl, Neural message passing for quantum chemistry, in *Proceedings of the 34th International Conference on Machine Learning* (2017), pp. 1263–1272

P. Singh and B. Raman, *Deep Learning Through the Prism of Tensors*, Studies in Big Data 162, https://doi.org/10.1007/978-981-97-8019-8

X. Glorot, Y. Bengio, Understanding the difficulty of training deep feedforward neural networks, in *International Conference on Artificial Intelligence and Statistics* (2010)

I. Goodfellow, J. Pouget-Abadie, M. Mirza, B. Xu, D. Warde-Farley, S. Ozair, A. Courville, Y. Bengio, Generative adversarial nets, in *Advances in Neural Information Processing Systems*, vol. 27 (2014)

M. Gori, G. Monfardini, F. Scarselli, A new model for learning in graph domains, in *Proceedings. 2005 IEEE International Joint Conference on Neural Networks*, vol. 2 (2005), pp. 729–734

J.-B. Grill, F. Strub, F. Altch'e, C. Tallec, P.H. Richemond, E. Buchatskaya, C. Doersch, B.Á. Pires, Z.D. Guo, M.G. Azar, B. Piot, K. Kavukcuoglu, R. Munos, M. Valko, Bootstrap your own latent: a new approach to self-supervised learning, in *Advances in Neural Information Processing Systems (NeurIPS)* (2020), pp. 21271–21284

W.L. Hamilton, R. Ying, J. Leskovec, Inductive representation learning on large graphs. *Advances in Neural Information Processing Systems (NeurIPS)* (2017), pp. 1024–1034

K. He, X. Zhang, S. Ren, J. Sun, Delving deep into rectifiers: surpassing human-level performance on ImageNet classification, in *Proceedings of the IEEE International Conference on Computer Vision* (2015a), pp. 1026–1034

K. He, X. Zhang, S. Ren, J. Sun, Deep residual learning for image recognition, in *Proceedings of the IEEE Conference on Computer Vision and Pattern Recognition* (2015b), pp. 770–778

K. He, H. Fan, Y. Wu, S. Xie, R.B. Girshick, Momentum contrast for unsupervised visual representation learning, in *2020 IEEE/CVF Conference on Computer Vision and Pattern Recognition (CVPR)* (2019), pp. 9726–9735

K. He, H. Fan, Y. Wu, S. Xie, R.B. Girshick, Momentum contrast for unsupervised visual representation learning, in *Proceedings of the IEEE/CVF Conference on Computer Vision and Pattern Recognition (CVPR)* (2020), pp. 9726–9735

G.E. Hinton, R.R. Salakhutdinov, Reducing the dimensionality of data with neural networks. Science **313**(5786), 504–507 (2006)

S. Hochreiter, J. Schmidhuber, Long short-term memory. Neural Comput. **9**(8), 1735–1780 (1997)

J.J. Hopfield, Neural networks and physical systems with emergent collective computational abilities. Proc. Natl. Acad. Sci. **79**(8), 2554–2558 (1982)

A.G. Howard, M. Zhu, B. Chen, D. Kalenichenko, W. Wang, T. Weyand, M. Andreetto, H. Adam, Mobilenets: efficient convolutional neural networks for mobile vision applications. arXiv, abs/1704.04861 (2017)

G. Huang, Z. Liu, K.Q. Weinberger, Densely connected convolutional networks, in *2017 IEEE Conference on Computer Vision and Pattern Recognition (CVPR)* (2016), pp. 2261–2269

F.N. Iandola, M.W. Moskewicz, K. Ashraf, S. Han, W.J. Dally, K. Keutzer, Squeezenet: AlexNet-level accuracy with 50x fewer parameters and <1mb model size. arXiv, abs/1602.07360 (2016)

J. Jost, *Riemannian Geometry and Geometric Analysis* (Springer, 2017)

D.P. Kingma, J. Ba, Adam: a method for stochastic optimization. CoRR, abs/1412.6980 (2014)

D.P. Kingma, M. Welling, Auto-encoding variational Bayes, in *International Conference on Learning Representations (ICLR)* (2014)

T.N. Kipf, M. Welling, Semi-supervised classification with graph convolutional networks. arXiv preprint arXiv:1609.02907 (2016)

T.N. Kipf, M. Welling, Semi-supervised classification with graph convolutional networks, in *International Conference on Learning Representations (ICLR)* (2017)

A. Krizhevsky, I. Sutskever, G.E. Hinton, ImageNet classification with deep convolutional neural networks. Commun. ACM **60**, 84–90 (2012)

Z. Lan, M. Chen, S. Goodman, K. Gimpel, P. Sharma, R. Soricut, Albert: a lite BERT for self-supervised learning of language representations. arXiv preprint arXiv:1909.11942 (2019)

Y. LeCun, L. Bottou, Y. Bengio, P. Haffner, Gradient-based learning applied to document recognition. Proc. IEEE **86**(11), 2278–2324 (1998)

J.M. Lee, *Introduction to Riemannian Manifolds* (Springer, 2018)

M. Lewis, Y. Liu, N. Goyal, M. Ghazvininejad, A.R. Mohamed, O. Levy, V. Stoyanov, L. Zettlemoyer, Bart: denoising sequence-to-sequence pre-training for natural language generation, translation, and comprehension, in *Annual Meeting of the Association for Computational Linguistics* (2019)

D.C. Liu, J. Nocedal, On the limited memory BFGS method for large scale optimization. Math. Progr. **45**(1), 503–528 (1989)

Y. Liu, M. Ott, N. Goyal, J. Du, M. Joshi, D. Chen, O. Levy, M. Lewis, L. Zettlemoyer, V. Stoyanov, Roberta: a robustly optimized BERT pretraining approach. arXiv preprint arXiv:1907.11692 (2019)

Z. Liu, Y. Lin, Y. Cao, H. Hu, Y. Wei, Z. Zhang, S. Lin, B. Guo, Swin transformer: hierarchical vision transformer using shifted windows, in *2021 IEEE/CVF International Conference on Computer Vision (ICCV)* (2021), pp. 9992–10002

T. Luong, H, Pham, C.D. Manning, Effective approaches to attention-based neural machine translation. arXiv, abs/1508.04025 (2015)

W.S. McCulloch, W. Pitts, A logical calculus of the ideas immanent in nervous activity. Bull. Math. Biol. **52**, 99–115 (1990)

L. McInnes, J. Healy, UMAP: uniform manifold approximation and projection for dimension reduction. arXiv, abs/1802.03426 (2018)

T. Mikolov, K. Chen, G.S. Corrado, J. Dean, Efficient estimation of word representations in vector space, in *Proceedings of the International Conference on Learning Representations (ICLR)* (2013)

M. Minsky, S. Papert, *Perceptrons: An Introduction to Computational Geometry* (MIT Press, 1969)

Y. Nesterov, A method for solving the convex programming problem with convergence rate $o(1/k^2)$. Proc. USSR Acad. Sci. **269**, 543–547 (1983)

P. Petersen, *Riemannian Geometry* (Springer, 2016)

A. Radford, K. Narasimhan, Improving language understanding by generative pre-training. OpenAI preprint (2018)

C. Raffel, N.M. Shazeer, A. Roberts, K. Lee, S. Narang, M. Matena, Y. Zhou, W. Li, P.J. Liu, Exploring the limits of transfer learning with a unified text-to-text transformer. J. Mach. Learn. Res. **21**, 140:1–140:67 (2019)

F. Rosenblatt, The perceptron: a probabilistic model for information storage and organization in the brain. Psychol. Rev. **65**(6), 386 (1958)

D.E. Rumelhart, G.E. Hinton, R.J. Williams, Learning representations by back-propagating errors. Nature **323**(6088), 533–536 (1986)

S. Sabour, N. Frosst, G.E. Hinton, Dynamic routing between capsules, in *Advances in Neural Information Processing Systems*, vol. 30 (2017)

V. Sanh, L. Debut, J. Chaumond, T. Wolf, Distilbert, a distilled version of BERT: smaller, faster, cheaper and lighter. arXiv preprint arXiv:1910.01108 (2019)

F. Scarselli, M. Gori, A.C. Tsoi, M. Hagenbuchner, G. Monfardini, The graph neural network model. IEEE Trans. Neural Netw. **20**(1), 61–80 (2009)

K. Simonyan, A. Zisserman, Very deep convolutional networks for large-scale image recognition. CoRR, abs/1409.1556 (2014)

I. Sutskever, O. Vinyals, Q.V. Le, Sequence to sequence learning with neural networks. arXiv, abs/1409.3215 (2014)

C. Szegedy, W. Liu, Y. Jia, P. Sermanet, S. Reed, D. Anguelov, D. Erhan, V. Vanhoucke, A. Rabinovich, Going deeper with convolutions, in *Proceedings of the IEEE Conference on Computer Vision and Pattern Recognition* (2015), pp. 1–9

M. Tan, Q.V. Le, Efficientnet: rethinking model scaling for convolutional neural networks. arXiv, abs/1905.11946 (2019)

J.B. Tenenbaum, V. de Silva, J.C. Langford, A global geometric framework for nonlinear dimensionality reduction. Science **290**(5500), 2319–2323 (2000)

L. van der Maaten, G. Hinton, Visualizing data using t-SNE. J. Mach. Learn. Res. **9**, 2579–2605 (2008)

A. Vaswani, N. Shazeer, N. Parmar, J. Uszkoreit, L. Jones, A.N. Gomez, L. Kaiser, I. Polosukhin, Attention is all you need, in *Advances in Neural Information Processing Systems*, vol. 30 (2017)

P. Velickovic, G. Cucurull, A. Casanova, A. Romero, P. Lio', Y. Bengio, Graph attention networks. arXiv, abs/1710.10903 (2017a)

P. Velickovic, G. Cucurull, A. Casanova, A. Romero, P. Lio', Y. Bengio, Graph attention networks. arXiv preprint arXiv:1710.10903 (2017b)

D.B. West, *Introduction to Graph Theory* (Pearson, 2000)

MIX
Papier aus verantwortungsvollen Quellen
Paper from responsible sources
FSC® C105338

If you have any concerns about our products,
you can contact us on
ProductSafety@springernature.com

In case Publisher is established outside the EU,
the EU authorized representative is:
Springer Nature Customer Service Center GmbH
Europaplatz 3, 69115 Heidelberg, Germany

Printed by Libri Plureos GmbH
in Hamburg, Germany